Basic MATHEMATICS for Occupational and Vocational Students

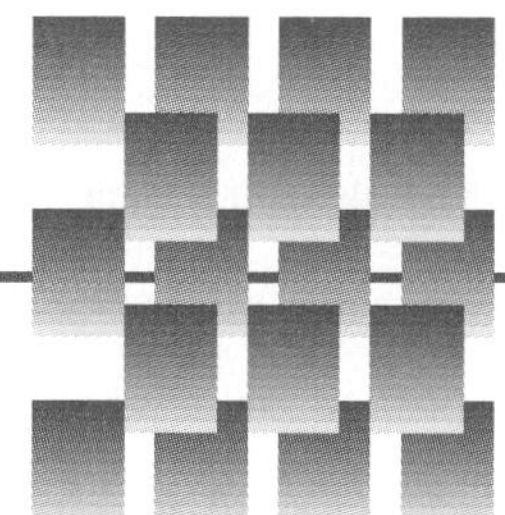

Basic MATHEMATICS for Occupational and Vocational Students

Richard C. Spangler

Basic Skills Division Chairman and Instructor, Retired
Tacoma Community College

Former Instructor, Bates Technical College
Tacoma, Washington

Upper Saddle River, New Jersey
Columbus, Ohio

Library of Congress Cataloging-in-Publication Data
Spangler, Richard C.
Basic mathematics for occupational and vocational students/Richard C. Spangler.
p. cm.
Includes index.
ISBN 0-13-081053-3
1. Arithmetic. I. Title.

QA107 .S596 2001
513--dc21 00-058466

Vice President and Publisher: Dave Garza
Editor in Chief: Stephen Helba
Associate Editor: Michelle Churma
Production Editor: Louise N. Sette
Production Supervision: York Production Services
Design Coordinator: Robin G. Chukes
Cover Designer: Jason Moore
Cover art: © Jason Moore
Production Manager: Brian Fox
Marketing Manager: Chris Bracken

This book was set by York Graphic Services, Inc. It was printed and bound by Banta Book Group. The cover was printed by Phoenix Color Corp.

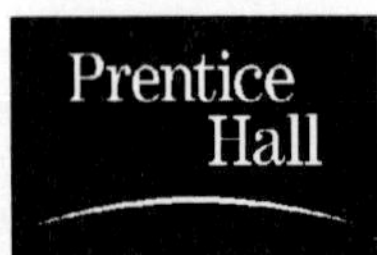

10 9 8 7 6 5 4 3 2 1
ISBN 0-13-081053-3

This text is dedicated to my dear wife,
Margaret, and my mathematics-loving children,
Mike, David, and Carole.

PREFACE

This book is designed for technical and vocational students who need a solid, thorough background in computational skills as applied to the technical and trade work and who desire mathematical success. It is suitable for use in any self-study situation or standard lecture mode. There are many illustrative examples and problems that will be of value both for self-instruction and for formal classwork. The text is designed to be a companion to *Mathematics for Technical and Vocational Students: A Worktext,* 2nd Edition.

Features

- A pretest is given at the beginning of each chapter. The results will tell the student which sections of the chapter to study. Those who need more than a refresher are encouraged to skip the pretest and study all of the chapter.
- The examples are plentiful and lead the students step by step to the solution.
- The practice checks are sequenced periodically with each section. After a new idea or technique is explained, students are directed to work a practice exercise in the margin. The practice exercises are self-paced and allow the student to become actively involved with the material before starting the problem set at the end of the unit. The solutions to the practice checks are conveniently located within the margins of each chapter.
- The section problem sets are graduated in degree of difficulty and afford the students more drill, practice, and reinforcement. The answers to odd-numbered problems appear in the back of the text.
- Technical, occupational, and vocational application problems are contained throughout the text.
- Space is provided for the students to work out the problems in each problem set.
- To help retention, most problem sets begin with review problems. Test Your Memory cumulative reviews are given at the close of Chapters 3, 6, 9, and 12 to further reinforce retention.
- Each chapter ends with a chapter overview, a self-test, and a chapter test to aid the students in their preparation for an instructor-given chapter test. These tests contain a variety of problems representative of those found in the chapter, and all answers are given in the back of the book.

In short, the book is designed for student understanding, retention, and success with short explanations, practice exercises, and constant review of skills.

Use of Calculators

No reference to the use of calculators is made in this basic text for two reasons:

1. Some instructors require that calculators not be used in basic mathematics.
2. For instructors who encourage their use, calculators are so widespread that any text reference to their use for solving problems or exercises would not be necessary.

Supplements

For the Instructor

- The Instructor's Manual includes answers to all the exercises, self-tests, and chapter tests. The manual also includes Just for Fun sections as an optional mental challenge for students.
- The Test Item File contains a quick assessment check, a pretest covering all chapters, and five free-response tests per chapter and two final examinations.
- The Prentice Hall Custom Test is an electronic test bank that allows instructors to create a variety of tests, including numerous randomized versions of the same test. The Custom Test can also be used for on-line testing.

For the Students

- The Student Solutions Manual consists of worked-out, step-by-step solutions to selected odd-numbered end-of-section exercises. This manual may be purchased as an optional resource by students.

Special Notes to the Learner

How to Be Successful in Mathematics

Most people find reading a mathematics or technical book very difficult, but, with the following tips, it doesn't have to be that way.

> *Read slowly and carefully.* Reading a mathematics or technical book is not like reading a novel. You can read and comprehend a page from an average novel in two or three minutes, but reading and comprehending a page from a mathematics or technical book could take you up to an hour. Don't be shocked if you find yourself rereading something several times, because every word and symbol is important.
>
> *Be actively involved.* Mathematics is not a spectator sport. Work out all examples step by step on paper. Both the practice exercises and the problem sets are part of this involvement. Be involved right from the start—the pay-off will be more learning power.
>
> *Seek help.* Even with the most careful reading and practice, some concepts will remain fuzzy. Don't be afraid to seek help—there are no "dumb" questions.
>
> *Take time to review.* Allow time for ideas to sink in. It helps to review material from time to time.

Success in mathematics comes with perseverance, patience, and doing.

Richard C. Spangler

CONTENTS

PART ONE

Whole Numbers

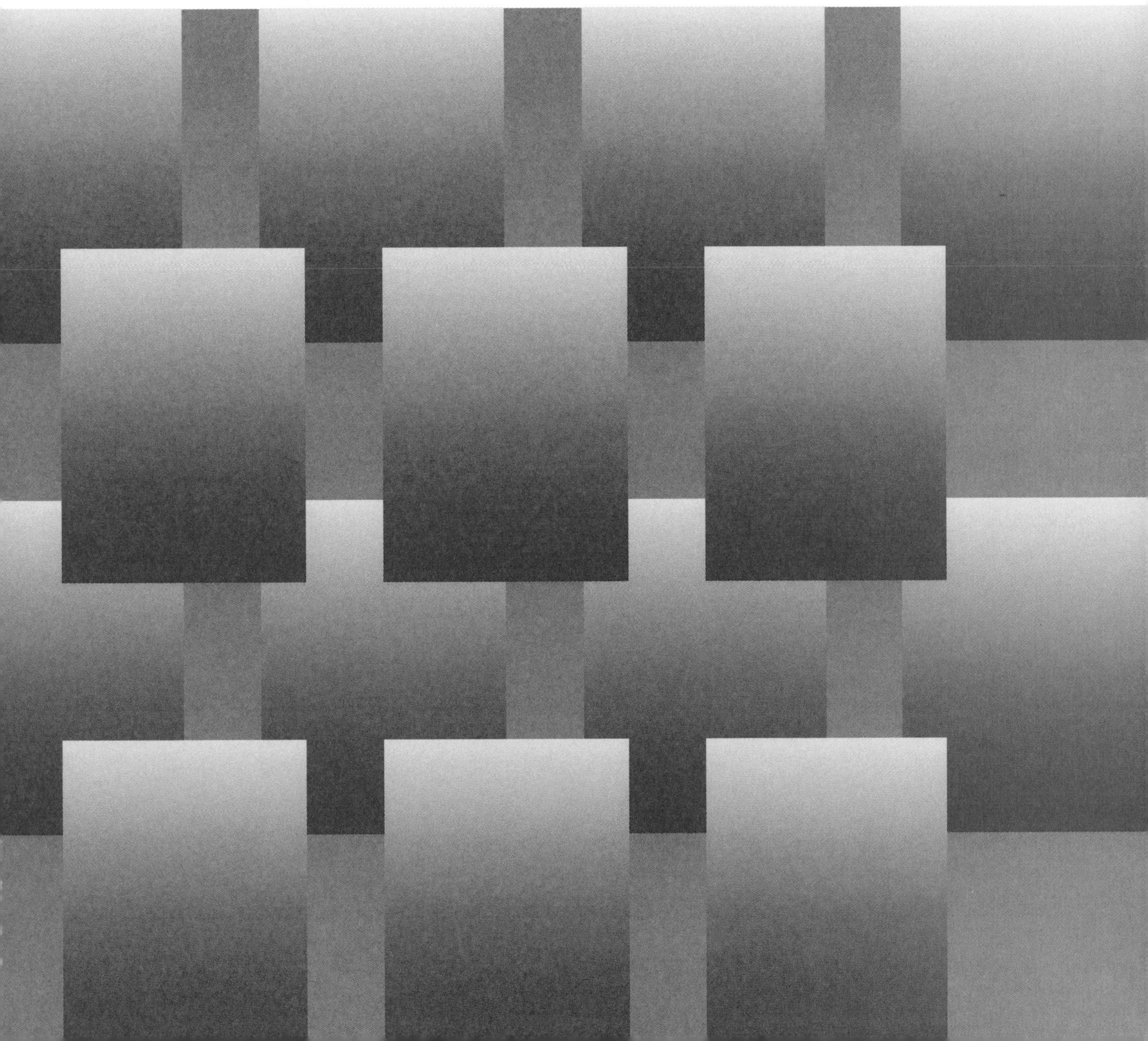

CHAPTER 1

Whole Numbers and the Place Value System

Choice: Either skip the Pretest and study all of Chapter 1 or take the Pretest and study only the section(s) you need to study.

Pretest

Directions. This Pretest will help you determine which sections to study in Chapter 1. If any of your answers are incorrect, or if you omitted any exercise, turn to the indicated section(s) and study the material; then take the Chapter Test.

1–1 **1.** Write a number for the number of corners in a square.

2. Name the smallest whole number.

3. Name the counting number on the number line that is marked by A.

4. Name the whole numbers between 73 and 76.

5. Name the second digit in 484,396.

1–2 **6.** Write an expanded number for 438.

7. Write a standard number for 3000 + 60 + 2.

1–3 **8.** Write a word name for 71,056.

9. What is the meaning of the 6 in the standard number 361,042?

1–4 **10.** Place $>$ or $<$ between 15,090 and 15,009 to make a true statement.

1–5 **11.** Round 18 to the nearest ten.

12. Round 644 to the nearest hundred.

13. Round 7532 to the nearest thousand.

The answers are in the margin on page 6.

1–1 COUNTING AND WHOLE NUMBERS

The Counting Numbers

When we want to find the number of objects, counting is required. Suppose you have a pile of one dollar bills, and you have to find how many there are. You would mentally count 1, 2, 3, 4, 5, and so forth, until you come to the last bill. The numbers that you mentally use in counting are called *counting* or *natural numbers*. The numerals that represent natural numbers are 1, 2, 3, 4, ... The three dots mean that the next number after 4 is 5, the next number after 5 is 6, and so on, following this pattern.

■ **Example 1**

Name the first 15 counting numbers.

Solution 1, 2, 3, 4, 5, 6, 7, 8, 9, 10, 11, 12, 13, 14, 15

■ **Example 2**

Name the natural number that is the same as the number of months in a year.

Solution 12

Practice check: Do exercises 1–3 in the margin.

1. Name the natural numbers between 203 and 210.
2. Name the first counting number.
3. Name the last natural number.

The answers are in the margin on page 7.

Number Line

One way to picture the natural numbers is with a number line. This is done by first drawing a line with equally spaced marks.

We shall call this point the first one. The arrow means that the line and points go on forever in this direction.

Figure 1–1

We name each point with a counting number, starting with the first point.

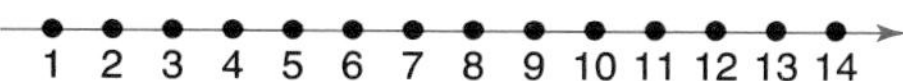

Figure 1–2

This is called a *natural number line*. We shall use number lines later on to help in comparing sizes and to add and subtract.

Whole Numbers

When we include 0 with the counting numbers, we have the *whole numbers*. The whole numbers are 0, 1, 2, 3 . . . . Here is a picture of the whole number line.

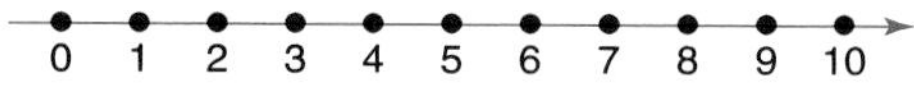

Figure 1–3

Even though 0 is not a natural number we can still use it to name a number of things or objects.

■ **Example 3**

Name how many square-headed bolts have six sides.

Solution 0

Digits

Consider the whole number 658. Each of the numbers in 658 is called a *digit*. Thus 6 is a digit, 5 is a digit, and 8 is a digit. In fact, 658 is a three-digit number.

Answers to Chapter 1 Pretest

1. 4 **2.** 0 **3.** 59 **4.** 74, 75 **5.** 8 **6.** 400 + 30 + 8 **7.** 3062 **8.** seventy-one thousand, fifty-six **9.** six ten-thousands **10.** 15,090 > 15,009 **11.** 20 **12.** 600 **13.** 80

4. Name the digits in 10456.

5. Name the third digit in 6409.

6. How many digits are in 40569?

The answers are in the margin on page 9.

■ Examples

4. A one-digit number is 8.

5. A two-digit number is 64.

6. A four-digit number is 9546.

7. The second digit of 2941 is 9.

8. The first digit of 68 is 6.

9. The fifth digit of 156049 is 4.

Practice check: Do exercises 4–6 in the margin.

Notice that, in our number system, the only digits we use to form any whole number are 0, 1, 2, 3, 4, 5, 6, 7, 8, and 9. These are the first ten whole numbers.

Standard Number

A *standard number* is written with one or more digits following one another. Some examples of standard numbers are

68 5 204 58,406 1598

■ Looking Back

You should now be able to do the following:

1. Write the counting numbers or any portion of the counting numbers.

2. Tell the difference between the counting numbers and the whole numbers.

3. Name the digits in any standard number.

4. Name the number of digits in any standard number.

■ Problems 1–1

1. Write a number for the number of weeks in a year.

2. Name the counting numbers.

3. What is the difference between the counting numbers and the whole numbers?

4. Name the counting numbers between 78 and 84.

5. Name the whole numbers from 50 through 63.

6. Name the whole number on the number line that is marked by *A*.

7. Name the natural number on the number line that is marked by A.

Answers to exercises 1–3

1. 204, 205, 206, 207, 208, 209
2. 1
3. There is no last natural number.

Name the digits in the number.

8. 201 **9.** 7 **10.** 6401 **11.** 98

Write the number of digits in each standard number.

12. 5 **13.** 1501 **14.** 61 **15.** 68,401

16. Name the second digit in 6041.

The answers to odd-numbered problems are in the back of the book.

1–2 EXPANDED NUMBERS

In section 1-1 we used counting numbers to indicate the number of objects. Count the number of bolts in the square.

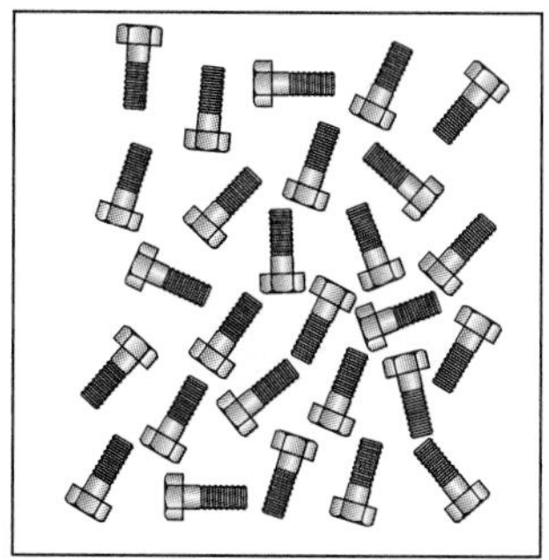

Figure 1–4

You are correct if you counted 27 bolts. The order in which we write our digits in standard numbers is based on ten or multiples of ten. We can divide our bolts in the following manner:

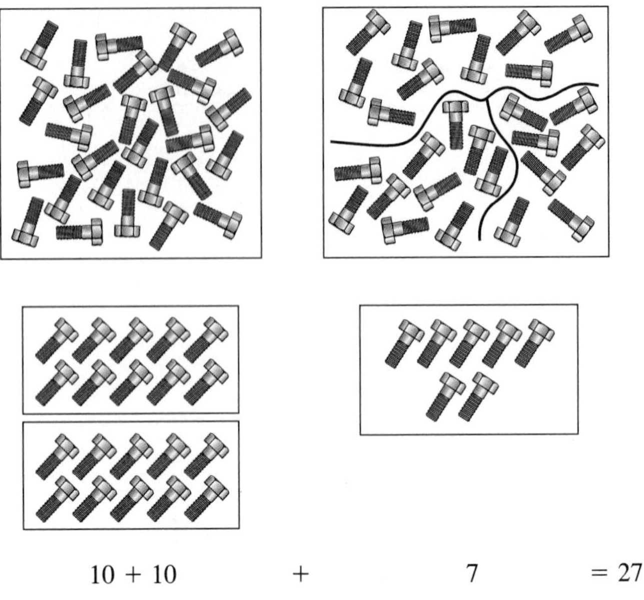

Figure 1–5

So 27 means 2 tens + 7 ones, or 20 + 7.

Consider the standard number 7864. By adding together 7000, 800, 60, and 4, we get the standard number 7864.

$$\begin{array}{r} 7000 \\ 800 \\ 60 \\ \underline{4} \\ 7864 \end{array}$$

This example could also be written as 7000 + 800 + 60 + 4 = 7864. We say this is the expanded number for 7864.

From this example, you should be able to draw some conclusions about the meaning of each digit in 7864.

The 7 names 7000 or 7 thousands.

The 8 names 800 or 8 hundreds.

The 6 names 60 or 6 tens.

The 4 names 4 or 4 ones.

Note: The symbol = is called the *equal symbol* and can be read as "is equal to." Thus, 7000 + 800 + 60 + 4 is equal to 7864.

You will use expanded numbers to help you understand addition, subtraction, multiplication, and division.

■ Example 1

Write an expanded number for 2346.

Solution An expanded number for 2346 is 2000 + 300 + 40 + 6.
An expanded number for 2346 can also be stated as 2 thousands + 3 hundreds + 4 tens + 6 ones.

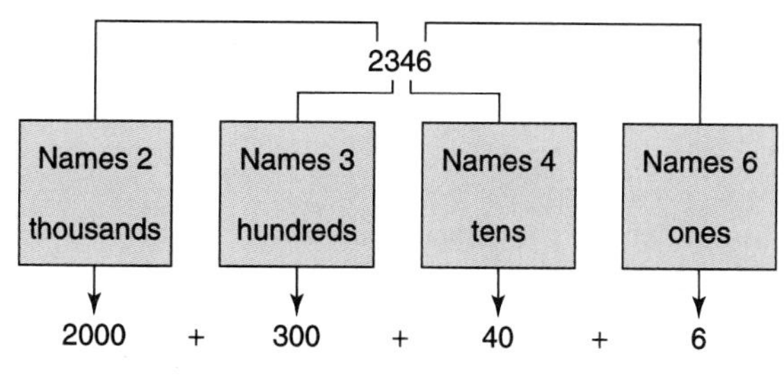

Figure 1–6

Answers to exercises 4–6

4. 1, 0, 4, 5, 6 **5.** 0 **6.** 5

■ **Example 2**

Write an expanded number for 52.

Solution 52 = 50 + 2, or 5 tens + 2 ones.

■ **Example 3**

Write an expanded number for 752.

Solution 752 = 700 + 50 + 2

Practice check: Do exercises 7–11 in the margin.

7. Write an expanded number for 73.

8. Write an expanded number for 168.

9. Write an expanded number for 1945.

10. Write an expanded number for 4249.

11. Write an expanded number for 19,578.

The answers are in the margin on page 10.

Now suppose we start with an expanded number. From this, let's write a standard number. Write a standard number for 8000 + 600 + 10 + 5. An easy way to convert is to place the expanded number in columns and then add.

$$\begin{array}{r} 8000 \\ 600 \\ 10 \\ \underline{5} \\ 8615 \end{array}$$

So, 8000 + 600 + 10 + 5 = 8615.

■ **Example 4**

Write the standard number for 70 + 5.

Solution 70 + 5 = 75

■ **Example 5**

Write the standard number for 600 + 40 + 1.

Solution 600 + 40 + 1 = 641

■ **Example 6**

Write the standard number for 30,000 + 9000 + 300 + 60 + 7.

Solution 30,000 + 9000 + 300 + 60 + 7 = 39,367

Practice check: Do exercises 12–14 in the margin.

12. Write the standard number for 600 + 90 + 2.

13. Write the standard number for 7000 + 800 + 10 + 6.

14. Write the standard number for 10,000 + 9000 + 600 + 70 + 9.

The answers are in the margin on page 10.

Answers to exercises 7–11.

7. 70 + 3 **8.** 100 + 60 + 8
9. 1000 + 900 + 40 + 5
10. 4000 + 200 + 40 + 9
11. 10,000 + 9000 + 500 + 70 + 8

Answers to exercises 12–14.

12. 692 **13.** 7816
14. 19,679

Zeros

Zeros are used as placeholders in standard numbers.

3 0 4 0

↗ Placeholder for 0 hundreds ↖ Placeholder for 0 ones

■ Example 7

Write an expanded number for 4806.

Solution

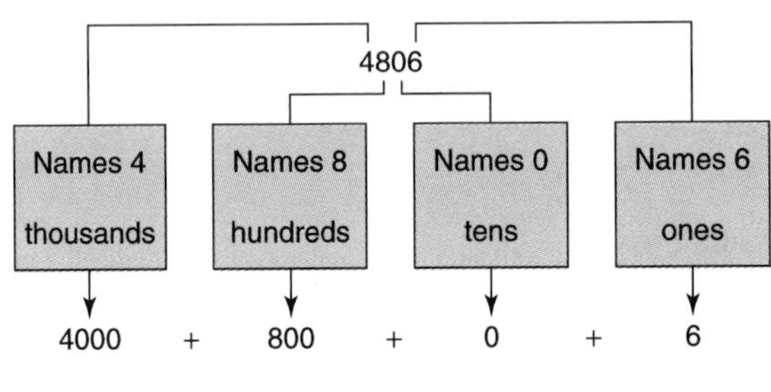

Figure 1–7

4000 + 800 + 6, or 4 thousands + 8 hundreds + 6 ones.

Notice we do not have to include the 0 tens in the expanded number.

■ Example 8

Write an expanded number for 6045.

Solution 6045 = 6000 + 40 + 5
We leave out 0 hundreds.

■ Example 9

Write an expanded number for 2004.

Solution 2004 = 2000 + 4
We leave out 0 hundreds and 0 tens.

Practice check: Do exercises 15–18 in the margin.

15. Write an expanded number for 605.

16. Write an expanded number for 5400.

17. Write an expanded number for 7050.

18. Write an expanded number for 60,402.

The answers are in the margin on page 13.

■ Example 10

Write the standard number for 5000 + 20 + 9.

Solution Think:

$$\begin{array}{r} 5000 \\ 20 \\ 9 \\ \hline 5029 \end{array} \qquad 5000 + 20 + 9 = 5029$$

■ Example 11

Write the standard number for 6000 + 500 + 7.

Solution Think:

Practice check: Do exercises 19–21 in the margin.

19. Write the standard number for 700 + 5.

20. Write the standard number for 9000 + 50.

21. Write the standard number for 50,000 + 600 + 8.

The answers are in the margin on page 13.

■ Looking Back

You should now be able to do the following:

1. Write an expanded number from a standard number.
2. Write a standard number from an expanded number.

■ Problems 1–2

Review

1. Name the whole number on the number line that is marked by *A*.

10 11 *A* 20 21

2. Name the fourth digit in 78,560.

Write expanded numbers.

3. 4526
4. 1389
5. 6821
6. 6677
7. 78
8. 56
9. 211
10. 653
11. 9090
12. 4040
13. 6601
14. 4390
15. 206
16. 309
17. 1022
18. 4303

Write standard numbers.

19. 3000 + 500 + 60 + 4
20. 8000 + 600 + 70 + 1

21. 9000 + 600 + 10 + 3

22. 4000 + 500 + 80 + 3

23. 40 + 3

24. 70 + 8

25. 300 + 50 + 8

26. 900 + 60 + 1

27. 6000 + 70 + 3

28. 5000 + 10 + 6

29. 2000 + 8

30. 1000 + 7

31. 2000 + 50

32. 9000 + 10

33. 6000 + 700

34. 1000 + 100

Write expanded numbers.

35. 26,577

36. 18,666

37. 23,406

38. 66,047

39. 30,009

40. 50,010

41. 89,000

42. 12,000

Write standard numbers.

43. 70,000 + 9000 + 700 + 20 + 6

44. 50,000 + 7000 + 300 + 70 + 2

45. 50,000 + 600 + 9

46. 10,000 + 900 + 1

47. 10,000 + 7000 + 5

48. 70,000 + 1000 + 9

49. 20,000 + 6

50. 90,000 + 5

The answers to review and odd-numbered exercises are in the back of the book.

Answers to exercises 15–18.

15. 600 + 5 **16.** 5000 + 400
17. 7000 + 50
18. 60,000 + 400 + 2

Answers to exercises 19–21.

19. 705 **20.** 9050
21. 50,608

1–3 WORD NAMES AND PLACE VALUE

Each number has a written name. For example, we write *seven* as the word name for 7. Thus two, fifty-two, and one hundred twenty-one are word names for the numbers 2, 52, and 121.

■ Examples

Number	*Word Name*
1. 8	eight
2. 11	eleven
3. 68	sixty-eight
4. 305	three hundred five

Practice check: Do exercises 22–25 in the margin.

Write word names:

22. 39 **23.** 63 **24.** 98 **25.** 15

The answers are in the margin on page 15.

Being able to read both small and large numbers is very important—either for job-related activities or for just keeping up with what is going on in the world. In order to read or write numbers of any size, you need to memorize the *place value periods. Periods* are separated by commas, and each has no more than three digits.

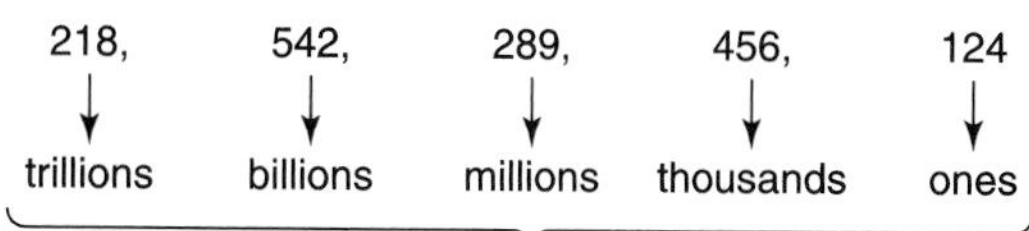

Figure 1–8

We see from this that

124 is in the *ones* period.
456 is in the *thousands* period.
289 is in the *millions* period.
542 is in the *billions* period.
218 is in the *trillions* period.

If you haven't already, it's important that you memorize these periods from right to left, even though you read numbers from left to right.

■ **Example 5**

In the number 78,456,789,292, in what period is 78? 456? 789? 292?

Solution 78 is in the billions period.
456 is in the millions period.
789 is in the thousands period.
292 is in the ones period.

Practice check: Do Exercises 26 and 27 in the margin.

26. In the number 92,541,021,698, in what period is 698? 92? 021? 541?

27. In the number 304,568,002,411,666, in what period is 002? 568? 666? 304? 411?

The answers are in the margin on page 17.

When reading large numbers, you attach the period name to the number of each period. Let's see how each period is read in the standard number 218,542,289,546,124.

218 is read, "two hundred eighteen *trillion*."
542 is read, "five hundred forty-two *billion*."
289 is read, "two hundred eighty-nine *million*."
546 is read, "five hundred forty-six *thousand*."
124 is read, "one hundred twenty-four."

Actually, the last number could have the word *ones* added, but this is not the usual practice. Notice that the periods are separated with commas.

■ **Example 6**

Write a word name for 18,248,405.

Solution Eighteen *million,* two hundred forty-eight *thousand,* four hundred five.

Practice check: Do exercises 28–33 in the margin.

Read to yourself, then write the word name for the following numbers:

28. 308 **29.** 6020

30. 23,468,235 **31.** 6,023,005

32. 105,004,016

33. 33,406,843,003

The answers are in the margin on page 17.

Given a word name, let's make a standard number.

■ **Example 7**

Write a standard number for six hundred fifty-eight million, two hundred nine thousand, ten.

Solution The standard number is 658,209,010.

■ **Example 8**

Write a standard number for two hundred five trillion, eighty-nine million, seven hundred twenty-six thousand, one hundred six.

Solution The standard number is 205,000,089,726,106.
Note there are no billions, so we must include three zeros as place holders.

Practice check: Do exercises 34–36 in the margin.

Write a standard number for the following:

34. Sixty thousand, forty-five.

35. Nine hundred sixty million, nine thousand.

36. Three hundred seventeen million, two hundred six thousand, four hundred eighty-four.

The answers are in the margin on page 17.

To understand the place value of each number within each period, study the place value chart on the next page.

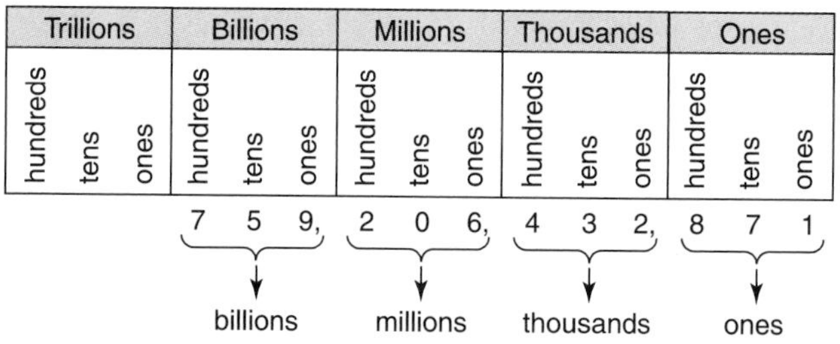

Trillions			Billions			Millions			Thousands			Ones		
hundreds	tens	ones	hundreds	tens	ones	hundreds	tens	ones	hundreds	tens	ones	hundreds	tens	ones
			7	5	9,	2	0	6,	4	3	2,	8	7	1
			billions			millions			thousands			ones		

Figure 1–9

Answers to exercises 22–25.

22. Thirty-nine **23.** Sixty-three
24. Ninety-eight **25.** Fifteen

Refer to the chart and the example of 759,206,432,871. The 8 means 8 hundreds, the 3 means 3 ten thousands, and the 6 means 6 millions.

There are 4 hundred thousands.
There are 2 hundred millions.
There are 5 ten billions.
There are 0 ten millions.

■ Examples

Read each number to yourself first, then state what the digit 5 means.

9. 359,409 — 5 ten thousands
10. 604,568 — 5 hundreds
11. 95,469,900 — 5 millions
12. 677,540,006 — 5 hundred thousands

Practice check: Do exercises 37–39 in the margin.

Read each number to yourself first, then state what the digit 7 means.

37. 401,276 **38.** 640,076,500,986

39. 10,098,587

The answers are in the margin on page 17.

We can also learn to identify the place values when given a standard number like 608,497,513.

■ Examples

In the number 608,497,513, what digit names the place value?

	Answer
13. ten millions	0
14. hundreds	5
15. one hundred millions	6
16. tens	1
17. ten thousands	9

Practice check: Do exercises 40–43 in the margin.

In the number 58,209, what digit names the following place value?

40. thousands **41.** tens

42. ten thousands **43.** ones

The answers are in the margin on page 17.

■ Looking Back

You should now be able to do the following:

1. State the period name of 298 and 19 when given a standard number like 19,504,298.
2. Write a word name when given a standard number.
3. Write a standard number when given a word name.

4. State the meaning of the digits 1, 5, 7, and 9 when given a standard number like 794,516,200.
5. Identify the hundreds digit, ten thousands digit, hundred millions digit, and so on when given a standard number like 794,516,200.

■ Problems 1–3

Review

1. Name the whole numbers between 105,689 an 105,693.
2. Name the digits in 209,541,233.
3. Write an expanded number for 5304.
4. Write a standard number for 7000 + 400 + 5.

In the number 477,980,247,604,369, name the period of the following groups of numbers.

5. 247
6. 369

Write word names.

7. 43
8. 204
9. 968
10. 666
11. 2,358
12. 80,341
13. 131,015
14. 268,305,400

Write a standard number.

15. Three hundred fifty-two.
16. One hundred three thousand nine.
17. Seven million.
18. Five hundred sixty-two million, fifteen thousand, two hundred five.

In the number 289,304,561, what digit names the following place value?

19. ones

20. ten millions

21. tens

In the standard number 412,536,987, state the meaning of

22. 2

23. 4

Write a word for the numbers in the following statements.

24. The population of California in 1990 was 29,760,021.

25. The public debt of the United States in 1994 was $4,692,800,000,000.

Write word names.

26. 68,204,980,460

27. 299,680,411,571,410

Write a standard number.

28. Four hundred two trillion, three thousand, four hundred five.

29. Twenty-eight trillion, five million, one hundred seventeen thousand.

Answers to exercises 26 and 27.

26. ones, billions, thousands, millions
27. millions, billions, ones, trillions, thousands.

Answers to exercises 28–33.

28. Three hundred eight.
29. Six thousand twenty.
30. Twenty-three million, four hundred sixty-eight thousand, two hundred thirty-five.
31. Six million, twenty-three thousand five.
32. One hundred five million, four thousand sixteen.
33. Thirty-three billion, four hundred six million, eight hundred forty-three thousand three.

Answers to exercises 34–36.

34. 60,045
35. 960,009,000
36. 317,206,484

Answers to exercises 37–39.

37. 7 tens **38.** 7 ten millions
39. 7 ones

Answers to exercises 40–43.

40. 8 **41.** 0 **42.** 5 **43.** 9

The answers to review and odd-numbered exercises are in the back of the book.

1–4 COMPARING NUMBERS (ORDER)

Everyday comparisons are made using numbers. If Jack works six days a week and Irene works four, we can say that Irene works fewer days because four is less than six.

On a number line, 4 is to the left of 6.

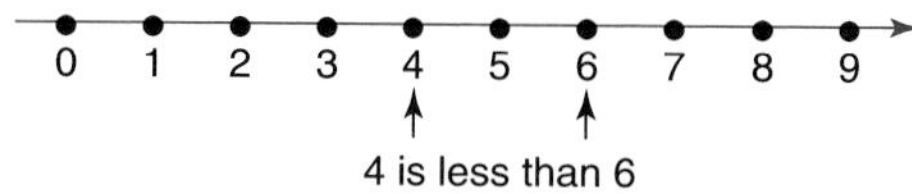

Figure 1–10

Since 6 is to the right of 4, we can also say that 6 is greater than 4. The symbol $<$ is used for "is less than," and the symbol $>$ is used for "is greater than." Thus, $4 < 6$ and $6 > 4$. The symbol always points to the smaller number.

■ Example 1

Place $<$ or $>$ between 341 and 352 to make a true statement.

Solution This number line shows that 341 is to the left of 352.

Figure 1–11

Thus, $341 < 352$. This is read, "341 is less than 352."

Since 352 is to the right of 341, we can also state that $352 > 341$, or "352 is greater than 341."

■ Example 2

Place $<$ and $>$ symbols between the following numbers to make true statements.

a. 6 14 b. 23 19 c. 153 168

d. 88 86 e. 2348 2356 f. 0 8

Solutions

a. $6 < 14$ b. $23 > 19$ c. $153 < 168$

d. $88 > 86$ e. $2348 < 2356$ f. $0 < 8$

Practice check: Do exercises 44–49 in the margin.

Place $<$ and $>$ symbols between the following numbers to make true statements.

44. 10 0 **45.** 56 65

46. 104 101 **47.** 66 86

48. 354 345 **49.** 1232 1090

The answers are in the margin on page 21.

■ Looking Back

Given two different whole numbers, you should be able to state which is larger and which is smaller by using the symbols $<$ and $>$.

■ Problems 1–4

Review

1. Write a number for the number of faces on a pair of dice.

2. Name the whole number on the number line that is marked by *A*.

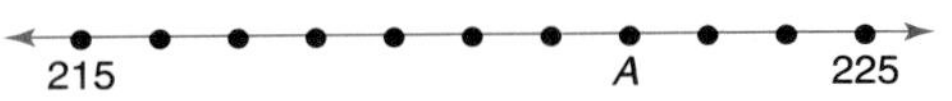

3. Write an expanded number for 7502.

4. Write a standard number for $5000 + 60 + 9$.

5. Write a word name for 519,604,533.

6. Write a standard number for nine hundred twenty-six thousand, five hundred eighty.

Place < or > symbols between the following numbers to make true statements.

7. 3 8 **8.** 8 4 **9.** 0 9 **10.** 16 0 **11.** 38 29

12. 20 30 **13.** 998 997 **14.** 664 662 **15.** 545 454

16. 787 878 **17.** 3468 3369 **18.** 1645 1654

19. 13,467 13,461 **20.** 56,741 56,480

21. 239,466 238,576 **22.** 690,426 690,421

The answers to review and odd-numbered exercises are in the back of the book.

1–5 ROUNDING

Sometimes exact answers aren't necessary. For example, the population of New York City is approximately 9 million people. Tom weighs about 160 pounds. These numbers are not exact because they have been rounded.

Rounding to the Nearest Ten

When we round to the nearest ten, our answers will be numbers like 20, 50, 630, 7750, and so on.

■ **Example 1** Round 38 to the nearest ten.

Solution

Figure 1–12

Notice that 38 is closer to 40 than to 30. Thus, 38 rounds *up* to 40.

■ **Example 2** Round 52 to the nearest ten.

Solution

Figure 1–13

This number is closer to 50 than to 60, so we round *down* to 50.
Another method gives the same result.

Rule

To round to the nearest ten:

Step 1 Locate the tens digit.

Step 2 Look at the digit just to the right of the tens digit.

a. If this digit is 5 or more, round up.

b. If this digit is less than 5, round down.

■ **Example 3** Round 34 to the nearest ten.

Solution **Step 1** Locate the tens digit.

Tens
↓
34
↗

Step 2 Look at the digit to the right. Less than 5.

Since 4 is less than 5, we round down to 30. Thus, 34 rounded to the nearest ten is 30.

■ **Example 4** Round 725 to the nearest ten.

Solution **Step 1** Locate the tens digit.

Tens
↓
725
↗

Step 2 Look at the digit to the right. 5 or more

Since the ones digit is 5 or more, we round up to 730. Thus, 725 rounded to the nearest ten is 730.

Practice check: Do exercises 50–63 in the margin.

Round to the nearest ten.

50. 27	**51.** 63	**52.** 85
53. 11	**54.** 79	**55.** 346
56. 989	**57.** 145	**58.** 35
59. 574	**60.** 98	**61.** 783
62. 54	**63.** 526	

The answers are in the margin on page 23.

Rounding to the Nearest Hundred

When we round to the nearest hundred, our answers will be numbers such as 100, 200, 400, 4300, 6500, and so on.

■ **Example 5** Round 362 to the nearest hundred.

Solution

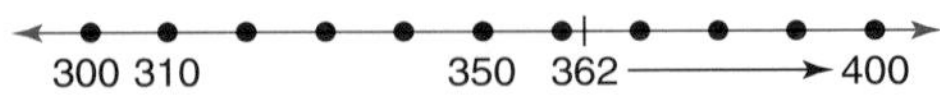

Figure 1–14

The number line indicates that 362 is closer to 400 than 300, so we round up to 400.

Rule

To Round to the Nearest Hundred:

Step 1 Locate the hundreds digit.

Step 2 Look at the digit just to the right of the hundred's digit.

a. If the digit is 5 or more, round up.

b. If the digit is less than 5, round down.

Answers to exercises 44–49.

44. $10 > 0$ **45.** $56 < 65$
46. $104 > 101$ **47.** $66 < 86$
48. $354 > 345$ **49.** $1232 > 1090$

■ **Example 6** Round 274 to the nearest hundred.

Solution **Step 1** Locate the hundreds digit.

Hundreds
↓
274
↗

Step 2 Look at the digit to the right. More than 5

We round up to 300.

■ **Example 7** Round 3409 to the nearest hundred.

Solution **Step 1** Locate the hundreds digit.

Hundreds
↓
3409
↗

Step 2 Look at the digit to the right. Less than 5

We round down to 3400.

Practice check: Do exercises 64–71 in the margin.

Round to the nearest hundred.

64. 3151 **65.** 751 **66.** 269
67. 188 **68.** 7436 **69.** 9468
70. 150 **71.** 4961

The answers are in the margin on page 23.

Rounding to the Nearest Thousand

When we round to the nearest thousand, our answers will be numbers such as 1000, 2000, 63,000, 4000, and so on.

■ **Example 8** Round 8246 to the nearest 1000.

Solution

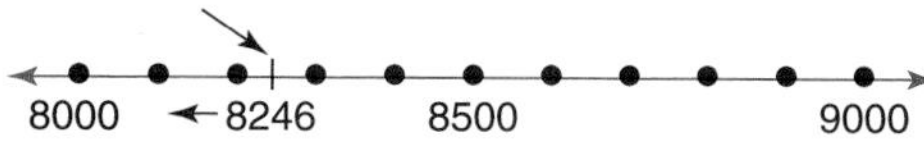

Figure 1–15

We find that 8246 is closer to 8000 than 9000, so we round down to 8000.

In general, the rules for rounding to the nearest thousand are the same as for tens and hundreds.

■ **Example 9** Round 3654 to the nearest thousand.

Solution **Step 1** Locate the thousands digit.

Thousands
↓
3654
↖

Step 2 Look at the digit to the right. 5 or more

We round up to 4000.

■ **Example 10** Round 34,581 to the nearest thousand.

Solution Thousands

↓

34,581

↗

5 or more

We round up to 35,000.

Practice check: Do exercises 72–79 in the margin.

Round to the nearest thousand.

72. 5647 **73.** 7391

74. 1581 **75.** 2915

76. 59,640 **77.** 78,400

78. 15,688 **79.** 987,651

The answers are in the margin on page 25.

■ Looking Back

You should now be able to do the following:

1. Round a whole number to the nearest ten.
2. Round a whole number to the nearest hundred.
3. Round a whole number to the nearest thousand.

Problems 1–5

Review

1. Name the whole number on the number line that is marked by A.

256 · · · · · · · A · · 266

2. Name the smallest whole number.

3. Write an expanded number for 5860.

4. Write a standard number for 90,000 + 5000 + 70 + 8.

5. Write a word name for 306,022.

6. In the number 670,411,333,001, state the meaning of 7.

7. Place $<$ or $>$ between 3168 and 3175 to make a true statement.

8. Place $<$ or $>$ between 16,708 and 16,717 to make a true statement.

Round to the nearest ten.

9. 18 **10.** 89 **11.** 842 **12.** 104 **13.** 225 **14.** 635

Round to the nearest hundred.

15. 258 **16.** 124 **17.** 4889 **18.** 6666 **19.** 708 **20.** 677

Round to the nearest thousand.

21. 7824 **22.** 500 **23.** 1450 **24.** 2918 **25.** 78,450 **26.** 25,651

Round to the nearest thousand.

27. 418,344 **28.** 801,509 **29.** 846,501,766 **30.** 19,682

31. The population in Florida in 1990 was 12,937,926. What was the population to the nearest thousand?

32. The annual salaries of three employees of the Jim Dandy Lock Shop are \$18,455, \$19,829, and \$19,271. Round each salary to the nearest hundred dollars.

Answers to exercises 50–63.

50. 30 **51.** 60
52. 90 **53.** 10
54. 80 **55.** 350
56. 990 **57.** 150
58. 40 **59.** 570
60. 100 **61.** 780
62. 50 **63.** 530

Answers to exercises 64–71.

64. 3200 **65.** 800
66. 300 **67.** 200
68. 7400 **69.** 9500
70. 200 **71.** 5000

The answers to review and odd-numbered exercises are in the back of the book.

CHAPTER 1 OVERVIEW

Summary

1–1 You learned the difference between natural and whole numbers.

1–2 You learned how to expand a standard number and how to write a standard number from an expanded number.

1–3 You learned the place value of periods and digits and how to write word names for numbers.

1–4 You learned how to signify that one number is larger or smaller than another number.

1–5 You learned how to round numbers to the nearest ten, hundred, or thousand.

Terms To Remember

	Page		***Page***
Counting numbers	4	Order	17
Digit	5	Place Value	13
Expanded numbers	7	Rounding	19
Natural numbers	4	Standard numbers	6
Number line	5	Whole numbers	5

Definitions

The natural or counting numbers begin with 1.
The whole numbers begin with 0.

Rules

- To round to the nearest ten:
  - **Step 1** Locate the tens digit.
  - **Step 2** Look at the digit just to the right of the tens digit
    - **a.** If this digit is 5 or more, round up.
    - **b.** If this digit is less than 5, round down.
- To round to the nearest hundred:
  - **Step 1** Locate the hundreds digit.
  - **Step 2** Look at the digit just to the right of the hundreds digit.
    - **a.** If the digit is 5 or more, round up.
    - **b.** If the digit is less than 5, round down.

Self-Test

The answers are in the back of the book.

1–1 **1.** Write a number for the number of weeks in a year.

2. Name the first whole number.

3. Name the natural number on the number line marked by A.

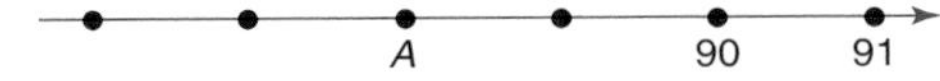

4. Name the counting numbers between 40 and 49.

5. Name the fourth digit in 468,098.

1–2 **6.** Write 7048 as an expanded number.

7. Write 6000 + 40 + 8 as a standard number.

Answers to exercises 72–79.

72. 6000	**73.** 7000
74. 2000	**75.** 3000
76. 60,000	**77.** 78,000
78. 16,000	**79.** 988,000

1–3 8. Write a word name for 18,406.

9. What is the meaning of the 4 in the standard number 348,094?

10. Place < or > symbols between 923 and 932 to make a true statement.

11. Round 3461 to the nearest hundred.

Chapter Test

This test will aid in your preparation for a possible chapter test given by your instructor. The answers are in the back of the book. Go back and review the appropriate section(s) if you missed any test items.

1–1 1. What is the difference between whole numbers and natural numbers?

2. Write a number for the number of minutes in an hour.

3. Name the first counting number.

4. Name the whole number on the number line that is marked by A.

5. Name the counting numbers between 371 and 383.

6. Name the third digit in 368,405.

1–2 Write as expanded numbers.

7. 784

8. 2040

Write as standard numbers.

9. 5000 + 600 + 70 + 1

10. 8000 + 60

1–3 **11.** Write a word name for 215,408,003.

12. Write a standard number of twenty-eight billion, four hundred thousand, six hundred eighty-two.

13. What is the meaning of the 5 in the standard number 546,711,003?

1–4 Place $<$ or $>$ symbols between the following numbers to make a true statement.

14. 846 748

15. 19,010 19,100

1–5 Round to the nearest ten.

16. 19

17. 645

18. 404

Round to the nearest hundred.

19. 348

20. 956

21. 2261

Round to the nearest thousand.

22. 5622

23. 1241

24. 9567

CHAPTER 2

Addition and Subtraction of Whole Numbers

Choice: Either skip the Pretest and study all of Chapter 2 or take the Pretest and study only the sections you need to study.

Pretest

Directions: This Pretest will help you determine which sections to study in Chapter 2. If any of your answers are incorrect, or if you omitted any exercise, turn to the indicated section(s) and study the material; then take the Chapter Test.

2–1 Push yourself to add quickly without counting. If you have to count to find any answer, study all of section 2–1.

1. $3 + 4$ **2.** $8 + 5$ **3.** $6 + 5$ **4.** $3 + 8$ **5.** $3 + 9$

6. $5 + 9$ **7.** $8 + 6$ **8.** $7 + 8$ **9.** $8 + 9$ **10.** $9 + 7$

11. $6 + 7$ **12.** $4 + 9$ **13.** $4 + 8$ **14.** $9 + 9$ **15.** $7 + 5$

2–2 Add using the short method.

16. $\begin{array}{r} 32 \\ +46 \\ \hline \end{array}$ **17.** $\begin{array}{r} 3241 \\ +3057 \\ \hline \end{array}$

2–3 Add using the short method.

18. $\begin{array}{r} 493 \\ +868 \\ \hline \end{array}$ **19.** $\begin{array}{r} 7846 \\ +5194 \\ \hline \end{array}$

2–4 **20.** Add and check: $56 + 704 + 7 + 84$.

2–5 **21.** Estimate the sum to the nearest hundred: $478 + 842 + 104 + 556$.

2–6 **22.** Ray bowled games of 240, 189, 222, 201, and 172. What was his total for five games?

2–7 Push yourself to subtract quickly without counting. If you have to count to find any answer, study all of section 2–7.

23. $\begin{array}{r} 17 \\ -9 \\ \hline \end{array}$ **24.** $\begin{array}{r} 15 \\ -8 \\ \hline \end{array}$ **25.** $\begin{array}{r} 16 \\ -9 \\ \hline \end{array}$ **26.** $\begin{array}{r} 18 \\ -9 \\ \hline \end{array}$

27. $\begin{array}{r} 14 \\ -8 \\ \hline \end{array}$ **28.** $\begin{array}{r} 13 \\ -9 \\ \hline \end{array}$ **29.** $\begin{array}{r} 12 \\ -8 \\ \hline \end{array}$ **30.** $\begin{array}{r} 15 \\ -9 \\ \hline \end{array}$

31. $\begin{array}{r} 11 \\ -4 \\ \hline \end{array}$ **32.** $\begin{array}{r} 13 \\ -6 \\ \hline \end{array}$ **33.** $\begin{array}{r} 17 \\ -8 \\ \hline \end{array}$ **34.** $\begin{array}{r} 12 \\ -4 \\ \hline \end{array}$

2–8 **35.** Subtract. $\begin{array}{r} 7842 \\ -4310 \\ \hline \end{array}$

2–9 **36.** Subtract. $\begin{array}{r} 8421 \\ -7925 \\ \hline \end{array}$

37. Subtract.

$$\begin{array}{r} 5001 \\ -789 \\ \hline \end{array}$$

38. Jim machined 8 gears on Monday and 16 gears on Thursday. How many more gears did Jim machine on Thursday than on Monday?

The answers are in the margin on page 33.

2–1 BASIC ADDITION FACTS

Turn to Problems 2–1 and do the additions mentally *without counting.* **This is important!** If you get at least 95 answers correct, go on to section 2–2. If not, study section 2–1.

Simple Addition

In Chapter 1 we discussed how to use the counting numbers when we wanted to find the number of objects or things. Adding is like counting. Adding is finding how many things are in two or more sets of objects. The following example is simple, but it illustrates a point: A machine shop prepared 8 tapered shanks on Monday for a customer and 9 tapered shanks on Tuesday. How many tapered shanks did the machine shop prepare altogether?

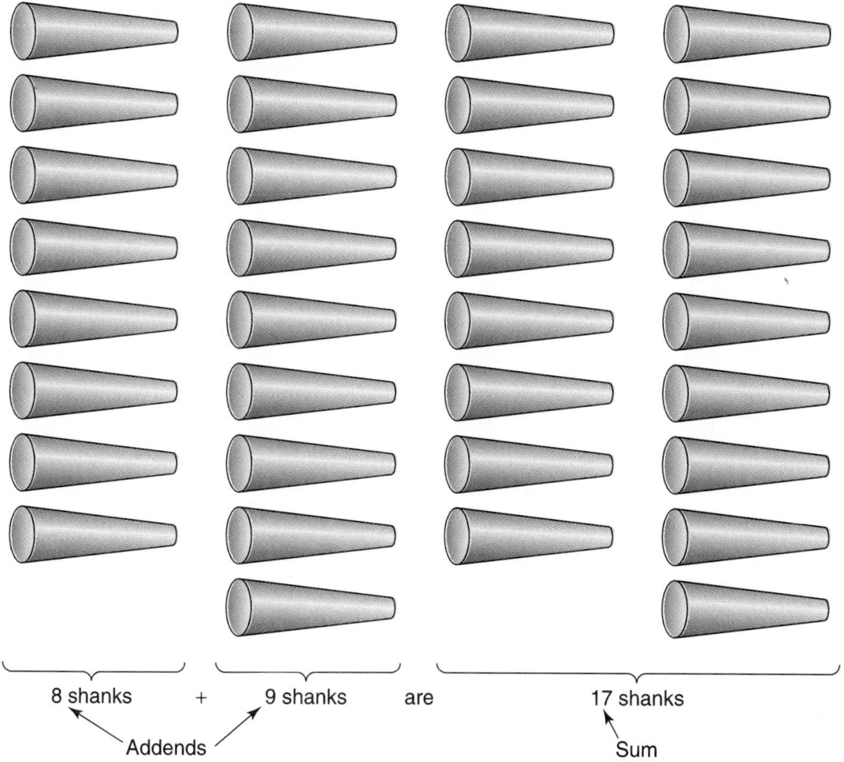

Figure 2–1

The numbers that are added are called *addends*. The result is called the *sum*.

Practice check: Do exercises 1–6 in the margin.

Find the sum by joining two sets and counting.

1. $2 + 7 =$ **2.** $3 + 8$

3. $4 + 4 =$ **4.** $5 + 6 =$

5. $6 + 2$ **6.** $7 + 4$

The answers are in the margin on page 33.

Addition can also be shown by using the number line.

■ Example 1

Add 6 and 3 using the number line.

Solution We start at 6 and make an arrow 3 units in length.

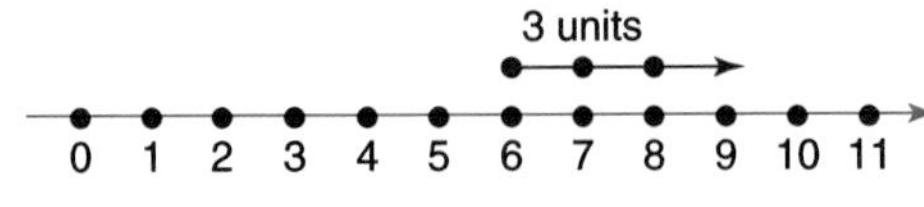

Figure 2–2

Thus, $6 + 3 = 9$.

Practice check: Do exercises 7–10 in the margin.

Illustrate the addition by drawing a number line.

7. $2 + 7$ **8.** $3 + 8$

9. $4 + 5$

10. Write the addition that is shown by the number line.

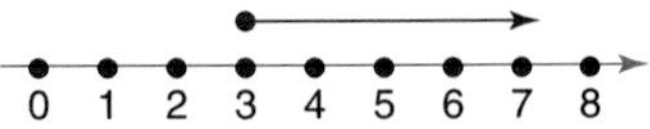

The answers are in the margin on page 33.

Addition of Zero

Joe picked 8 onions from the garden on Tuesday, and none on Wednesday. How many onions did he pick in the two days?

8	+	0	=	8
↑		↑		↑
Tuesday		Wednesday		Total picked

Rule

When adding zero and another number, the sum is the other number.

■ Examples

2. $0 + 8 = 8$ **3.** $15 + 0 = 15$ **4.** $0 + 0 = 0$

$$\begin{array}{r} 0 \\ +8 \\ \hline 8 \end{array} \qquad \begin{array}{r} 15 \\ +0 \\ \hline 15 \end{array} \qquad \begin{array}{r} 0 \\ +0 \\ \hline 0 \end{array}$$

Practice check: Do exercises 11–14 in the margin.

Add.

11. $10 + 0$ **12.** $0 + 2$

13. $88 + 0$ **14.** $0 + 27$

The answers are in the margin on page 33.

To add quickly, you should memorize the addition facts. Use the table to help you. Since adding 0 is easy, these facts are not shown.

+	1	2	3	4	5	(6)	7	8	9
1	2	3	4	5	6	7	8	9	10
2	3	4	5	6	7	8	9	10	11
3	4	5	6	7	8	9	10	11	12
(4)	5	6	7	8	9	(10)	11	12	13
5	6	7	8	9	10	11	12	13	14
6	7	8	9	10	11	12	13	14	15
7	8	9	10	11	12	13	14	15	16
8	9	10	11	12	13	14	15	16	17
9	10	11	12	13	14	15	16	17	18

■ Looking Back

You should now be able to add any two digits. It is important for you to memorize the addition facts before doing Problems 2–1.

■ Problems 2–1

Do the following additions mentally without counting. Force yourself to be quick. Read the answers to someone while he/she checks them in the back of the book.

1. 3 +2	**2.** 2 +2	**3.** 0 +9	**4.** 3 +4	**5.** 4 +6
6. 1 +1	**7.** 5 +8	**8.** 3 +5	**9.** 0 +8	**10.** 4 +7
11. 5 +9	**12.** 1 +2	**13.** 3 +3	**14.** 4 +5	**15.** 0 +7
16. 5 +7	**17.** 6 +9	**18.** 1 +3	**19.** 2 +4	**20.** 3 +6
21. 0 +6	**22.** 4 +8	**23.** 4 +4	**24.** 1 +4	**25.** 5 +6
26. 6 +8	**27.** 0 +5	**28.** 2 +5	**29.** 3 +7	**30.** 1 +5
31. 4 +9	**32.** 5 +5	**33.** 0 +4	**34.** 6 +7	**35.** 7 +9
36. 1 +6	**37.** 2 +6	**38.** 3 +8	**39.** 0 +3	**40.** 6 +6
41. 7 +8	**42.** 1 +7	**43.** 2 +7	**44.** 3 +9	**45.** 0 +2
46. 7 +7	**47.** 8 +9	**48.** 1 +8	**49.** 2 +8	**50.** 8 +8

51. $\begin{array}{r} 0 \\ +1 \\ \hline \end{array}$ **52.** $\begin{array}{r} 2 \\ +9 \\ \hline \end{array}$ **53.** $\begin{array}{r} 9 \\ +9 \\ \hline \end{array}$ **54.** $\begin{array}{r} 1 \\ +9 \\ \hline \end{array}$ **55.** $\begin{array}{r} 2 \\ +3 \\ \hline \end{array}$

56. $\begin{array}{r} 6 \\ +4 \\ \hline \end{array}$ **57.** $\begin{array}{r} 8 \\ +0 \\ \hline \end{array}$ **58.** $\begin{array}{r} 7 \\ +5 \\ \hline \end{array}$ **59.** $\begin{array}{r} 4 \\ +2 \\ \hline \end{array}$ **60.** $\begin{array}{r} 5 \\ +0 \\ \hline \end{array}$

61. $\begin{array}{r} 9 \\ +4 \\ \hline \end{array}$ **62.** $\begin{array}{r} 9 \\ +7 \\ \hline \end{array}$ **63.** $\begin{array}{r} 3 \\ +0 \\ \hline \end{array}$ **64.** $\begin{array}{r} 7 \\ +2 \\ \hline \end{array}$ **65.** $\begin{array}{r} 9 \\ +8 \\ \hline \end{array}$

66. $\begin{array}{r} 1 \\ +0 \\ \hline \end{array}$ **67.** $\begin{array}{r} 7 \\ +4 \\ \hline \end{array}$ **68.** $\begin{array}{r} 1 \\ +1 \\ \hline \end{array}$ **69.** $\begin{array}{r} 5 \\ +4 \\ \hline \end{array}$ **70.** $\begin{array}{r} 9 \\ +6 \\ \hline \end{array}$

71. $\begin{array}{r} 6 \\ +3 \\ \hline \end{array}$ **72.** $\begin{array}{r} 4 \\ +1 \\ \hline \end{array}$ **73.** $\begin{array}{r} 5 \\ +2 \\ \hline \end{array}$ **74.** $\begin{array}{r} 6 \\ +1 \\ \hline \end{array}$ **75.** $\begin{array}{r} 9 \\ +3 \\ \hline \end{array}$

76. $\begin{array}{r} 6 \\ +6 \\ \hline \end{array}$ **77.** $\begin{array}{r} 8 \\ +1 \\ \hline \end{array}$ **78.** $\begin{array}{r} 9 \\ +2 \\ \hline \end{array}$ **79.** $\begin{array}{r} 8 \\ +5 \\ \hline \end{array}$ **80.** $\begin{array}{r} 9 \\ +5 \\ \hline \end{array}$

81. $\begin{array}{r} 1 \\ +3 \\ \hline \end{array}$ **82.** $\begin{array}{r} 6 \\ +5 \\ \hline \end{array}$ **83.** $\begin{array}{r} 7 \\ +3 \\ \hline \end{array}$ **84.** $\begin{array}{r} 2 \\ +6 \\ \hline \end{array}$ **85.** $\begin{array}{r} 8 \\ +7 \\ \hline \end{array}$

86. $\begin{array}{r} 8 \\ +2 \\ \hline \end{array}$ **87.** $\begin{array}{r} 9 \\ +9 \\ \hline \end{array}$ **88.** $\begin{array}{r} 4 \\ +3 \\ \hline \end{array}$ **89.** $\begin{array}{r} 5 \\ +3 \\ \hline \end{array}$ **90.** $\begin{array}{r} 2 \\ +1 \\ \hline \end{array}$

91. $\begin{array}{r} 8 \\ +4 \\ \hline \end{array}$ **92.** $\begin{array}{r} 8 \\ +6 \\ \hline \end{array}$ **93.** $\begin{array}{r} 5 \\ +1 \\ \hline \end{array}$ **94.** $\begin{array}{r} 7 \\ +6 \\ \hline \end{array}$ **95.** $\begin{array}{r} 8 \\ +3 \\ \hline \end{array}$

96. $\begin{array}{r} 7 \\ +1 \\ \hline \end{array}$ **97.** $\begin{array}{r} 7 \\ +7 \\ \hline \end{array}$ **98.** $\begin{array}{r} 8 \\ +8 \\ \hline \end{array}$ **99.** $\begin{array}{r} 9 \\ +1 \\ \hline \end{array}$ **100.** $\begin{array}{r} 0 \\ +0 \\ \hline \end{array}$

Memorize these facts until you can get all of Problems 2–1 correct.

The answers to the problems are in the back of the book.

The Commutative Law of Addition

Notice the pattern.

$3 + 2 = 5 \qquad 9 + 2 = 11$

$2 + 3 = 5 \qquad 2 + 9 = 11$

$8 + 4 = 12 \qquad 3 + 4 = 7$

$4 + 8 = 12 \qquad 4 + 3 = 7$

When we add 3 and 2, we get the same sum as 2 + 3. Thus, we can say that 3 + 2 = 2 + 3. This shows that the order in which we add numbers does not affect the sum. This law is called the commutative (*come mew ta tive*) law of addition. By using the commutative law, you can cut your memorizing in half. Notice that 9 + 7 is the same as 7 + 9.

2–2 ADDITION WITHOUT "CARRYING"

Turn to Problems 2–2 and do exercises 21–42 mentally without counting. If you get at least 20 answers correct, go on to section 2–3. If not, study section 2–2. To add numbers with two or more digits, we first add ones, then tens, then hundreds, and so on. First, we shall illustrate addition by using expanded notation.

■ **Example 1**

Add 56 and 42 using expanded notation.

Solution **Step 1** Expand each number and place them in column form.

5 tens + 6 ones
4 tens + 2 ones

Step 2 Add the ones.

5 tens + 6 ones
4 tens + 2 ones
8 ones

Step 3 Add the tens.

5 tens + 6 ones
4 tens + 2 ones
9 tens + 8 ones

Step 4 Change to a standard number.

9 tens + 8 ones = 98

Practice check: Do exercises 15–17 in the margin.

This is not the method we will use to add numbers. This expanded method is used to show the meaning of addition. To add we use the short method.

■ **Example 2**

Add 56 and 42 using the short method.

Solution **Step 1** Write the numbers in columns.

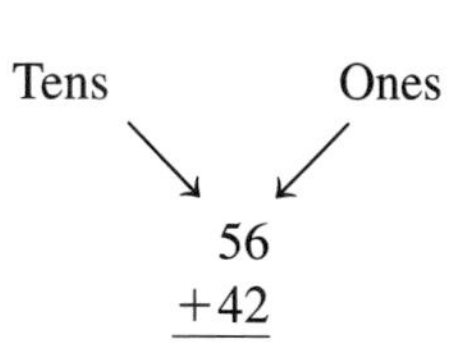

Answers to Chapter 2 Pretest

1. 7 **2.** 13 **3.** 11 **4.** 11 **5.** 12 **6.** 14 **7.** 14 **8.** 15 **9.** 17 **10.** 16 **11.** 13 **12.** 13 **13.** 12 **14.** 18 **15.** 12 **16.** 78 **17.** 6298 **18.** 1361 **19.** 13,040

Answers to Chapter 2 Pretest

20. 851 **21.** 2000 **22.** 1024 **23.** 8 **24.** 7 **25.** 7 **26.** 9 **27.** 6 **28.** 4 **29.** 4 **30.** 6 **31.** 7 **32.** 7 **33.** 9 **34.** 8 **35.** 3532 **36.** 496 **37.** 4212 **38.** 8

Answers to exercises 1–6

1. 9 **2.** 11 **3.** 9 **4.** 11 **5.** 8 **6.** 11

Answers to exercises 7–10

7.

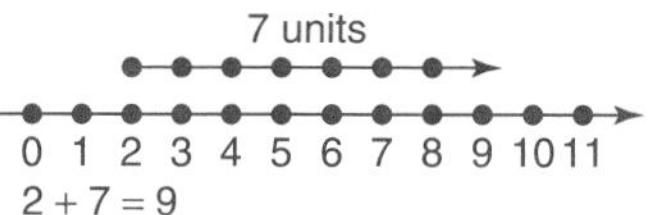

8.

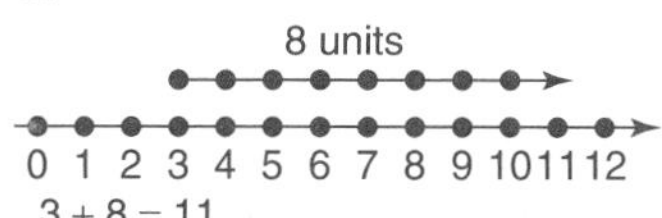

9.

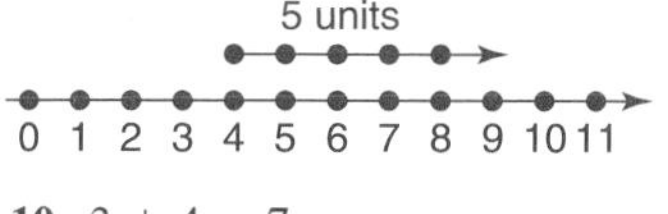

10. 3 + 4 = 7

Answers to exercises 11–14

11. 10 **12.** 2 **13.** 88 **14.** 27

Add using expanded notation.

15. 44 and 32 **16.** 12 and 65

17. 73 and 16

The answers are in the margin on page 34.

Answers to exercises 15–17

15. 76 **16.** 77 **17.** 89

Add using the short method.

18. $\begin{array}{r} 53 \\ +34 \\ \hline \end{array}$ **19.** $\begin{array}{r} 29 \\ +50 \\ \hline \end{array}$

20. 84 + 15 = **21.** 45 + 22 =

The answers are in the margin on page 36.

Add using the short method.

22. $\begin{array}{r} 305 \\ +493 \\ \hline \end{array}$ **23.** $\begin{array}{r} 613 \\ +315 \\ \hline \end{array}$

24. 116 + 712 =

25. 290 + 402 =

26. $\begin{array}{r} 4206 \\ +2731 \\ \hline \end{array}$ **27.** $\begin{array}{r} 2330 \\ +5456 \\ \hline \end{array}$

28. 6018 + 3911 =

29. 1455 + 6223 =

The answers are in the margin on page 36.

Step 2 Add ones. $\begin{array}{r} 56 \\ +42 \\ \hline 8 \end{array}$ (Think: 6 + 2 = 8)

Step 3 Add tens. $\begin{array}{r} 56 \\ +42 \\ \hline 98 \end{array}$ (Think: 5 + 4 = 9)

We must be careful to put the ones and tens in straight columns.

Practice check: Do exercises 18–21 in the margin.

In addition, we add from right to left.

■ Example 3

Add 1403 and 7503 using the short method.

Solution

$\begin{array}{r} 1403 \\ +7503 \\ \hline 6 \end{array}$ ↑ Add ones.

$\begin{array}{r} 1403 \\ +7503 \\ \hline 06 \end{array}$ ↑ Add tens.

$\begin{array}{r} 1403 \\ +7503 \\ \hline 906 \end{array}$ ↑ Add hundreds.

$\begin{array}{r} 1403 \\ +7503 \\ \hline 8906 \end{array}$ ↑ Add thousands.

Be sure to keep the columns very straight.

Practice check: Do exercises 22–29 in the margin.

■ Looking Back

You should now be able to add two numbers that do not require "carrying."

■ Problems 2–2

Review

1. 3 + 4 **2.** 5 + 8 **3.** 3 + 5 **4.** 4 + 7

5. 5 + 9 **6.** 5 + 7 **7.** 6 + 9 **8.** 4 + 8

9. 5 + 6 **10.** 6 + 8 **11.** 4 + 9 **12.** 6 + 7

13. 7 + 9 **14.** 3 + 8 **15.** 7 + 8 **16.** 3 + 9

17. $8 + 9$ **18.** $2 + 9$ **19.** $9 + 6$ **20.** $9 + 4$

Add using the short method.

21. $\begin{array}{r} 28 \\ +1 \\ \hline \end{array}$ **22.** $\begin{array}{r} 15 \\ +3 \\ \hline \end{array}$ **23.** $\begin{array}{r} 34 \\ +53 \\ \hline \end{array}$ **24.** $\begin{array}{r} 26 \\ +43 \\ \hline \end{array}$

25. $\begin{array}{r} 81 \\ +15 \\ \hline \end{array}$ **26.** $\begin{array}{r} 56 \\ +32 \\ \hline \end{array}$ **27.** $29 + 50$ **28.** $71 + 26$

29. $\begin{array}{r} 622 \\ +71 \\ \hline \end{array}$ **30.** $\begin{array}{r} 354 \\ +34 \\ \hline \end{array}$ **31.** $\begin{array}{r} 450 \\ +329 \\ \hline \end{array}$ **32.** $\begin{array}{r} 102 \\ +213 \\ \hline \end{array}$

33. $\begin{array}{r} 123 \\ +666 \\ \hline \end{array}$ **34.** $\begin{array}{r} 222 \\ +456 \\ \hline \end{array}$ **35.** $432 + 67$ **36.** $874 + 21$

37. $\begin{array}{r} 5441 \\ +230 \\ \hline \end{array}$ **38.** $\begin{array}{r} 7340 \\ +341 \\ \hline \end{array}$ **39.** $\begin{array}{r} 4315 \\ +2182 \\ \hline \end{array}$ **40.** $\begin{array}{r} 8430 \\ +1438 \\ \hline \end{array}$

41. $1084 + 815$ **42.** $6441 + 530$ **43.** $\begin{array}{r} 8430 \\ +4538 \\ \hline \end{array}$ **44.** $\begin{array}{r} 8776 \\ +7013 \\ \hline \end{array}$

45. $\begin{array}{r} 43{,}244 \\ +23{,}145 \\ \hline \end{array}$ **46.** $\begin{array}{r} 54{,}834 \\ +44{,}155 \\ \hline \end{array}$ **47.** $62{,}781 + 210$

48. $81{,}005 + 972$

The answers to odd-numbered problems are in the back of the book.

Answers to exercises 18–21

18. 87 **19.** 79 **20.** 84 +15 = 99

21. 45 +22 = 67

Answers to exercises 22–29

22. 798 **23.** 928 **24.** 828
25. 692 **26.** 6937 **27.** 7786
28. 6018 +3911 = 9929 **29.** 1455 +6223 = 7678

2–3 ADDITION WITH "CARRYING"

We shall use expanded notation to show the meaning of "carrying" in addition, so that you will understand the short method.

"Carrying" Tens

■ Example 1

Add 36 and 28.

Solution

$$\begin{array}{rcl} 36 & = & 3 \text{ tens} + \ \ 6 \text{ ones} \\ \underline{+28} & = & \underline{2 \text{ tens} + \ \ 8 \text{ ones}} \\ & & 5 \text{ tens} + 14 \text{ ones} \end{array}$$

↑ Add the tens. ↑ Add the ones.

Look at the 14 ones. Notice that 14 ones = 1 ten + 4 ones. We cannot have more than 9 ones in the ones column (we have 14), so we have to "carry" the extra 1 ten over into the tens column.

$$\begin{array}{l} 3 \text{ tens} + \ \ 6 \text{ ones} \\ \underline{2 \text{ tens} + \ \ 8 \text{ ones}} \\ 5 \text{ tens} + 14 \text{ ones} \end{array}$$

5 tens + 1 ten + 4 ones

6 tens + 4 ones = 64

■ Example 2

Add 38 and 55 using expanded notation.

Solution **Step 1** Add ones. (Think: 8 ones + 5 ones = 13 ones.)

$$\begin{array}{rcl} \text{"Carry"} \longrightarrow & & 1 \text{ ten} \\ 38 & = & 3 \text{ tens} + 8 \text{ ones} \\ \underline{+55} & = & \underline{5 \text{ tens} + 5 \text{ ones}} \\ & & 3 \text{ ones} \longleftarrow \text{"Keep"} \end{array}$$

We "carry" 1 ten and "keep" 3 ones

Step 2 Add tens.

$$\begin{array}{l} 1 \text{ ten} \\ 3 \text{ tens} + 8 \text{ ones} \\ \underline{5 \text{ tens} + 5 \text{ ones}} \\ 9 \text{ tens} + 3 \text{ ones} = 93 \end{array}$$

When we add for "real," we use the short method.

■ Example 3

Add 27 and 48 using the short method.

Solution **Step 1** Add ones. (Think: 7 + 8 = 15.)

"Carry" 1 ten. ⟶ 1

$$\begin{array}{r} 27 \\ +48 \\ \hline 5 \end{array}$$ ⟵ Keep 5 ones

Step 2 Add tens. (Think: 1 + 2 + 4 = 7.)

$$\begin{array}{r} 1 \\ 27 \\ +48 \\ \hline 75 \end{array}$$

Practice check: Do exercises 30–32 in the margin.

Add using the short method.

30. $\begin{array}{r} 27 \\ +38 \\ \hline \end{array}$ **31.** $\begin{array}{r} 33 \\ +19 \\ \hline \end{array}$ **32.** $\begin{array}{r} 47 \\ +23 \\ \hline \end{array}$

The answers are in the margin on page 39.

"Carrying" Hundreds

■ **Example 4**

Add 681 and 273 using expanded notation.

Solution **Step 1** Add ones.

6 hundreds + 8 tens + 1 one
2 hundreds + 7 tens + 3 ones
———————————————
4 ones
(No "carrying" needed.)

Step 2 Add tens. (Think: 8 tens + 7 tens = 15 tens = 1 hundred + 5 tens.)

We need to "keep" 5 tens and "carry" 1 hundred.

"Carry" ⟶ 1 hundred
6 hundreds + 8 tens + 1 one
2 hundreds + 7 tens + 3 ones
———————————————
+ 5 tens + 4 ones
↑
"Keep"

Step 3 Add hundreds

1 hundred
6 hundreds + 8 tens + 1 one
2 hundreds + 7 tens + 3 ones
———————————————
9 hundreds + 5 tens + 4 ones = 954

■ **Example 5**

Add 567 and 371 using the short method.

Solution **Step 1** Add ones. (Think: 7 + 1 = 8. No "carrying" needed.)

$$\begin{array}{r} 567 \\ +371 \\ \hline 8 \end{array}$$

Step 2 Add tens. (Think: $6 + 7 = 13$. "Keep" 3 tens, "carry" 1 hundred.)

$$\text{"Carry"} \longrightarrow \begin{array}{r} 1 \\ 567 \\ +371 \\ \hline 38 \end{array}$$

↑ "Keep"

Step 3 Add hundreds. (Think: $1 + 5 + 3 = 9$.)

$$\begin{array}{r} 1 \\ 567 \\ +371 \\ \hline 938 \end{array}$$

To check, we add in reverse order.

$$\begin{array}{r} 1 \\ 371 \\ +567 \\ \hline 938 \end{array}$$

Practice check: Do exercises 33–36 in the margin.

Add and check using the short method.

33. $\begin{array}{r} 382 \\ +495 \\ \hline \end{array}$ **34.** $\begin{array}{r} 163 \\ +574 \\ \hline \end{array}$

35. $\begin{array}{r} 746 \\ +193 \\ \hline \end{array}$ **36.** $\begin{array}{r} 350 \\ +483 \\ \hline \end{array}$

The answers are in the margin on page 40.

"Carrying" Thousands

■ **Example 6**

Add 6419 and 2730 using expanded notation.

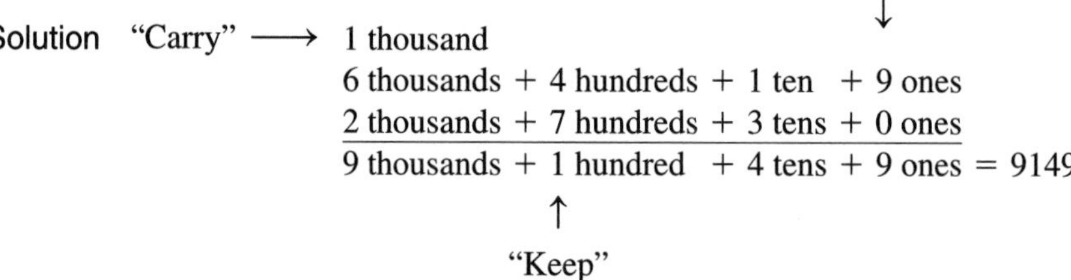

■ **Example 7**

Add 4832 and 8346 and check using the short method.

Solution

$$\begin{array}{r} \\ 4832 \\ +8346 \\ \hline 8 \end{array} \qquad \begin{array}{r} \\ 4832 \\ +8346 \\ \hline 78 \end{array} \qquad \begin{array}{r} 1 \\ 4832 \\ +8346 \\ \hline 178 \end{array} \qquad \begin{array}{r} 1 \\ 4832 \\ +8346 \\ \hline 13178 \end{array}$$

↑	↑	↑	↑
Add ones.	Add tens.	Add hundreds.	Add thousands.
		"Keep" 1 hundred; "carry" 1 thousand.	Write 3 in thousands place and 1 in 10 thousands place.

Check:
```
    1
 8346
+4832
13178
```

Practice check: Do exercises 37 and 38 in the margin.

Combined "Carrying"

■ Example 8

Add 7433 and 9771 and check using the short method.

Solution

	1	11	11
7433	7433	7433	7433
+9771	+9771	+9771	+9771
4	04	204	17204
↑	↑	↑	↑
Add ones.	Add tens.	Add hundreds.	Add thousands.
	"Keep" 0 tens; "carry" 1 hundred.	"Keep" 2 hundreds; "carry" 1 thousand.	

Check:
```
   11
 9771
+7433
17204
```

Practice check: Do exercises 39 and 40 in the margin.

■ Looking Back

You should now be able to add and check any two whole numbers.

■ Problems 2–3

Add and check using the short method.

1. 46 +7 **2.** 53 +9 **3.** 15 +69 **4.** 18 +73

5. 99 +1 **6.** 99 +2 **7.** 467 +392 **8.** 391 +446

9. 490 +813 **10.** 512 +690 **11.** 999 +11 **12.** 999 +1

Answers to exercises 30–32

30. 65 **31.** 52 **32.** 70

Add and check using the short method.

37. 4881 +6903 **38.** 8476 +7821

The answers are in the margin on page 40.

Add and check using the short method.

39. 6783 +5726 **40.** 8349 +2957

The answers are in the margin on page 40.

13. $\begin{array}{r} 709 \\ +201 \\ \hline \end{array}$

14. $\begin{array}{r} 609 \\ +101 \\ \hline \end{array}$

15. $\begin{array}{r} 7624 \\ +2196 \\ \hline \end{array}$

16. $\begin{array}{r} 4773 \\ +3149 \\ \hline \end{array}$

17. $\begin{array}{r} 5889 \\ +2652 \\ \hline \end{array}$

18. $\begin{array}{r} 3936 \\ +2765 \\ \hline \end{array}$

19. $\begin{array}{r} 9941 \\ +8079 \\ \hline \end{array}$

20. $\begin{array}{r} 6599 \\ +5711 \\ \hline \end{array}$

21. $\begin{array}{r} 9999 \\ +1 \\ \hline \end{array}$

22. $\begin{array}{r} 9999 \\ +101 \\ \hline \end{array}$

Add and check.

23. $\begin{array}{r} 56{,}968 \\ +34{,}887 \\ \hline \end{array}$

24. $\begin{array}{r} 77{,}438 \\ +89{,}674 \\ \hline \end{array}$

25. $\begin{array}{r} 194{,}604 \\ +299{,}410 \\ \hline \end{array}$

26. $\begin{array}{r} 380{,}491 \\ +894{,}609 \\ \hline \end{array}$

27. $\begin{array}{r} 999{,}999 \\ +1 \\ \hline \end{array}$

28. $\begin{array}{r} 999{,}999 \\ +10{,}001 \\ \hline \end{array}$

The answers to odd-numbered exercises are in the back of the book.

Answers to exercises 33–36

33. 877 **34.** 737 **35.** 939 **36.** 833

Answers to exercises 37 and 38

37. 11784 **38.** 16297

Answers to exercises 39 and 40

39. 12509 **40.** 11306

2–4 ADDITION OF MORE THAN TWO NUMBERS

Column Addition

When we do column addition, the sum found in each step is added mentally to the next number. This takes much practice.

■ **Example 1**

Add. $\begin{array}{r} 5 \\ 8 \\ 6 \\ +9 \\ \hline \end{array}$

Solution We mentally add two numbers at a time, beginning at the top.

Step 1

$$\begin{array}{r} \left.\begin{array}{r} 5 \\ 8 \end{array}\right\} \\ 6 \\ +9 \\ \hline \end{array} \longrightarrow \begin{array}{r} 5 \\ +8 \\ \hline 13 \end{array}$$

Step 2 Mentally add 13 to the next number, 6.

$$\begin{array}{r} 5 \\ 8 \\ 6 \\ +9 \\ \hline \end{array} \longrightarrow \begin{array}{r} 13 \\ +6 \\ \hline 19 \end{array}$$

Step 3 Mentally add 19 to the last number, 9.

$$\begin{array}{r} 5 \\ 8 \\ 6 \\ +9 \\ \hline \end{array} \longrightarrow \begin{array}{r} 19 \\ +9 \\ \hline 28 \end{array}$$

You must take care to do this process mentally. Review your basic addition facts if necessary. To make certain your sum is correct, you add it again from bottom to top.

Practice check: Do exercises 41–44 in the margin.

Add and check mentally.

41.	**42.**	**43.**	**44.**
5	2	7	9
8	4	1	4
3	9	5	3
4	8	6	2

Redo this exercise until you can add quickly to get the correct sum. The answers are in the margin on page 43.

Example 2 shows addition that involves more than one column of numbers.

■ Example 2

Add 4162, 9608, and 4173 using the short method.

Solution **Step 1** Add ones.

$$\begin{array}{r} 1 \\ 4162 \\ 9608 \\ 4173 \\ \hline 3 \end{array} \begin{array}{l} \longleftarrow \text{“Carry”} \\ \\ \\ \\ \longleftarrow \text{“Keep”} \end{array}$$

Note: $2 + 8 + 3 = 13$. We carry 1 ten and keep 3 ones.

Step 2 Add tens.

$$\begin{array}{r} 11 \\ 4162 \\ 9608 \\ 4173 \\ \hline 43 \end{array} \begin{array}{l} \text{“Carry”} \\ \\ \\ \\ \text{“Keep”} \end{array}$$

Note: $1 + 6 + 0 + 7 = 14$. We carry 1 hundred and keep 4 tens.

Step 3 Add hundreds.

$$\begin{array}{r} 11 \\ 4162 \\ 9608 \\ 4173 \\ \hline 943 \end{array}$$

Add and check.

45.
```
 14
 73
+36
```

46.
```
 394
 925
 368
+560
```

47.
```
 4330
 9426
 5831
+7304
```

The answers are in the margin on page 44.

Step 4 Add thousands.

```
   11
 4162
 9608
 4173
17943
```

Check by adding, starting at the bottom.

Practice check: Do exercises 45–47 in the margin.

■ Looking Back

You should now be able to add and check more than two whole numbers.

■ Problems 2–4

Add and check.

1.
```
 6
 5
 9
+7
```

2.
```
 5
 8
 4
+6
```

3.
```
 9
 4
 3
+7
```

4.
```
 9
 2
 7
+8
```

5.
```
 46
 78
+53
```

6.
```
 59
 22
+41
```

7.
```
 78
 41
+83
```

8.
```
 43
 71
+41
```

9.
```
 432
 149
+571
```

10.
```
 326
 418
+369
```

11.
```
 321
 469
+741
```

12.
```
 304
 691
+439
```

13.
```
 3041
+6182
```

14.
```
 4293
+6294
```

15.
```
 3004
+7148
```

16.
```
 9005
+4888
```

17.
```
  258
   16
 1082
+315
```

18.
```
   29
 4101
   61
+980
```

19.
```
 2116
   82
  641
+5019
```

20.
```
  389
   16
 4302
  +43
```

Arrange vertically and add.

SAMPLE: 51 + 6141 + 18 + 301 *Solution*

$$\begin{array}{r} 51 \\ 6141 \\ 18 \\ \underline{301} \\ 6511 \end{array}$$

Answers to exercises 41–44

41. 20 **42.** 23 **43.** 19 **44.** 18

21. 29 + 3041 + 29 + 461

22. 304 + 39 + 4005 + 17

23. 1941 + 63 + 305 + 6052

24. 30 + 18 + 410 + 3643

25.
$$\begin{array}{r} 3148 \\ 6932 \\ 4193 \\ \underline{+1484} \end{array}$$

26.
$$\begin{array}{r} 5668 \\ 3204 \\ 9832 \\ \underline{+1098} \end{array}$$

27.
$$\begin{array}{r} 6{,}241{,}803 \\ 763 \\ 401{,}039 \\ 9{,}840 \\ \underline{+19} \end{array}$$

28.
$$\begin{array}{r} 31 \\ 104{,}316 \\ 9{,}041{,}098 \\ 462 \\ \underline{+7{,}058} \end{array}$$

The answers to odd-numbered exercises are in the back of the book.

2–5 ESTIMATING SUMS

A Review of Rounding

Rule

To round to the nearest ten, hundred, or thousand:

Step 1 Locate the place value number to be rounded to (tens, hundreds, or thousands).

Step 2 Look at the digit to the right of this number.

- **a.** If the digit is 5 or more, round up.
- **b.** If the digit is less than 5, round down.

■ **Example 1**

Round 648 to the nearest ten.

Solution 648 ← 5 or more

↓

650 Raise 4 tens to 5 tens

Answers to exercises 45–47

45. 123 **46.** 2247 **47.** 26,891

■ **Example 2**

Round 6251 to the nearest hundred.

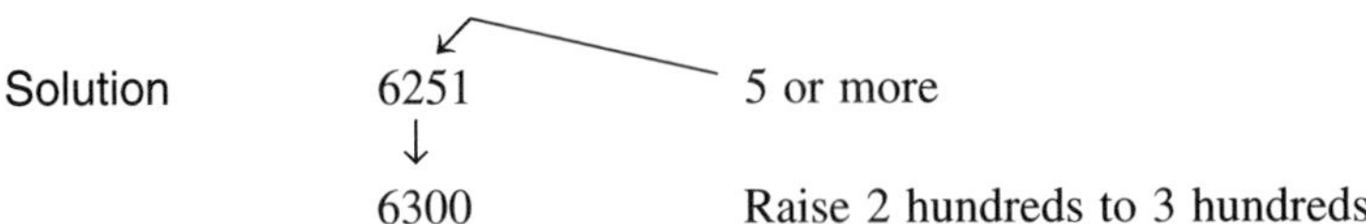

■ **Example 3**

Round 15,268 to the nearest thousand.

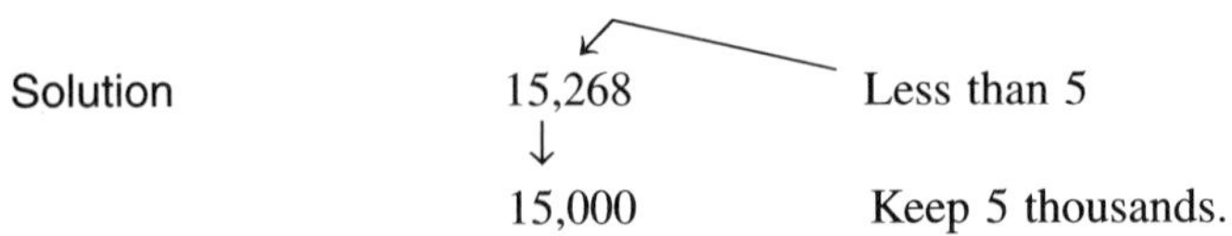

Practice check: Do exercises 48–50 in the margin.

48. Round 274 to the nearest ten.

49. Round 4579 to the nearest hundred.

50. Round 90,562 to the nearest thousand.

The answers are in the margin on page 47.

Estimating

Sometimes when we want to be less than accurate, when we want to find out "about how many," or when we don't care for an exact answer, we estimate. Examples of estimates: the population of the world is about 4 billion; the air line distance between New York and Moscow is about 5000 miles.

We estimate sums by first rounding each addend. Notice how much simpler the addition problem becomes.

■ **Example 4**

Estimate the sum by first rounding each addend to the nearest ten.

Solution

	Nearest ten
84	80
68	70
31	30
+15	+20
	200

The sum is about 200.

■ **Example 5**

Estimate the sum by first rounding each addend to the nearest hundred.

Solution

	Nearest hundred
645	600
780	800
231	200
+859	+900
	2500

The sum is about 2500.

■ **Example 6**

Estimate the sum by first rounding each addend to the nearest thousand.

Solution

	Nearest thousand
6083	6000
2560	3000
1718	2000
+8346	+8000
	19,000

The sum is about 19,000.

By rounding the addends in example 6 to the nearest ten, the estimate will be closer to the exact answer, but again the person must decide how close the guess needs to be.

Rule

To Estimate a Sum:

Step 1 Round all addends either to the nearest ten, hundred, or thousand.

Step 2 Find the sum of the rounded addends.

Practice check: Do exercises 51–53 in the margin.

51. Estimate the sum by first rounding each addend to the nearest ten

66
12
84
+35

52. Estimate the sum by first rounding each addend to the nearest hundred.

782 + 216 + 930 + 156

53. Give three estimates of the sum by first rounding each addend to the nearest ten, hundred, and thousand.

5710 + 1099 + 7520 + 3361

The answers are in the margin on page 47.

■ Looking Back

You should now be able to estimate the sum to the nearest ten, hundred, or thousand when you are given two or more whole numbers.

■ Problems 2–5

Review

Add and check.

1. 5268 +7321

2. 7081 +2399

3. 15 + 4684 + 32 + 980

4. 560 + 81 + 2041 + 18

Round to the nearest ten, then hundred, then thousand.

5. 6582

6. 3815

Estimate the sums by rounding each addend to the nearest ten. Then find the exact sums and compare.

7. 65 31 44 +76

8. 24 87 65 +96

9. 67 91 24 +37

10. 85 37 23 +19

Estimate the sums by rounding each addend to the nearest hundred. Then find the exact sums and compare.

11. 273
189
726
+508

12. 829
684
729
+432

13. 701
142
387
+495

14. 708
554
621
+386

Estimate the sums by rounding each addend to the nearest thousand. Then find the exact sums and compare.

15. 4237
6489
5076
+1684

16. 1467
3015
9864
+2046

17. 1288
3074
4385
+7907

18. 6872
8360
6009
+3785

Estimate the sums by rounding each addend to the nearest ten and then to the nearest hundred. By comparing the estimates, which sums do you find incorrect?

19. 386
747
+593
1726

20. 829
495
387
+726
2137

21. 2,927
4,385
3,074
+1,288
11,674

22. 8,360
3,496
3,785
+5,842
21,483

Give three estimates of the sum by rounding each addend to the nearest ten, hundred, and thousand. Then find the exact answer and compare.

23. 18,745
10,674
60,485
+12,008

24. 94,036
87,468
34,805
+58,009

The answers to review and odd-numbered exercises are in the back of the book.

2–6 APPLIED PROBLEMS

Solving word problems helps you to apply your computational skills to real-life situations. First read the problem very carefully. It is important to note that the answer to a word problem is not complete unless it is in the form of a statement.

■ **Example 1**

Sam Markette is a manager of a supermarket. In one evening his stockers stacked 325 cans of beans, 189 cans of peas, 241 cans of corn, and 68 cans of tomatoes on the shelves. How many cans of vegetables did the stockers stack?

Solution We need to find the total number of cans that were stacked, so we add.

$$\begin{array}{r} 325 \\ 189 \\ 241 \\ \underline{68} \\ 823 \end{array}$$

The men stacked 823 cans of vegetables.

Answers to exercises 48–50

48. 270 **49.** 4600 **50.** 91,000

Answers to exercises 51–53

51. 200 **52.** 2100
53. 17,690 nearest ten
17,700 nearest hundred
18,000 nearest thousand

■ Example 2

The attendance at a basketball tournament on three nights was 2341, 1986, and 3067. Find the total attendance.

Solution To find the total attendance we add.

$$\begin{array}{r} 2341 \\ 1986 \\ \underline{3067} \\ 7394 \end{array}$$

The total attendance was 7,394.

■ Looking Back

You should now be able to solve applied word problems involving addition.

■ Problems 2–6

Write a statement for the answer.

1. **Sales.** A salesperson received the following bonuses during a certain week: $87 on Monday, $66 on Wednesday, and $93 on Friday. Find the total bonus for the week.

2. **Welding.** Ned works in a welding shop that manufactured the following amounts of items in a certain week: 570 lb, 264 lb, 382 lb, 101 lb, and 236 lb. Find the total number of pounds of items made during that week.

3. **Business.** A service station in Jacksonville sold the following number of gallons of gas: Monday, 843; Tuesday, 689; Wednesday, 704; Thursday, 961; Friday, 756. How many gallons were sold during the five days?

4. **Management.** During March, taxi 23 of the Green Balm Taxi Company traveled the following numbers of miles: first week, 756; second week, 971; third week, 389; fourth week, 609. How many miles did the taxi travel during the month?

5. **Accounting.** Sam Winthrop paid his four employees monthly salaries of $1250, $1826, $1394, and $832. What was his total monthly payroll?

6. **Sales.** During May, Agatha Cradelmeyer's Apparel sold 385 skirts, 1056 blouses, and 847 slacks. Find the total number of skirts, blouses, and slacks sold.

7. **Accounting.** An auditorium has 878 seats on the main floor and 224 seats in the balcony. How many tickets would have to be sold for a full house?

8. **Marketing.** An appliance outlet sold 43,086 kitchen stoves during the first quarter of the year and 55,149 kitchen stoves during the second quarter. How many units were sold during the first 6 months?

9. **Clerical.** The census for a certain town has 21,560 persons under 25 years old, 17,981 between 25 and 62 years, and 9472 over 62 years old. How many people live in that town?

10. **Agriculture.** During one day a farmer delivered four truckloads of grain to the grain elevator. The weights of the loads were 3471 lb, 2968 lb, 3348 lb, and 2746 lb. About how many pounds of grain did the farmer deliver that day to the nearest hundred?

The answers to odd-numbered exercises are in the back of the book.

2–7 BASIC SUBTRACTION FACTS

Turn to Problems 2–7 and do the subtraction mentally using the addition facts without counting. This is very important! If you get at least 80 answers correct, go on to section 2–8. If not, study section 2–7.

In subtraction of whole numbers, we start with a number and take some of it away. The number that is left is called the difference. For example, suppose we have six triangles and take two of them away.

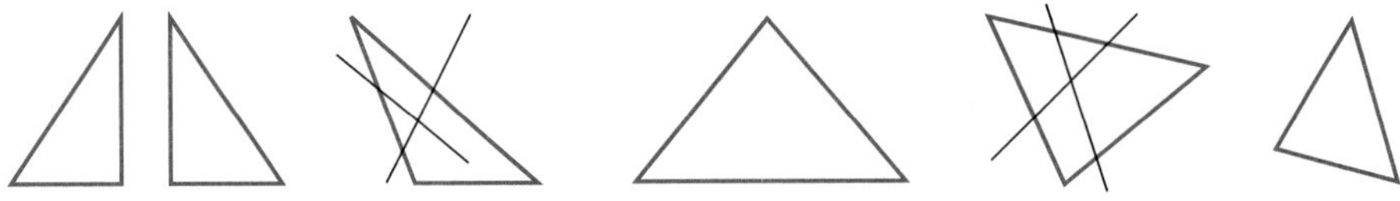

Figure 2–3

Obviously we have four triangles left.

This can be written as $6 - 2 = 4$, or

$$\begin{array}{rl} 6 & \longleftarrow \text{Minuend} \\ \underline{-2} & \longleftarrow \text{Subtrahend} \\ 4 & \longleftarrow \text{Difference} \end{array}$$

We can say, “6 take away 2 is 4,” or “2 from 6 is 4.”

Subtraction can also be illustrated using the number line.

■ **Example 1**

Find the difference: $7 - 3$.

Solution

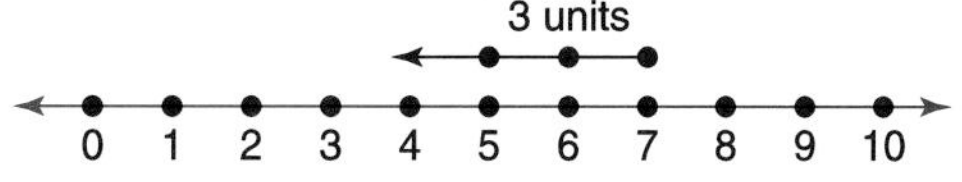

Figure 2–4

Begin at 7 and draw the arrow three units to the *left*. The difference is 4.

Practice check: Do exercises 54–59 in the margin.

Do the following subtractions using the number line.

54. $8 - 3$ **55.** $5 - 3$
56. $9 - 7$ **57.** $5 - 4$
58. $8 - 5$ **59.** $7 - 3$

The answers are in the margin on page 51.

Subtraction is really undoing addition.

■ **Example 2**

Subtract.

$$\begin{array}{r} 9 \\ -6 \\ \hline \end{array}$$

Solution Compare these examples

$$6 + 3 = 9 \quad \text{and} \quad 9 - 6 = 3$$

Knowing that $6 + 3$ is 9, we can get the difference by looking at the 6 and thinking, "What number plus 6 is 9?"

The difference is 3, and $9 - 6 = 3$.

■ **Example 3**

Subtract.

$$\begin{array}{r} 13 \\ -7 \\ \hline \end{array}$$

Solution $\begin{array}{r} 13 \\ -7 \\ \hline \end{array}$ Look at the 7 and think: *What number plus 7 is 13?* $\begin{array}{r} 13 \\ -7 \\ \hline 6 \end{array}$

The difference is 6.

■ **Example 4**

Subtract.

$$15 - 0$$

Solution $\begin{array}{r} 15 \\ -0 \\ \hline \end{array}$ Look at the 0 and think: *What number plus 0 is 15?* $\begin{array}{r} 15 \\ -0 \\ \hline 15 \end{array}$

The difference is 15.

■ **Example 5**

Subtract.

$$5 - 5$$

Subtract using the basic addition facts to get the difference.

60. $\begin{array}{r} 15 \\ -9 \\ \hline \end{array}$ **61.** $\begin{array}{r} 10 \\ -7 \\ \hline \end{array}$ **62.** $\begin{array}{r} 9 \\ -2 \\ \hline \end{array}$

63. $\begin{array}{r} 11 \\ -8 \\ \hline \end{array}$

The answers are in the margin on page 52.

Solution $\begin{array}{r} 5 \\ -5 \\ \hline \end{array}$ Look at the 5 and think: *What number plus 5 is 5?* $\begin{array}{r} 5 \\ -5 \\ \hline 0 \end{array}$

The difference is 0.

Practice check: Do exercises 60–63 in the margin.

■ Looking Back

You should now be able to find the difference such as 11 − 6, and 17 − 8 using the basic addition facts.

Before doing Problems 2–7, go back to Problems 2–1 and review the basic addition facts.

■ Problems 2–7

Subtract using the basic addition facts to get the difference. Force yourself to be quick. Read the answers to someone while he/she checks them in the back of the book.

1. $\begin{array}{r} 10 \\ -1 \\ \hline \end{array}$	**2.** $\begin{array}{r} 18 \\ -9 \\ \hline \end{array}$	**3.** $\begin{array}{r} 11 \\ -2 \\ \hline \end{array}$	**4.** $\begin{array}{r} 16 \\ -8 \\ \hline \end{array}$	**5.** $\begin{array}{r} 9 \\ -8 \\ \hline \end{array}$	**6.** $\begin{array}{r} 17 \\ -9 \\ \hline \end{array}$	**6.** $\begin{array}{r} 14 \\ -7 \\ \hline \end{array}$
8. $\begin{array}{r} 2 \\ -2 \\ \hline \end{array}$	**9.** $\begin{array}{r} 12 \\ -3 \\ \hline \end{array}$	**10.** $\begin{array}{r} 9 \\ -2 \\ \hline \end{array}$	**11.** $\begin{array}{r} 8 \\ -1 \\ \hline \end{array}$	**12.** $\begin{array}{r} 15 \\ -8 \\ \hline \end{array}$	**13.** $\begin{array}{r} 12 \\ -6 \\ \hline \end{array}$	**14.** $\begin{array}{r} 3 \\ -0 \\ \hline \end{array}$
15. $\begin{array}{r} 11 \\ -3 \\ \hline \end{array}$	**16.** $\begin{array}{r} 8 \\ -6 \\ \hline \end{array}$	**17.** $\begin{array}{r} 7 \\ -1 \\ \hline \end{array}$	**18.** $\begin{array}{r} 16 \\ -9 \\ \hline \end{array}$	**19.** $\begin{array}{r} 13 \\ -7 \\ \hline \end{array}$	**20.** $\begin{array}{r} 4 \\ -4 \\ \hline \end{array}$	**21.** $\begin{array}{r} 10 \\ -5 \\ \hline \end{array}$
22. $\begin{array}{r} 13 \\ -9 \\ \hline \end{array}$	**23.** $\begin{array}{r} 6 \\ -5 \\ \hline \end{array}$	**24.** $\begin{array}{r} 10 \\ -7 \\ \hline \end{array}$	**25.** $\begin{array}{r} 7 \\ -5 \\ \hline \end{array}$	**26.** $\begin{array}{r} 5 \\ -0 \\ \hline \end{array}$	**27.** $\begin{array}{r} 14 \\ -8 \\ \hline \end{array}$	**28.** $\begin{array}{r} 11 \\ -6 \\ \hline \end{array}$
29. $\begin{array}{r} 5 \\ -4 \\ \hline \end{array}$	**30.** $\begin{array}{r} 8 \\ -4 \\ \hline \end{array}$	**31.** $\begin{array}{r} 12 \\ -8 \\ \hline \end{array}$	**32.** $\begin{array}{r} 9 \\ -6 \\ \hline \end{array}$	**33.** $\begin{array}{r} 6 \\ -4 \\ \hline \end{array}$	**34.** $\begin{array}{r} 4 \\ -3 \\ \hline \end{array}$	**35.** $\begin{array}{r} 15 \\ -9 \\ \hline \end{array}$
36. $\begin{array}{r} 12 \\ -7 \\ \hline \end{array}$	**37.** $\begin{array}{r} 7 \\ -7 \\ \hline \end{array}$	**38.** $\begin{array}{r} 9 \\ -5 \\ \hline \end{array}$	**39.** $\begin{array}{r} 6 \\ -3 \\ \hline \end{array}$	**40.** $\begin{array}{r} 3 \\ -2 \\ \hline \end{array}$	**41.** $\begin{array}{r} 14 \\ -9 \\ \hline \end{array}$	**42.** $\begin{array}{r} 11 \\ -7 \\ \hline \end{array}$
43. $\begin{array}{r} 8 \\ -0 \\ \hline \end{array}$	**44.** $\begin{array}{r} 8 \\ -5 \\ \hline \end{array}$	**45.** $\begin{array}{r} 13 \\ -9 \\ \hline \end{array}$	**46.** $\begin{array}{r} 2 \\ -1 \\ \hline \end{array}$	**47.** $\begin{array}{r} 10 \\ -6 \\ \hline \end{array}$	**48.** $\begin{array}{r} 7 \\ -4 \\ \hline \end{array}$	**49.** $\begin{array}{r} 9 \\ -0 \\ \hline \end{array}$

50. $\begin{array}{r} 4 \\ -2 \\ \hline \end{array}$	**51.** $\begin{array}{r} 5 \\ -3 \\ \hline \end{array}$	**52.** $\begin{array}{r} 6 \\ -6 \\ \hline \end{array}$	**53.** $\begin{array}{r} 10 \\ -9 \\ \hline \end{array}$	**54.** $\begin{array}{r} 11 \\ -9 \\ \hline \end{array}$	**55.** $\begin{array}{r} 9 \\ -1 \\ \hline \end{array}$	**56.** $\begin{array}{r} 17 \\ -8 \\ \hline \end{array}$
57. $\begin{array}{r} 12 \\ -9 \\ \hline \end{array}$	**58.** $\begin{array}{r} 9 \\ -7 \\ \hline \end{array}$	**59.** $\begin{array}{r} 8 \\ -7 \\ \hline \end{array}$	**60.** $\begin{array}{r} 15 \\ -7 \\ \hline \end{array}$	**61.** $\begin{array}{r} 11 \\ -8 \\ \hline \end{array}$	**62.** $\begin{array}{r} 8 \\ -2 \\ \hline \end{array}$	**63.** $\begin{array}{r} 7 \\ -6 \\ \hline \end{array}$
64. $\begin{array}{r} 16 \\ -7 \\ \hline \end{array}$	**65.** $\begin{array}{r} 13 \\ -6 \\ \hline \end{array}$	**66.** $\begin{array}{r} 13 \\ -4 \\ \hline \end{array}$	**67.** $\begin{array}{r} 6 \\ -1 \\ \hline \end{array}$	**68.** $\begin{array}{r} 10 \\ -3 \\ \hline \end{array}$	**69.** $\begin{array}{r} 7 \\ -2 \\ \hline \end{array}$	**70.** $\begin{array}{r} 14 \\ -6 \\ \hline \end{array}$
71. $\begin{array}{r} 11 \\ -5 \\ \hline \end{array}$	**72.** $\begin{array}{r} 5 \\ -1 \\ \hline \end{array}$	**73.** $\begin{array}{r} 12 \\ -4 \\ \hline \end{array}$	**74.** $\begin{array}{r} 9 \\ -3 \\ \hline \end{array}$	**75.** $\begin{array}{r} 6 \\ -2 \\ \hline \end{array}$	**76.** $\begin{array}{r} 4 \\ -1 \\ \hline \end{array}$	**77.** $\begin{array}{r} 15 \\ -6 \\ \hline \end{array}$
78. $\begin{array}{r} 12 \\ -5 \\ \hline \end{array}$	**79.** $\begin{array}{r} 9 \\ -4 \\ \hline \end{array}$	**80.** $\begin{array}{r} 3 \\ -1 \\ \hline \end{array}$	**81.** $\begin{array}{r} 14 \\ -5 \\ \hline \end{array}$	**82.** $\begin{array}{r} 11 \\ -4 \\ \hline \end{array}$	**83.** $\begin{array}{r} 8 \\ -3 \\ \hline \end{array}$	**84.** $\begin{array}{r} 13 \\ -5 \\ \hline \end{array}$
85. $\begin{array}{r} 10 \\ -4 \\ \hline \end{array}$	**86.** $\begin{array}{r} 7 \\ -3 \\ \hline \end{array}$	**87.** $\begin{array}{r} 5 \\ -2 \\ \hline \end{array}$	**88.** $\begin{array}{r} 7 \\ -7 \\ \hline \end{array}$			

The answers are in the back of the book.

2–8 SUBTRACTION WITHOUT "BORROWING"

Turn to Problems 2–8 and do exercises 21–46 mentally without counting. If you get at least 23 answers correct, go on to section 2–9. If not, study section 2–8.

To subtract numbers with two or more digits, we subtract ones, then tens, then hundreds, and so on. First, we will illustrate subtraction by using expanded notation.

■ Example 1

Subtract 72 from 93.

Solution **Step 1** Expand each numeral and place them in column form.

$$\begin{array}{l} 9 \text{ tens} + 3 \text{ ones} \\ 7 \text{ tens} + 2 \text{ ones} \end{array}$$

Answers to exercises 54–59

54.

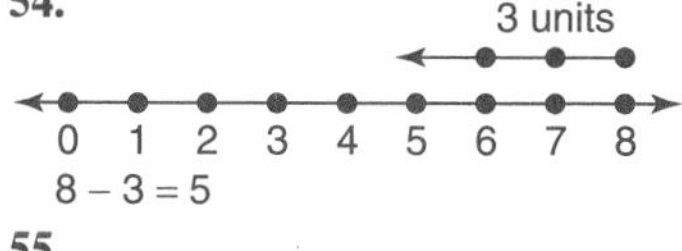

55.

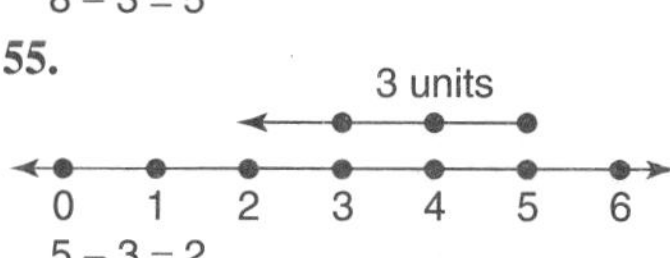

56.

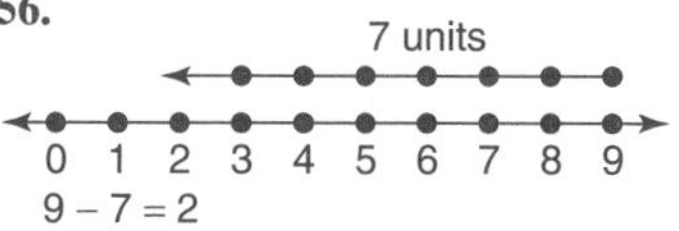

57.

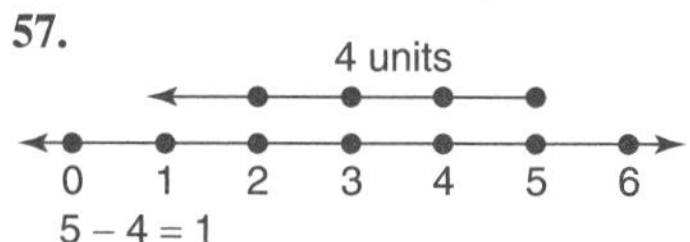

58.

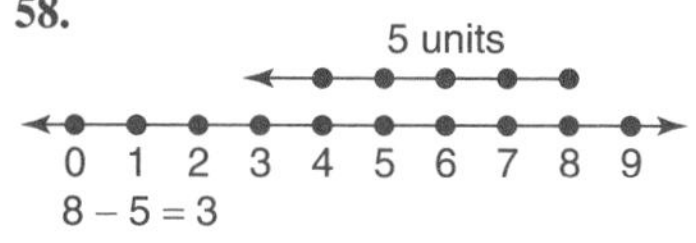

59.

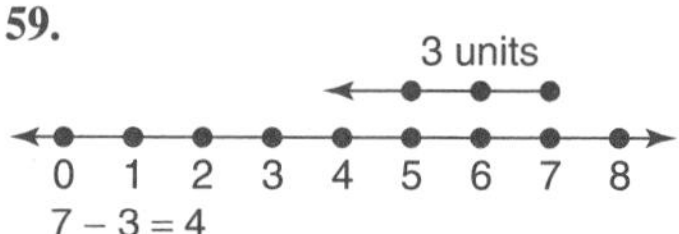

Answers to exercises 60–63

60. 6 **61.** 3
62. 7 **63.** 3

Step 2 Subtract the ones.

$$\begin{array}{r} 9 \text{ tens} + 3 \text{ ones} \\ \underline{7 \text{ tens} + 2 \text{ ones}} \\ 1 \text{ one} \end{array} \qquad (\textit{Think: } 3 - 2 = 1.)$$

Step 3 Subtract the tens.

$$\begin{array}{r} 9 \text{ tens} + 3 \text{ ones} \\ \underline{7 \text{ tens} + 2 \text{ ones}} \\ 2 \text{ tens} + 1 \text{ one} \end{array} \qquad (\textit{Think: } 9 - 7 = 2.)$$

Thus, $93 - 72 = 21$

In order to subtract rapidly we use the short method.

■ **Example 2**

Subtract 1431 from 4953 using the short method.

Solution

4953	4953	4953	4953
−1431	−1431	−1431	−1431
2	22	522	3522
↑	↑	↑	↑
Subtract ones.	Subtract tens.	Subtract hundreds.	Subtract thousands.

The difference is 3522.

Practice check: Do exercises 64–66 in the margin.

Subtract using the short method.

64. $\begin{array}{r} 9752 \\ \underline{5340} \end{array}$ **65.** $\begin{array}{r} 6495 \\ \underline{2323} \end{array}$

66. $\begin{array}{r} 8964 \\ \underline{3431} \end{array}$

The answers are in the margin on page 54.

■ Looking Back

You should now be able to subtract two whole numbers when no "borrowing" is involved.

■ Problems 2–8

Try to push yourself to subtract as quickly as possible.

Review

1. $\begin{array}{r} 9 \\ \underline{-2} \end{array}$ **2.** $\begin{array}{r} 8 \\ \underline{-6} \end{array}$ **3.** $\begin{array}{r} 7 \\ \underline{-5} \end{array}$ **4.** $\begin{array}{r} 8 \\ \underline{-4} \end{array}$

5. $\begin{array}{r} 9 \\ \underline{-6} \end{array}$ **6.** $\begin{array}{r} 6 \\ \underline{-4} \end{array}$ **7.** $\begin{array}{r} 9 \\ \underline{-5} \end{array}$ **8.** $\begin{array}{r} 6 \\ \underline{-3} \end{array}$

9. $\begin{array}{r} 8 \\ \underline{-5} \end{array}$ **10.** $\begin{array}{r} 5 \\ \underline{-3} \end{array}$ **11.** $\begin{array}{r} 9 \\ \underline{-7} \end{array}$ **12.** $\begin{array}{r} 8 \\ \underline{-7} \end{array}$

13. $\begin{array}{r} 8 \\ -2 \\ \hline \end{array}$ 14. $\begin{array}{r} 7 \\ -2 \\ \hline \end{array}$ 15. $\begin{array}{r} 9 \\ -3 \\ \hline \end{array}$ 16. $\begin{array}{r} 6 \\ -2 \\ \hline \end{array}$

17. $\begin{array}{r} 9 \\ -4 \\ \hline \end{array}$ 18. $\begin{array}{r} 8 \\ -3 \\ \hline \end{array}$ 19. $\begin{array}{r} 7 \\ -3 \\ \hline \end{array}$ 20. $\begin{array}{r} 5 \\ -2 \\ \hline \end{array}$

Subtract using the short method.

21. $\begin{array}{r} 47 \\ -5 \\ \hline \end{array}$ 22. $\begin{array}{r} 69 \\ -6 \\ \hline \end{array}$ 23. $\begin{array}{r} 79 \\ -48 \\ \hline \end{array}$ 24. $\begin{array}{r} 37 \\ -13 \\ \hline \end{array}$

25. $\begin{array}{r} 59 \\ -31 \\ \hline \end{array}$ 26. $\begin{array}{r} 88 \\ -25 \\ \hline \end{array}$ 27. $\begin{array}{r} 557 \\ -346 \\ \hline \end{array}$ 28. $\begin{array}{r} 682 \\ -140 \\ \hline \end{array}$

29. $\begin{array}{r} 495 \\ -173 \\ \hline \end{array}$ 30. $\begin{array}{r} 797 \\ -543 \\ \hline \end{array}$ 31. $\begin{array}{r} 598 \\ -260 \\ \hline \end{array}$ 32. $\begin{array}{r} 427 \\ -104 \\ \hline \end{array}$

33. $\begin{array}{r} 6419 \\ -3208 \\ \hline \end{array}$ 34. $\begin{array}{r} 9876 \\ -6404 \\ \hline \end{array}$ 35. $\begin{array}{r} 7469 \\ -3343 \\ \hline \end{array}$ 36. $\begin{array}{r} 9504 \\ -4301 \\ \hline \end{array}$

37. $\begin{array}{r} 4936 \\ -1235 \\ \hline \end{array}$ 38. $\begin{array}{r} 8557 \\ -1121 \\ \hline \end{array}$ 39. $\begin{array}{r} 86134 \\ -70022 \\ \hline \end{array}$ 40. $\begin{array}{r} 86593 \\ -61061 \\ \hline \end{array}$

41. $\begin{array}{r} 97999 \\ -10 \\ \hline \end{array}$ 42. $\begin{array}{r} 89999 \\ -111 \\ \hline \end{array}$ 43. $\begin{array}{r} 49394 \\ -6173 \\ \hline \end{array}$ 44. $\begin{array}{r} 69754 \\ -2434 \\ \hline \end{array}$

45. $\begin{array}{r} 34256 \\ -153 \\ \hline \end{array}$ 46. $\begin{array}{r} 67574 \\ -253 \\ \hline \end{array}$

Answers to exercises 64–66

64. 4412 **65.** 4172 **66.** 5533

If you missed any of the assigned exercises, go back and review the subtraction facts in Section 2–7.

The answers to review and odd-numbered exercises are in the back of the book.

2–9 SUBTRACTION WITH "BORROWING"

"Borrowing" from the Tens Place

Consider this subtraction:

$$\begin{array}{r} 54 \\ -17 \\ \hline \end{array}$$

Notice that we cannot subtract 7 from 4. In order to subtract we must use another name for 54. Consider the expanded number for 54:

$$54 = 5 \text{ tens} + 4 \text{ ones}$$

Suppose we take or "borrow" 1 ten of the 5 tens, change it to 10 ones, and add it to the 4 ones.

$$\begin{aligned} 54 &= 5 \text{ tens} + 4 \text{ ones} \\ &= 4 \text{ tens} + 1 \text{ ten} + 4 \text{ ones} \\ &= 4 \text{ tens} + 10 \text{ ones} + 4 \text{ ones} \\ &= 4 \text{ tens} + 14 \text{ ones} \end{aligned}$$

The subtraction in expanded notation becomes

$$\begin{array}{rcccc} 54 & = & 5 \text{ tens} + 4 \text{ ones} & = & 4 \text{ tens} + 14 \text{ ones} \\ \underline{-17} & = & \underline{1 \text{ ten} \ + 7 \text{ ones}} & = & \underline{1 \text{ ten} \ + \ \ 7 \text{ ones}} \end{array}$$

↑ Notice we can now subtract 7 from 14

$$\begin{array}{l} 4 \text{ tens} + 14 \text{ ones} \\ \underline{1 \text{ ten} \ + \ \ 7 \text{ ones}} \\ 3 \text{ tens} + \ \ 7 \text{ ones} = 37 \end{array}$$

The difference is 37.

■ **Example 1**

Subtract 24 from 41 using expanded notation.

Solution

Take 1 ten away. Add 1 ten or 10 ones to 1 one.

$$\begin{array}{l} 4 \text{ tens} + 1 \text{ one} \\ \underline{2 \text{ tens} + 4 \text{ ones}} \end{array} \qquad \begin{array}{l} 3 \text{ tens} + 11 \text{ ones} \\ \underline{2 \text{ tens} + \ \ 4 \text{ ones}} \\ 1 \text{ ten} \ + \ \ 7 \text{ ones} = 17 \end{array}$$

The difference is 17.

Practice check: Do exercises 67–69 in the margin.

Subtract using expanded notation.

67. $\begin{array}{r} 67 \\ -48 \\ \hline \end{array}$ **68.** $\begin{array}{r} 71 \\ -53 \\ \hline \end{array}$ **69.** $\begin{array}{r} 50 \\ -23 \\ \hline \end{array}$

The answers are in the margin on page 57.

The expanded method was used to show the meaning of "borrowing" in subtraction. For calculating quickly it is best to use the short method.

■ **Example 2**

Subtract 28 from 51 using the short method.

Solution **Step 1** We cannot subtract 8 from 1, so we rename 51.

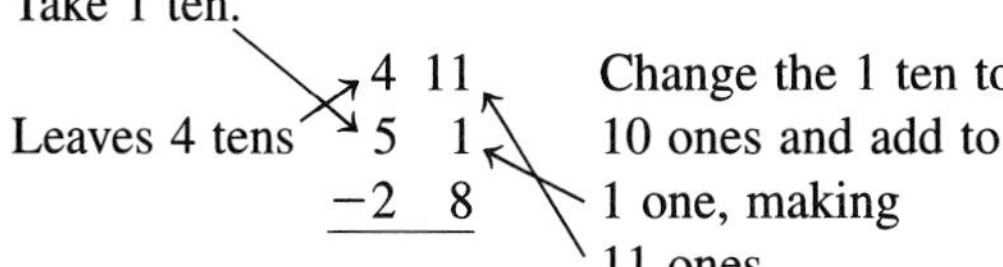

Now we can subtract.

Step 2

$$\begin{array}{rr} 4 & 11 \\ \not{5} & \not{1} \\ -2 & 8 \\ \hline & 3 \end{array}$$

Subtract ones (11 − 8 = 3).

Step 3

$$\begin{array}{rr} 4 & 12 \\ \not{5} & \not{1} \\ -2 & 8 \\ \hline 2 & 3 \end{array}$$

Subtract tens (4 − 2 = 2).

■ **Example 3**

Subtract 39 from 85 using the short method.

Solution

$$\begin{array}{rr} 7 & 15 \\ \not{8} & \not{5} \\ -3 & 9 \\ \hline 4 & 6 \end{array}$$

Practice check: Do exercises 70–72 in the margin.

Subtract using the short method.

70. $\begin{array}{r} 67 \\ -48 \\ \hline \end{array}$ **71.** $\begin{array}{r} 71 \\ -53 \\ \hline \end{array}$ **72.** $\begin{array}{r} 50 \\ -23 \\ \hline \end{array}$

The answers are in the margin on page 57.

"Borrowing" from the Hundreds Place

■ **Example 4**

Subtract 284 from 476 using expanded notation.

Solution

4 hundreds + 7 tens + 6 ones
2 hundreds + 8 tens + 4 ones

We cannot subtract 8 tens from 7 tens, so we rename.

Take 1 hundred. Add 10 tens to 7 tens.

3 17
~~4~~ hundreds + ~~7~~ tens + 6 ones
2 hundreds + 8 tens + 4 ones
1 hundred + 9 tens + 2 ones = 192

■ **Example 5**

Subtract 284 from 476 using the short method.

Solution **Step 1**

```
  476
 -284
    2 ← Subtract 4 from 6 and get 2
```

Step 2 We cannot subtract 8 from 7.

```
  476
 -284
```

So we rename 4 as 3 and 7 as 17, and then subtract.

```
 3 17
 4  7 6
 2  8 4
    9 2
```

Step 3 Complete the subtraction.

```
  3 17
  4  7 6
 -2  8 4
  1  9 2
```

■ **Example 6**

Subtract 583 from 747 using the short method.

Solution

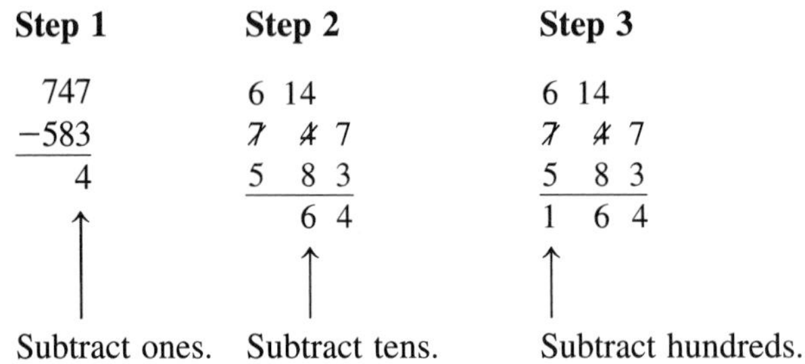

Practice check: Do exercises 73–75 in the margin.

Subtract using the short method.

73. 645 − 391

74. 868 − 573

75. 431 − 250

The answers are in the margin on page 59.

"Borrowing" Two or More Times

■ **Example 7**

Subtract 2877 from 5463 using the short method.

Solution **Step 1** We cannot subtract 7 from 3.

```
  5463
 -2877
```

So we rename 6 as 5 and 3 as 13 and then subtract.

```
      5 13
  5 4 6̸  3̸
 −2 8 7  7
         6
```

Subtract ones (13 − 7 = 6).

Step 2 We cannot subtract 7 from 5. So we rename 4 as 3 and 5 as 15 and then subtract.

```
     15
   3 5̸ 13
 5 4̸ 6̸ 3̸
−2 8 7 7
     8 6
```

Subtract tens (15 − 7 = 8).

Step 3 We cannot subtract 8 from 3. So we rename 5 as 4 and 3 as 13 and then subtract.

```
    13 15
 4  3̸  5̸ 13
 5̸  4̸  6̸  3̸
−2  8  7  7
    5  8  6
```

Subtract hundreds (13 − 8 = 5).

Step 4 We complete the subtraction.

```
    13 15
 4  3̸  5̸ 13
 5̸  4̸  6̸  3̸
−2  8  7  7
 2  5  8  6
```

Subtract thousands (4 − 2 = 2).

Subtract 2877 from 5463 using the short method without looking at the explanation.

■ Example 8

Subtract 3576 from 9352 using the short method.

Solution

```
    12 14
 8  2̸  4̸ 12
 9̸  3̸  5̸  2̸
−3  5  7  6
 5  7  7  6
```

We check by adding:

```
  5776
+3576
  9352
```

Practice check: Do exercises 76–78 in the margin.

■ Looking Back

You should now be able to subtract one whole number from another when "borrowing" is required.

Answers to exercises 67–69

67.
6 tens + 7 ones = 5 tens + 17 ones
4 tens + 8 ones = 4 tens + 8 ones
1 ten + 9 ones
= 19

68.
7 tens + 1 one = 6 tens + 11 ones
5 tens + 3 ones = 5 tens + 3 ones
1 ten + 8 ones
= 18

69.
5 tens + 0 ones = 4 tens + 10 ones
2 tens + 3 ones = 2 tens + 3 ones
2 tens + 7 ones
= 27

Answers to exercises 70–72

70. 19 **71.** 18 **72.** 27

Subtract using the short method. Check by adding.

76. 5417 − 3769

77. 6433 − 4667

78. 8943 − 6975

The answers are in the margin on page 59.

■ Problems 2–9

Try to push yourself to subtract as quickly as possible without counting.

Review

1. $\begin{array}{r} 18 \\ -9 \\ \hline \end{array}$ **2.** $\begin{array}{r} 11 \\ -2 \\ \hline \end{array}$ **3.** $\begin{array}{r} 17 \\ -9 \\ \hline \end{array}$ **4.** $\begin{array}{r} 12 \\ -3 \\ \hline \end{array}$ **5.** $\begin{array}{r} 15 \\ -8 \\ \hline \end{array}$ **6.** $\begin{array}{r} 11 \\ -3 \\ \hline \end{array}$

7. $\begin{array}{r} 16 \\ -9 \\ \hline \end{array}$ **8.** $\begin{array}{r} 13 \\ -7 \\ \hline \end{array}$ **9.** $\begin{array}{r} 13 \\ -9 \\ \hline \end{array}$ **10.** $\begin{array}{r} 14 \\ -8 \\ \hline \end{array}$ **11.** $\begin{array}{r} 11 \\ -6 \\ \hline \end{array}$ **12.** $\begin{array}{r} 12 \\ -8 \\ \hline \end{array}$

13. $\begin{array}{r} 15 \\ -9 \\ \hline \end{array}$ **14.** $\begin{array}{r} 12 \\ -7 \\ \hline \end{array}$ **15.** $\begin{array}{r} 14 \\ -9 \\ \hline \end{array}$ **16.** $\begin{array}{r} 11 \\ -7 \\ \hline \end{array}$ **17.** $\begin{array}{r} 17 \\ -8 \\ \hline \end{array}$ **18.** $\begin{array}{r} 13 \\ -8 \\ \hline \end{array}$

Subtract using the short method. Check by adding.

19. $\begin{array}{r} 63 \\ -49 \\ \hline \end{array}$ **20.** $\begin{array}{r} 94 \\ -18 \\ \hline \end{array}$ **21.** $\begin{array}{r} 55 \\ -27 \\ \hline \end{array}$ **22.** $\begin{array}{r} 46 \\ -28 \\ \hline \end{array}$ **23.** $\begin{array}{r} 70 \\ -56 \\ \hline \end{array}$

24. $\begin{array}{r} 80 \\ -42 \\ \hline \end{array}$ **25.** $\begin{array}{r} 734 \\ -442 \\ \hline \end{array}$ **26.** $\begin{array}{r} 626 \\ -362 \\ \hline \end{array}$ **27.** $\begin{array}{r} 673 \\ -595 \\ \hline \end{array}$ **28.** $\begin{array}{r} 547 \\ -468 \\ \hline \end{array}$

29. $\begin{array}{r} 813 \\ -648 \\ \hline \end{array}$ **30.** $\begin{array}{r} 763 \\ -387 \\ \hline \end{array}$ **31.** $\begin{array}{r} 8142 \\ -6957 \\ \hline \end{array}$ **32.** $\begin{array}{r} 2851 \\ -1972 \\ \hline \end{array}$ **33.** $\begin{array}{r} 4816 \\ -3947 \\ \hline \end{array}$

34. $\begin{array}{r} 5362 \\ -3874 \\ \hline \end{array}$ **35.** $\begin{array}{r} 9640 \\ -3767 \\ \hline \end{array}$ **36.** $\begin{array}{r} 5620 \\ -2834 \\ \hline \end{array}$

Subtract. Check by adding.

37. 34,983 − 25,998

38. 716,538 − 423,979

39. 758,627 − 75,998

40. 837,573 − 690,978

The answers to review and odd-numbered exercises are in the back of the book.

2–10 SUBTRACTION AND ZEROS

You have to be extra careful in renaming when the number to be subtracted contains zeros.

Answers to exercises 73–75

73. 254 **74.** 295 **75.** 181

Answers to exercises 76–78

76. 1648 **77.** 1766 **78.** 1968

■ Example 1

Subtract 542 from 701 using the short method.

Solution **Step 1** We cannot subtract 2 from 1.

```
 701
−542
```

Since there are 0 tens, we look to the 7 hundreds to begin renaming.

We rename 7 as 6 and 0 as 10.

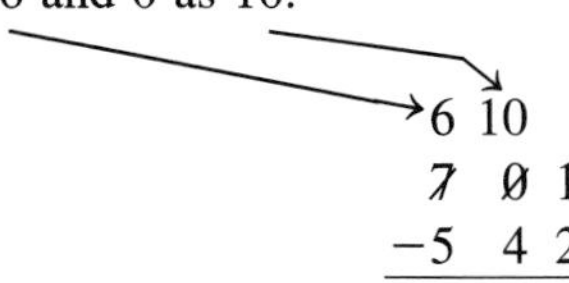

```
 6 10
 7  0 1
−5  4 2
```

(the 7 and the 0 are crossed out)

Step 2 We still cannot subtract 2 from 1, so we rename 10 as 9 and 1 as 11.

```
    9
 6 10 11
 7  0  1
−5  4  2
```

(the 10, 7, 0 and 1 are crossed out)

Step 3 We complete the subtraction.

```
    9
 6 10 11     Check
 7  0  1       159
−5  4  2      +542
 1  5  9       701
```

■ Example 2

Subtract 158 from 604 using the short method.

Solution

```
    9
 5 10 14     Check
 6  0  4       446
−1  5  8      +158
 4  4  6       604
```

(the 10, 6, 0 and 4 are crossed out)

Subtract using the short method.

a. 701 − 542   b. 604 − 158

Practice check: Do exercises 79–82 in the margin.

Subtract using the short method.

79. 905 − 718   **80.** 301 − 76

81. 7043 − 6885   **82.** 9041 − 683

The answers are in the margin on page 61.

More on Zeros

■ Example 3

Subtract 2845 from 5004 using the short method.

Solution **Step 1** We cannot subtract 5 from 4.

$$\begin{array}{r} 5004 \\ -2845 \\ \hline \end{array}$$

Since there are 0 tens and 0 hundreds, we look to the 5 thousands to begin renaming. We rename 5 as 4 and 0 as 10.

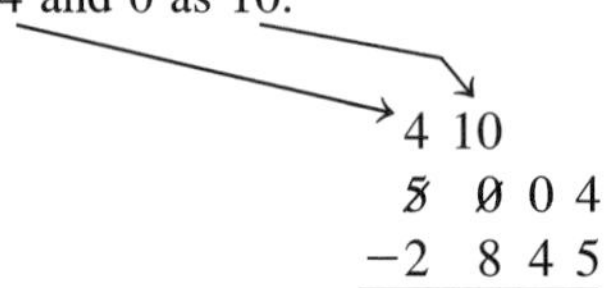

Next we rename 10 as 9 and 0 as 10.

$$\begin{array}{rrrr} & 9 & & \\ \not{4} & \not{10} & 10 & \\ \not{5} & \not{0} & \not{0} & 4 \\ -2 & 8 & 4 & 5 \\ \hline \end{array}$$

Next we rename 10 as 9 and 4 as 14.

$$\begin{array}{rrrr} & 9 & 9 & \\ 4 & \not{10} & \not{10} & 14 \\ \not{5} & \not{0} & \not{0} & \not{4} \\ -2 & 8 & 4 & 5 \\ \hline \end{array}$$

Step 2 We complete the subtraction.

$$\begin{array}{rrrr} & 9 & 9 & \\ 4 & \not{10} & \not{10} & 14 \\ \not{5} & \not{0} & \not{0} & \not{4} \\ -2 & 8 & 4 & 5 \\ \hline 2 & 1 & 5 & 9 \end{array} \qquad \begin{array}{r} \text{Check} \\ 2159 \\ +2845 \\ \hline 5004 \end{array}$$

■ Example 4

Subtract 3468 from 7000 using the short method.

Solution **Step 1** We cannot subtract 8 from 0.

$$\begin{array}{r} 7000 \\ -3468 \\ \hline \end{array}$$

Since there are 0 tens and 0 hundreds, we look to the 7 thousands to begin renaming.

Answers to exercises 79–82

79. 187 **80.** 225 **81.** 158
82. 8358

We rename 7 as 6 and 0 as 10.

```
 6 10
 7̸  0̸ 0 0
−3  4 6 8
```

Next we rename 10 as 9 and 0 as 10.

```
     9
 6 1̸0̸ 10
 7̸  0̸  0̸ 0
−3  4  6 8
```

Next we rename 10 as 9 and 0 as 10.

```
    9  9
 6 1̸0̸ 1̸0̸ 10
 7̸  0̸  0̸  0̸
−3  4  6  8
```

Step 2 We complete the subtraction.

```
    9  9
 6 1̸0̸ 1̸0̸ 10      Check
 7̸  0̸  0̸  0̸        3532
−3  4  6  8       +3468
 3  5  3  2        7000
```

Subtract using the short method.

a. 5004 − 2845 b. 7000 − 3468

Practice check: Do exercises 83–85 in the margin.

Subtract using the short method. Check your work.

83. 5004 − 3165 **84.** 7000 − 3260

85. 4000 − 1584

The answers are in the margin on page 63.

■ Looking Back

You should now be able to subtract one whole number from another when zeros occur as digits.

■ Problems 2–10

Subtract and check.

Review

1. 5623 − 2410 **2.** 8948 − 5626 **3.** 67 − 29 **4.** 34 − 18

5. 643 − 289 **6.** 433 − 274 **7.** 4728 − 2749 **8.** 5427 − 4688

Subtract using the short method. Check.

9. $\begin{array}{r} 50 \\ -38 \\ \hline \end{array}$ **10.** $\begin{array}{r} 30 \\ -17 \\ \hline \end{array}$ **11.** $\begin{array}{r} 60 \\ -25 \\ \hline \end{array}$ **12.** $\begin{array}{r} 80 \\ -63 \\ \hline \end{array}$

13. $\begin{array}{r} 370 \\ -88 \\ \hline \end{array}$ **14.** $\begin{array}{r} 140 \\ -52 \\ \hline \end{array}$ **15.** $\begin{array}{r} 202 \\ -156 \\ \hline \end{array}$ **16.** $\begin{array}{r} 809 \\ -595 \\ \hline \end{array}$

17. $\begin{array}{r} 9640 \\ -756 \\ \hline \end{array}$ **18.** $\begin{array}{r} 3460 \\ -579 \\ \hline \end{array}$ **19.** $\begin{array}{r} 7604 \\ -5437 \\ \hline \end{array}$ **20.** $\begin{array}{r} 4607 \\ -2368 \\ \hline \end{array}$

21. $\begin{array}{r} 5406 \\ -3888 \\ \hline \end{array}$ **22.** $\begin{array}{r} 6305 \\ -4637 \\ \hline \end{array}$ **23.** $\begin{array}{r} 6065 \\ -3496 \\ \hline \end{array}$ **24.** $\begin{array}{r} 3015 \\ -1638 \\ \hline \end{array}$

25. $\begin{array}{r} 9004 \\ -6416 \\ \hline \end{array}$ **26.** $\begin{array}{r} 5006 \\ -2398 \\ \hline \end{array}$ **27.** $\begin{array}{r} 9001 \\ -4324 \\ \hline \end{array}$ **28.** $\begin{array}{r} 3008 \\ -1119 \\ \hline \end{array}$

29. $\begin{array}{r} 3000 \\ -641 \\ \hline \end{array}$ **30.** $\begin{array}{r} 6000 \\ -589 \\ \hline \end{array}$ **31.** $\begin{array}{r} 8000 \\ -6461 \\ \hline \end{array}$ **32.** $\begin{array}{r} 5000 \\ -3243 \\ \hline \end{array}$

Subtract and check.

33. $\begin{array}{r} 60{,}040 \\ -27{,}284 \\ \hline \end{array}$ **34.** $\begin{array}{r} 40{,}080 \\ -27{,}298 \\ \hline \end{array}$ **35.** $\begin{array}{r} 496{,}000 \\ -57{,}470 \\ \hline \end{array}$ **36.** $\begin{array}{r} 298{,}000 \\ -89{,}256 \\ \hline \end{array}$

37. $\begin{array}{r} 500{,}000 \\ -215{,}628 \\ \hline \end{array}$ **38.** $\begin{array}{r} 800{,}000 \\ -698{,}461 \\ \hline \end{array}$

The answers to review and odd-numbered exercises are in the back of the book.

2–11 APPLIED PROBLEMS

Answers to exercises 83–85

83. 1839 **84.** 3740 **85.** 2416

To solve word problems, read the problem carefully and decide what you need to find out. Be sure the answer to the word problem is in the form of a statement.

■ Example 1

A certain lift truck weighs 1862 pounds without a load. If the total weight of lift truck and load is 2321 pounds, how many pounds does the load weigh?

Solution We know that 1862 pounds of the total 2321 pounds is the weight of the lift truck. Thus, 1862 pounds plus what is 2321 pounds. The what is the weight of the load. We find the answer by subtracting 2321 and 1862.

$$\begin{array}{rrrr} & 12 & 11 & \\ 1 & \not{2} & \not{1} & 11 \\ \not{2} & \not{3} & \not{2} & \not{1} \\ -1 & 8 & 6 & 2 \\ \hline & 4 & 5 & 9 \end{array}$$

The load weighs 459 lb.

■ Example 2

Mr. Ramsey bought a new car that cost \$14,085. He got \$1995 for his trade-in. How much does he still owe?

Solution Think: \$1995 plus what is \$14,085? The what is how much he still owes. We find the answer by subtracting \$14,085 and \$1995.

$$\begin{array}{rrr} & 9 & \\ 3 & \not{10} & 18 \\ \not{4} & \not{0} & \not{8}5 \\ 1 & 9 & 95 \\ \hline 12 & 1 & 90 \end{array}$$

He still owes \$12,190.

■ Looking Back

You should now be able to solve applied problems involving subtraction.

■ Problems 2–11

Subtract and check.

Review

1. $\begin{array}{r} 6431 \\ -2846 \\ \hline \end{array}$ **2.** $\begin{array}{r} 3321 \\ -1683 \\ \hline \end{array}$ **3.** $\begin{array}{r} 5502 \\ -2615 \\ \hline \end{array}$ **4.** $\begin{array}{r} 3208 \\ -1919 \\ \hline \end{array}$

5. $\begin{array}{r} 8015 \\ -3137 \\ \hline \end{array}$ **6.** $\begin{array}{r} 9068 \\ -5674 \\ \hline \end{array}$ **7.** $\begin{array}{r} 6000 \\ -4321 \\ \hline \end{array}$ **8.** $\begin{array}{r} 3000 \\ -1908 \\ \hline \end{array}$

Solve. Write a statement for the answer.

9. **Auto Mechanic.** John Jacobs found that his car repair bill was $946. If he paid $315, what was his balance?

10. **Construction.** A loaded truck of dry wall weighs 9561 lb. If the truck empty weighs 8352 lb, find the weight of the dry wall.

11. **Generic.** Find the missing dimension.

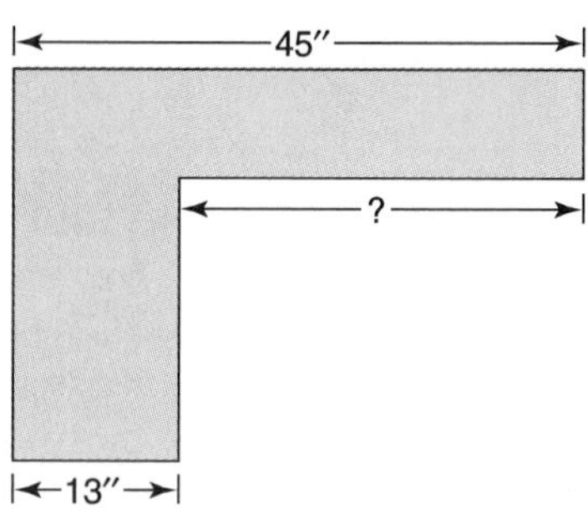

12. **Machinist.** A machinist has a steel bar weighing 37 lb. If he uses 15 lb to make tapered steel shafts, how many pounds are left?

13. **Electrician.** A circuit has 17 amps. If a parallel circuit of 8 amps is removed, what is the remaining current in the circuit?

14. **Contractor.** A contractor has 6200 board feet of hardwood flooring. If he lays 3468 board feet on a new house, how many board feet does he have left? If he needs 3612 board feet for his next job, how many board feet will he have to order to complete the second job?

15. **Business.** A new desktop computer system cost $3200 but was reduced to $2515. How much was it reduced?

16. **Constructon.** A 2×6 board 20 ft long is cut into five pieces: 3 ft, 4 ft, 5 ft, and 7 ft long. How long is the fifth piece, assuming there is no waste from the cutting?

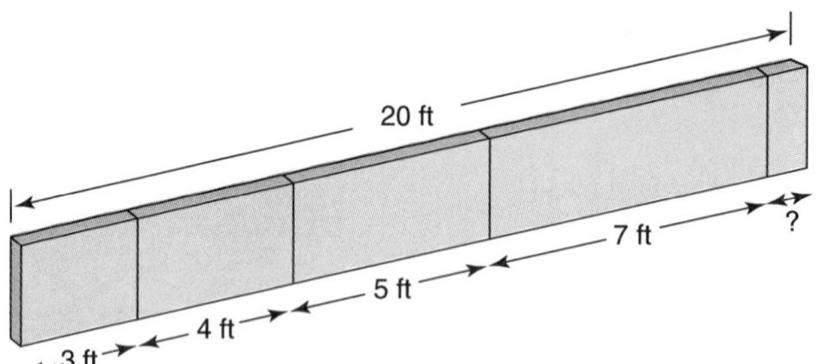

17. Metal Worker. A metal worker requested 192 thread cutting screws from the supply cage for his job. At the end of the day he returned 32 screws. How many did he use?

18. Generic. Find the missing length.

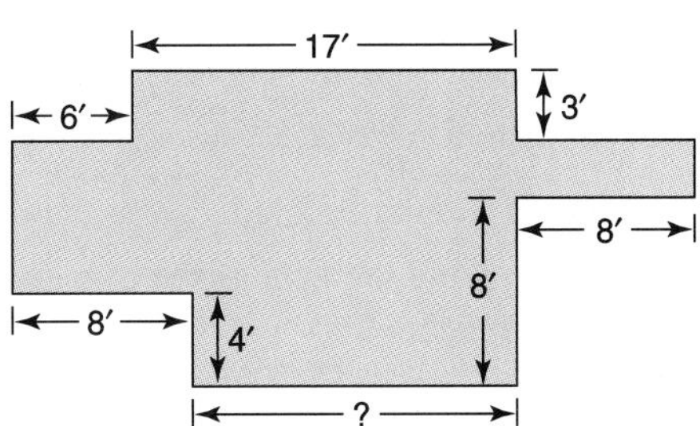

The answers to review and odd-numbered exercises are in the back of the book.

CHAPTER 2 OVERVIEW

Summary

2–1 You learned to do the basic addition facts mentally without counting.

2–2 You learned to add two numbers that do not require "carrying."

2–3 You learned to add any two whole numbers when "carrying" is necessary, and to check the result.

2–4 You learned to add more than two whole numbers, and check the result.

2–5 You learned to estimate sums to the nearest ten, hundred, or thousand.

2–6 You learned to solve applied problems involving addition and whole numbers.

2–7 You learned to give the basic subtraction facts mentally without counting.

2–8 You learned to subtract when no "borrowing" is involved and to check the result.

2–9 You learned to subtract when "borrowing" is required and to check the result.

2–10 You learned to subtract when zeros occur as digits and to check the result.

2–11 You learned to solve applied problems involving subtraction and whole numbers.

Terms To Remember

	Page		***Page***
Addends	30	Difference	48
Basic Addition Facts	29	Minuend	48
Basic Subtraction Facts	48	Number Line	49
"Borrowing" in Subtraction	54	Rounding	43
"Carrying" in Addition	36	Subtrahend	48
		Sum	30

Laws

- Commutative Law of Addition
  Let a and b be natural numbers.
  Then $a + b = b + a$.

Rules

- When adding zero and another number, the sum is the other number.
- To estimate a sum:

  Step 1 Round each addend to the nearest ten, hundred, or thousand.

  Step 2 Find the sum of the rounded addends.

Self-Test

2–1 Push yourself to add quickly without counting.

1. $5 + 9$ **2.** $8 + 6$ **3.** $8 + 8$ **4.** $7 + 9$

5. $7 + 6$ **6.** $3 + 8$ **7.** $5 + 6$ **8.** $9 + 3$

2–2 Add using the short method.

9. $\begin{array}{r} 56 \\ \underline{43} \end{array}$ **10.** $\begin{array}{r} 6243 \\ \underline{3245} \end{array}$

2–3 Add using the short method.

11. $\begin{array}{r} 56 \\ \underline{35} \end{array}$ **12.** $\begin{array}{r} 7378 \\ \underline{6817} \end{array}$

2–4 Add and check.

13. $\begin{array}{r} 73 \\ 46 \\ \underline{61} \end{array}$ **14.** $\begin{array}{r} 5699 \\ \underline{3002} \end{array}$

2–5 Estimate the sum to the nearest ten. Then find the exact sum and compare.

15. $\begin{array}{r} 76 \\ 23 \\ 51 \\ \underline{35} \end{array}$

2–6 16. A welder worked 40 hours on Monday, 38 hours on Tuesday, 40 hours on Wednesday, 34 hours on Thursday, and 42 hours on Friday. Find the total number of hours worked.

2–7 Push yourself to subtract without counting.

17. 18 − 9 **18.** 13 − 7 **19.** 10 − 6 **20.** 11 − 8

21. 13 − 4 **22.** 16 − 8 **23.** 14 − 8 **24.** 15 − 6

Subtract and check.

2–8 25. $\begin{array}{r} 8637 \\ -4215 \\ \hline \end{array}$

2–9 26. $\begin{array}{r} 5438 \\ -3779 \\ \hline \end{array}$

2–10

27. $\begin{array}{r} 8053 \\ -6576 \\ \hline \end{array}$ **28.** $\begin{array}{r} 5003 \\ -2364 \\ \hline \end{array}$

2–11

29. A contractor has 5 rolls of cable, each with 1000 feet in a roll. The electricians use 3615 feet of cable on the job. How many feet of cable are left?

The answers are in the back of the book.

Chapter Test

Directions: This test will aid in your preparation for a possible chapter test given by your instructor. The answers are in the back of the book. Go back and review in the appropriate section(s) if you missed any test items.

2–1 Push yourself to add quickly without counting.

1. 5 + 8 **2.** 9 + 5 **3.** 4 + 3 **4.** 9 + 6

5. 8 + 4 **6.** 5 + 6 **7.** 8 + 6 **8.** 9 + 4

9. 7 + 6 **10.** 7 + 9 **11.** 8 + 3 **12.** 8 + 7

13. 9 + 8 **14.** 9 + 9 **15.** 4 + 5 **16.** 9 + 3

2–2 Add using the short method.

17. 43
35

18. 201
312

19. 5134
2812

2–3 Add using the short method.

20. 73
38

21. 386
747

22. 6489
5076

2–4 Add and check.

23. 46
59
78

24. 432
326
469

25. 4682
7109

26. 49
5603
82
456

2–5 Estimate the sum to the nearest ten. Then find the exact sum and compare.

27. 56
13
67
42

Estimate the sum to the nearest hundred. Then find the exact sum and compare.

28. 372
107
928
807

Estimate the sum to the nearest thousand. Then find the exact sum and compare.

29. $\begin{array}{r} 7324 \\ 8821 \\ 6741 \\ 2786 \\ \hline \end{array}$

2–6 **30.** A roofer laid 1355 composition shingles on Monday, 1405 on Tuesday, 1520 on Wednesday, and 780 on Thursday. How many composition shingles did he lay altogether?

31. The numbers of passengers carried each week in May by the Green Balm Taxi Company were 3106, 3568, 2981, and 2603. How many passengers were carried in May?

2–7 Push yourself to subtract quickly without counting.

32. $\begin{array}{r} 18 \\ -9 \\ \hline \end{array}$ **33.** $\begin{array}{r} 17 \\ -8 \\ \hline \end{array}$ **34.** $\begin{array}{r} 15 \\ -7 \\ \hline \end{array}$ **35.** $\begin{array}{r} 16 \\ -7 \\ \hline \end{array}$ **36.** $\begin{array}{r} 13 \\ -4 \\ \hline \end{array}$

37. $\begin{array}{r} 14 \\ -6 \\ \hline \end{array}$ **38.** $\begin{array}{r} 12 \\ -4 \\ \hline \end{array}$ **39.** $\begin{array}{r} 15 \\ -6 \\ \hline \end{array}$ **40.** $\begin{array}{r} 9 \\ -4 \\ \hline \end{array}$ **41.** $\begin{array}{r} 11 \\ -4 \\ \hline \end{array}$

42. $\begin{array}{r} 13 \\ -5 \\ \hline \end{array}$ **43.** $\begin{array}{r} 17 \\ -9 \\ \hline \end{array}$

Subtract and check.

2–8 **44.** $\begin{array}{r} 7528 \\ -4305 \\ \hline \end{array}$ **45.** $\begin{array}{r} 3614 \\ -2503 \\ \hline \end{array}$

2–9 **46.** $\begin{array}{r} 6531 \\ -2955 \\ \hline \end{array}$ **47.** $\begin{array}{r} 4347 \\ -1689 \\ \hline \end{array}$

2–10

48. $\begin{array}{r} 70 \\ -36 \\ \hline \end{array}$

49. $\begin{array}{r} 7507 \\ -3658 \\ \hline \end{array}$

50. $\begin{array}{r} 4002 \\ -1625 \\ \hline \end{array}$

2–11

51. A contractor has 4580, 4 × 12 wall boards in storage. If 2044 wall boards are used on a construction sight, how many are left in storage?

52. One U.S. state has an average of 1067 controlled forest fires annually. In 1998 it controlled 1654 fires. How many more controlled fires did the state have in 1998 than the average?

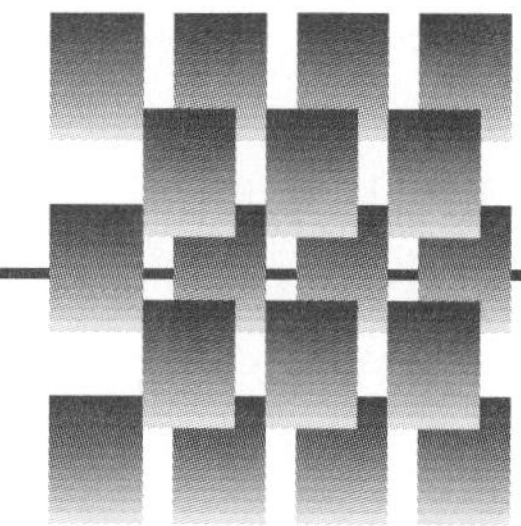

CHAPTER 3

Multiplication and Division of Whole Numbers

Choice: Either skip the Pretest and study all of Chapter 3 or take the Pretest and study only the sections you need to study.

Pretest

Directions: This Pretest will help you determine which sections to study in Chapter 3. If any of your answers are incorrect, or if you omitted any exercise, turn to the indicated section(s) and study the material; then take the Chapter Test.

3–1 Push yourself to multiply quickly without counting.

1. 7×9 **2.** 4×8 **3.** 6×7 **4.** 8×6

5. 5×3 **6.** 9×4 **7.** 4×3 **8.** 6×7

9. 5×9 **10.** 9×8

Multiply.

3–2 **11.** $\begin{array}{r} 502 \\ \underline{\times 100} \end{array}$ **12.** 7000×5

3–3 **13.** $\begin{array}{r} 3346 \\ \underline{\times 5} \end{array}$

3–4 **14.** $\begin{array}{r} 708 \\ \times 60 \\ \hline \end{array}$

3–5 **15.** $\begin{array}{r} 462 \\ \times 74 \\ \hline \end{array}$

16. $\begin{array}{r} 4704 \\ \times 305 \\ \hline \end{array}$

3–6 **17.** Randy Corporation bought 35 desks for $529 apiece. How much did all the desks cost?

3–7 Push yourself to divide quickly without counting.

18. $28 \div 4$ **19.** $\frac{48}{8}$ **20.** $54 \div 6$ **21.** $\frac{24}{6}$

22. $63 \div 9$ **23.** $\frac{64}{8}$ **24.** $72 \div 9$ **25.** $\frac{24}{3}$

26. $35 \div 7$ **27.** $\frac{56}{7}$

3–8 Divide.

28. $4\overline{)804}$

3–9 **29.** $5\overline{)396}$

30. $3\overline{)4067}$

3–10 Divide and check.

31. $56\overline{)3847}$

32. $73\overline{)14825}$

3–11

33. Lloyd earned $455 in one week. If he worked 35 hours that week, how much an hour did he earn?

The answers are in the margin on page 75.

3–1 BASIC MULTIPLICATION FACTS

Turn to Problem Set 3–1 and do the multiplications mentally, *without counting.* **This is important!** If you get at least 90 answers correct, go on to Section 3–2. If not, study Section 3–1.

When multiplying we begin with two numbers called *factors* that result in a third number called the *product.* For example,

$$6 \times 8 = 48$$

Factor ↗ (6), Multiplication sign (×), Factor ↑ (8), Product ↖ (48)

Suppose you purchased seven 1″ by 5″ bolts at $5 apiece. The clerk at the cash register could punch the $5 key seven times or the multiple of 7 key to find the total price.

When we add seven 5s, the sum is 35.

$$5 + 5 + 5 + 5 + 5 + 5 + 5 = 35$$

We can get the same amount by using multiplication.

$$7 \times 5 = 35$$

The cost for 7 bolts is $35.

You can see that multiplication is a short cut for addition when the same number is being added many times over. The first number, 7, tells us how many times the second number, 5, is used as an addend. We memorize that $7 \times 5 = 35$ This is called a *basic multiplication fact.*

■ Example 1

Show with repeated addition the meaning of 5×3.

$$3 + 3 + 3 + 3 + 3 = 15$$

$$5 \times 3 = 15 \qquad \textit{Basic fact}$$

■ Example 2

Show with repeated addition the meaning of 6×1.

Solution We list 1 as an addend 6 times.

$$1 + 1 + 1 + 1 + 1 + 1 = 6$$

$$6 \times 1 = 6 \qquad \textit{Basic fact}$$

■ Example 3

Show with repeated addition the meaning of 4×0.

$$0 + 0 + 0 + 0 = 0$$

$$4 \times 0 = 0 \qquad \textit{Basic fact}$$

Practice check: Do exercises 1–6 in the margin.

Show with repeated addition the meaning of

1. 3×5 **2.** 1×6

3. 2×9 **4.** 9×2

5. 4×5 **6.** 5×4

The answers are in the margin on page 75.

To multiply quickly, you should memorize the multiplication facts. Use the table here to help you. Since multiplication by 0 and 1 are easy, these facts are not shown.

×	2	3	4	5	(6)	7	8	9
2	4	6	8	10	12	14	16	18
3	6	9	12	15	18	21	24	27
(4)	8	12	16	20	(24)	28	32	36
5	10	15	20	25	30	35	40	45
6	12	18	24	30	36	42	48	54
7	14	21	28	35	42	49	56	63
8	16	24	32	40	48	56	64	72
9	18	27	32	45	54	63	72	81

The multiplication fact $6 \times 4 = 24$ is shown.

■ Looking Back

You should now be able to multiply any two digits. It is important for you to memorize the multiplication facts before doing Problems 3–1.

■ Problems 3–1

Do the following multiplications mentally without counting. Force yourself to be quick. Read the answers to someone while he/she checks them in the back of the book.

1. $\begin{array}{r} 3 \\ \times 1 \\ \hline \end{array}$

2. $5 \times 8 =$

3. $\begin{array}{r} 1 \\ \times 0 \\ \hline \end{array}$

4. $4 \times 3 =$

5. $\begin{array}{r} 6 \\ \times 3 \\ \hline \end{array}$

6. $9 \times 3 =$

7. $\begin{array}{r} 7 \\ \times 5 \\ \hline \end{array}$

8. $2 \times 1 =$

9. $\begin{array}{r} 6 \\ \times 1 \\ \hline \end{array}$

10. $2 \times 6 =$

11. $\begin{array}{r} 1 \\ \times 7 \\ \hline \end{array}$

12. $6 \times 6 =$

13. $\begin{array}{r} 8 \\ \times 3 \\ \hline \end{array}$

14. $2 \times 0 =$

15. $\begin{array}{r} 4 \\ \times 4 \\ \hline \end{array}$

16. $3 \times 0 =$

17. $\begin{array}{r} 7 \\ \times 8 \\ \hline \end{array}$

18. $9 \times 7 =$

19. $\begin{array}{r} 1 \\ \times 5 \\ \hline \end{array}$

20. $4 \times 8 =$

21. $\begin{array}{r} 5 \\ \times 6 \\ \hline \end{array}$

22. $0 \times 2 =$

23. $\begin{array}{r} 4 \\ \times 6 \\ \hline \end{array}$

24. $6 \times 0 =$

25. $\begin{array}{r} 7 \\ \times 4 \\ \hline \end{array}$

26. $5 \times 2 =$

27. $\begin{array}{r} 9 \\ \times 4 \\ \hline \end{array}$

28. $8 \times 8 =$

29. $\begin{array}{r} 8 \\ \times 9 \\ \hline \end{array}$

30. $0 \times 1 =$

31. $\begin{array}{r} 9 \\ \times 5 \\ \hline \end{array}$

32. $6 \times 4 =$

33. $\begin{array}{r} 1 \\ \times 9 \\ \hline \end{array}$

34. $6 \times 5 =$

35. $\begin{array}{r} 1 \\ \times 4 \\ \hline \end{array}$

36. $9 \times 2 =$

Answers to Pretest

1. 63 **2.** 32 **3.** 42 **4.** 48 **5.** 15 **6.** 36 **7.** 12 **8.** 42 **9.** 45 **10.** 72 **11.** 50,200 **12.** 35,000 **13.** 16,730 **14.** 42,480 **15.** 34,188 **16.** 1,434,720 **17.** $18,515 **18.** 7 **19.** 6 **20.** 9 **21.** 4 **22.** 7 **23.** 8 **24.** 8 **25.** 8 **26.** 5 **27.** 8 **28.** 201 **29.** 79 R 1 **30.** 1355 R 2 **31.** 68 R 39 **32.** 203 R 6 **33.** $13

Answers to exercises 1–6

1. $3 \times 5 = 5 + 5 + 5 = 15$
2. $1 \times 6 = 6 = 6$
3. $2 \times 9 = 9 + 9 = 18$
4. $9 \times 2 = 2 + 2 + 2 + 2 + 2 + 2 + 2 + 2 + 2 = 18$
5. $4 \times 5 = 5 + 5 + 5 + 5 = 20$
6. $5 \times 4 = 4 + 4 + 4 + 4 + 4 = 20$

37. $\begin{array}{r} 5 \\ \times 5 \\ \hline \end{array}$

38. $0 \times 6 =$

39. $\begin{array}{r} 9 \\ \times 0 \\ \hline \end{array}$

40. $4 \times 0 =$

41. $\begin{array}{r} 5 \\ \times 1 \\ \hline \end{array}$

42. $5 \times 9 =$

43. $\begin{array}{r} 4 \\ \times 2 \\ \hline \end{array}$

44. $3 \times 7 =$

45. $\begin{array}{r} 8 \\ \times 4 \\ \hline \end{array}$

46. $2 \times 8 =$

47. $\begin{array}{r} 8 \\ \times 2 \\ \hline \end{array}$

48. $2 \times 7 =$

49. $\begin{array}{r} 7 \\ \times 0 \\ \hline \end{array}$

50. $0 \times 5 =$

51. $\begin{array}{r} 9 \\ \times 1 \\ \hline \end{array}$

52. $2 \times 3 =$

53. $\begin{array}{r} 8 \\ \times 5 \\ \hline \end{array}$

54. $1 \times 3 =$

55. $\begin{array}{r} 7 \\ \times 6 \\ \hline \end{array}$

56. $2 \times 2 =$

57. $\begin{array}{r} 7 \\ \underline{\times 9} \end{array}$

58. $5 \times 0 =$

59. $\begin{array}{r} 4 \\ \underline{\times 1} \end{array}$

60. $0 \times 4 =$

61. $\begin{array}{r} 8 \\ \underline{\times 7} \end{array}$

62. $5 \times 7 =$

63. $\begin{array}{r} 8 \\ \underline{\times 6} \end{array}$

64. $4 \times 7 =$

65. $\begin{array}{r} 0 \\ \underline{\times 0} \end{array}$

66. $2 \times 4 =$

67. $\begin{array}{r} 3 \\ \underline{\times 4} \end{array}$

68. $2 \times 5 =$

69. $\begin{array}{r} 0 \\ \underline{\times 3} \end{array}$

70. $3 \times 6 =$

71. $\begin{array}{r} 1 \\ \underline{\times 1} \end{array}$

72. $9 \times 6 =$

73. $\begin{array}{r} 3 \\ \underline{\times 8} \end{array}$

74. $7 \times 3 =$

75. $\begin{array}{r} 3 \\ \underline{\times 3} \end{array}$

76. $4 \times 5 =$

77. $\begin{array}{r} 5 \\ \underline{\times 3} \end{array}$

78. $7 \times 1 =$

79. $\begin{array}{r} 7 \\ \underline{\times 7} \end{array}$

80. $6 \times 7 =$

81. $\begin{array}{r} 9 \\ \underline{\times 9} \end{array}$

82. $5 \times 4 =$

83. $\begin{array}{r} 3 \\ \underline{\times 2} \end{array}$

84. $0 \times 8 =$

85. $\begin{array}{r} 8 \\ \underline{\times 0} \end{array}$

86. $1 \times 6 =$

87. $\begin{array}{r} 9 \\ \underline{\times 8} \end{array}$

88. $2 \times 9 =$

89. $\begin{array}{r} 1 \\ \underline{\times 2} \end{array}$

90. $3 \times 9 =$

91. $\begin{array}{r} 3 \\ \underline{\times 5} \end{array}$

92. $6 \times 2 =$

93. $\begin{array}{r} 4 \\ \times 9 \\ \hline \end{array}$

94. $1 \times 8 =$

95. $\begin{array}{r} 6 \\ \times 9 \\ \hline \end{array}$

96. $7 \times 2 =$

97. $\begin{array}{r} 8 \\ \times 1 \\ \hline \end{array}$

98. $0 \times 7 =$

99. $\begin{array}{r} 6 \\ \times 8 \\ \hline \end{array}$

100. $0 \times 9 =$

Memorize these facts until you can get all of Problems 3–1 correct.

The answers are in the back of the book.

3–2 MULTIPLICATION OF 10, 100, OR 1000, OR MULTIPLES OF 10, 100, OR 1000

Multiplying a Number and 10, 100, or 1000

From section 3–1 we learned that

$$4 \times 10 = 10 + 10 + 10 + 10 = 40,$$

and

$$6 \times 100 = 100 + 100 + 100 + 100 + 100 + 100 = 600$$

Discover the pattern for writing this product directly:

$$10 \times 4 = 40 \qquad 32 \times 100 = 3200$$

$$23 \times 10 = 230 \qquad 1000 \times 3 = 3000$$

$$100 \times 8 = 800 \qquad 1000 \times 98 = 98{,}000$$

Rule

In general, the product of a number and 10, 100, or 1000 is that number followed respectively by one, two, or three zeros.

■ **Example 1**

Multiply 36 by 10 by writing the answer directly.

Solution All we need to do is *mentally* multiply 1×36. Write down the product 36, followed by *one* zero.

$$\begin{array}{r} 36 \\ \times 10 \\ \hline 360 \end{array}$$

■ **Example 2**

Find the product of 39 and 100 directly.

Solution $\begin{array}{r} 39 \\ \times 100 \\ \hline 3900 \end{array}$ Write 39 followed by *two* zeros.

■ **Example 3**

Find the product of 53 and 1000 directly.

Solution $\begin{array}{r} 53 \\ \times 1000 \\ \hline 53{,}000 \end{array}$ Write 53 followed by *three* zeros.

Practice check: Do exercises 7–15 in the margin.

7. $\begin{array}{r} 10 \\ \times 9 \\ \hline \end{array}$ **8.** $10 \times 14 =$

9. $\begin{array}{r} 989 \\ \times 10 \\ \hline \end{array}$ **10.** $\begin{array}{r} 100 \\ \times 8 \\ \hline \end{array}$

11. $17 \times 100 =$ **12.** $\begin{array}{r} 82 \\ \times 100 \\ \hline \end{array}$

13. $9 \times 1000 =$ **14.** $\begin{array}{r} 1000 \\ \times 21 \\ \hline \end{array}$

15. $1000 \times 604 =$

The answers are in the margin on page 81.

Multiplying a Number and Multiples of 10, 100, or 1000

■ **Example 4**

Find the product of 7×40.

To write the answer directly, all we need to do is *mentally* multiply 7 by 4, then write down the product followed by one 0. (We are multiplying a multiple of 10, and it has one 0.) Thus, the product is 280.

■ **Example 5**

Find the product of 6×300 directly.

Solution $\begin{array}{r} 6 \\ \times 300 \\ \hline 1800 \end{array}$ Mentally multiply 6 by 3 and write the product 18, followed by *two* zeros.

■ **Example 6**

Find the product of 9×8000 directly.

Solution $\begin{array}{r} 8000 \\ \times 9 \\ \hline 72{,}000 \end{array}$ Mentally multiply 8 by 9, write the product 72, followed by *three* zeros.

Practice check: Do exercises 16–21 in the margin.

Write the products directly.

16. $3 \times 20 =$ **17.** $\begin{array}{r} 80 \\ \times 3 \\ \hline \end{array}$

18. $\begin{array}{r} 500 \\ \times 5 \\ \hline \end{array}$ **19.** $\begin{array}{r} 7 \\ \times 800 \\ \hline \end{array}$

20. $8000 \times 4 =$ **21.** $\begin{array}{r} 9 \\ \times 5000 \\ \hline \end{array}$

The answers are in the margin on page 81.

Rule

In general, to find the product of a number and a multiple of 10, we write the product of the two nonzero whole numbers followed by the number of zeros in the multiple of 10.

■ Looking Back

You should now be able to

1. Multiply 10, 100, or 1000 by any whole number.
2. Multiply multiples of 10, 100, or 1000 by 1, 2, 3, 4, 5, 6, 7, 8 or 9.

■ Problems 3–2

Multiply.

1. $\begin{array}{r} 10 \\ \times 7 \\ \hline \end{array}$

2. $15 \times 100 =$

3. $\begin{array}{r} 6 \\ \times 10 \\ \hline \end{array}$

4. $\begin{array}{r} 517 \\ \times 10 \\ \hline \end{array}$

5. $\begin{array}{r} 705 \\ \times 1000 \\ \hline \end{array}$

6. $10 \times 234 =$

7. $\begin{array}{r} 105 \\ \times 100 \\ \hline \end{array}$

8. $\begin{array}{r} 1000 \\ \times 25 \\ \hline \end{array}$

9. $\begin{array}{r} 891 \\ \times 10 \\ \hline \end{array}$

10. $10 \times 23 =$

11. $\begin{array}{r} 98 \\ \times 1000 \\ \hline \end{array}$

12. $\begin{array}{r} 7 \\ \times 100 \\ \hline \end{array}$

13. $\begin{array}{r} 800 \\ \times 6 \\ \hline \end{array}$

14. $5000 \times 8 =$

15. $\begin{array}{r} 5 \\ \times 700 \\ \hline \end{array}$

16. $\begin{array}{r} 48 \\ \times 10 \\ \hline \end{array}$

17. $50 \times 6 =$

18. $1000 \times 576 =$

19. $\begin{array}{r} 30 \\ \times 2 \\ \hline \end{array}$

20. $\begin{array}{r} 56 \\ \times 100 \\ \hline \end{array}$

21. $20 \times 7 =$

22. $\begin{array}{r} 7 \\ \times 200 \\ \hline \end{array}$

23. $\begin{array}{r} 100 \\ \times 461 \\ \hline \end{array}$

24. $\begin{array}{r} 7 \\ \times 6000 \\ \hline \end{array}$

25. $400 \times 3 =$

26. $\begin{array}{r} 100 \\ \times 8 \\ \hline \end{array}$

27. $\begin{array}{r} 1000 \\ \times 8 \\ \hline \end{array}$

28. $5 \times 1000 =$

29. $\begin{array}{r} 3 \\ \times 40 \\ \hline \end{array}$

30. $\begin{array}{r} 3000 \\ \times 9 \\ \hline \end{array}$

31. Printer. A litho-offset press can produce 3000 impressions per hour. If the press ran for 5 hours, how many impressions did it produce?

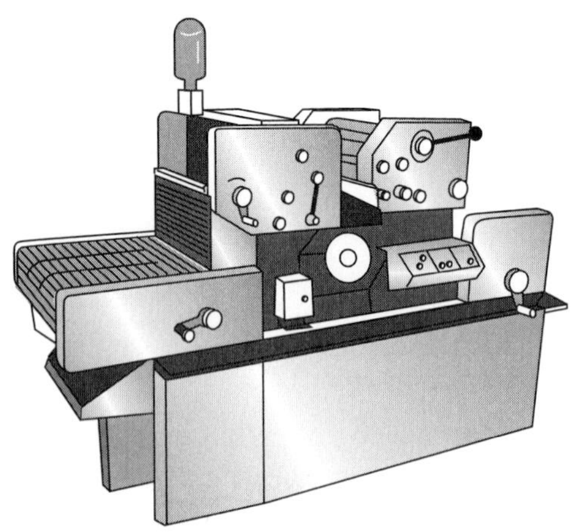

32. Welders. A certain job requires 1000 pieces of bar stock. How long would the bar stock have to be if each piece is 11 inches long?

33. Masons. A brick layer can lay an average of 400 bricks a day. How many bricks could he lay in 5 days?

34. Diesel Mechanics. How many cc will a fuel pump deliver in 400 strokes if it delivers 8 cc of fuel per stroke?

35. Aviation. The liquid pressure on a surface is 9 psi. What is the total force if the surface area is 100 sq. in.

The answers to odd-numbered problems are in the back of the book.

3–3 MULTIPLYING BY ONE DIGIT

Multiplication without "Carrying"

Find the product of 4×21.

One way to show how the product 4×21 can be obtained is the vertical expanded method.

$$\begin{array}{rcr} 21 & = & 20 + 1 \\ \times 4 & & \times 4 \\ \hline & & 4 \\ & & 80 \\ \hline & & 84 \end{array}$$

Expand 21.

(Step 1: $1 \times 4 = 4$.)
(Step 2: $20 \times 4 = 80$.)
(Step 3: Add.)

Using this method for finding products is a correct way, but it is long and tedious in most cases. It is best to use the short method.

■ **Example 1**

Multiply 132 and 3 using the short method.

Solution **Step 1** Multiply ones

$$\begin{array}{r} 132 \\ \times 3 \\ \hline 6 \end{array}$$

Step 2 Multiply tens

$$\begin{array}{r} 132 \\ \times 3 \\ \hline 96 \end{array}$$

Step 3 Multiply hundreds

$$\begin{array}{r} 132 \\ \times 3 \\ \hline 396 \end{array}$$

Notice that we distribute the 3 first to the ones, then to the tens, and finally to the hundreds.

Practice check: Do exercises 22–30 in the margin.

Multiplication with "Carrying"

Find the product of 7 and 73 using the horizontal expanded method.

$7 \times 73 = 7 \times (70 + 3)$	Replace 73 by 70 + 3.
$= (7 \times 70) + (7 \times 3)$	Multiplying the 70 and 3 by 7.
$= 490 + 21$	Multiply.
$= 511$	

Another way to show how the product 7×73 can be obtained is the vertical expanded method.

$73 = 70 + 3$	Expand 73.
$\times 7 = \quad \times 7$	
21	(Step 1: $3 \times 7 = 21$.)
490	(Step 2: $70 \times 7 = 490$.)
511	(Step 3: Add.)

For calculating quickly, however, we use the short method.

■ Example 2

Multiply 73 and 7 using the short method.

Solution **Step 1** Multiply ones. $3 \times 7 = 21$

$$\begin{array}{r} 2 \\ 73 \\ \times 7 \\ \hline 1 \end{array}$$

Write the 1 in the ones place and the 2 above the 7 in the tens place as a "carry" number.

The "carry" number

Step 2 Multiply tens. 7×7 and add 2 $[(7 \times 7) + 2 = 51]$

$$\begin{array}{r} 2 \\ 73 \\ \times 7 \\ \hline 511 \end{array}$$

Write the 1 in the the tens place and the 5 in the hundreds place.

Multiply 73 and 7 using the short method without looking at the steps above.

Practice check: Do exercises 31–34 in the margin.

Answers to exercises 7–15

7. 90 **8.** 140 **9.** 9890 **10.** 800 **11.** 1700 **12.** 8200 **13.** 9000 **14.** 21,000 **15.** 604,000

Answers to exercises 16–21

16. 60 **17.** 240 **18.** 2500 **19.** 5600 **20.** 32,000 **21.** 45,000

Multiply.

22. 2×14 **23.** 5×7

24. 1×11 **25.** $22 \times 3 =$

26. $\begin{array}{r} 34 \\ \times 2 \\ \hline \end{array}$ **27.** $\begin{array}{r} 12 \\ \times 4 \\ \hline \end{array}$

28. $\begin{array}{r} 234 \\ \times 2 \\ \hline \end{array}$ **29.** $\begin{array}{r} 121 \\ \times 4 \\ \hline \end{array}$

30. $\begin{array}{r} 223 \\ \times 3 \\ \hline \end{array}$

The answers are in the margin on page 83.

Multiply using the short method.

31. $\begin{array}{r} 46 \\ \times 3 \\ \hline \end{array}$ **32.** $\begin{array}{r} 74 \\ \times 5 \\ \hline \end{array}$

33. $\begin{array}{r} 29 \\ \times 8 \\ \hline \end{array}$ **34.** $\begin{array}{r} 63 \\ \times 4 \\ \hline \end{array}$

The answers are in the margin on page 83.

■ **Example 3**

Multiply 4286 and 6.

Solution Expanded method:

$$\begin{array}{rcrl} 4286 & = & 4000 + 200 + 80 + 6 & \\ \underline{\times 6} & = & \underline{\times 6} & \\ & & 36 & (6 \times 6) \\ & & 480 & (80 \times 6) \\ & & 1200 & (200 \times 6) \\ & & \underline{2400} & (4000 \times 6) \\ & & 25716 & \end{array}$$

Solution Short method:

Step 1 Multiply ones. $(6 \times 6 = 36)$

$$\begin{array}{r} 3 \\ 4286 \\ \uparrow \\ \underline{\times 6} \\ 6 \end{array}$$

Write the 6 in the ones place and the 3 above 8 tens as a "carry" number.

Step 2 Multiply tens. $(8 \times 6) + 3 = 51$ The "carry" number

$$\begin{array}{r} 53 \\ 4286 \\ \nwarrow \\ \underline{\times 6} \\ 16 \end{array}$$

Write the 1 in the tens place; "carry" the 5

Step 3 Multiply hundreds. $(2 \times 6) + 5 = 17$ The "carry" number

$$\begin{array}{r} 153 \\ 4286 \\ \nwarrow \\ \underline{\times 6} \\ 716 \end{array}$$

Write the 7 in the hundreds place; "carry" the 1.

Step 4 Multiply thousands. $(4 \times 6) + 1 = 25$ The "carry" number

$$\begin{array}{r} 153 \\ 4286 \\ \nwarrow \\ \underline{\times 6} \\ 25716 \end{array}$$

Multiply 4286 and 6 using the short method without looking at the steps above.

Practice check: Do exercises 35–40 in the margin.

Multiply.

35. 754 ×4

36. 816 ×7

37. 468 ×9

38. 6642 ×3

39. 4832 ×5

40. 9164 ×9

The answers are in the margin on page 84.

Multiplication Involving Zero

When zeros are involved, we must be careful to note that the product of any number and zero is zero.

Answers to exercises 22–30

22. 28 **23.** 35 **24.** 11
25. 66 **26.** 68 **27.** 48
28. 468 **29.** 484 **30.** 669

Answers to exercises 31–34

31. 138 **32.** 370 **33.** 232
34 252

■ Example 4

Multiply 506 and 60.

Solution Expanded method:

$$\begin{array}{rl} 500 + 6 & \\ \times 6 & \\ \hline 36 & (6 \times 6) \\ 3000 & (500 \times 6) \\ \hline 3036 & \end{array}$$

Solution Short method:

Step 1 Multiply ones.

$$\begin{array}{r} 3 \\ 506 \\ \times 6 \\ \hline 6 \end{array}$$

Write 6 in the ones place; "carry" the 3.

The "carry" number.

Step 2 Multiply tens. $(0 \times 6) + 3 = 3$

$$\begin{array}{r} 3 \\ 506 \\ \times 6 \\ \hline 36 \end{array}$$

Note: Remember $0 \times 6 = 0$. Write the 3 in the tens place.

Step 3 Multiply hundreds. $5 \times 6 = 30$

$$\begin{array}{r} 3 \\ 506 \\ \times 6 \\ \hline 3036 \end{array}$$

■ Example 5

Multiply 2009 and 8.

Solution Expanded method:

$$\begin{array}{rl} 2000 + 9 & \\ \times 8 & \\ \hline 72 & (9 \times 8) \\ 16000 & (2000 \times 8) \\ \hline 16072 & \end{array}$$

Answers to exercises 35–40

35. 3016 **36.** 5712 **37.** 4212
38. 19,926 **39.** 24,160
40. 82,476

Solution Short method:

Step 1	**Step 2**	**Step 3**	**Step 4**
$\begin{array}{r} 7 \\ 2009 \\ \underline{\times 8} \\ 2 \end{array}$	$\begin{array}{r} 7 \\ 2009 \\ \underline{\times 8} \\ 72 \end{array}$	$\begin{array}{r} 7 \\ 2009 \\ \underline{\times 8} \\ 072 \end{array}$	$\begin{array}{r} 7 \\ 2009 \\ \underline{\times 8} \\ 16072 \end{array}$
$9 \times 8 = 72$	$(0 \times 8) + 7 = 7$	$0 \times 8 = 0$	$2 \times 8 = 16$

Practice check: Do exercises 41–44 in the margin.

Multiply.

41. $\begin{array}{r} 607 \\ \underline{\times 5} \end{array}$ **42.** $\begin{array}{r} 2075 \\ \underline{\times 3} \end{array}$

43. $\begin{array}{r} 4006 \\ \underline{\times 7} \end{array}$ **44.** $\begin{array}{r} 7408 \\ \underline{\times 9} \end{array}$

The answers are in the margin on page 86.

■ Looking Back

You should now be able to multiply any whole number by 0, 1, 2, 3, 4, 5, 6, 7, 8, or 9.

Problems 3–3

Review

1. $\begin{array}{r} 500 \\ \underline{\times 6} \end{array}$ **2.** $\begin{array}{r} 1000 \\ \underline{\times 23} \end{array}$ **3.** $\begin{array}{r} 80 \\ \underline{\times 9} \end{array}$

4. $\begin{array}{r} 368 \\ \underline{\times 100} \end{array}$ **5.** $\begin{array}{r} 7 \\ \underline{\times 6000} \end{array}$ **6.** $\begin{array}{r} 10 \\ \underline{\times 15} \end{array}$

Multiply.

7. $\begin{array}{r} 63 \\ \underline{\times 8} \end{array}$ **8.** $\begin{array}{r} 84 \\ \underline{\times 7} \end{array}$ **9.** $\begin{array}{r} 59 \\ \underline{\times 4} \end{array}$ **10.** $\begin{array}{r} 29 \\ \underline{\times 5} \end{array}$

11. $\begin{array}{r} 43 \\ \underline{\times 9} \end{array}$ **12.** $\begin{array}{r} 65 \\ \underline{\times 5} \end{array}$ **13.** $\begin{array}{r} 484 \\ \underline{\times 8} \end{array}$ **14.** $\begin{array}{r} 596 \\ \underline{\times 9} \end{array}$

15. $\begin{array}{r} 799 \\ \underline{\times 6} \end{array}$ **16.** $\begin{array}{r} 489 \\ \underline{\times 6} \end{array}$ **17.** $\begin{array}{r} 902 \\ \underline{\times 5} \end{array}$ **18.** $\begin{array}{r} 304 \\ \underline{\times 4} \end{array}$

19. $\begin{array}{r} 754 \\ \underline{\times 3} \end{array}$ **20.** $\begin{array}{r} 856 \\ \underline{\times 3} \end{array}$ **21.** $\begin{array}{r} 409 \\ \underline{\times 6} \end{array}$ **22.** $\begin{array}{r} 604 \\ \underline{\times 7} \end{array}$

23. $\begin{array}{r} 192 \\ \times 8 \\ \hline \end{array}$

24. $\begin{array}{r} 163 \\ \times 8 \\ \hline \end{array}$

25. $\begin{array}{r} 3216 \\ \times 5 \\ \hline \end{array}$

26. $\begin{array}{r} 4321 \\ \times 6 \\ \hline \end{array}$

27. $\begin{array}{r} 5108 \\ \times 6 \\ \hline \end{array}$

28. $\begin{array}{r} 7305 \\ \times 7 \\ \hline \end{array}$

29. $\begin{array}{r} 9814 \\ \times 2 \\ \hline \end{array}$

30. $\begin{array}{r} 8146 \\ \times 3 \\ \hline \end{array}$

31. $\begin{array}{r} 6023 \\ \times 7 \\ \hline \end{array}$

32. $\begin{array}{r} 7018 \\ \times 9 \\ \hline \end{array}$

33. $\begin{array}{r} 4444 \\ \times 4 \\ \hline \end{array}$

34. $\begin{array}{r} 6666 \\ \times 6 \\ \hline \end{array}$

35. $\begin{array}{r} 9002 \\ \times 8 \\ \hline \end{array}$

36. $\begin{array}{r} 5006 \\ \times 2 \\ \hline \end{array}$

37. $\begin{array}{r} 9988 \\ \times 7 \\ \hline \end{array}$

38. $\begin{array}{r} 8989 \\ \times 5 \\ \hline \end{array}$

39. $\begin{array}{r} 43{,}623 \\ \times 4 \\ \hline \end{array}$

40. $\begin{array}{r} 68{,}416 \\ \times 5 \\ \hline \end{array}$

41. $\begin{array}{r} 78{,}043 \\ \times 5 \\ \hline \end{array}$

42. $\begin{array}{r} 12{,}045 \\ \times 6 \\ \hline \end{array}$

43. $\begin{array}{r} 10{,}804 \\ \times 8 \\ \hline \end{array}$

44. $\begin{array}{r} 40{,}609 \\ \times 8 \\ \hline \end{array}$

45. $\begin{array}{r} 468{,}205 \\ \times 6 \\ \hline \end{array}$

46. $\begin{array}{r} 348{,}306 \\ \times 2 \\ \hline \end{array}$

47. Forestry. A cat skinner worked for 254 hours at $9.00 per hour. Compute the labor costs.

48. Police Science. The Freedom City P.D. receives about 24,893 fingerprint cards per year. How many would be on file 4 years from now?

49. Painting. At $25.30 per gallon, find the cost of 7 gallons of primer.

50. Construction. If one man can lay 125 square feet of tiles per hour, how many tiles can 4 men lay in an 8-hour shift?

The answers to odd-numbered problems are in the back of the book.

Answers to exercises 41–44

41. 3035 **42.** 6225
43. 28,042 **44.** 66,672

3–4 MULTIPLYING MULTIPLES OF 10, 100, OR 1000 BY MULTIPLES OF 10, 100, OR 1000

Notice the pattern:

1. $400 \times 10 =$ 4000

↑↑ ↑ ↑↑↑

Total of 3 zeros in factors — Three zeros in product

2. $700 \times 200 =$ 140,000

↑↑ ↑↑ ↑↑↑↑

Total of 4 zeros in factors — Four zeros in product

3. $50 \times 30 =$ 1500

↑ ↑ ↑↑

Total of 2 zeros in factors — Two zeros in product

Rule

When we multiply with factors that are multiples of 10, we write down the product of the two nonzero whole numbers and then write down after it the total number of zeros that appear in both factors.

■ **Example 1**

Multiply 300 and 70 by writing the answer directly.

Solution **Step 1** Multiply 7×3 and write the product 21.

$$\begin{array}{r} 300 \\ \times 70 \\ \hline 21 \end{array}$$

Step 2 Write three 0s following the 21.

$$\begin{array}{r} 300 \\ \times 70 \\ \hline 21{,}000 \end{array}$$

■ **Example 2**

Multiply 8000 and 300 by writing the answer directly.

Solution **Step 1** Multiply 3×8 and write the product 24.

$$\begin{array}{r} 8000 \\ \times 300 \\ \hline 24 \end{array}$$

Step 2 Write five 0s following the 24.

$$\begin{array}{r} 8000 \\ \times 300 \\ \hline 2{,}400{,}000 \end{array}$$

Practice check: Do exercises 45–52 in the margin.

Consider 856 × 5000.

$$\begin{aligned} 856 \times 5000 &= 856 \times (5 \times 1000) && \text{Replace 5000 by } 5 \times 1000. \\ &= (856 \times 5) \times 1000 \\ &= 4280 \times 1000 && \text{Multiply.} \\ &= 4{,}280{,}000 && \text{Multiply.} \end{aligned}$$

Answers to this type of exercise can also be found using the following method:

Rule

When we multiply by a factor that is a multiple of 10, we write down as many zeros that appear in the multiple of 10 and then write down the product of the two nonzero whole numbers.

Multiply by writing the answers directly.

45. 60 ×20 **46.** 10 ×20

47. 400 ×300 **48.** 100 ×10

49. 100 ×200 **50.** 5000 ×600

51. 5000 ×100 **52.** 8000 ×3000

The answers are in the margin on page 89.

■ Example 3

Multiply 841 and 50.

Solution **Step 1**

```
  841
  ×50
    0
```

Write a 0 below the 0 of the second factor.

Step 2

```
   841
   ×50
42,050
```

Multiply 841 by 5.

■ Example 4 Multiply 356 and 3000.

Solution **Step 1**

```
 356
×300
  00
```

Write a 0 below each 0 in the second factor.

Step 2

```
    356
   ×300
106,800
```

Multiply 356 by 3.

Practice check: Do exercises 53–58 in the margin.

Multiply by writing the answer directly.

53. 23 ×60 **54.** 568 ×20

55. 86 ×300 **56.** 254 ×800

57. 609 ×4000 **58.** 5068 ×5000

The answers are in the margin on page 89.

■ Looking Back

You should now be able to

1. Multiply multiples of 10, 100, or 1000 by multiples of 10, 100, or 1000.
2. Multiply any whole number by multiples of 10, 100, or 1000.

Problems 3–4

Review

1. 89×6

2. 561×5

3. 608×9

4. 4162×4

5. 6503×8

6. 8005×7

Multiply.

7. 50×10

8. 10×80

9. 100×60

10. 2000×100

11. 80×40

12. 500×70

13. 30×600

14. 900×300

15. 6000×80

16. 900×5000

17. 3000×500

18. 9000×3000

19. 56×40

20. 31×60

21. 289×50

22. 804×60

23. 98×600

24. 56×300

25. 681×400

26. 305×700

27. 571 ×4000

28. 7452 ×9000

29. 5063 ×3000

30. 9999 ×5000

Answers to exercises 45–52

45. 1200 **46.** 200 **47.** 120,000
48. 1000 **49.** 20,000
50. 3,000,000 **51.** 500,000
52. 24,000,000

31. 3005 ×8000

32. 64,469 ×5000

33. 71,503 ×80,000

34. 30,491 ×60,000

Answers to exercises 53–58

53. 1380 **54.** 11,360
55. 25,800 **56.** 203,200
57. 2,436,000 **58.** 25,340,000

35. 560,248 ×700,000

36. Agriculture. Find the value of a 3000-acre ranch at $856.00 per acre.

37. Fire Science. Firemen put 3000 gallons of water into a burning building. If water weighs 8 pounds per gallon, how many pounds of water were used?

38. Clerical. The suburb of Jacksonville has 800 families. Find the annual income of the community if the annual average of each family is $22,560.

39. Plumbing. The cost of 4′ copper straight pipe is $2.56 per foot. Find the cost of 100 of these pipes.

40. Electrician. An electrician uses 2000 inches of copper wire to wrap a coil. How many inches of wire would be needed to complete a contract for 25 coils?

If you missed any of the problems, rework them. Be careful about your basic multiplication facts and follow the rules on pages 86 and 87.

The answers to odd-numbered problems are in the back of the book.

3–5 MULTIPLYING BY TWO DIGITS OR MORE

Multiplication by Two Digits

Consider 46×26. We use the horizontal expanded method to help us find the product.

$46 \times 26 = 46 \times (20 + 6)$	Replace 26 with 20 + 6.
$= (46 \times 20) + (46 \times 6)$	Multiplying the 20 and 6 by 46.
$= 920 + 276$	Multiply.
$= 1196$	Add.

Thus, $46 \times 26 = 1196$

Another way to show how the product 46×26 can be obtained is the vertical expanded method.

46 =	46	
×26 =	20 + 6	Expand 26.
	276	$(46 \times 6 = 276)$
	920	$(46 \times 20 = 920)$
	1196	Add 276 and 920.

For calculating quickly, we use the short method.

■ **Example 1**

Multiply 46 and 26 using the short method.

Solution We first find the product of 46×6.

$$\begin{array}{r} 46 \\ \times 26 \\ \hline \end{array}$$

Step 1 $6 \times 6 = 36$

$$\begin{array}{r} 3 \\ 46 \\ \uparrow \\ \times 26 \\ \hline 6 \end{array}$$

Step 2 $(4 \times 6) + 3 = 27$

$$\begin{array}{r} 46 \\ \nwarrow \\ \times 26 \\ \hline 276 \end{array}$$

Next we find the product 46×20.

$$\begin{array}{r} 46 \\ \times 26 \\ \hline \end{array}$$

To do this, first write a 0 in the ones place, since we are multiplying 46 by 20.

Step 1

$$\begin{array}{r} 3 \\ 46 \\ \times 26 \\ \hline 276 \\ 0 \end{array}$$

Step 2 $6 \times 2 = 12$

$$\begin{array}{r} 1 \\ \not{3} \\ 46 \\ \nearrow \\ \times 26 \\ \hline 276 \\ 20 \end{array}$$

1 ← Cross out the 3. Write the new "carry" number above 3.

Step 3 $(4 \times 2) + 1 = 9$

$$\begin{array}{r} 1 \\ \not{3} \\ 46 \\ \uparrow \\ \times 26 \\ \hline 276 \\ 920 \end{array}$$

Add.

```
   1
   3
  46
 ×26
 276    (46 × 6) ⟶ Partial product
 920    (46 × 20) → Partial product
1196 ⟶ Product
```

Multiply 46 and 26 using the short method without looking at the steps above.

Practice check: Do exercises 59–62 in the margin.

Multiply.

59. 27 ×15 **60.** 38 ×25 **61.** 74 ×33

62. 63 ×59

The answers are in the margin on page 93.

■ Example 2

Multiply 842 and 67 using the short method.

Solution **Step 1** We first find the product 842 × 7.

```
 842
 ×67
5894
```

Step 2 Next we find the product 842 × 60.

```
                     842
                     ×67
Write a zero first, 5894
then multiply.     50520
```

Note:

```
  842
  ×60
50520
```

To find the product of a number and a multiple of 10, we write the product of the two nonzero whole numbers (842 and 6) followed by the number of zeros in the multiple of 10. (In this case one zero.)

Step 3 Add

```
  842
 × 67
 5894    (842 × 7) ⟶ Partial product
50520    (842 × 60) ⟶ Partial product
56414 ⟶ Product
```

Multiply 842 and 67 using the short method without looking at the steps above.

Practice check: Do exercises 63–65 in the margin.

Multiply.

63. 146 ×38 **64.** 624 ×53 **65.** 985 ×29

The answers are in the margin on page 93.

■ Example 3

Multiply 642 and 218 using the short method.

Solution **Step 1** We find the product 642 × 8

```
 642
×218
5136
```

Step 2 We find the product 642 × 10

```
                          642
                         ×218
                         5136
Write zero first; then   6420
multiply.
```

Step 3 We find the product 642 × 200.

```
   642
  ×218
  5136
  6420
128400
```

Fill in zeros; then multiply.

Step 4 Add.

```
   642
  ×218
  5136    (642 × 8) ——→ Partial product
  6420    (642 × 10) ——→ Partial product
128400    (642 × 200) ——→ Partial product
139956  ——————————→ Product
```

Multiply 642 and 218 using the short method without looking at the steps above.

Practice check: Do exercises 66–68 in the margin.

Multiply.

66. 543 ×284 **67.** 564 ×813

68. 729 ×165

The answers are in the margin on page 95.

■ Example 4

Multiply 6438 and 291 using the short method.

Solution

```
   6438
   ×291
   6438     (Step 1: 6438 × 1.)
 579420     (Step 2: 6438 × 90.)
1287600     (Step 3: 6438 × 200.)
1863458
```

The product is 1,863,458.

■ Example 5

Multiply 9746 and 2318 using the short method.

Solution

```
    9746
  × 2318
   77968    (Step 1: 9746 × 8.)
   97460    (Step 2: 9746 × 10.)
 2923800    (Step 3: 9746 × 300.)
19492000    (Step 4: 9746 × 2000.)
22591228
```

The product is 22,591,228.

Practice check: Do exercises 69–72 in the margin.

Multiply.

69. 2183 ×632 **70.** 8312 ×328

71. 4891 ×8439 **72.** 2846 ×7184

The answers are in the margin on page 95.

Multiplication with Zeros

■ Example 6

Multiply 6305 and 86 using the short method.

Solution

Step 1

```
 6305
  ×86
37830 ← (6305 × 6)
```

Step 2

```
  6305
   ×86
 37830
504400 ← (6305 × 80)
```

Write in one zero first; then multiply by 8

Step 3 Add.

```
  6305
   ×86
 37830
504400
542230
```

The product is 542,230.

■ Example 7

Multiply 604 and 840 using the short method.

Solution

Step 1

```
  604
 ×840
24160 ← (604 × 40)
```

Write in one zero first, then multiply by 4.

Step 2

```
   604
  ×840
 24160
483200 ← (604 × 800)
```

Write in two 0s, then multiply by 8.

Step 3 Add

```
   604
  ×840
 24160
483200
507360
```

The product is 507,360.

Practice check: Do exercises 73–75 in the margin.

■ Example 8

Multiply 4803 and 704 using the short method.

Solution

Step 1

```
 4803
 ×704 (4803 × 4)
19212
```

Step 2

```
   4803
   ×704
  19212
3362100 (4803 × 700)
```

Write in two zeros first, then multiply by 7.

Step 3 Add.

```
   4803
   ×704
  19212
3362100
3381312
```

The product is 3,381,312.

■ Example 9

Multiply 5008 and 3070 using the short method.

Solution **Step 1**

```
  5008
 ×3070
350560  (5008 × 70)
```

Step 2

```
    5008
   ×3070
  350560
15024000  (5008 × 3000)
15374560
```

The product is 15,374,560.

Example 10

Multiply 50067 and 9004 using the short method.

Solution **Step 1**

```
 50067
 ×9004
200268  (50067 × 4)
```

Step 2

```
    50067
    ×9004
   200268
450603000  (50067 × 9000)
450803268
```

The product is 450,803,268.

Practice check: Do exercises 76–80 in the margin.

Answers to exercises 59–62

59. 405 **60.** 950 **61.** 2442 **62.** 3717

Answers to exercises 63–65

63. 5548 **64.** 33,072 **65.** 28,565

Multiply.

73. 4304 ×29

74. 5019 ×70

75. 908 ×360

The answers are in the margin on page 95.

Multiply.

76. 7053 ×402

77. 2008 ×706

78. 7002 ×4060

79. 4030 ×6007

80. 70,402 ×3,008

The answers are in the margin on page 95.

Looking Back

You should now be able to multiply any two whole numbers.

Problems 3–5

Review

1. 700×3

2. $4 \times 3000 =$

3. 353×6

4. 6032×9

5. 84×60

6. 8000×407

Multiply.

7. 62×34

8. 65×57

9. 71×39

10. 84×73

11. 563×29

12. 713×56

13. 225×62

14. 918×32

15. 704×56

16. 506×29

17. 223×416

18. 786×435

19. 266×499

20. 818×732

21. 2308×64

22. 9806×47

23. 7046×27

24. 3035×48

25. 7809×20

26. 4067×70

27. 309×770

28. 704×420

29. 8143×283

30. 6712×431

31. $\begin{array}{r} 5487 \\ \times 149 \\ \hline \end{array}$

32. $\begin{array}{r} 6777 \\ \times 393 \\ \hline \end{array}$

33. $\begin{array}{r} 406 \\ \times 308 \\ \hline \end{array}$

34. $\begin{array}{r} 902 \\ \times 406 \\ \hline \end{array}$

35. $\begin{array}{r} 6007 \\ \times 502 \\ \hline \end{array}$

36. $\begin{array}{r} 8003 \\ \times 709 \\ \hline \end{array}$

37. $\begin{array}{r} 2368 \\ \times 4916 \\ \hline \end{array}$

38. $\begin{array}{r} 8716 \\ \times 4394 \\ \hline \end{array}$

39. $\begin{array}{r} 3090 \\ \times 4006 \\ \hline \end{array}$

40. $\begin{array}{r} 7002 \\ \times 4070 \\ \hline \end{array}$

41. $\begin{array}{r} 50406 \\ \times 8004 \\ \hline \end{array}$

42. $\begin{array}{r} 70908 \\ \times 2006 \\ \hline \end{array}$

43. $\begin{array}{r} 64{,}158 \\ \times 7{,}892 \\ \hline \end{array}$

44. $\begin{array}{r} 84{,}171 \\ \times 5{,}747 \\ \hline \end{array}$

45. $\begin{array}{r} 861{,}472 \\ \times 2{,}384 \\ \hline \end{array}$

46. $\begin{array}{r} 242{,}853 \\ \times 8{,}462 \\ \hline \end{array}$

47. $\begin{array}{r} 743{,}284 \\ \times 73{,}843 \\ \hline \end{array}$

48. $\begin{array}{r} 816{,}246 \\ \times 51{,}284 \\ \hline \end{array}$

49. $\begin{array}{r} 500{,}608 \\ \times 70{,}801 \\ \hline \end{array}$

50. $\begin{array}{r} 700{,}707 \\ \times 20{,}406 \\ \hline \end{array}$

51. $\begin{array}{r} 7{,}042{,}007 \\ \times 610{,}705 \\ \hline \end{array}$

52. $\begin{array}{r} 6{,}038{,}006 \\ \times 850{,}609 \\ \hline \end{array}$

Answers to exercises 66–68

66. 154,212 **67.** 458,532
68. 120,285

Answers to exercises 69–72

69. 1,379,656 **70.** 2,726,336
71. 41,275,149 **72.** 21,164,064

Answers to exercises 73–75

73. 124,816 **74.** 351,330
75. 326,880

Answers to exercises 76–80

76. 2,835,306 **77.** 1,417,648
78. 28,428,120 **79.** 24,208,210
80. 211,769,216

The answers to review and odd-numbered exercises are in the back of the book.

3–6 APPLIED PROBLEMS

Remember, the solution of a word problem is not complete unless the answer is in the form of a statement.

■ Example 1

Suppose a car weighs 4062 pounds. What is the total weight of 168 cars of this type?

Solution We multiply to find the total weight.

$$\begin{array}{r} 4062 \\ \times 168 \\ \hline 32496 \\ 243720 \\ 406200 \\ \hline 682416 \end{array}$$

The actual total weight is 682,416 lb.

Solve.

81. Berg Grocery ordered 405 cases of Santiam Beans. What was the total cost of the beans if each case cost $30.00?

82. Margaret ordered 105 cellular telephones for her electronics store. If she realizes a profit of $135.00 each, how much money did she make if she sold all of them?

83. A carton contains 16 bottles of pop. If each bottle contains 12 ounces, how many ounces are there in the carton?

The answers are in the margin on page 98.

■ Example 2

The warehouse of Buy-Rite Grocers delivered 648 dozen eggs. How many eggs were shipped?

Solution We multiply by 12 to find the number of eggs, since there are 12 eggs in one dozen.

$$\begin{array}{r} 648 \\ \times 12 \\ \hline 1296 \\ 648 \\ \hline 7776 \end{array}$$

There are 7776 eggs in 648 dozen.

Practice check: Do exercises 81–83 in the margin.

■ Looking Back

You should now be able to solve word problems using multiplication.

Problems 3–6

Review

1. $\begin{array}{r} 7653 \\ \times 8 \\ \hline \end{array}$ **2.** $\begin{array}{r} 5007 \\ \times 9 \\ \hline \end{array}$ **3.** $\begin{array}{r} 86 \\ \times 17 \\ \hline \end{array}$ **4.** $\begin{array}{r} 382 \\ \times 39 \\ \hline \end{array}$ **5.** $\begin{array}{r} 684 \\ \times 217 \\ \hline \end{array}$

6. $\begin{array}{r} 4653 \\ \times 682 \\ \hline \end{array}$ **7.** $\begin{array}{r} 5533 \\ \times 3164 \\ \hline \end{array}$ **8.** $\begin{array}{r} 503 \\ \times 73 \\ \hline \end{array}$ **9.** $\begin{array}{r} 7609 \\ \times 74 \\ \hline \end{array}$ **10.** $\begin{array}{r} 5013 \\ \times 60 \\ \hline \end{array}$

11. $\begin{array}{r} 306 \\ \times 802 \\ \hline \end{array}$ **12.** $\begin{array}{r} 5006 \\ \times 804 \\ \hline \end{array}$ **13.** $\begin{array}{r} 7002 \\ \times 8040 \\ \hline \end{array}$ **14.** $\begin{array}{r} 60502 \\ \times 4009 \\ \hline \end{array}$

Solve. Write a statement for the answer.

15. Welder. If an electric welding unit weighs 325 pounds, what is the total weight of 15 units?

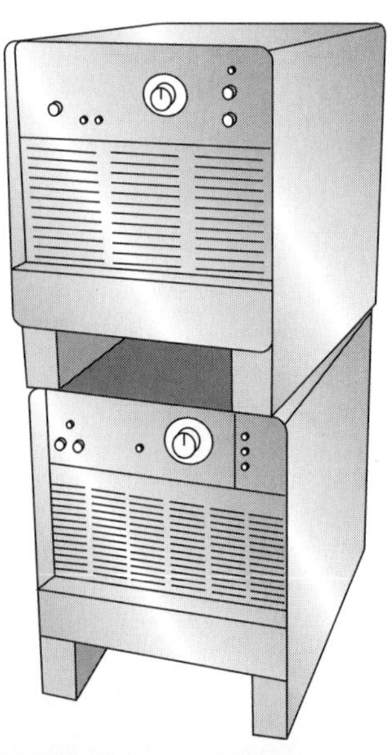

16. Management. An auto parts store ordered 3046 cases of high-quality engine oil at $18 a case. Find the total cost.

17. Secretarial. Sadie Jackson works 205 days per year as a secretary. If she types an average of 52 pages per day, how many pages does she type per year?

18. Construction. There are 250 cedar shingles in a bundle. How many cedar shingles are there in 452 bundles?

19. Mason. A mason saves $206 per month. How much will he save by the end of 3 years?

20. Drafting. The cost to excavate 2500 cu yd for a diversion dam is $15 per cu yd. Find the total cost for the excavation.

21. Construction. How many square feet of turf needs to be planted in a field $150' \times 50'$?

22. Welder. A welder examines the cost of rolls of wire and tape. He finds the cost of wire is $73.10 per roll and the cost of tape is $6.52 per roll. Find the total cost of 8 rolls of wire and 10 rolls of tape.

23. Firefighting. A water pump pumped 675 gallons of water per minute at a fire. How many gallons would be pumped in 3 hours?

24. Machinist. One lathe can produce 1620 brass bushings in one hour. How many bushings can 6 lathes produce during a 40-hour week?

25. Clerical. Flett Dairy received the following gallons of whole milk from 6 dairy farms: 3540 gallons, 2354 gallons, 3170 gallons, 450 gallons, 4471 gallons, and 1160 gallons. How many gallons were delivered to Flett Dairy?

26. Accounting. A certain foreign country has a fleet of 72 oil tankers. Each tanker has a capacity of 3,305,684 gallons of oil. How many gallons can be transported, if all tankers are at sea at full capacity?

The answers to odd-numbered problems are in the back of the book.

Answers to exercises 81–83

81. \$12,150 **82.** \$14,175
83. 192 oz

3–7 BASIC DIVISION FACTS

Before reading any further, turn to page 74 and review the basic multiplication facts.

Turn to Problems 3–7 and do the division mentally, *without counting.* **This is important!** If you get at least 90 answers correct, go on to section 3–8. If not, study section 3–7.

Members of a machine shop entered a shop pool. At the end the pot held \$35.00 with 5 equal winners. One method for dividing the money would be to give a dollar to each person one at a time until all of the \$35.00 was used up. How many dollars did each one get?

We need to break or divide 35 into 5 groups until all 35 dollars are used up.

Person A		Person B		Person C		Person D		Person E	
\$7	+	\$7	+	\$7	+	\$7	+	\$7	= \$35

Using this method, the \$35 would be used up when each person receives 7 dollars.

We can be more efficient and obtain the same result using division.

$$35 \div 5 = 7$$

↑ Division sign

This is read, "35 divided by 5 is 7," and is a basic division fact.

When dividing, we begin with two numbers called the *dividend* and *divisor.* The result of the division is called the *quotient.*

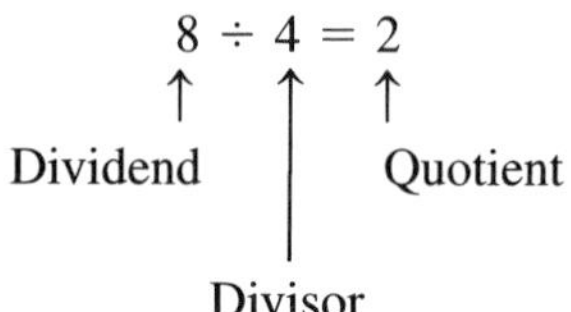

Memorize these names.

Another way to write the same division is $\frac{8}{4} = 2$. This division fact is related to the multiplication fact, $4 \times 2 = 8$ or $2 \times 4 = 8$. We can use the multiplication facts to find the related division facts.

■ Example 1

Find the quotient of $20 \div 4$. The problem can be restated as $\frac{20}{4} = ?$ We use $4 \times ? = 20$

Solution Think: "What number times 4 is 20?"
Of course, $4 \times 5 = 20$, so $20 \div 4 = 5$.

■ Example 2

$27 \div 9 = ?$ This can be restated as $\frac{27}{9} = ?$ We use $9 \times ? = 27$

Solution Think: "What number times 9 is 27?"
Since $9 \times 3 = 27$, then $27 \div 9 = 3$.

Practice check: Do exercises 84–89 in the margin.

Divide by using the basic multiplication facts.

84. $24 \div 3 = ?$ **85.** $16 \div 2 = ?$

86. $42 \div 7 = ?$ **87.** $25 \div 5 = ?$

88. $36 \div 9 = ?$ **89.** $81 \div 9 = ?$

The answers are in the margin on page 101.

Dividing by 1

■ Example 3

$8 \div 1 = ?$ The problem can be restated as $\frac{8}{1} = ?$ We use $1 \times ? = 8$

Solution Think: "What number times 1 is 8?"
Since $8 \times 1 = 8$, then $8 \div 1 = 8$

Rule

Anytime we divide a number by 1, the quotient is always the dividend.

Dividing by 0

In our number system we cannot divide by 0. Let's try it to see what happens.

■ Example 4

$5 \div 0 = ?$ The problem can be restated as $\frac{5}{0} = ?$ We use $0 \times ? = 5$

Solution Think: "What number times 0 is 5?" Let's try to find one.

$$5 \times 0 = 0 \qquad 7 \times 0 = 0 \qquad 11 \times 0 = 0$$

Since any number times 0 is still 0, there is *no number divided by* 0 that will give 5 for a quotient.

Suppose we divide 0 by 0.

■ Example 5

$0 \div 0 = ?$ Restating the problem we have $\frac{0}{0} = ?$ since $0 \times ? = 0$

Solution Think: "What number times 0 is 0?"

$6 \times 0 = 0$

$7 \times 0 = 0$

$12 \times 0 = 0$

$108 \times 0 = 0$ All of these are true.

. . . We could go on forever.

In fact the quotients that would make $0 \div 0$ or $\frac{0}{0}$ true are countless. Since there is *no one answer* to $0 \div 0$, we cannot divide 0 by 0.

Rule

In our number system, it is not possible to divide by 0.

Practice check: Do exercises 90–96 in the margin.

Divide.

90. $9 \div 1 = ?$ **91.** $11 \div 0 = ?$

92. $\frac{56}{0} = ?$ **93.** $29 \div 1 = ?$

94. $\frac{260}{0} = ?$ **95.** $\frac{381}{1} = ?$

96. $\frac{0}{0} = ?$

The answers are in the margin on page 101.

Dividing a Number by Itself

■ **Example 6**

$5 \div 5 = ?$

Solution Think "What number times 5 is 5?"
Since $1 \times 5 = 5$, then $5 \div 5 = 1$.

Rule

Any number, except 0, divided by itself is equal to 1.

Dividing 0 by a Number Other Than Zero

■ **Example 7**

$0 \div 7 = ?$

Solution Think "What number times 7 is 0?"
Since $0 \times 7 = 0$, then $0 \div 7 = 0$.

Rule

Zero divided by any number other than zero is zero.

Note: $0 \div 11$ is different from $11 \div 0$. Remember, we can't divide by 0, but we can divide a nonzero number into 0.

Practice check: Do exercises 97–104 in the margin.

97. $\frac{6}{6} = ?$ **98.** $\frac{0}{23} = ?$

99. $0 \div 17 = ?$

100. $26 \div 26 = ?$

101. $29 \div 0 = ?$

102. $\frac{106}{106} = ?$ **103.** $\frac{31}{0} = ?$

104. $0 \div 238 = ?$

The answers are in the margin on page 103.

■ Looking Back

You should now be able to do the following:

1. Use the basic multiplication facts to find quotients of such divisions as $18 \div 2$ and $\frac{45}{9}$.
2. Divide a number by 1 to get the same number.
3. Divide a number, except 0, by itself to get a quotient of 1.
4. Divide zero by a number other than zero, to get zero for a quotient.

■ Problems 3–7

Review the basic multiplication facts beginning on page 74. Then do the following divisions mentally without counting. Force yourself to be quick. Say the answers to someone while he/she checks them in the back of the book.

1. $3 \div 1 = ?$ **2.** $\frac{40}{5} = ?$ **3.** $1 \div 0 = ?$

4. $12 \div 4 = ?$ **5.** $18 \div 6 = ?$ **6.** $\frac{27}{9} = ?$

7. $35 \div 7 = ?$

8. $\frac{2}{1} = ?$

9. $6 \div 1 = ?$

10. $12 \div 2 = ?$

11. $\frac{7}{1} = ?$

12. $36 \div 6 = ?$

13. $24 \div 8 = ?$

14. $0 \div 2 = ?$

15. $16 \div 4 = ?$

16. $\frac{3}{0} = ?$

17. $\frac{56}{7} = ?$

18. $63 \div 9 = ?$

19. $\frac{5}{1} = ?$

20. $32 \div 4 = ?$

21. $30 \div 5 = ?$

22. $\frac{0}{2} = ?$

23. $24 \div 4 = ?$

24. $0 \div 6 = ?$

25. $28 \div 7 = ?$

26. $\frac{10}{5} = ?$

27. $36 \div 9 = ?$

28. $64 \div 8 = ?$

29. $\frac{72}{8} = ?$

30. $0 \div 1 = ?$

31. $\frac{45}{9} = ?$

32. $24 \div 6 = ?$

33. $\frac{9}{1} = ?$

34. $30 \div 6 = ?$

35. $4 \div 1 = ?$

36. $18 \div 9 = ?$

Answers to exercises 84–89

84. 8, since $8 \times 3 = 24$
85. 8, since $8 \times 2 = 16$
86. 6, since $6 \times 7 = 42$
87. 5, since $5 \times 5 = 25$
88. 4, since $4 \times 9 = 36$
89. 9, since $9 \times 9 = 81$

Answers to exercises 90–96

90. 9 **91.** not possible
92. not possible **93.** 29
94. not possible **95.** 381
96. not possible

---

37. $25 \div 5 = ?$

38. $\frac{6}{0} = ?$

39. $9 \div 0 = ?$

40. $\frac{0}{4} = ?$

41. $5 \div 1 = ?$

42. $45 \div 9 = ?$

43. $\frac{8}{4} = ?$

44. $21 \div 3 = ?$

45. $32 \div 8 = ?$

46. $\frac{16}{2} = ?$

47. $16 \div 8 = ?$

48. $\frac{0}{7} = ?$

49. $0 \div 5 = ?$

50. $9 \div 1 = ?$

51. $6 \div 2 = ?$

52. $\frac{40}{8} = ?$

53. $3 \div 1 = ?$

54. $42 \div 7 = ?$

55. $\frac{4}{4} = ?$

56. $\frac{63}{7} = ?$

57. $0 \div 0 = ?$

58. $\frac{4}{1} = ?$

59. $\frac{0}{4} = ?$

60. $56 \div 8 = ?$

61. $\frac{9}{9} = ?$

62. $35 \div 5 = ?$

63. $\frac{48}{8} = ?$

64. $28 \div 4 = ?$

65. $7 \div 7 = ?$

66. $8 \div 2 = ?$

Answers to exercises 97–104

97. 1 **98.** 0 **99.** 0 **100.** 1
101. not possible **102.** 1
103. not possible **104.** 0

67. $\frac{12}{3} = ?$

68. $\frac{10}{2} = ?$

69. $\frac{3}{0} = ?$

70. $18 \div 3 = ?$

71. $1 \div 1 = ?$

72. $\frac{54}{9} = ?$

73. $24 \div 3 = ?$

74. $\frac{21}{7} = ?$

75. $3 \div 3 = ?$

76. $\frac{20}{4} = ?$

77. $15 \div 5 = ?$

78. $7 \div 1 = ?$

79. $\frac{49}{7} = ?$

80. $42 \div 6 = ?$

81. $\frac{20}{5} = ?$

82. $\frac{6}{3} = ?$

83. $\frac{0}{8} = ?$

84. $\frac{6}{1} = ?$

85. $72 \div 9 = ?$

86. $\frac{18}{2} = ?$

87. $2 \div 1 = ?$

88. $27 \div 3 = ?$

89. $\frac{15}{3} = ?$

90. $12 \div 6 = ?$

91. $\frac{36}{4} = ?$ **92.** $\frac{8}{1} = ?$ **93.** $54 \div 6 = ?$

94. $\frac{14}{7} = ?$ **95.** $8 \div 1 = ?$ **96.** $0 \div 7 = ?$

97. $48 \div 6 = ?$ **98.** $\frac{9}{0} = ?$ **99.** $6 \div 6 = ?$

The answers are given in the back of the book.

3–8 DIVIDING BY ONE DIGIT WITHOUT REGROUPING

Dividing Powers of 10, 100, or 1000

Consider $20 \div 2$. We can write this division as $\frac{20}{2}$ or $2\overline{)20}$. This tells us to find how many 2s we can use as addends to make a sum of 20. One way we could find out is by repeated subtraction.

$$
\begin{array}{rl}
20 & \\
\underline{-2} & 1 \\
18 & \\
\underline{-2} & 2 \\
16 & \\
\underline{-2} & 3 \\
14 & \\
\underline{-2} & 4 \\
12 & \\
\underline{-2} & 5 \\
10 & \\
\underline{-2} & 6 \\
8 & \\
\underline{-2} & 7 \\
6 & \\
\underline{-2} & 8 \\
4 & \\
\underline{-2} & 9 \\
2 & \\
\underline{-2} & 10 \\
0 &
\end{array}
$$

There are ten 2s in 20. This method would take too long and take up too much paper on larger numbers.

There is a shorter way! On divisions such as these,-we shall use the form $2\overline{)20}$.

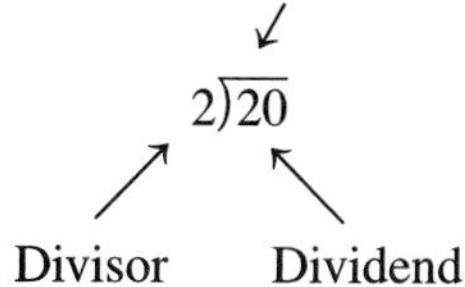

Think: "What number times 2 is 20?" Since

$10 \times \quad = 20$, then $2\overline{)20}$ with 10 written above. Put the 1 over the 2 (tens place).

$20 \div 2 = 10$

■ **Example 1**

Divide $2\overline{)40}$.

Solution $\begin{array}{r} 20 \\ 2\overline{)40} \end{array}$ because $20 \times 2 = 40$. Thus, $40 \div 2 = 20$.

■ **Example 2**

Divide $2\overline{)160}$.

Solution Place the 8 over the 6 (tens place).

$\begin{array}{r} 80 \\ 2\overline{)160} \end{array}$ because $80 \times 2 = 160$. Thus, $160 \div 2 = 80$.

Let's look at Example 2 again: $2\overline{)160}$ The quotient of course is 80. One way to get the "8" is to mentally divide 16 by 2. Once you have the "8" then it is easy to find the quotient 80, since $80 \times 2 = 160$.

Try this method on Example 3.

■ **Example 3**

Divide $7\overline{)210}$.

Solution **Step 1** Mentally divide 21 by 7. $21 \div 7 = 3$

Step 2 Write the quotient. $\begin{array}{r} 30 \\ 7\overline{)210} \end{array}$, since $30 \times 7 = 210$.
Thus, $210 \div 7 = 30$.

Practice check: Do exercises 105–111 in the margin.

Divide.

105. $2\overline{)60}$ **106.** $3\overline{)90}$

107. $6\overline{)180}$ **108.** $9\overline{)270}$

109. $2\overline{)100}$ **110.** $7\overline{)140}$

111. $5\overline{)250}$

The answers are in the margin on page 107.

Dividing without Regrouping

■ **Example 4**

Divide $2\overline{)46}$ using the expanded method.

Solution **Step 1** Expand the dividend. $2\overline{)46} = 2\overline{)40 + 6}$

Step 2 Divide 40 by 2. $\begin{array}{l} \;\;20 \\ 2\overline{)40 + 6} \end{array}$

Step 3 Divide 6 by 2. $\begin{array}{l} \;\;20 + 3\text{, or } 23 \\ 2\overline{)40 + 6} \end{array}$

■ Example 5

Divide $3\overline{)363}$ using the expanded method.

Solution **Step 1** Expand the dividend. $3\overline{)363} = 3\overline{)300 + 60 + 3}$.

Step 2 Divide 300, 60, and 3 by 3. $3\overline{)363} = \begin{array}{r} 100 + 20 + 1 = 121 \\ 3\overline{)300 + 60 + 3} \end{array}$

Practice check: Do exercises 112–114 in the margin.

Divide using the expanded method.

112. $2\overline{)64}$ **113.** $3\overline{)693}$

114. $2\overline{)4862}$

The answers are in the margin on page 109.

We have shown the expanded method to help you understand the division process. It would be very complicated and impractical to always divide in this way. Example 6 shows a short method for division that can be used for all types of division exercises.

■ Example 6

Divide $2\overline{)46}$ using the short method.

Solution **Step 1** Divide 4 by 2.

$$\begin{array}{r} 2 \\ 2\overline{)46} \end{array}$$

Place the 2 over the 4.

Step 2 Multiply 2 by 2 and subtract.

$$\begin{array}{r} 2 \\ 2\overline{)46} \\ \underline{4} \end{array} \leftarrow (2 \times 2)$$

Step 3 Bring down the 6.

$$\begin{array}{r} 2 \\ 2\overline{)46} \\ \underline{4} \\ 06 \end{array}$$

Step 4 Divide 6 by 2.

$$\begin{array}{r} 23 \\ 2\overline{)46} \\ \underline{4} \\ 06 \end{array}$$

Place the 3 over the 6.

Step 5 Multiply 3 by 2 and subtract.

$$\begin{array}{r} 23 \\ 2\overline{)46} \\ \underline{4} \\ 06 \\ \underline{6} \\ 0 \end{array} \leftarrow (3 \times 2)$$

The quotient is 23.

■ Example 7

Divide $3\overline{)363}$ using the short method.

Solution **Step 1** Divide.
$3 \div 3 = 1$

$$\begin{array}{r} 1 \\ 3\overline{)363} \end{array}$$

Place the 1 over the 3.

Step 2 Multiply and subtract.

$$\begin{array}{r} 1 \\ 3\overline{)363} \\ \underline{3} \\ 0 \end{array} \leftarrow (1 \times 3)$$

Step 3 Bring down the 6.

$$\begin{array}{r} 1 \\ 3\overline{)363} \\ \underline{3} \\ 06 \end{array}$$

Step 4 Divide.
$6 \div 3 = 2$

$$\begin{array}{r} 12 \\ 3\overline{)363} \\ \underline{3} \\ 06 \end{array}$$ Place the 2 above the 6.

Step 5 Multiply and subtract.

$$\begin{array}{r} 12 \\ 3\overline{)363} \\ \underline{3} \\ 06 \\ \underline{6} \leftarrow (2 \times 3) \\ 0 \end{array}$$

Step 6 Bring down the 3.

$$\begin{array}{r} 12 \\ 3\overline{)363} \\ \underline{3} \\ 06 \\ \underline{6} \\ 03 \end{array}$$

Step 7 Divide.
$3 \div 3 = 1$

$$\begin{array}{r} 121 \\ 3\overline{)363} \\ \underline{3} \\ 06 \\ \underline{6} \\ 03 \end{array}$$ Place the 1 over the 3.

Step 8 Multiply and subtract.

$$\begin{array}{r} 121 \\ 3\overline{)363} \\ \underline{3} \\ 06 \\ \underline{6} \\ 03 \\ \underline{3} \leftarrow (1 \times 3) \\ 0 \end{array}$$

The quotient is 121.

Divide $3\overline{)363}$ using the short method without looking at the steps above.

Rule

The steps for dividing using the short method are divide, multiply, subtract, and bring down. In order for you to divide quickly, these steps should be memorized.

Practice check: Do exercises 115–117 in the margin.

■ Looking Back

You should now be able to divide such problems as $2\overline{)482}$ using the short method.

■ Problems 3–8

Divide using the short method.

1. $2\overline{)62}$ **2.** $2\overline{)24}$ **3.** $2\overline{)28}$ **4.** $2\overline{)86}$ **5.** $3\overline{)36}$

Answers to exercises 105–111

105. $2\overline{)60}$ = 30 since $30 \times 2 = 60$

106. $3\overline{)90}$ = 30 since $30 \times 3 = 90$

107. $6\overline{)180}$ = 30 since $30 \times 6 = 180$

108. $9\overline{)270}$ = 30 since $30 \times 9 = 270$

109. $2\overline{)100}$ = 50 since $50 \times 2 = 100$

110. $7\overline{)140}$ = 20 since $20 \times 7 = 140$

111. $5\overline{)250}$ = 50 since $50 \times 5 = 250$

Divide. Use the short method.

115. $2\overline{)64}$ **116.** $3\overline{)693}$

117. $2\overline{)4862}$

The answers are in the margin on page 109.

6. $3\overline{)69}$ **7.** $3\overline{)93}$ **8.** $3\overline{)63}$ **9.** $4\overline{)48}$ **10.** $4\overline{)84}$

11. $5\overline{)55}$ **12.** $8\overline{)88}$ **13.** $2\overline{)426}$ **14.** $2\overline{)286}$ **15.** $3\overline{)696}$

16. $3\overline{)396}$ **17.** $4\overline{)884}$ **18.** $4\overline{)484}$ **19.** $2\overline{)6842}$ **20.** $2\overline{)8462}$

21. $3\overline{)3693}$ **22.** $3\overline{)6396}$ **23.** $4\overline{)4488}$ **24.** $4\overline{)8448}$ **25.** $2\overline{)42,648}$

26. $4\overline{)84,480}$

27. **Diesel.** An alternator generates 90 amps over a period of 5 hours. What is the output of the alternator?

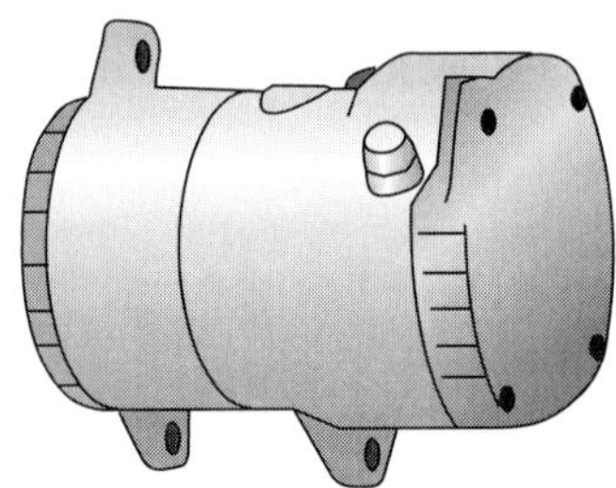

28. Marketing. During a six-hour sale 8 calculators were sold for $208.00. Find the price for each calculator.

29. Police Science. Car 28 of the Richfield Police Department traveled 558 miles in a 9 day period. Find the average number of miles traveled per day.

30. Mason. A mason completes 350 square feet of concrete blocks in 7 hours. What is the average number of square feet of concrete blocks completed in 1 hour?

The answers to odd-numbered problems are in the back of the book.

3–9 DIVIDING BY ONE DIGIT WITH REGROUPING AND REMAINDERS

In all of the examples in section 3–8, the divisor could be divided into the first digit of the dividend. This will not always be the case.

■ **Example 1**

Divide 148 ÷ 2.

Solution To understand the short-cut method, we begin with the expanded method. Expanded method:

$$2\overline{)148} = \overset{50 + 20 + 4 = 74}{2\overline{)100 + 40 + 8}}$$

Dividing each addend by 2.

We can shorten the expanded method further.

$$\overset{70 + 4 = 74}{2\overline{)140 + 8}}$$

↗ Combining 100 + 40.

Solution Short method:

Step 1 Look at the dividend. Choose the first digit or digits that are equal to or greater than the divisor. This digit or combination of digits is called the first *partial dividend.*

$2\overline{)\mathbf{148}}$

$2\overline{)\mathbf{14}\,8}$

Step 2 Divide the first partial dividend by the divisor (14 ÷ 2 = 7). The answer is called a *partial quotient.* Place it above the last digit of the first partial dividend.

$\mathbf{2}\overline{)\mathbf{14}\,8}$

$\overset{\mathbf{7}}{2\overline{)148}}$

Step 3 Multiply the divisor by the partial quotient (2 × 7 = 14). Place the product under the first partial dividend.

$\overset{7}{2\overline{)148}}$ with **14** below

Step 4 Subtract the product from the first partial dividend.

$\overset{7}{2\overline{)148}}$, 14, **0**

(*Note:* If the remainder is more than or equal to the divisor, the partial quotient is wrong.)

Step 5 Bring down the next digit in the dividend. The remainder, if any, and this digit together are called the second partial dividend.

$\overset{7}{2\overline{)148}}$, 14↓, **8**

Step 6 Divide the second partial dividend by the divisor (8 ÷ 2 = 4). Place this new partial quotient directly above the digit th\at was brought down from the dividend.

$\overset{\mathbf{74}}{\mathbf{2}\overline{)148}}$, 14, **8**

Answers to exercises 112–114

112. $\overset{30 + 2 = 32}{2\overline{)60 + 4}}$

113. $\overset{200 + 30 + 1 = 231}{3\overline{)600 + 90 + 3}}$

114. $\overset{2000+400+30+1 = 2431}{2\overline{)4000+800+60+2}}$

Answers to exercises 115–117

115. 32 **116.** 231 **117.** 2431

You continue the process found in Steps 3–6 (divide, multiply, subtract, and bring down) until all of the dividend digits have been brought down and the final remainder is less than the divisor

```
   74
2)148
  14
   8
   8
   0
```

The quotient is 74.

Practice check: Do exercises 118–120 in the margin.

Divide.

118. $3\overline{)249}$ **119.** $6\overline{)486}$

120. $2\overline{)1842}$

The answers are in the margin on page 112.

Remainders other than Zero

Thus far the division examples have shown a final remainder of 0. This means that the divisor could be divided into the dividend evenly. When this is not true, the final remainder is greater than 0 but less than the divisor.

■ Example 2

Divide $5\overline{)3558}$.

Solution **Step 1**

```
   7
5)3558
```

Divide 35 by 5.
Write 7 in the hundreds place.

Step 2

```
   7
5)3558
  35↓  ←— (7 × 5)
   5
```

Multiply, subtract, and bring down.

Step 3

```
   71
5)3558
  35
   5
```

Divide 5 by 5.
Write 1 in the tens place.

Step 4

```
   71
5)3558
  35 |
   5 |
   5↓  ←— (1 × 5)
    8
```

Multiply 1 × 5, subtract, and bring down.

Step 5

```
   711
5)3558
  35 |
   5 |
   5↓
    8
```

Divide 5 by 5. There is only one group of 5s in 8. Write the 1 in the ones place.

Step 6

```
   711
5)3558
  35
   5
   5
    8
    5  ←— (1 × 5)
    3  ←— remainder
```

Multiply 1 × 5, and subtract.

Since all of the digits in the divisor have been brought down and 3 is less than 5, the division is finished.

The quotient is 711 R 3 (R means remainder).

Practice check: Do exercises 121–123 in the margin.

Divide.

121. $3\overline{)278}$ **122.** $6\overline{)1869}$

123. $8\overline{)4089}$

The answers are in the margin on page 112.

■ **Example 3**

Divide $3\overline{)2867}$

Solution **Step 1**

$$\begin{array}{r} 9 \\ 3\overline{)2867} \\ \underline{27} \\ 16 \end{array}$$

1 Divide 28 by 3. There are 9 groups of 3s in 28.
2 Multiply 9 × 3.
3 Subtract and bring down the 6.

Step 2

$$\begin{array}{r} 95 \\ 3\overline{)2867} \\ \underline{27} \\ 16 \\ \underline{15} \\ 17 \end{array}$$

1 Divide 16 by 3. There are 5 groups of 3s in 16.
2 Multiply 5 × 3.
3 Subtract and bring down the 7.

Step 3

$$\begin{array}{r} 955 \\ 3\overline{)2867} \\ \underline{27} \\ 16 \\ \underline{15} \\ 17 \\ \underline{15} \\ 2 \end{array}$$

1 Divide 17 by 3. There are 5 groups of 3s in 17.
2 Multiply 5 × 3.
3 Subtract.

The quotient is 955 R 2.

Divide $3\overline{)2867}$ using the short method without looking at the steps above.

Practice check: Do exercises 124–126 in the margin.

Divide.

124. $3\overline{)5264}$ **125.** $7\overline{)9361}$

126. $4\overline{)3356}$

The answers are in the margin on page 112.

Quotients and Zeros

■ **Example 4**

Divide $3\overline{)2252}$

Solution **Step 1**

$$\begin{array}{r} 7 \\ 3\overline{)2252} \\ \underline{21} \\ 15 \end{array}$$

1 Divide 22 by 3.
2 Multiply 7 × 3.
3 Subtract and bring down the 5.

Step 2

$$\begin{array}{r} 75 \\ 3\overline{)2252} \\ \underline{21} \\ 15 \\ \underline{15} \\ 2 \end{array}$$

1 Divide 15 by 3.
2 Multiply 5 × 3.
3 Subtract and bring down the 2.

Notice that the ones place is empty, so we continue dividing even though 2 is less than 3.

Answers to exercises 118–120

118. 83 **119.** 81 **120.** 921

Answers to exercises 121–123

121. 92 R 2 **122.** 311 R 3
123. 511 R 1

Answers to exercises 124–126

124. 1754 R 2 **125.** 1337 R 2
126. 839

Step 3

```
     750 ←
  3)2252    1   Divide 2 by 3.
    21
     15
     15
  →   2
      0 ← 2     Multiply 0 × 3.
      2 ← 3     Subtract.
```

The quotient is 750 R 2.

■ Example 5

Divide $4\overline{)2432}$

Solution **Step 1**

```
     6 ←
  4)2432    1   Divide 24 by 4.
    24   ← 2    Multiply 6 × 4.
     3   ← 3    Subtract and bring down the 3.
```

Step 2

```
     60 ←
  4)2432    1   Divide 3 by 4.
    24
  →  3
     0   ← 2    Multiply 6 × 4.
    32   ← 3    Subtract and bring down the 2.
```

Step 3

```
     608 ←
  4)2432    1   Divide 32 by 4.
    24
     3
     0
  → 32
    32   ← 2    Multiply 6 × 4.
     0   ← 3    Subtract.
```

The quotient is 608.

Divide $3\overline{)2252}$ and $4\overline{)2432}$ using the short method without looking at the steps above.

■ Example 6

Channel iron is delivered in 21-ft lengths. If a section must be cut into 4-ft lengths, how many pieces can be obtained from each length?

Solution Since we need to find the number of lengths in each 21-ft length, we divide 21 by 4.

```
     5
  4)21
    20
     1
```

There are 5 pieces with 1 ft left over in each channel iron.

Practice check: Do exercises 127–129 in the margin.

Divide.

127. $3\overline{)2723}$ **128.** $7\overline{)61740}$

129. $5\overline{)321046}$

The answers are in the margin on page 115.

Checking

To check division, we multiply the divisor by the quotient and add the remainder to get the dividend.

■ Example 7

Divide and check $8\overline{)3047}$.

Solution

$$\begin{array}{r} 380 \\ 8\overline{)3047} \\ \underline{24} \\ 64 \\ \underline{64} \\ 7 \\ \underline{0} \\ 7 \end{array} \qquad \text{Check: } \begin{array}{rl} 380 & \text{Quotient} \\ \underline{\times 8} & \text{Divisor} \\ 3040 & \\ \underline{+7} & \text{Remainder} \\ 3047 & \text{Dividend} \end{array}$$

The quotient is 380 R 7.

Practice check: Do exercises 130 and 131 in the margin.

Divide.

130. $6\overline{)7204}$ **131.** $9\overline{)73805}$

The answers are in the margin on page 115.

Here is a summary of the steps for the short-cut method.

Rule

Short-Cut Division Method:

Step 1 Divide the divisor into a partial dividend.

Step 2 Multiply the divisor by the partial quotient.

Step 3 Subtract the product of Step 2 from the partial dividend of Step 1.

Step 4 Bring down the next digit in the dividend.

Step 5 Repeat Steps 1 to 4 until all of the digits in the dividend have been brought down and the remainder in Step 3 is less than the divisor.

■ Looking Back

You should be able to divide whole numbers where the divisor is a one-digit number.

■ Problems 3–9

Divide and check.

1. $3\overline{)279}$ **2.** $2\overline{)148}$ **3.** $2\overline{)186}$ **4.** $3\overline{)246}$

5. $6\overline{)2468}$ **6.** $7\overline{)3579}$ **7.** $6\overline{)2154}$ **8.** $8\overline{)5763}$

9. $3\overline{)14595}$

10. $2\overline{)6008}$

11. $5\overline{)38403}$

12. $7\overline{)28568}$

13. $8\overline{)37580}$

14. $3\overline{)7582}$

15. $8\overline{)644}$

16. $6\overline{)87302}$

17. $7\overline{)23583}$

18. $8\overline{)7348}$

19. $6\overline{)92405}$

20. $7\overline{)96058}$

21. $9\overline{)705070}$

22. $9\overline{)604900}$

23. $6\overline{)5009560}$

24. $7\overline{)3004067}$

25. **Printer.** A pressman earned $900 in 5 days. How much does he earn each day?

26. **Welding.** Seven oil tanks are constructed in a welding shop. If the total number of gallons in the 7 tanks is 385 gallons, what is the capacity of each tank?

27. **Mason.** Working together on a job, 5 bricklayers lay 5175 bricks per day. On the average, how many bricks does each worker lay per day?

28. **Aeronautics.** A jet passenger plane travels 2296 miles in 5 hours. What is the average speed per hour?

29. **Forestry.** A certain National Forest has 8 plots containing 356 acres. Determine the average number of acres per plot.

30. **Electrician.** How many 9-inch-long pieces of insulation wire can be cut from a 500-foot roll of wire? (12 inches = 1 foot)

The answers to odd-numbered problems are in the back of the book.

3–10 DIVIDING BY TWO OR MORE DIGITS

Finding partial quotients get harder when the divisor contains two or more digits. We can use multiplication to help find the correct partial quotient.

■ Example 1

Divide $26\overline{)58}$.

Solution Think: "How many 26s are there in 58 ones?"
We can see that there is at least 1. Are there 2? Let's see.

$$\begin{array}{r} 26 \\ \underline{\times 2} \\ 52 \end{array}$$ Yes, there are two 26s in 58.

Are there three?

$$\begin{array}{r} 26 \\ \underline{\times 3} \\ 78 \end{array}$$ No, there are not!

Divide.

$$\begin{array}{r} 2 \\ 26\overline{)58} \\ \underline{52} \\ 6 \end{array}$$ *The quotient is* 2 *R* 6.

■ Example 2

Divide $18\overline{)76}$.

Step 1 Think: "How many 18s are there in 76?"
We guess there are 6.

$$\begin{array}{r} 18 \\ \underline{\times 6} \\ 108 \end{array}$$ Too much.

Let's try 4.

$$\begin{array}{r} 18 \\ \underline{\times 4} \\ 72 \end{array}$$ Eureka!

Step 2 Complete the division.

$$\begin{array}{r} 4 \\ 18\overline{)76} \\ \underline{72} \\ 4 \end{array}$$ The quotient is 4 R 4.

Practice check: Do exercises 132–135 in the margin.

■ Example 3

Divide $28\overline{)170}$

Solution **Step 1** $28\overline{)170}$ Think: "How many 28s are there in 170?"
We guess there are 8.

$$\begin{array}{r} 28 \\ \underline{\times 8} \\ 224 \end{array}$$ Too much.

Answers to exercises 127–129

127. 907 R 2 **128.** 8820
129. 64,209 R 1

Answers to exercises 130 and 131

130. 1200 R 4 **131.** 8200 R 5

Divide and check.

132. $17\overline{)70}$ **133.** $22\overline{)93}$

134. $33\overline{)88}$ **135.** $19\overline{)73}$

The answers are in the margin on page 117.

Let's try 5.
$$\begin{array}{r} 28 \\ \times 5 \\ \hline 140 \end{array}$$

We divide.
$$\begin{array}{r} 5 \\ 28\overline{)170} \\ \underline{140} \\ 30 \end{array}$$
The remainder after subtraction is greater than the divisor. This means that our partial quotient is too small.

Let's try 6.
$$\begin{array}{r} 28 \\ \times 6 \\ \hline 168 \end{array}$$
Eureka!

Step 2 Complete the division.
$$\begin{array}{r} 6 \\ 28\overline{)170} \\ \underline{168} \\ 2 \end{array}$$
The quotient is 6 R 2.

Practice check: Do exercises 136–138 in the margin.

Divide and check.

136. $26\overline{)234}$ **137.** $46\overline{)380}$

138. $60\overline{)305}$

The answers are in the margin on page 118.

■ **Example 4**

Divide $38\overline{)4672}$

Solution **Step 1**
$$\begin{array}{r} 1 \\ \mathbf{38}\overline{)\mathbf{46}72} \\ \underline{38} \\ 87 \end{array}$$
1 Divide 46 by 38.
2 Multiply 1 × 38.
3 Subtract and bring down.

Step 2
$$\begin{array}{r} 12 \\ \mathbf{38}\overline{)4672} \\ \underline{38} \\ \mathbf{87} \\ \underline{76} \\ 112 \end{array}$$
1 Divide 87 by 38.
2 Multiply 2 × 38.
3 Subtract and bring down.

Step 3
$$\begin{array}{r} 122 \\ \mathbf{38}\overline{)4672} \\ \underline{38} \\ 87 \\ \underline{76} \\ \mathbf{112} \\ \underline{76} \\ 36 \end{array}$$
1 Divide 112 by 38.
2 Multiply 2 × 38.
3 Subtract.

Step 4 Check.
$$\begin{array}{r} 122 \\ \times 38 \\ \hline 976 \\ \underline{3660} \\ 4636 \\ +36 \\ \hline 4672 \end{array}$$

The quotient is 122 R 36.

■ **Example 5**

Divide $73\overline{)38466}$

Solution

```
     526     We first divide 384 by 73      Check:    526
 73)38466    and place the 5 over the                 ×73
    365      4 (hundreds place).                     1578
     196                                            36820
     146                                            38398
      506                                            +68
      438                                           38466
       68
```

The quotient is 526 R 68

Practice check: Do exercises 139–143 in the margin.

■ Example 6

Divide 427)5416

Solution **Step 1**

```
        1
427)5416    1   Divide 541 by 427.
    427  ←— 2   Multiply 1 × 427.
    1146 ←— 3   Subtract and bring down.
```

Step 2

```
       12
427)5416    1   Divide 1146 by 427.
    427
    1146
     854 ←— 2   Multiply 2 × 427.
     292 ←— 3   Subtract.
```

Check:

```
 427
 ×12
 854
4270
5124
+292
5416
```

The quotient is 12 R 292

Divide 427)5416 using the short method without looking at the steps above.

Practice check: Do exercises 144 and 145 in the margin.

Again, you must be extra careful when zeros are involved in the quotient.

■ Example 7

Divide 152)91709

Solution **Step 1**

```
        6
152)91709    1   Divide 917 by 152.
    912  ←— 2    Multiply 6 × 152.
     50  ←— 3    Subtract and bring down.
```

Answers to exercises 132–135

132. 4 R 2 **133.** 4 R 5
134. 2 R 22 **135.** 3 R 16

Divide and check.

139. 32)887 **140.** 49)683
141. 25)1956 **142.** 38)5430
143. 85)42461

The answers are in the margin on page 118.

Divide and check.

144. 285)1564 **145.** 321)8032

The answers are in the margin on page 118.

Answers to exercises 136–138

136. 9 **137.** 8 R 12 **138.** 5 R 5

Answers to exercises 139–143

139. 27 R 23 **140.** 13 R 46
141. 78 R 6 **142.** 142 R 34
143. 499 R 46

Answers to exercises 144 and 145

144. 5 R 139 **145.** 25 R 7

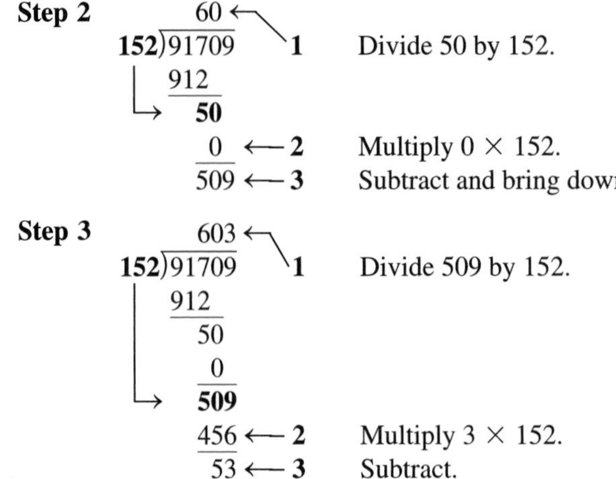

Check:
```
   152
  ×603
   456
 91200
 91656
   +53
 91709
```
The quotient is 603 R 53.

Divide 152)91709 using the short method without looking at the steps above.

■ Example 8

A beverage company manufactures 789 liter bottles of ginger ale per day. How many 12-liter bottle cartons can be filled? How many will be left?

Solution We divide

```
    65
12)789
   72
    69
    60
     9
```

Check

```
  65
 ×12
 130
 650
 780
  +9
 789
```

65 cartons with 9 bottles left.

Practice check: Do exercises 146 and 147 in the margin.

Divide and check.

146. 341)78946 **147.** 203)62156

The answers are in the margin on page 120.

■ Looking Back

You should now be able to divide whole numbers by two- or three-digit numbers

■ Problems 3–10

Divide and check.

Review

1. 3)346 **2.** 7)2804 **3.** 5)2504

4. $9\overline{)19304}$ **5.** $6\overline{)23004}$ **6.** $9\overline{)144320}$

7. $29\overline{)66}$ **8.** $31\overline{)73}$ **9.** $51\overline{)98}$ **10.** $18\overline{)62}$

11. $13\overline{)108}$ **12.** $84\overline{)336}$ **13.** $68\overline{)476}$ **14.** $39\overline{)185}$

15. $46\overline{)789}$ **16.** $47\overline{)880}$ **17.** $26\overline{)893}$ **18.** $17\overline{)294}$

19. $96\overline{)7872}$ **20.** $43\overline{)1935}$ **21.** $18\overline{)1567}$ **22.** $38\overline{)2706}$

23. $39\overline{)5630}$ **24.** $56\overline{)7846}$ **25.** $16\overline{)4384}$ **26.** $40\overline{)8703}$

27. $73\overline{)29631}$ **28.** $27\overline{)13070}$ **29.** $33\overline{)56958}$ **30.** $32\overline{)23470}$

31. $47\overline{)20036}$ **32.** $18\overline{)49986}$ **33.** $87\overline{)35005}$ **34.** $50\overline{)61050}$

35. $640\overline{)1930}$ **36.** $285\overline{)9572}$ **37.** $745\overline{)52150}$ **38.** $475\overline{)75436}$

39. $768\overline{)522490}$ **40.** $306\overline{)25120}$ **41.** $601\overline{)433727}$ **42.** $679\overline{)539805}$

43. $400\overline{)208600}$ **44.** $589\overline{)417012}$ **45.** $625\overline{)398125}$ **46.** $8304\overline{)64842}$

47. $5280\overline{)155130}$ **48.** $6080\overline{)1935000}$ **49.** $3246\overline{)25466737}$

Answers to exercises 146–147

146. 231 R 175 **147.** 306 R 38

50. Painting. How many gallons of primer at $21.00 a gallon can be bought for $780?

The answers to odd-numbered problems are in the back of the book.

3–11 APPLIED PROBLEMS

When you solve applied problems, be sure to write a statement for your answer.

■ **Example 1**

Sam Jackson's annual salary is $11,148. What is his monthly salary?

Solution Since there are 12 months in a year, we divide $11,148 by 12.

```
     929
12)11148
   108
     34
     24
     108
     108
       0
```

His monthly salary is $929.

■ **Example 2**

A student had the following test scores in arithmetic: 85%, 92%, 73%, 67%, 82%, 100%, 75%, and 90%. What was the average score?

Solution To find the average, we add all the scores together and divide by the number of scores.

Step 1 Add the scores.

```
 85 ⎫
 92 ⎪
 73 ⎪
 67 ⎪
 82 ⎬ 8 scores
100 ⎪
 75 ⎪
 90 ⎪
---  ⎪
664 ⎭
```

Step 2 Divide by the number of scores.

```
   83
8)664
  64
   24
   24
    0
```

The average score is 83%.

Practice check: Do exercise 148 in the margin.

■ Looking Back

You should now be able to solve word problems involving division.

■ Problems 3–11

Divide and check.

Review

1. $7\overline{)30461}$

2. $5\overline{)73004}$

3. $28\overline{)1694}$

4. $18\overline{)24658}$

5. $302\overline{)410843}$

6. $984\overline{)6604321}$

7. Architect. Architect Martha Wick earned a weekly salary of $600. If she worked 40 hours that week, how much an hour did she earn?

8. Automotive. A tank of kerosene contains 550 gallons. If 71 gallons are used each day, how many days will it take to use up the kerosene?

9. Finance. Bill Lindstrom owes the bank $1335. He plans to pay it back in 15 equal monthly installments. How much will be returned each month?

10. Technician. Harry Moses assembled 43,070 mini-radio units in one year. How many radio units did he average per day? (Assume 365 days in a year.)

11. Engineer Technician. A rocket-powered automobile traveled 74,180 feet for a test run. How many miles did it travel if there are 5280 feet in one mile?

12. Agriculture. A farm hand was told not to spend over $2632 on baling wire. Each roll of wire cost $22. How many rolls of wire could he buy and how much money was left over?

13. Fire Science. In 1998 sixteen firemen in Fire House Number 18 fought 992 fires. What was the average number of fires fought per person if duplication is not considered?

14. Aeronautics. A jet traveled 3240 miles in 6 hours. What was its average speed per hour?

Solve.

148. An automobile traveled a distance of 1248 miles at an average rate of 52 mph. How many hours did it take for the automobile to travel that distance?

The answer is in the margin on page 123.

15. **Clerical.** Fryer Bakery uses 73 lb of wheat flour a week. How many weeks will the flour last if they have 482 lb on hand?

16. **Finance.** Sally Buyrite bought a Ford LTD for $3516 twelve years ago. If she sold it for $1140, what was the yearly depreciation?

The answers to review and odd-numbered exercises are in the back of the book.

CHAPTER 3 OVERVIEW

Summary

3–1 You learned to give the basic multiplication facts mentally without counting.

3–2 You learned to multiply 10, 100, or 1000 by any whole number and to multiply multiples of 10, 100, or 1000 by a one-digit number.

3–3 You learned to multiply any whole number by a one-digit number.

3–4 You learned to multiply multiples of 10, 100, or 1000 by any whole number.

3–5 You learned to multiply by numbers containing several digits.

3–6 You learned to solve applied problems involving multiplication of whole numbers.

3–7 You learned to give the basic division facts mentally without counting.

3–8 You learned to divide by one-digit numbers without regrouping.

3–9 You learned to divide by one-digit numbers using regrouping and remainders.

3–10 You learned to divide by numbers containing two or more digits.

3–11 You learned to solve applied problems involving division of whole numbers.

Terms To Remember

	Page		*Page*
Basic Multiplication Facts	74	Partial Product	91
Dividend	98	Product	73
Divisor	98	Quotient	98
Factor	73	Remainder	110
Partial Dividend	109	Repeated Division	104

Rules

- In general, the product of a number and 10, 100, or 1000 is that number followed respectively by one, two, or three zeros.
- In general, to find the product of a number and a multiple of 10, we write the product of the two nonzero whole numbers followed by the number of zeros in the multiple of 10.
- When we multiply with factors that are multiples of 10, we write down the product of the two nonzero whole numbers and then write down after it the total number of zeros that appear in both factors.
- When we multiply by a factor that is a multiple of 10, we write down as many zeros that appear in the multiple of 10 and then write down the product of the two nonzero whole numbers.

- Anytime we divide a number by 1, the quotient is always the dividend.
- In our number system, it is not possible to divide by 0.
- Any number, except 0, divided by itself is equal to 1.
- Zero divided by any number other than zero is zero.
- The steps for dividing using the short method are divide, multiply, subtract, and bring down. In order for you to divide quickly, these steps should be memorized.
- Short-Cut Division Method

Step 1 Divide the divisor into a partial dividend.
Step 2 Multiply the divisor by the partial quotient.
Step 3 Subtract the product of Step 2 from the partial dividend of Step 1.
Step 4 Bring down the next digit in the dividend.
Step 5 Repeat Steps 1 to 4 until all of the digits in the dividend have been brought down and the remainder in Step 3 is less than the divisor.

Answers to exercises 148

148. 52)1248 = 24 hours
$$\begin{array}{r} 24 \\ 52\overline{)1248} \\ \underline{104} \\ 208 \\ \underline{208} \\ 0 \end{array}$$

Self-Test

3–1, 3–2 Multiply.

1. 10 × 8

2. 806 × 1000

3. 89 × 100 = ?

4. 6000 × 4 = ?

5. 5 × 4000

6. 5 × 70

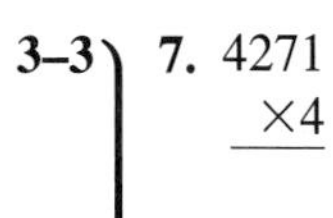

3–3, 3–4

7. 4271 × 4

8. 3009 × 7

9. 607 × 40

10. 506 × 7000

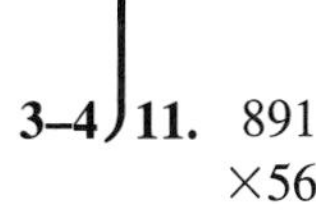

11. 891 × 56

12. 3146 × 1891

13. 5309 × 33

14. 7005 × 6070

3–6 **15.** A salesperson sold 56 cars for $13,650 each. What was the total amount of sales?

16. A lot in the shape of a rectangle measured 238 meters by 106 meters. Find the area of the lot. (Area is found by multiplying length by the width.)

3–7, 3–8 Divide.

17. $3\overline{)6393}$

3–9 Divide.

18. $3\overline{)685}$ 19. $6\overline{)830}$ 20. $7\overline{)502}$ 21. $3\overline{)23468}$

3–10 22. $72\overline{)2646}$ 23. $52\overline{)47369}$ 24. $285\overline{)9609}$

25. $144\overline{)10800}$ 26. $805\overline{)329425}$ 27. $258\overline{)191409}$

3–11 28. Green Balm Taxi #28 traveled 582 miles on Monday, 708 miles on Tuesday, 467 miles on Wednesday, 0 miles on Thursday, and 358 miles on Friday. What was the average number of miles traveled each day during the five-day period?

29. A jet flew a distance of 8175 miles at an average rate of 545 miles per hour. How many hours did it take the jet to fly this distance?

Chapter Test

Directions: This test will aid in your preparation for a possible chapter test given by your instructor. The answers are in the back of the book. Go back and review in the appropriate section(s) if you missed any test items.

3–1, 3–2 Multiply.

1. $\begin{array}{r} 10 \\ \times 7 \\ \hline \end{array}$

2. $\begin{array}{r} 717 \\ \times 1000 \\ \hline \end{array}$

3. 100×67

4. 5×8000

5. $\begin{array}{r} 9 \\ \times 80 \\ \hline \end{array}$

6. $\begin{array}{r} 4 \\ \times 5000 \\ \hline \end{array}$

3–3, 3–4

7. $\begin{array}{r} 5381 \\ \times 3 \\ \hline \end{array}$

8. $\begin{array}{r} 4008 \\ \times 8 \\ \hline \end{array}$

9. $\begin{array}{r} 703 \\ \times 70 \\ \hline \end{array}$

3–5

10. $\begin{array}{r} 409 \\ \times 6000 \\ \hline \end{array}$

11. $\begin{array}{r} 782 \\ \times 69 \\ \hline \end{array}$

12. $\begin{array}{r} 4218 \\ \times 1971 \\ \hline \end{array}$

13. $\begin{array}{r} 4208 \\ \times 44 \\ \hline \end{array}$

14. $\begin{array}{r} 8001 \\ \times 7080 \\ \hline \end{array}$

Solve.

3–6 15. How many sheets of laser printer paper are in 86 packages if each package contains 500 sheets (1 ream)?

Divide and check.

3–9

16. $2\overline{)6486}$

17. $3\overline{)794}$

18. $7\overline{)408}$

3–10

19. $6\overline{)34579}$

20. $83\overline{)3735}$

21. $41\overline{)56478}$

22. $245\overline{)20580}$

23. $904\overline{)238534}$

24. $349\overline{)282108}$

Solve.

3–11 25. During an automoble trip across the United States, a retired couple traveled 2573 miles while using 83 gallons of gasoline. How many miles did they travel on each gallon of gasoline?

Test Your Memory

These problems review Chapters 1–3.

1. How many digits are there in 706,470?

2. Write a standard number for 80,000 + 500 + 6.

3. Write an expanded number for 507,320.

4. Write a word name for 23,506,007.

5. Which number is larger, 642,776 or 645,772?

6. Round to the nearest ten, then hundred, then thousnd: 75,362.

Mixed Practice: Solve.

7. $\begin{array}{r} 649 \\ +208 \\ \hline \end{array}$ **8.** $\begin{array}{r} 8905 \\ +6435 \\ \hline \end{array}$ **9.** $\begin{array}{r} 798 \\ -425 \\ \hline \end{array}$ **10.** $\begin{array}{r} 9836 \\ -4324 \\ \hline \end{array}$

11. $\begin{array}{r} 6000 \\ \times 4 \\ \hline \end{array}$ **12.** $\begin{array}{r} 8243 \\ \times 6 \\ \hline \end{array}$ **13.** $2\overline{)642}$ **14.** $15\overline{)765}$

15. $\begin{array}{r} 86 \\ 43 \\ 29 \\ +57 \\ \hline \end{array}$

16. $\begin{array}{r} 200 \\ \times 46 \\ \hline \end{array}$

17. $3\overline{)194}$

18. $\begin{array}{r} 5344 \\ 71 \\ 643 \\ +4069 \\ \hline \end{array}$

19. $\begin{array}{r} 646 \\ -584 \\ \hline \end{array}$

20. $\begin{array}{r} 5163 \\ -3894 \\ \hline \end{array}$

21. $\begin{array}{r} 647 \\ \times 40 \\ \hline \end{array}$

22. $\begin{array}{r} 4173 \\ \times 362 \\ \hline \end{array}$

23. $6\overline{)7340}$

24. $\begin{array}{r} 6204 \\ -4386 \\ \hline \end{array}$

25. $\begin{array}{r} 709 \\ \times 5000 \\ \hline \end{array}$

26. $\begin{array}{r} 134 \\ \times 2 \\ \hline \end{array}$

27. $5\overline{)33040}$

28. $\begin{array}{r} 3047 \\ \times 4 \\ \hline \end{array}$

29. $\begin{array}{r} 4539 \\ 8605 \\ 7189 \\ +2433 \\ \hline \end{array}$

30. $\begin{array}{r} 3005 \\ -1388 \\ \hline \end{array}$

31. $\begin{array}{r} 4318 \\ \times 3216 \\ \hline \end{array}$

32. $26\overline{)13498}$

33. $470\overline{)310409}$

34. $\begin{array}{r} 7021 \\ \times 89 \\ \hline \end{array}$

35. $\begin{array}{r} 4003 \\ \times 604 \\ \hline \end{array}$

36. $539\overline{)708562}$

37. $\begin{array}{r} 7002 \\ \times 7040 \\ \hline \end{array}$

38. $308\overline{)182604}$

39. An automobile supply store took an inventory of their drill bits. They have 108 $\frac{1}{2}$-in. drill bits, 79 $\frac{3}{4}$-in. drill bits, 24 1-in. drill bits, and 98 2-in. drill bits. How many drill bits do they have in stock?

40. An electrical contractor used 205 ft of conduit for a job. If the inventory of conduit was 989 ft in his warehouse before the job, how much did he have left after he finished the job?

41. A customer's natural gas bill shows that a total of 96 therms were used. Of this total 64 therms were used for hot water and the remainder for cooking. How many therms were used for cooking?

42. Five loads of gravel were delivered to a construction sight. They were as follows: 3468 lb, 4179 lb, 3898 lb, 3018 lb, and 1864 lb. Find the total weight.

43. Find the missing dimension.

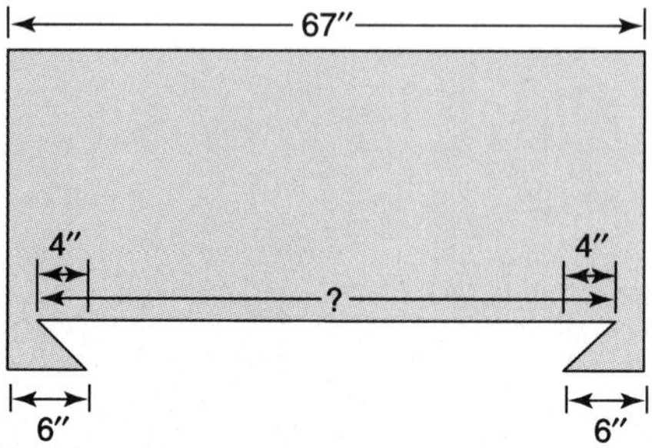

44. A microwave oven can be bought on credit for $15 down and 12 monthly payments of $11. What is the total cost of the microwave?

45. A container weighs 167 lb. The number of gallons it contains can be found by dividing its weight in pounds by 8.35. How many gallons does the container contain?

46. A trucker uses 56 gallons of diesel when traveling 1344 miles. Find how many miles the truck travels per gallon.

47. An aluminum casting weighs 1520 lb. If aluminum costs $1.00 a pound, find the cost of 7 of these castings.

The answers are in the back of the book.

PART TWO

FRACTIONS AND MIXED NUMBERS

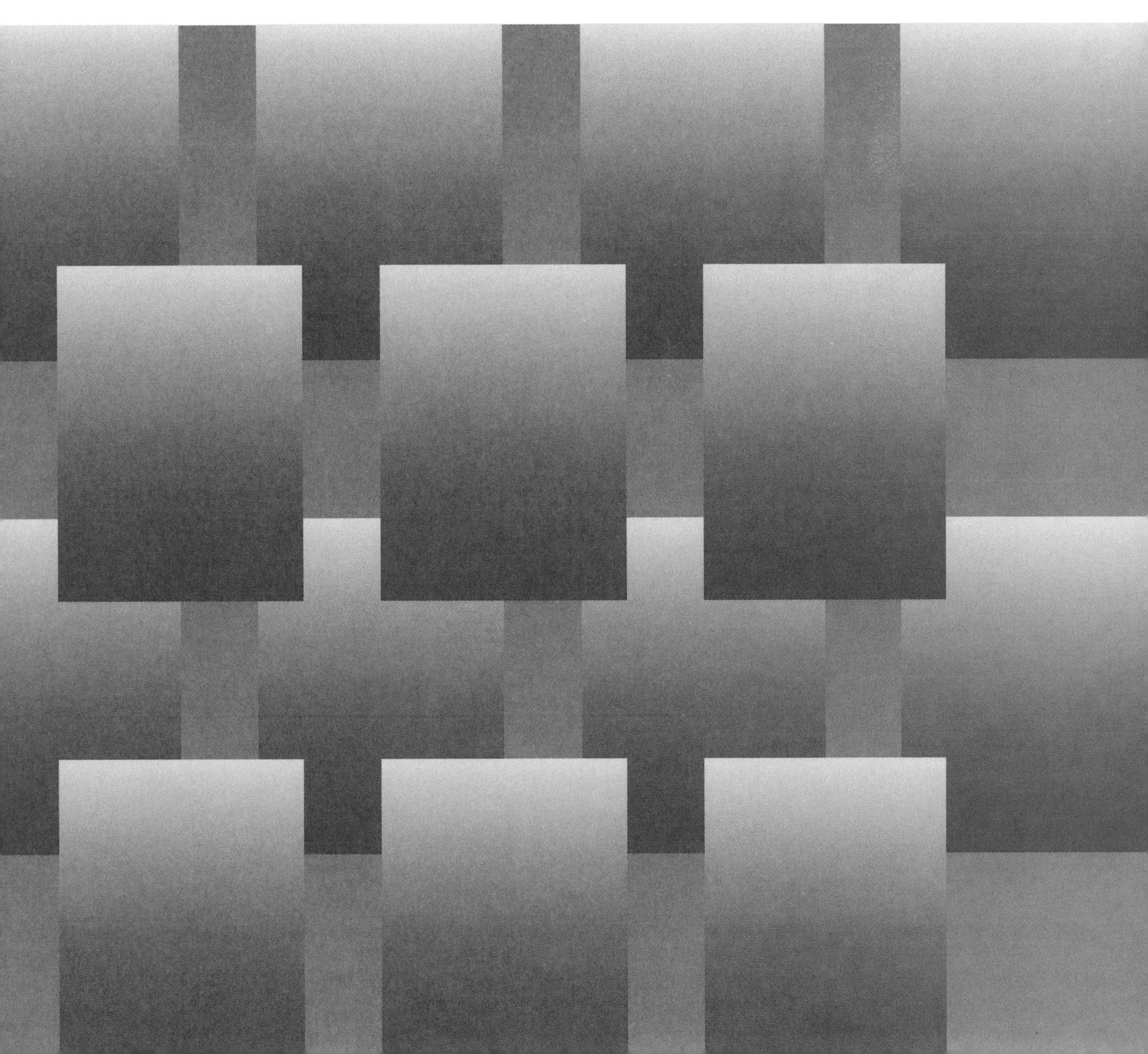

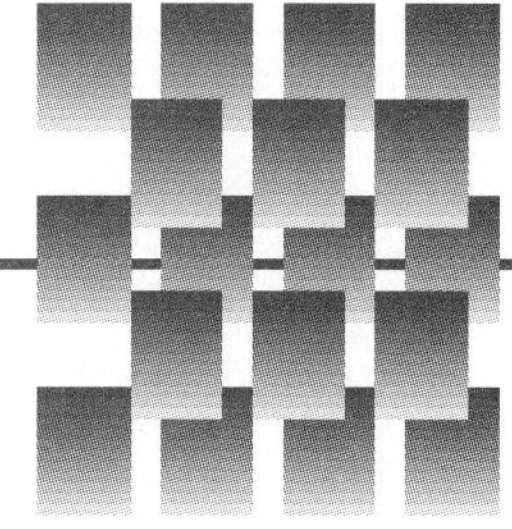

CHAPTER 4

Multiplication and Division of Fractions

Choice: Either skip the Pretest and study all of Chapter 4 or take the Pretest and study only the sections you need to study.

Pretest

Directions: This Pretest will help you determine which sections to study in Chapter 4. If any of your answers are incorrect, or if you omitted any exercise, turn to the indicated section(s) and study the material; then take the Chapter Test.

4–1 **1.** Name a fraction that the shaded part of the object suggests.

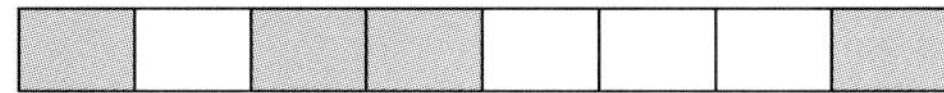

2. Name the point on the number line.

4–2 **3.** Find the factors of 24.

4–3 **4.** Find the prime factorization of 36.

4–4 **5.** Find the missing numerator: $\frac{7}{8} = \frac{?}{24}$

6. Reduce $\frac{56}{72}$ to lowest terms.

4–5 7. Multiply: $\dfrac{3}{8} \times \dfrac{4}{9}$

4–6 8. Divide and simplify: $\dfrac{7}{8} \div \dfrac{4}{14}$

4–7 9. Simplify: $\dfrac{\frac{5}{6}}{\frac{10}{27}}$

4–8 10. An exotic dish at the Cellars Broadway calls for $\frac{1}{16}$ cup of chopped exotic mushrooms per serving. How many requests for the dish can be filled if the chef has only $\frac{3}{4}$ cup of the mushrooms?

The answers are in the margin on page 134.

4–1 THE MEANING OF FRACTIONS

A Part of the Whole

In Chapters 1–3 we studied addition, subtraction, multiplication, and division with whole numbers. The whole numbers are 0, 1, 2, 3, 4, 5, and so on. There is no last whole number. We can count forever!

Suppose a whole pie is cut into five pieces so that each piece is exactly the same size.

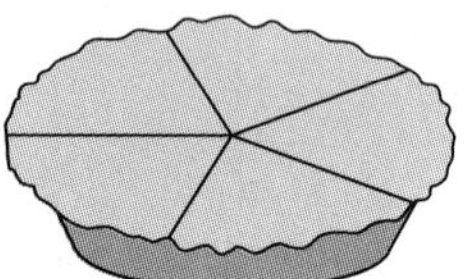

Figure 4–1

Since each piece is a part of a whole thing, we need a name for *a part of a whole.*

Imagine one of the five pieces, ready to eat on your plate. We express this piece as $\frac{1}{5}$. (We say, "one-fifth.") We call this number a fraction, so we have $\frac{1}{5}$ of the whole pie.

■ Example 1

Let's consider dividing the whole object below into 10 equal parts.

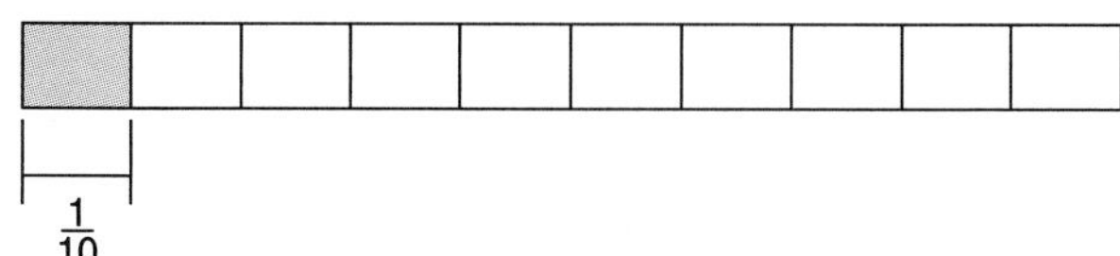

Figure 4–2

The shaded part represents $\frac{1}{10}$ (one-tenth) of the object. One part out of 10 parts is shaded. Use this as a memory aid.

$$\frac{1}{10} \begin{array}{l} \leftarrow \text{One part is shaded} \\ \leftarrow \text{The whole thing has 10 parts.} \end{array}$$

Each part is called $\frac{1}{10}$ of the whole.

Practice check: Do exercises 1–4 in the margin.

Name a fraction that the shaded part of the object suggests.

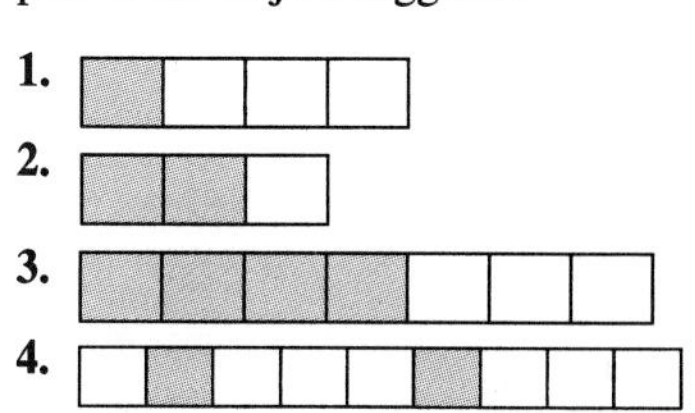

The answers are in the margin on page 134.

Let's consider the portion of a whole number line between 0 and 1.

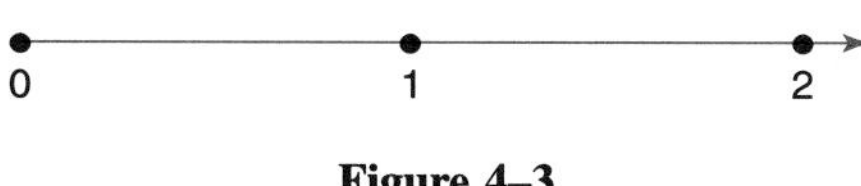

Figure 4–3

■ Example 2

Divide the part between 0 and 1 into two equal pieces and place a mark at that point.

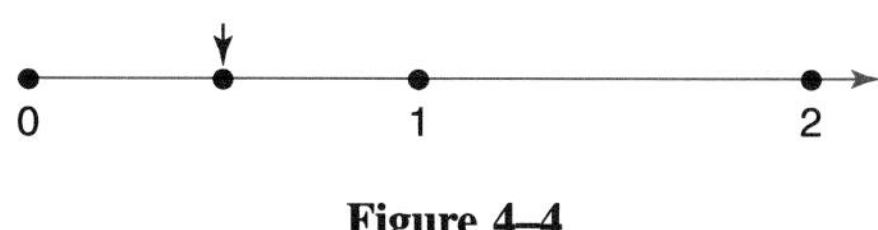

Figure 4–4

Moving from the 0 mark to the first mark is 1 part out of 2 parts between 0 and 1. We call this first point $\frac{1}{2}$ (one-half).

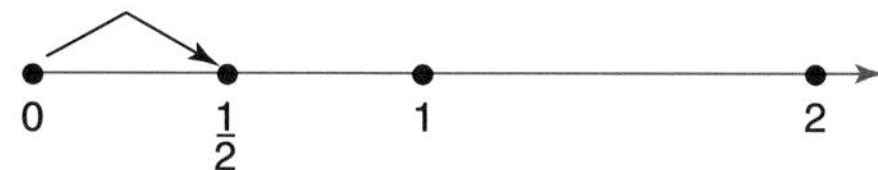

Figure 4–5

■ Example 3

Divide the part between 0 and 1 into four equal pieces and place marks at the points. (*Note:* To divide the space into four equal parts, we use 3 marks.)

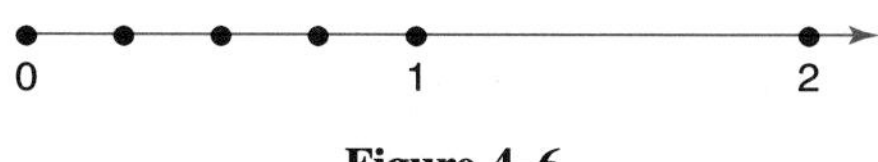

Figure 4–6

Answers to Chapter 4 Pretest

Circle the answers you missed or omitted. Cross them off after you've studied the corresponding section(s), then take the Chapter Test.

1. $\frac{4}{8}$ **2.** $18\frac{2}{6}$

3. 1, 2, 3, 4, 6, 8, 12, 24
4. $2 \times 2 \times 3 \times 3$ **5.** 21

6. $\frac{7}{9}$ **7.** $\frac{1}{6}$ **8.** $\frac{49}{16}$

9. $\frac{9}{4}$ **10.** 12 servings

Answers to exercises 1–4

1. $\frac{1}{4}$ **2.** $\frac{2}{3}$

3. $\frac{4}{7}$ (four-sevenths)

4. $\frac{2}{9}$ (two-ninths)

Moving from the 0 mark to the first mark is 1 part out of 4 parts between 0 and 1. We call this first point $\frac{1}{4}$ (one-fourth).

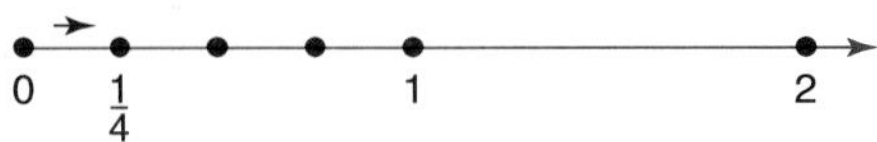

Figure 4–7

Moving from the 0 mark to the second mark is 2 parts out of 4 parts between 0 and 1. We call this second point $\frac{2}{4}$ (two-fourths).

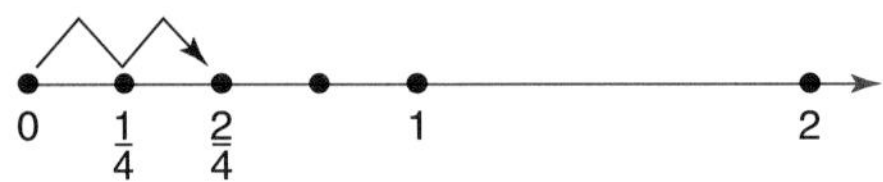

Figure 4–8

Moving from the 0 mark to the third mark is 3 parts out of 4 parts between 0 and 1. We call this third part $\frac{3}{4}$ (three-fourths).

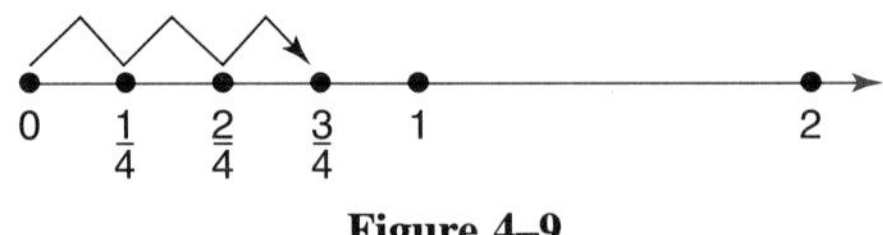

Figure 4–9

■ Example 4

Find the point $\frac{3}{5}$ on a number line.

Solution **Step 1** Draw a portion of the whole number line.

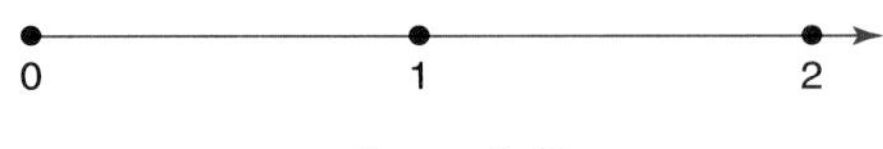

Figure 4–10

Step 2 Divide the part between 0 and 1 into 5 equal portions. Notice that we use 4 marks to make the division.

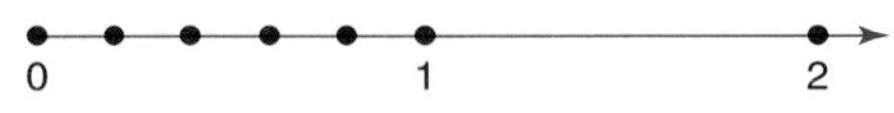

Figure 4–11

Step 3 From the 0 mark count over 3 marks. The third point is $\frac{3}{5}$ (three-fifths).

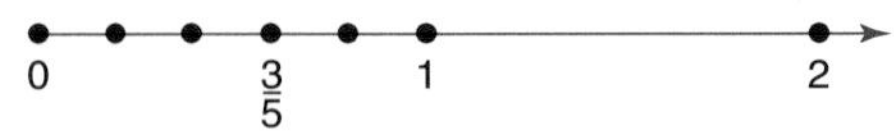

Figure 4–12

Practice check: Do exercises 5–7 in the margin.

Construct a number line for each exercise.

5. Find the point $\frac{2}{3}$ on a number line.

6. Find the point $\frac{4}{7}$ on a number line.

7. Find the point $\frac{5}{6}$ on a number line.

The answers are in the margin on page 136.

A Portion of a Group

We can also name a fractional portion of a group. The individual members of the group may or may not be different.

■ **Example 5**

Write a fractional number that tells what portion of the group is shaded.

     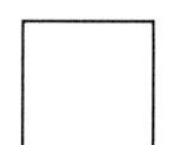  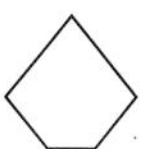

Figure 4–13

Solution

$\frac{3}{8}$ 3 ← There are 3 objects shaded.
8 ← There are 8 objects in the group.

■ **Example 6**

Write a fractional number that tells what portion of the group are pears.

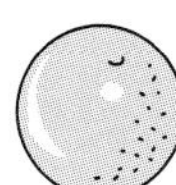   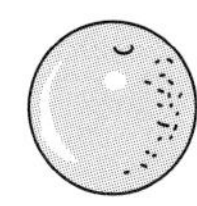  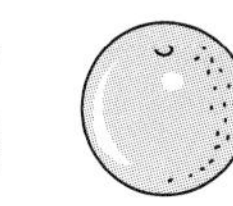 

Figure 4–14

Solution

2 ← There are two pears.
7 ← There are 7 in the group.

Numerators and Denominators

Consider the fraction $\frac{4}{7}$ (four-sevenths). There are special names for the top and bottom number in a fraction. The top number is called the *numerator.* The bottom number is called the *denominator.*

4 ← Numerator
7 ← Denominator

Here is a memory aid: To remember which one is the denominator, think of the word *deep*. The denominator is *deep* down, or the denominator is the bottom number. The other number then, is the numerator.

Practice check: Do exercises 8–11 in the margin.

Tell which is the numerator and denominator of the following fractions.

8. $\frac{6}{8}$ 9. $\frac{1}{4}$

10. $\frac{7}{9}$ 11. $\frac{3}{7}$

The answers are in the margin on page 136.

Answers to exercises 5–7

5.

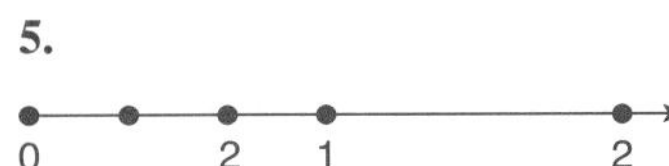

6.

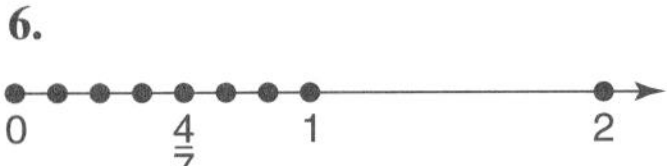

7.

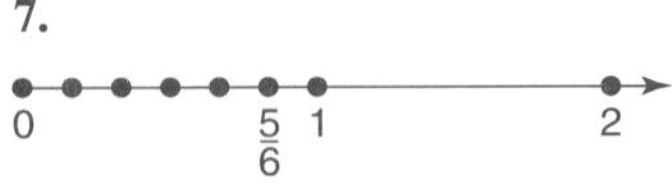

Answers to exercises 8–11

	Numerator	Denominator
8.	6	8
9.	1	4
10.	7	9
11.	3	7

Fractions That Name One Whole

Suppose we consider an object divided into three equal pieces.

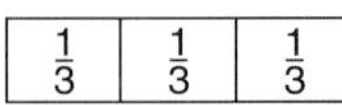

Figure 4–15

Each piece names $\frac{1}{3}$ of the whole. If we considered all 3 pieces together we would have $\frac{3}{3}$ or 1 whole.

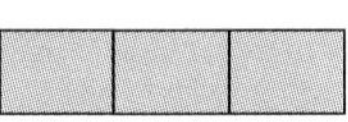

Figure 4–16

$\frac{3}{3}$ ← Three parts are shaded.
← The whole thing has 3 parts.

A number line will give us the same results.

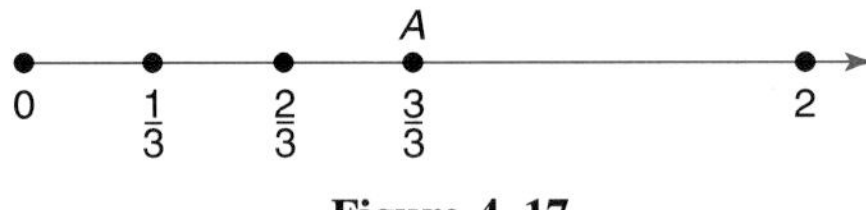

Figure 4–17

Point *A* is also 1, thus $\frac{3}{3} = 1$.

Rule

A fraction that has the same numerator and denominator is another name for 1 whole except when the numerator and denominator are 0.

We should remember from Chapter 3 that $\frac{0}{0}$ is not a number. $\frac{0}{5-5}$ is not a number, since $\frac{0}{5-5} = \frac{0}{0}$.

Practice check: Do exercises 12–17 in the margin.

Simplify the fractions.

12. $\frac{1}{1}$ **13.** $\frac{4}{4}$

14. $\frac{18}{18}$ **15.** $\frac{36}{36}$

16. $\frac{477}{477}$ **17.** $\frac{1060}{1060}$

The answers are in the margin on page 138.

Fractions That Name 0

Consider $\frac{0}{3}$.

$\frac{0}{3}$ ← We have none of the 3 pieces.
← The whole is divided into 3 equal parts.

Since we have no part of the whole, then $\frac{0}{3}$ is a fraction that names 0.

Rule

A fraction that has a numerator of zero and a whole number other than zero for a denominator is another name for 0.

Remember that $\frac{1}{0}$ is not a number since we cannot divide by 0.

Practice check: Do exercises 18–21 in the margin.

Simplify the fractions.

18. $\frac{0}{7}$ 19. $\frac{0}{19}$

20. $\frac{23-23}{82}$ 21. $\frac{0}{1684}$

The answers are in the margin on page 138.

Fractions That Name Whole Numbers

Consider the fraction $\frac{3}{1}$.

$$\frac{3 \leftarrow \text{This says we have 3 whole pieces.}}{1 \leftarrow \text{This says the whole is in one piece.}}$$

It has not been divided into pieces.

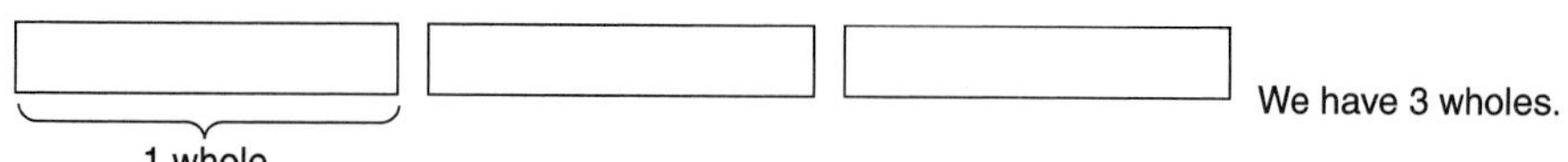

Figure 4–18

Thus, $\frac{3}{1} = 3$.

Rule

A fraction having a denominator of 1 is always equal to the numerator.

Practice check: Do exercises 22–32 in the margin.

Simplify.

22. $\frac{6}{1}$ 23. $\frac{18}{1}$ 24. $\frac{39}{1}$

25. $\frac{681}{1}$ 26. $\frac{7}{7}$ 27. $\frac{0}{5}$

28. $\frac{116}{1}$ 29. $\frac{15}{15}$ 30. $\frac{389}{389}$

31. $\frac{0}{62}$ 32. $\frac{3061}{1}$

The answers are in the margin on page 138.

Numerals for Whole Numbers and Fractions

Suppose we have two whole objects exactly alike; each divided into 3 equal parts. Consider one whole object and one piece of the second.

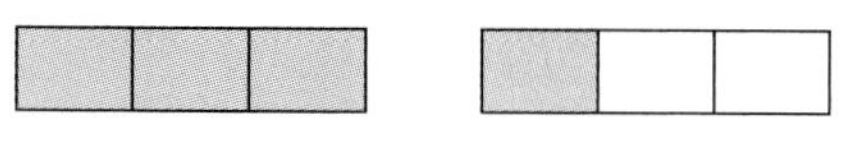

Figure 4–19

We have 4 pieces and since each piece is named $\frac{1}{3}$, we have the fraction $\frac{4}{3}$.

Figure 4–20

$$\frac{4 \leftarrow \text{Four pieces are shaded.}}{3 \leftarrow \text{The whole has 3 pieces.}}$$

Notice that $\frac{4}{3}$ is the same as 1 whole and $\frac{1}{3}$. We write this as $1\frac{1}{3}$ (one and one-third). Thus, $\frac{4}{3} = 1\frac{1}{3}$.

Answers to exercises 12–17

12. 1 **13.** 1 **14.** 1 **15.** 1
16. 1 **17.** 1

Answers to exercises 18–21

18. 0 **19.** 0 **20.** 0 **21.** 0

Answers to exercises 22–32

22. 6 **23.** 18 **24.** 39 **25.** 681
26. 1 **27.** 0 **28.** 116 **29.** 1
30. 1 **31.** 0 **32.** 3061

■ Example 7

Write a fractional number for the shaded part.

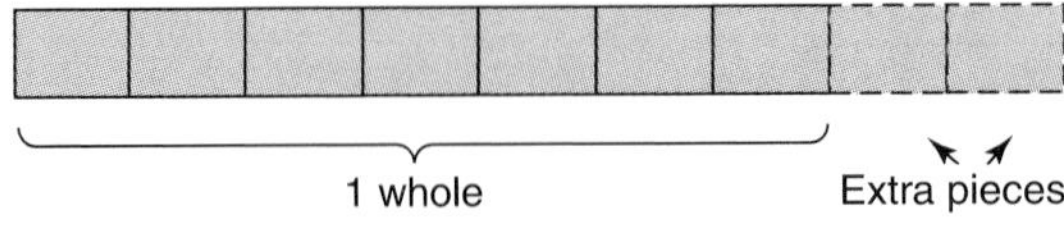

Figure 4–21

Solution

$\frac{9}{7}$ 9 ← Nine pieces are shaded.
7 ← The whole has 7 pieces.

■ Example 8

Draw an object that represents $2\frac{3}{4}$.

Solution Each piece is $\frac{1}{4}$ of a whole.

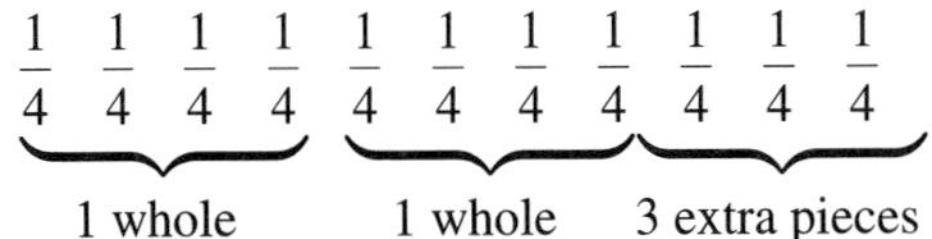

$\frac{1}{4}$ $\frac{1}{4}$ $\frac{1}{4}$ $\frac{1}{4}$ $\frac{1}{4}$ $\frac{1}{4}$ $\frac{1}{4}$ $\frac{1}{4}$ $\frac{1}{4}$ $\frac{1}{4}$ $\frac{1}{4}$

1 whole 1 whole 3 extra pieces

Notice that there are eleven $\frac{1}{4}$ pieces, so $2\frac{3}{4} = \frac{11}{4}$.

Practice check: Do exercises 33–35 in the margin.

Write a fractional number for the shaded part.

33.

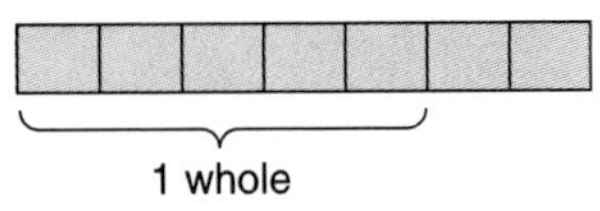

34.

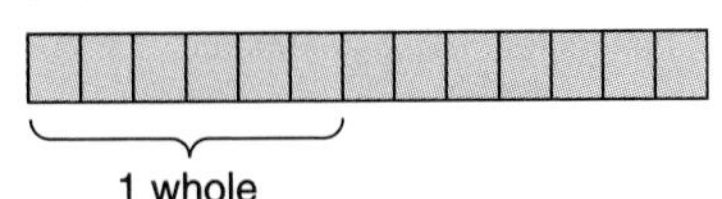

35. Draw an object like the one above that represents $1\frac{5}{8}$.

The answers are in the margin on page 141.

■ Example 9

Find $1\frac{2}{5}$ on a number line.

Solution This number is between 1 and 2.

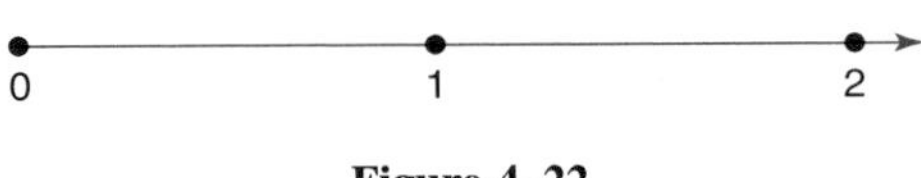

Figure 4–22

Step 1 Divide the portion between 1 and 2 into 5 equal parts.

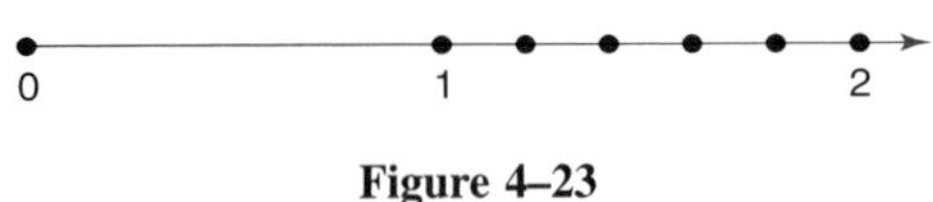

Figure 4–23

Each divided portion is $\frac{1}{5}$ of the whole between 1 and 2.

Step 2 From 1 count over 2 dots.

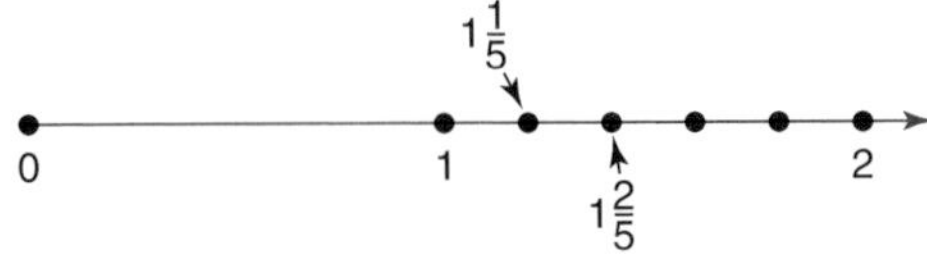

Figure 4–24

Practice check: Do exercises 36–39 in the margin.

Find the following numbers on a number line.

36. $3\frac{1}{4}$ **37.** $2\frac{3}{7}$

38. $6\frac{5}{6}$ **39.** $12\frac{5}{8}$

The answers are in the margin on page 141.

■ Looking Back

You should now be able to

1. Name and pictorially represent a part of one whole with a fractional number.
2. Name the numerator and denominator of a fraction.
3. Give fractional names for 1.
4. Give fractional names for 0.
5. Write fractions that name whole numbers.
6. Write and pictorially represent numbers that represent a combination of whole numbers and fractions.

■ Problems 4–1

Name a fraction that the shaded part of the object suggests.

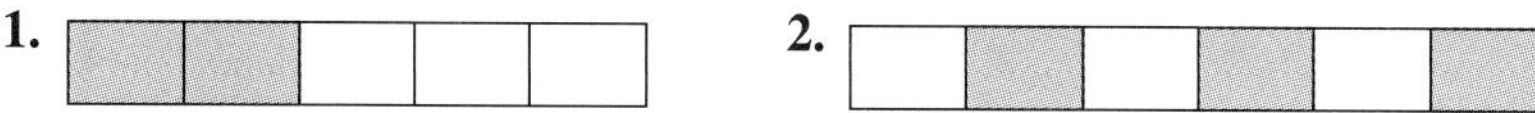

3.

4.

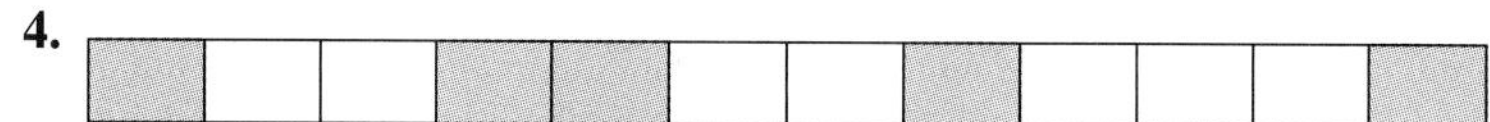

Shade in the portion of an object that is named by the fraction.

5. $\frac{5}{6}$ **6.** $\frac{3}{5}$ **7.** $\frac{2}{9}$ **8.** $\frac{6}{7}$

Find the following points on a number line.

9. $\frac{3}{4}$ **10.** $\frac{2}{5}$ **11.** $\frac{3}{6}$ **12.** $\frac{5}{8}$

Name the point on the number line.

13.

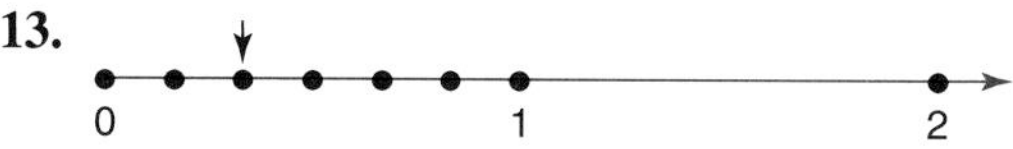

14.

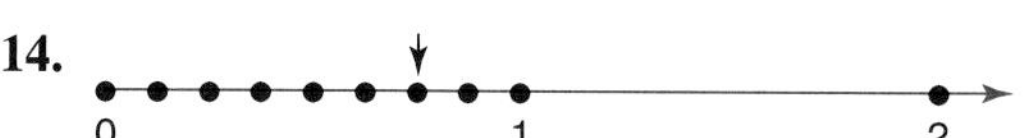

15. 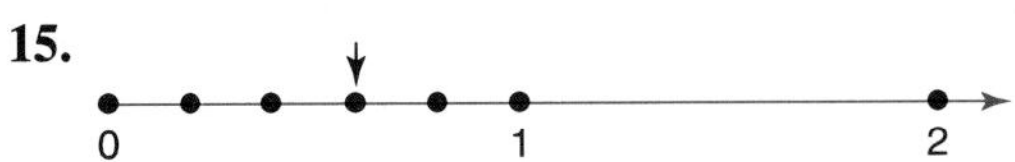

16.

17. Write a fractional number that tells what portion of the group are circles.

Name the numerator and denominator of each fraction.

18. $\frac{6}{8}$ 19. $\frac{7}{9}$ 20. $\frac{68}{56}$ 21. $\frac{172}{468}$

Simplify the fractions.

22. $\frac{16}{16}$ 23. $\frac{0}{31}$ 24. $\frac{0}{6}$ 25. $\frac{79}{79}$

26. $\frac{15}{0}$ 27. $\frac{26}{26}$ 28. $\frac{115}{115}$ 29. $\frac{98}{0}$

30. $\frac{0}{102}$ 31. $\frac{1056}{1056}$ 32. $\frac{0}{56}$ 33. $\frac{30}{30}$

Simplify the fractions.

34. $\frac{23 - 23}{0}$ 35. $\frac{23 - 17}{13 - 7}$ 36. $\frac{32 + 68}{2306 - 2206}$ 37. $\frac{56 - 38}{203 - 203}$

38. $\frac{23 \times 0}{684 \times 368}$ 39. $\frac{32 \times 48}{16 \times 96}$

Write a fractional number for the shaded part.

40.

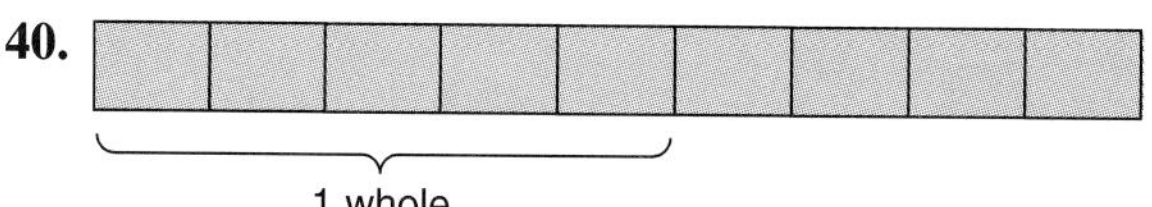

41.

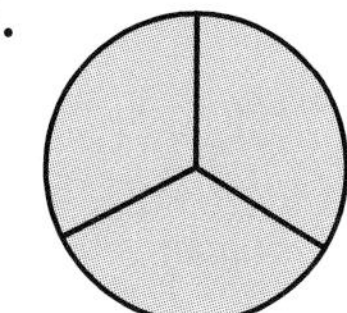

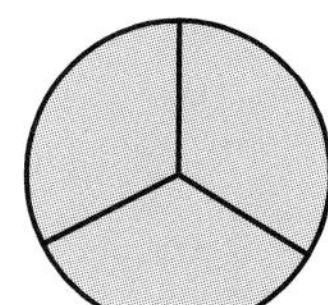

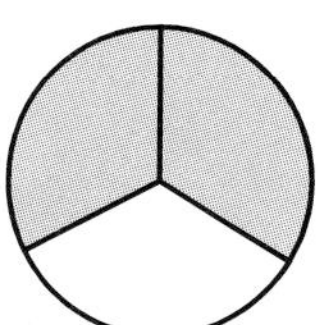

Find the following numbers on a number line.

42. $5\frac{3}{7}$

43. $18\frac{2}{5}$

Name the point on the number line.

44.

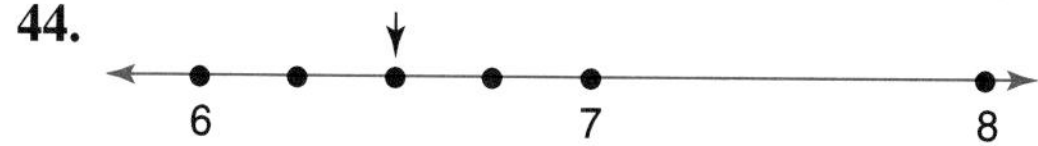

45.

46.

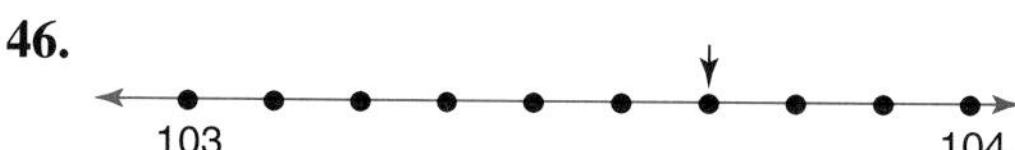

The answers to the odd-numbered exercises are in the back of the book.

Answers to exercises 33–35

33. $\frac{7}{5}$ or $1\frac{2}{5}$ 34. $\frac{13}{6}$ or $2\frac{1}{6}$

35.

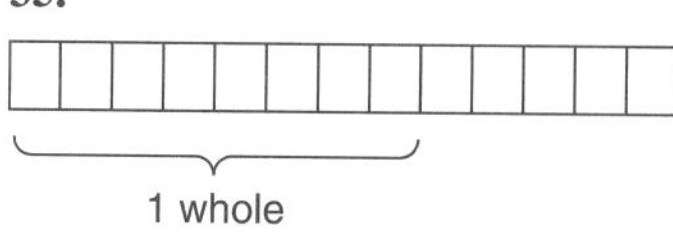

Answers to exercises 36–39

36.

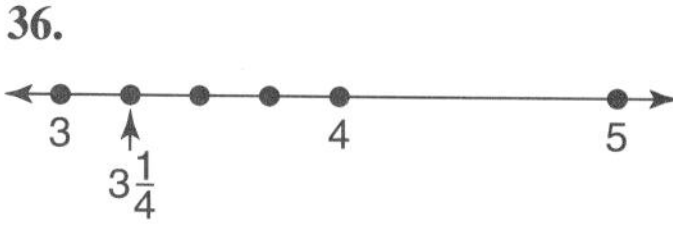

37.

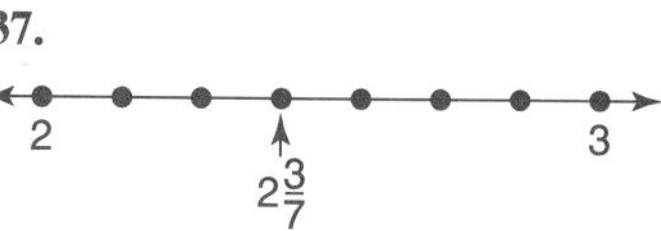

38.

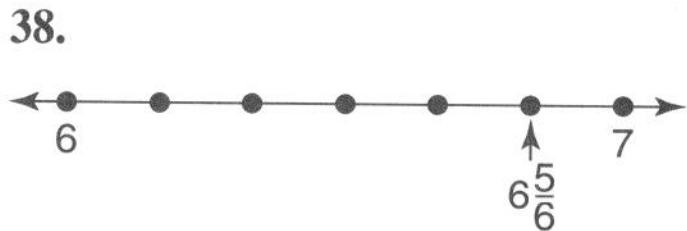

39.

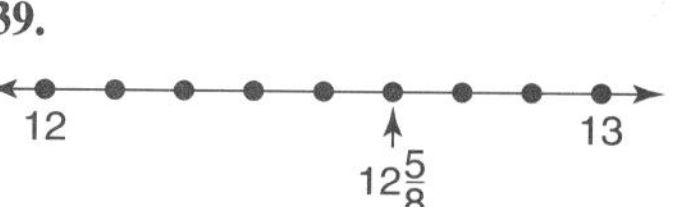

4–2 DIVISIBILITY AND FACTORS

Divisibility

One number is divisible by another number if the remainder is zero when the two numbers are divided.

■ Examples

1. 6 is divisible by 3.

$$\begin{array}{r} 2 \\ 3\overline{)6} \\ \underline{6} \\ 0 \end{array} \leftarrow \text{Remainder is zero.}$$

2. 25 is divisible by 5.

$$\begin{array}{r} 5 \\ 5\overline{)25} \\ \underline{25} \\ 0 \end{array} \leftarrow \text{Remainder is zero.}$$

3. 37 is *not* divisible by 4.

$$\begin{array}{r} 9 \\ 4\overline{)37} \\ \underline{36} \\ 1 \end{array} \leftarrow \text{Remainder is not zero.}$$

For our work later on, it will help if we know whether certain numbers are divisible by 2, 3, or 5.

Divisibility by 2

Rule

A number is divisible by 2 if its ones digit is divisible by 2.

A number whose ones digit is 0, 2, 4, 6, or 8 is an even number.

■ Example 4

Is 386 divisible by 2?

Solution Yes. 6 is divisible by 2, so 386 is divisible by 2.

■ Example 5

Is 97 divisible by 2?

Solution No. 7 is not divisible by 2, so 97 is not divisible by 2.

Memorize the rule.

Practice check: Do exercises 40–49 in the margin.

State whether each number is or is not divisible by 2.

40. 64 **41.** 38 **42.** 69

43. 107 **44.** 200 **45.** 682

46. 3066 **47.** 5785

48. 16,341 **49.** 356,348

The answers are in the margin on page 144.

Divisibility by 3

Rule

A number is divisible by 3 if the sum of its digits is divisible by 3.

■ **Example 6**

Is 237 divisible by 3?

Solution Yes. To determine whether 237 is divisible by 3, we add the digits: $2 + 3 + 7 = 12$.

12 is divisible by 3, so 237 is divisible by 3.

Notice that any ordering of the digits 2, 3, and 7 will also be divisible by 3. Examples: 273, 732, 723, 372, and 327.

■ **Example 7**

Is 728 divisible by 3?

Solution Again, we add the digits: $7 + 2 + 8 = 17$.

17 is not divisible by 3, so 728 is not divisible by 3.

Memorize the rule.

Practice check: Do exercises 50–59 in the margin.

State whether each number is or is not divisible by 3.

50. 57 **51.** 38
52. 306 **53.** 942
54. 543 **55.** 7008
56. 21,380 **57.** 2122
58. 50,389 **59.** 150,716

The answers are in the margin on page 144.

Divisibility by 5

Rule

A number is divisible by 5 if its ones digit is 0 or 5.

■ **Example 8**

Is 650 divisible by 5?

Solution Yes. The ones digit is zero, so 650 is divisible by 5.

■ **Example 9**

Is 168,452 divisible by 5?

Solution No. The ones digit is not 0 or 5, so 168,452 is not divisible by 5.

Memorize the rule.

Practice check: Do exercises 60–68 in the margin.

A summary of the rules for divisibility by 2, 3, and 5 is given below.

By 2: the last digit must be divisible by 2.
By 3: the sum of the digits must be divisible by 3.
By 5: the last digit must be 0 or 5.

Practice check: Do exercises 69–78 in the margin.

State whether each number is or is not divisible by 5.

60. 80 **61.** 57 **62.** 365
63. 89 **64.** 570 **65.** 1385
66. 12,366 **67.** 35,490
68. 117,344

The answers are in the margin on page 144.

State whether the following are divisible by 2, 3, or 5.

69. 632 **70.** 87 **71.** 770
72. 5705 **73.** 4914 **74.** 814
75. 6834 **76.** 960
77. 45,618 **78.** 51,750

The answers are in the margin on page 144.

Answers to exercises 40–49

40. yes **41.** yes **42.** no **43.** no **44.** yes **45.** yes **46.** yes **47.** no **48.** no **49.** yes

Answers to exercises 50–59

50. yes **51.** no **52.** yes **53.** yes **54.** yes **55.** yes **56.** no **57.** no **58.** no **59.** no

Answers to exercises 60–68

60. yes **61.** no **62.** yes **63.** no **64.** yes **65.** yes **66.** no **67.** yes **68.** no

Answers to exercises 69–78

69. by 2 **70.** by 3 **71.** by 2 and 5 **72.** by 5 **73.** by 2 and 3 **74.** by 2 **75.** by 2 and 3 **76.** by 2, 3 and 5 **77.** by 2 and 3 **78.** by 2, 3, and 5

Factors

In our discussion of factors, we shall use the natural numbers, 1, 2, 3, and so on. When multiplying, we begin with two numbers called factors that result in a third number called the product.

$$4 \times 3 = 12$$

↗ ↑ ↖
Factor Factor Product

So 4 is a factor of 12 and 3 is a factor of 12 because $4 \times 3 = 12$. Notice that 12 is divisible by 3 and 4.

■ Example 10

Find the factors of 12.

Solution

$1 \times 12 = 12$
↑ ↗
1 and 12 are factors of 12.

$2 \times 6 = 12$
↖ ↑
2 and 6 are factors of 12

$3 \times 4 = 12$
↖ ↗
3 and 4 are factors of 12.

Thus the factors of 12 are 1, 2, 3, 4, 6, and 12.

All natural numbers that will divide 12 evenly (with a zero remainder) are factors of 12. Notice that 1 and the number itself are always factors of a number.

■ Example 11

Find all the factors of 18.

Solution We mentally test natural numbers from 1 through 18 using divisibility tests where possible.

Factors	***Not Factors***
1 (Always a factor)	4 (Won't divide 18 evenly)
2 (18 is an even number.)	5 (18 does not end in 0 or 5.)
3 ($1 + 8 = 9$, and 9 is divisible by 3.)	7 (Won't divide 18 evenly)
6 ($6 \times 3 = 18$)	8 (Won't divide 18 evenly)
9 ($9 \times 2 = 18$)	10–17 (None of these numbers will divide 18 evenly.)
18 (The number itself is always a factor.)	

The factors of 18 are 1, 2, 3, 6, 9, and 18.

■ Example 12

Find all the factors of 24.

Solution When we mentally test the natural numbers for divisibility from 1 through 24, we find that the factors are

1, 2, 3, 4, 6, 8, 12, and 24.

Rule

The factors of a number are 1 and itself, and all the natural numbers between that evenly divide the number.

Practice check: Do exercises 79–86 in the margin.

Find the factors of the following numbers.

79. 3 **80.** 6 **81.** 7

82. 10 **83.** 17 **84.** 21

85. 28 **86.** 32

The answers are in the margin on page 147.

■ Looking Back

You should now be able to do the following:

1. Use a test to find whether a number is divisible by 2, 3, or 5.

2. List the factors of a natural number.

■ Problems 4–2

Review

1. Name a fraction that the shaded part of the object suggests.

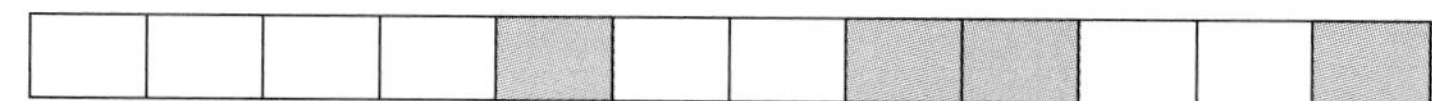

2. Draw a number line and locate the point $\frac{3}{7}$.

3. Name the numerator of $\frac{29}{17}$.

Simplify.

4. $\dfrac{18}{18}$ **5.** $\dfrac{56}{1}$ **6.** $\dfrac{0}{8}$ **7.** $\dfrac{361}{361}$ **8.** $\dfrac{15}{0}$

Write a fractional number for the shaded part.

9.

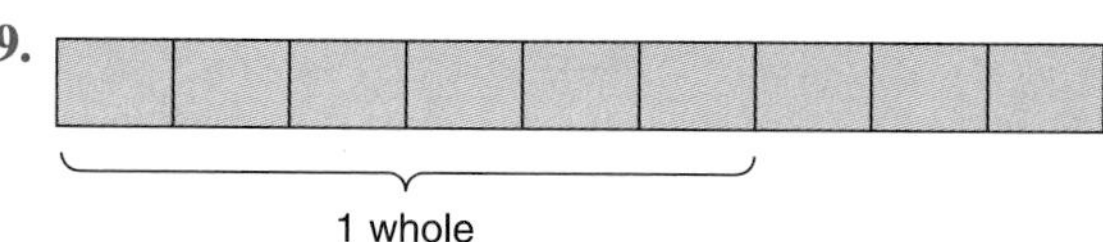

10. Name the point on the number line.

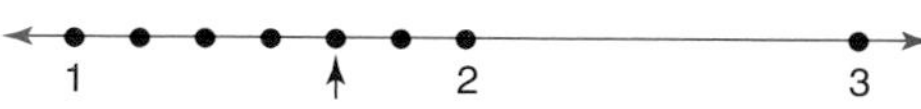

State whether the following numbers are divisible by 2, 3, or 5.

11. 21 **12.** 15 **13.** 66 **14.** 47 **15.** 70 **16.** 36

17. 99 **18.** 55 **19.** 87 **20.** 29 **21.** 238 **22.** 480

23. 389 **24.** 575 **25.** 106 **26.** 735 **27.** 603 **28.** 832

29. 947 **30.** 330 **31.** 6042 **32.** 3118 **33.** 1005 **34.** 8742

35. 2683 **36.** 31,041 **37.** 68,320 **38.** 13,485

39. 405,246 **40.** 841,305

Find the factors of the following numbers.

41. 15 **42.** 14 **43.** 19 **44.** 28 **45.** 22 **46.** 26

47. 25 **48.** 13 **49.** 1 **50.** 16 **51.** 19 **52.** 33

53. 2 **54.** 8 **55.** 44 **56.** 23 **57.** 43 **58.** 13

59. 30 **60.** 36 **61.** 54 **62.** 7 **63.** 67 **64.** 18

65. 38 **66.** 27 **67.** 100 **68.** 66 **69.** 88 **70.** 150

The answers to review and odd-numbered exercises are in the back of the book.

4–3 PRIME NUMBERS AND FACTORING

Prime and Composite Numbers

Example 1 will help you understand the meaning of a prime number.

Answers to exercises 79–86

79. 1, 3 **80.** 1, 2, 3, 6 **81.** 1, 7 **82.** 1, 2, 5, 10 **83.** 1, 17 **84.** 1, 3, 7, 21 **85.** 1, 2, 4, 7, 14, 28 **86.** 1, 2, 4, 8, 16, 32

■ **Example 1**

List all the factors of (a) 3, (b) 19, (c) 6, and (d) 21.

Solution **a.** The factors of 3 are 1 and 3.
b. The factors of 19 are 1 and 19.
c. The factors of 6 are 1, 2, 3, and 6.
d. The factors of 21 are 1, 3, 7, 21.

From this example we observe the following:

1. The numbers 3 and 19 have only two different factors: 1 and the number itself.
2. The numbers 6 and 21 have more than two different factors.

Rule

All natural numbers that have only two different factors, 1 and the number itself, are called prime numbers.

The numbers 2, 17, 29, 37, and 53 are examples of prime numbers. Each one of these numbers has only two different factors, 1 and itself. To decide whether a number is prime, we make sure the only factors are 1 and the number itself.

Practice check: Do exercise 87 in the margin.

87. From the following list of natural numbers, find the prime numbers.

29, 4, 35, 66, 57, 7, 94, 17, 91, 21, 79, 44

The answers are in the margin on page 149.

■ **Example 2**

Is 27 a prime number?

Solution Because 3 and 9 also divide 27 evenly, the answer is no.

Rule

All natural numbers that have more than two different factors are called composite numbers.

The numbers 6, 14, 26, 33, and 54 are composite numbers. Each one of these numbers has more than two different factors. A number is composite if we can find at least one other factor besides 1 and itself.

Practice check: Do exercise 88 in the margin.

88. From the following list of natural numbers, find the composite numbers.

13, 29, 48, 8, 16, 89, 67, 97, 22, 36, 5, 69, 75, 96, 31

The answers are in the margin on page 149.

What about the natural number 1? Is it prime or composite? It is neither, because it only has itself as a factor. To be a prime number it would have to have two different factors. To be a composite number it would have to have more than two different factors.

Rule

The first six prime numbers are 2, 3, 5, 7, 11, and 13.

Memorize this list. We will need them in the next section.

Practice check: Do exercise 89 in the margin.

89. Decide whether the following numbers are prime, composite, or neither.

2, 29, 40, 9, 21, 63, 51, 5, 1, 7, 32, 16, 11, 3, 36, 19

The answers are in the margin on page 150.

Prime Factorization

To factor a natural number means to write the number as a product of factors.

■ Example 3

Factor 36.

Solution

$36 = 6 \times 6$ $36 = 2 \times 18$ $36 = 4 \times 9$

$36 = 1 \times 36$ $36 = 3 \times 12$ $36 = 2 \times 2 \times 9$

These are some factorizations of 36. Others are

$36 = 4 \times 3 \times 3$ $36 = 6 \times 2 \times 3$ $36 = 3 \times 3 \times 2 \times 2$

Notice that the factors are a mixture of prime and composite numbers, except where 1 is a factor (1 is neither prime nor composite).

■ Example 4

List some factorizations of 54.

Solution

$54 = 2 \times 27$ $54 = 6 \times 9$ $54 = 2 \times 3 \times 9$

$54 = 6 \times 3 \times 3$ $54 = 2 \times 3 \times 3 \times 3$

Practice check: Do exercises 90 and 91 in the margin.

90. List 3 different factorizations of 28.

91. List 3 different factorizations of 42.

The answers are in the margin on page 150.

A special kind of factorization is prime factorization where all of the factors are prime numbers. Here are three different factorizations of 54.

$54 = 6 \times 9$ (↑ ↑) Both factors are composite numbers.

$54 = 2 \times 3 \times 9$ (↑ ↗ ↑) These factors are prime. This factor is composite.

Prime factorization

$54 = 2 \times 3 \times 3 \times 3$ (↑ ↑ ↑ ↗) All factors are prime.

The prime factorization of 54 is $2 \times 3 \times 3 \times 3$.

■ Example 5

Find the prime factorization of 18.

Solution We start dividing 18 by 2, the smallest prime number that will divide 18. For convenience, we introduce the "upside-down, short division" method below.

$2\overline{)18}$

9 ← Since 9 is not prime, we keep dividing.

$$\begin{array}{r} 2\underline{)18} \\ 3\underline{)\ 9} \\ \mathbf{3} \end{array}$$ ← Since 3 is prime, we are finished.

Thus, the prime factorization of 18 is $2 \times 3 \times 3$, because $2 \times 3 \times 3 = 18$, and every factor is a prime number. The order of the factors doesn't matter, so the prime factorization could also be written as $3 \times 2 \times 3$ or $3 \times 3 \times 2$.

Answer to exercise 87

29, 7, 17, 79, 91

Answer to exercise 88

48, 8, 16, 22, 36, 69, 75, 96

■ Example 6

Find the prime factorization of 78.

Solution Since 78 is an even number, we can start by dividing by 2, the smallest prime number.

$$\begin{array}{r} 2\underline{)78} \\ 3\underline{)39} \\ \mathbf{13} \end{array}$$ ← Since 13 is prime, we are finished.

The prime factorization of 78 is $2 \times 3 \times 13$.

Practice check: Do exercises 92–97 in the margin.

Find the prime factorization of the following numbers.

92. 6 **93.** 10 **94.** 12

95. 27 **96.** 38 **97.** 52

The answers are in the margin on page 150.

■ Example 7

Find the prime factorization of 112.

Solution We start dividing 112 by 2, the smallest prime number.

$$\begin{array}{r} 2\underline{)112} \\ 2\underline{)56} \\ 2\underline{)28} \\ 2\underline{)14} \\ 7 \end{array}$$ ← Since 7 is prime, we are finished.

The prime factorization of 112 is $2 \times 2 \times 2 \times 2 \times 7$.

The order of the factors doesn't matter, so the prime factorization could also be written as $2 \times 7 \times 2 \times 2 \times 2$ or $2 \times 2 \times 7 \times 2 \times 2$.

■ Example 8

Find the prime factorization of 29.

Solution This is a prime number, so it stands alone in factored form.

Rule

To find the prime factorization of a number:

Step 1 *Start with the smallest prime number that will divide the number.*

Step 2 *Divide each resulting quotient by a prime number.*

Step 3 *Stop when a resulting quotient is a prime number.*

Step 4 *The factorization is the product of all the prime number divisors and the last prime number quotient.*

Practice check: Do exercises 98–101 in the margin.

Find the prime factorization of the following numbers.

98. 100 **99.** 120 **100.** 396

101. 37

The answers are in the margin on page 150.

Answers to exercise 89

89. 2, prime; 29, prime; 40, composite; 9, composite; 21, composite; 63, composite; 51, composite; 5, prime; 1, neither; 7, prime; 32, composite; 16, composite; 11, prime; 3, prime; 36, composite; 19, prime

Answers to exercises 90 and 91

90. $2 \times 14, 4 \times 7, 2 \times 2 \times 7$
91. $2 \times 21, 6 \times 7, 2 \times 3 \times 7$

Answers to exercises 92–97

92. $6 = 2 \times 3$ **93.** $10 = 2 \times 5$
94. $12 = 2 \times 2 \times 3$
95. $27 = 3 \times 3 \times 3$
96. $38 = 2 \times 19$
97. $52 = 2 \times 2 \times 13$

Answers to exercises 98–101

98. $2 \times 2 \times 5 \times 5$
99. $2 \times 2 \times 2 \times 3 \times 5$
100. $2 \times 2 \times 3 \times 3 \times 11$
101. 37

■ Looking Back

Given any whole number greater than 1, you should now be able to find its prime factorization.

■ Problems 4–3

Review

1. Draw a number line and locate the point $\frac{4}{7}$.

Simplify.

2. $\dfrac{15}{15}$ **3.** $\dfrac{183}{1}$

4. Write a fractional number for the shaded part.

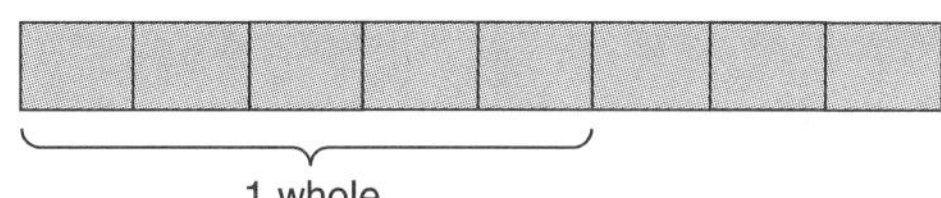

5. State the rule for divisibility by 3.

6. List all the factors of 44.

State whether the following numbers are prime, composite, or neither.

7. 12 **8.** 8 **9.** 1 **10.** 5 **11.** 17 **12.** 21

13. 75 **14.** 1 **15.** 13 **16.** 36 **17.** 46 **18.** 2

19. 59 **20.** 44 **21.** 7 **22.** 97 **23.** 68 **24.** 77

25. 27 **26.** 18 **27.** 57 **28.** 13 **29.** 38 **30.** 32

Find the prime factorizations of the following numbers.

31. 4 **32.** 8 **33.** 10 **34.** 12 **35.** 13 **36.** 14

37. 20 **38.** 19 **39.** 24 **40.** 22 **41.** 35 **42.** 30

43. 47 **44.** 38 **45.** 54 **46.** 56 **47.** 58 **48.** 63

49. 74 **50.** 72 **51.** 81 **52.** 79 **53.** 93 **54.** 82

55. 99 **56.** 88 **57.** 115 **58.** 116 **59.** 135 **60.** 144

61. 286 **62.** 195 **63.** 126 **64.** 132 **65.** 121 **66.** 255

Find the prime factorizations.

67. 475 **68.** 672 **69.** 1044 **70.** 1434 **71.** 5389 **72.** 3483

The answers to review and odd-numbered exercises are in the back of the book.

4–4 EQUAL FRACTIONS

Changing to Higher Terms

Consider $\frac{2}{5}$.

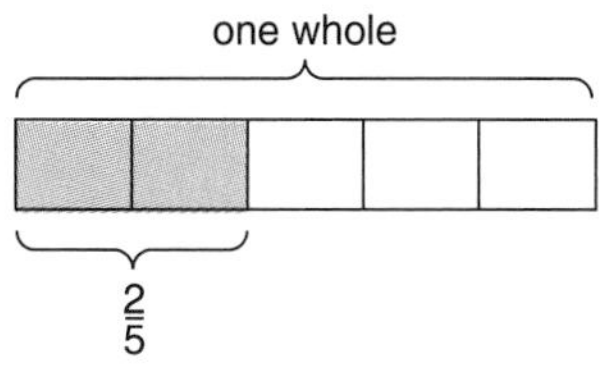

Figure 4–25

Let's divide each of the five pieces in half.

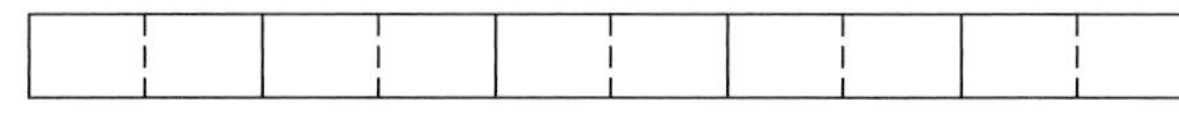

Figure 4–26

Now the whole is divided into 10 parts and each part is $\frac{1}{10}$ of the whole.

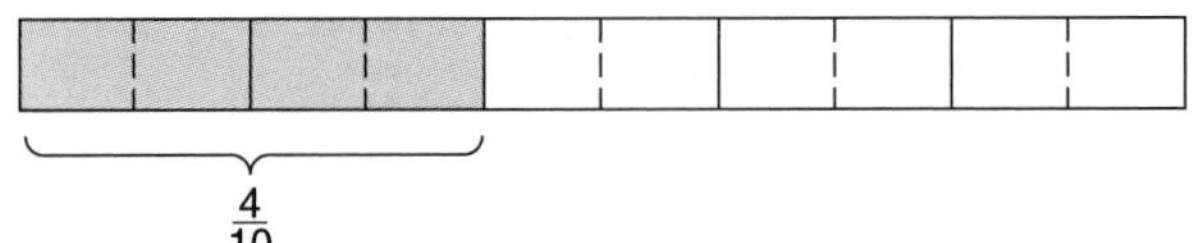

Figure 4–27

The shaded part that represents $\frac{4}{10}$ is the same size as the shaded part the represents $\frac{2}{5}$, so $\frac{2}{5}$ and $\frac{4}{10}$ represent the same amount, or $\frac{2}{5} = \frac{4}{10}$. We can get the same results by multiplying both numerator and denominator of $\frac{2}{5}$ by 2:

$$\frac{2 \times 2}{5 \times 2} = \frac{4}{10}$$

Rule

We can multiply the numerator and denominator of a fraction by any whole number greater than 0 to obtain another fraction that is equal to the first. The resulting fraction is in higher terms.

■ Example 1

Multiply the numerator and denominator of $\frac{3}{7}$ by 3 to obtain a fraction that is equal to $\frac{3}{7}$.

Solution $$\frac{3}{7} = \frac{3 \times 3}{7 \times 3} = \frac{9}{21}$$

Practice check: Do exercises 102–105 in the margin.

Multiply the numerator and denominator of the following fractions by the number in parentheses to obtain a fraction that is equal to the one given.

102. $\frac{2}{5}$ (5) **103.** $\frac{3}{5}$ (3)

104. $\frac{17}{21}$ (6) **105.** $\frac{37}{58}$ (2)

The answers are in the margin on page 154.

Suppose we need to form a fraction equal to another fraction when only the denominator is given. How do we find the numerator?

■ Example 2

$$\frac{2}{3} = \frac{?}{6}$$

Solution **Step 1** We ask the question

$$\frac{2}{3} = \frac{?}{6}$$

What times 3 makes 6? The answer of course is 2.

Step 2 This tells us to multiply both numerator and denominator of $\frac{2}{3}$ by 2 to obtain $\frac{4}{6}$.

$$\frac{2}{3} = \frac{2 \times 2}{3 \times 2} = \frac{4}{6}$$

The missing numerator is 4, and the fraction equal to $\frac{2}{3}$ is $\frac{4}{6}$.

■ **Example 3**

$$\frac{7}{8} = \frac{?}{40}$$

Solution **Step 1** We ask the question

$$\frac{7}{8} = \frac{?}{40}$$

What times 8 makes 40? The answer is 5.

Step 2 This tells us to multiply both numerator and denominator by 5 to obtain $\frac{35}{40}$.

The missing numerator is 35, and the fraction equal to $\frac{7}{8}$ is $\frac{35}{40}$.

Practice check: Do exercises 106–109 in the margin.

Change the following fractions to higher terms by finding the missing numerators.

106. $\frac{2}{7} = \frac{?}{14}$ **107.** $\frac{1}{4} = \frac{?}{28}$

108. $\frac{7}{9} = \frac{?}{27}$ **109.** $\frac{18}{23} = \frac{?}{69}$

The answers are in the margin on page 154.

Reducing to Lowest Terms

We can reduce fractions to lowest terms by reversing the previous procedure. Instead of multiplying, we can divide the numerator and denominator by the same number.

■ **Example 4**

Reduce $\frac{6}{8}$ to lowest terms.

Solution Both 6 and 8 are even numbers, so they are divisible by 2.

$$\frac{6}{8} = \frac{6 \div 2}{8 \div 2} = \frac{3}{4}$$

We divide 6 and 8 by 2 to get the reduced fraction of $\frac{3}{4}$. $\frac{3}{4}$ cannot be reduced any further because 3 and 4 are not both divisible by any whole number greater than 1.

■ **Example 5**

Reduce $\frac{30}{45}$ to lowest terms.

Solution **Step 1** We first note that 30 and 45 are both divisible by 3.

$$\frac{30}{45} = \frac{30 \div 3}{45 \div 3} = \frac{10}{15}$$

We divide 30 and 45 by 3 to get a reduced fraction of $\frac{10}{15}$.

This fraction is not yet reduced to lowest terms, since 10 and 15 are divisible by 5.

Step 2

$$\frac{10}{15} = \frac{10 \div 5}{15 \div 5} = \frac{2}{3}$$

We cannot reduce $\frac{2}{3}$ any further.

We can shorten the work of reducing by *canceling.*

Answers to exercises 102–105

102. $\frac{2}{3} = \frac{2 \times 5}{3 \times 5} = \frac{10}{15}$

103. $\frac{3}{5} = \frac{3 \times 3}{5 \times 3} = \frac{9}{15}$

104. $\frac{17}{21} = \frac{17 \times 6}{21 \times 6} = \frac{102}{126}$

105. $\frac{37}{58} = \frac{37 \times 2}{58 \times 2} = \frac{74}{116}$

Answers to exercises 106–109

106. $\frac{4}{14}$ **107.** $\frac{7}{28}$ **108.** $\frac{21}{27}$

109. $\frac{54}{69}$

■ **Example 6**

Reduce $\frac{6}{8}$ to lowest terms.

Solution

$$\frac{6}{8} = \frac{\overset{3}{\cancel{6}}}{\underset{4}{\cancel{8}}} = \frac{3}{4}$$ Mentally divide 6 and 8 by 2.

■ **Example 7**

Reduce $\frac{30}{45}$ to lowest terms.

Solution **Step 1**

$$\frac{30}{45} = \frac{\overset{10}{\cancel{30}}}{\underset{15}{\cancel{45}}}$$ Mentally divide 30 and 45 by 3.

Step 2

$$\frac{30}{45} = \frac{\overset{\overset{2}{\cancel{10}}}{\cancel{30}}}{\underset{\underset{3}{\cancel{15}}}{\cancel{45}}}$$ Divide 10 and 15 by 5.

Thus, $\frac{30}{45} = \frac{2}{3}$.

You can also reduce a fraction by first factoring the numerator and denominator into prime factors and then canceling common factors.

■ **Example 8**

Reduce $\frac{30}{45}$ to lowest terms.

Solution **Step 1** $\frac{30}{45} = \frac{2 \times 3 \times 5}{3 \times 3 \times 5}$ Factoring numerator and denominator

Step 2 $\frac{30}{45} = \frac{2 \times \cancel{3} \times \cancel{5}}{3 \times \cancel{3} \times \cancel{5}}$ Canceling common factors

Thus, $\frac{30}{45} = \frac{2}{3}$.

■ **Example 9**

Reduce $\frac{210}{270}$ to lowest terms.

Solution **Step 1** $\dfrac{210}{270} = \dfrac{2 \times 3 \times 5 \times 7}{2 \times 3 \times 3 \times 3 \times 5}$ Factoring numerator and denominator

Step 2 $\dfrac{210}{270} = \dfrac{\cancel{2} \times \cancel{3} \times \cancel{5} \times 7}{\cancel{2} \times \cancel{3} \times 3 \times 3 \times \cancel{5}}$ Canceling common factors

Thus, $\dfrac{210}{270} = \dfrac{7}{9}$.

Use whichever reducing method is easier for you.

Practice check: Do exercises 110–114 in the margin.

Reduce the following fractions to lowest terms.

110. $\dfrac{3}{18}$ **111.** $\dfrac{15}{45}$ **112.** $\dfrac{10}{25}$

113. $\dfrac{18}{24}$ **114.** $\dfrac{68}{85}$

The answers are in the margin on page 157.

■ Looking Back

You should now be able to raise a fraction to higher terms and reduce a fraction to lowest terms.

■ Problems 4–4

Do as many exercises as suggested by your instructor.

Review

1. Name the denominator of $\frac{16}{31}$.

2. Draw a number line and locate $5\frac{4}{6}$.

3. a. Is 680 divisible by 2?

b. State the rule for divisibility by 2?

4. Find the factors of 35.

5. List the first five prime numbers.

6. Find the prime factorization of 44.

Find the missing numerator.

7. $\frac{3}{4} = \frac{?}{8}$

8. $\frac{5}{6} = \frac{?}{12}$

9. $\frac{7}{9} = \frac{?}{36}$

10. $\frac{7}{8} = \frac{?}{32}$

11. $\frac{4}{7} = \frac{?}{42}$

12. $\frac{1}{8} = \frac{?}{48}$

13. $\frac{5}{8} = \frac{?}{48}$

14. $\frac{5}{7} = \frac{?}{21}$

15. $\frac{3}{4} = \frac{?}{68}$

16. $\frac{2}{3} = \frac{?}{39}$

17. $\frac{16}{25} = \frac{?}{75}$

18. $\frac{5}{8} = \frac{?}{32}$

19. $\frac{8}{8} = \frac{?}{56}$

20. $\frac{2}{9} = \frac{?}{72}$

21. $\frac{9}{13} = \frac{?}{52}$

22. $\frac{1}{12} = \frac{?}{48}$

23. $\frac{8}{11} = \frac{?}{55}$

24. $\frac{9}{13} = \frac{?}{52}$

25. $\frac{3}{22} = \frac{?}{66}$

26. $\frac{0}{7} = \frac{?}{14}$

27. $\frac{5}{16} = \frac{?}{64}$

28. $\frac{4}{15} = \frac{?}{45}$

29. $\frac{19}{24} = \frac{?}{96}$

30. $\frac{6}{17} = \frac{?}{51}$

Reduce the following fractions to lowest terms if possible.

31. $\frac{2}{4}$

32. $\frac{5}{10}$

33. $\frac{3}{18}$

34. $\frac{6}{8}$

35. $\frac{6}{20}$

36. $\frac{3}{9}$

37. $\frac{12}{16}$

38. $\frac{6}{10}$

39. $\frac{18}{24}$

40. $\frac{10}{16}$

41. $\frac{4}{16}$

42. $\frac{12}{18}$

43. $\frac{6}{21}$

44. $\frac{24}{40}$

45. $\frac{29}{40}$

46. $\frac{6}{48}$

47. $\frac{36}{81}$ **48.** $\frac{9}{15}$ **49.** $\frac{9}{12}$ **50.** $\frac{8}{30}$

51. $\frac{24}{72}$ **52.** $\frac{18}{32}$ **53.** $\frac{36}{54}$ **54.** $\frac{56}{64}$

55. $\frac{22}{44}$ **56.** $\frac{10}{40}$ **57.** $\frac{36}{90}$ **58.** $\frac{38}{57}$

59. $\frac{32}{45}$ **60.** $\frac{52}{91}$

Answers to exercises 110–114

110. $\frac{1}{6}$ **111.** $\frac{1}{3}$ **112.** $\frac{2}{5}$ **113.** $\frac{3}{4}$

114. $\frac{4}{5}$

Find the missing numerator.

61. $\frac{4}{7} = \frac{?}{210}$ **62.** $\frac{2}{21} = \frac{?}{147}$ **63.** $\frac{2}{13} = \frac{?}{169}$ **64.** $\frac{3}{13} = \frac{?}{91}$

65. $\frac{7}{11} = \frac{?}{132}$ **66.** $\frac{7}{8} = \frac{?}{1000}$ **67.** $\frac{5}{7} = \frac{?}{147}$ **68.** $\frac{5}{8} = \frac{?}{192}$

Reduce the following fractions to lowest terms if possible.

69. $\frac{84}{210}$ **70.** $\frac{22}{132}$ **71.** $\frac{150}{735}$ **72.** $\frac{105}{135}$

73. $\frac{147}{168}$ **74.** $\frac{34}{119}$ **75.** $\frac{91}{455}$ **76.** $\frac{680}{765}$

The answers to review and odd-numbered exercises are in the back of the book.

4–5 MULTIPLYING FRACTIONS

It is possible to picture the product of two fractions. Let us multiply $\frac{1}{4}$ and $\frac{2}{3}$ by using an illustration.

Step 1 Using a rectangle, first we show $\frac{2}{3}$.

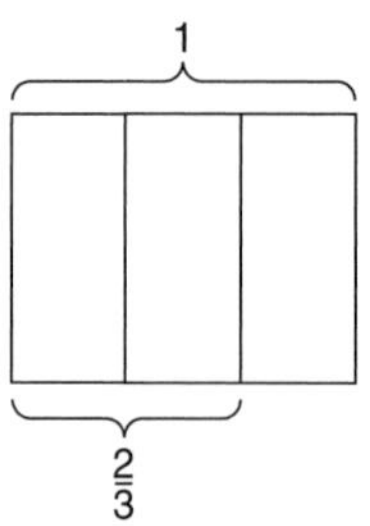

Figure 4–28

Step 2 Next, we divide the rectangle into 4 equal parts and show $\frac{1}{4}$.

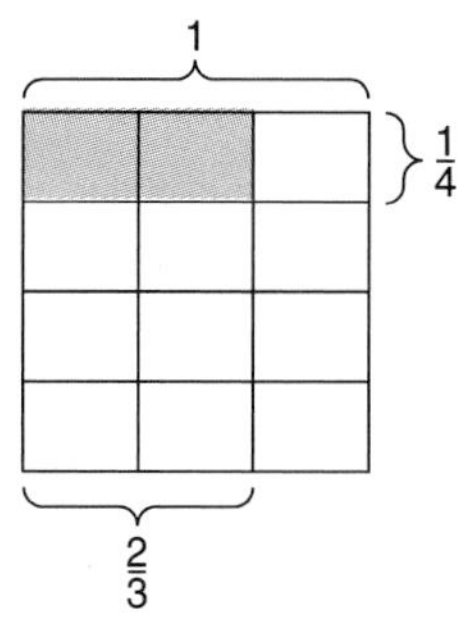

Figure 4–29

Step 3 State the product. Notice that the whole now has 12 equal parts, and 2 of them have overlapping shading. The product is $\frac{2}{12}$.

Thus, $\frac{1}{4} \times \frac{2}{3} = \frac{2}{12}$.

We can show the same results by simply multiplying the numerators and the denominators.

$$\frac{1}{4} \times \frac{2}{3} = \frac{1 \times 2}{4 \times 3} = \frac{2}{12}.$$

Rule

To multiply fractional numbers, we multiply the numerators together and multiply the denominators together.

■ Example 1

Multiply.

$$\frac{3}{5} \times \frac{2}{7}$$

Solution We multiply the numerators together and the denominators together.

$$\frac{3}{5} \times \frac{2}{7} = \frac{3 \times 2}{5 \times 7} = \frac{6}{35}$$

Do this step mentally.

Practice check: Do exercises 115–117 in the margin.

Sometimes the product must be reduced. We should always check to see if the fractional number is reduced to lowest terms.

■ Example 2

Multiply.

$$\frac{2}{3} \times \frac{7}{10}$$

Solution $\frac{2}{3} \times \frac{7}{10} = \frac{2 \times 7}{3 \times 10} = \frac{14}{30} = \frac{\overset{7}{\cancel{14}}}{\underset{15}{\cancel{30}}} = \frac{7}{15}$

Do this step mentally. (↑ $\frac{2 \times 7}{3 \times 10}$)

The product must be reduced. (→ $\frac{14}{30}$)

We divide both numerator and denominator by 2.

Thus, $\frac{2}{3} \times \frac{7}{10} = \frac{7}{15}$.

Practice check: Do exercises 118–121 in the margin.

You can save yourself a lot of work by reducing the terms of the fractions before you multiply the numerators and denominators. This is called canceling. It is important to remember that canceling requires division of a pair of numbers, one in the numerator and one in the denominator.

■ Example 3

Multiply.

$$\frac{2}{3} \times \frac{7}{10}$$

Solution $\frac{2}{3} \times \frac{7}{10} = \frac{2 \times 7}{3 \times 10}$ Both 2 and 10 are divisible by 2.

$= \frac{\overset{1}{\cancel{2}} \times 7}{3 \times \underset{5}{\cancel{10}}}$ Divide 2 and 10 by 2.

$= \frac{1 \times 7}{3 \times 5}$

$= \frac{7}{15}$

■ Example 4

Multiply.

$$\frac{4}{18} \times \frac{3}{10}$$

Multiply.

115. $\frac{3}{8} \times \frac{1}{5}$ **116.** $\frac{4}{5} \times \frac{7}{9}$

117. $\frac{5}{7} \times \frac{8}{9}$

The answers are in the margin on page 161.

Multiply. Reduce the results.

118. $\frac{5}{12} \times \frac{3}{4}$ **119.** $\frac{3}{10} \times \frac{1}{3}$

120. $\frac{8}{9} \times \frac{3}{4}$ **121.** $\frac{5}{6} \times \frac{9}{25}$

The answers are in the margin on page 161.

Solution $\frac{4}{18} \times \frac{3}{10} = \frac{4 \times 3}{18 \times 10}$ 4 and 10 are divisible by 2.

$$= \frac{\overset{2}{\cancel{4}} \times 3}{18 \times \underset{5}{\cancel{10}}}$$ Divide 4 and 10 by 2.

$$= \frac{\overset{2}{\cancel{4}} \times 3}{18 \times \underset{5}{\cancel{10}}}$$ 3 and 18 are divisible by 3.

$$= \frac{\overset{2}{\cancel{4}} \times \overset{1}{\cancel{3}}}{\underset{6}{\cancel{18}} \times \underset{5}{\cancel{10}}}$$ Divide 3 and 18 by 3.

$$= \frac{\overset{2}{\cancel{4}} \times \overset{1}{\cancel{3}}}{\underset{6}{\cancel{18}} \times \underset{5}{\cancel{10}}}$$ 2 and 6 are divisible by 2.

$$= \frac{\overset{\overset{1}{\cancel{2}}}{\cancel{4}} \times \overset{1}{\cancel{3}}}{\underset{\underset{3}{\cancel{6}}}{\cancel{18}} \times \underset{5}{\cancel{10}}}$$ Divide 2 and 6 by 2.

$$= \frac{1 \times 1}{3 \times 5} = \frac{1}{15}$$

Use canceling to multiply without looking at the steps above: $\frac{4}{18} \times \frac{3}{10}$

Practice check: Do exercises 122–125 in the margin.

Multiply. Use canceling.

122. $\frac{8}{9} \times \frac{3}{4}$ **123.** $\frac{10}{21} \times \frac{14}{15}$

124. $\frac{7}{10} \times \frac{5}{7}$ **125.** $\frac{9}{15} \times \frac{10}{24}$

The answers are in the margin on page 162.

Multiplying Three Fractions

To get the product of three or more fractions, we show the numerators multiplied together and the denominators multiplied together, then we cancel.

■ Example 5

Multiply.

$$\frac{3}{5} \times \frac{10}{21} \times \frac{7}{12}.$$

Solution $\frac{3}{5} \times \frac{10}{21} \times \frac{7}{12} = \frac{3 \times \overset{2}{\cancel{10}} \times 7}{\underset{1}{\cancel{5}} \times 21 \times 12}$ Divide 10 and 5 by 5.

$$= \frac{3 \times \overset{2}{10} \times \overset{1}{\cancel{7}}}{\underset{1}{5} \times \underset{3}{\cancel{21}} \times 12}$$ Divide 7 and 21 by 7.

$$= \frac{\overset{1}{\cancel{3}} \times \overset{2}{\cancel{10}} \times \overset{1}{\cancel{7}}}{\underset{1}{\cancel{5}} \times \underset{3}{\cancel{21}} \times \underset{4}{\cancel{12}}}$$ Divide 3 and 12 by 3.

$$= \frac{\overset{1}{\cancel{3}} \times \overset{\overset{1}{\cancel{2}}}{\cancel{10}} \times \overset{1}{\cancel{7}}}{\underset{1}{\cancel{5}} \times \underset{3}{\cancel{21}} \times \underset{\underset{2}{\cancel{4}}}{\cancel{12}}}$$ Divide 2 and 4 by 2.

$$= \frac{1}{6}$$

Answers to exercises 115–117

115. $\frac{3}{40}$ **116.** $\frac{28}{45}$ **117.** $\frac{40}{63}$

Answers to exercises 118–121

118. $\frac{5}{16}$ **119.** $\frac{1}{10}$ **120.** $\frac{2}{3}$

121. $\frac{3}{10}$

Multiply without looking at the steps above: $\frac{3}{5} \times \frac{10}{21} \times \frac{7}{12}$.

Practice check: Do exercises 126–128 in the margin.

Multiply.

126. $\frac{3}{7} \times \frac{1}{5} \times \frac{2}{3}$

127. $\frac{5}{8} \times \frac{4}{7} \times \frac{3}{10}$

128. $\frac{5}{12} \times \frac{3}{16} \times \frac{4}{5}$

The answers are in the margin on page 162.

■ Looking Back

You should now be able to multiply fractions using canceling. If you feel unsure about this area, go back and restudy the material before doing Problems 4–5.

■ Problems 4–5

Do as many exercises as suggested by your instructor.

Review

1. Simplify $\frac{5}{0}$.

2. Write a fractional number for the shaded part.

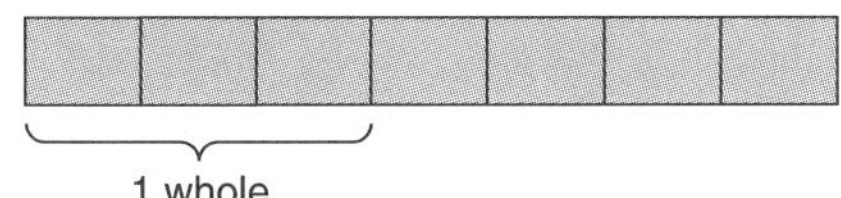

3. a. Is 853 divisible by 3?

b. State the rule for divisibility by 3.

4. Find the prime factorization of 105.

Answers to exercises 122–125

122. $\frac{2}{3}$ **123.** $\frac{4}{9}$ **124.** $\frac{1}{2}$ **125.** $\frac{1}{4}$

Answers to exercises 126–128

126. $\frac{2}{35}$ **127.** $\frac{3}{28}$ **128.** $\frac{1}{16}$

5. Find the missing numerator. $\frac{7}{9} = \frac{?}{36}$

6. Reduce $\frac{72}{66}$.

Multiply. Use canceling where possible.

7. $\frac{2}{3} \times \frac{4}{5}$

8. $\frac{4}{5} \times \frac{2}{3}$

9. $\frac{3}{8} \times \frac{5}{7}$

10. $\frac{3}{7} \times \frac{5}{6}$

11. $\frac{5}{6} \times \frac{5}{8}$

12. $\frac{3}{4} \times \frac{7}{11}$

13. $\frac{2}{7} \times \frac{3}{10}$

14. $\frac{6}{7} \times \frac{11}{12}$

15. $\frac{5}{9} \times \frac{7}{10}$

16. $\frac{4}{7} \times \frac{3}{8}$

17. $\frac{2}{7} \times \frac{1}{2}$

18. $\frac{5}{12} \times \frac{3}{4}$

19. $\frac{4}{9} \times \frac{7}{8}$

20. $\frac{2}{5} \times \frac{5}{7}$

21. $\frac{6}{6} \times \frac{7}{8}$

22. $\frac{2}{2} \times \frac{5}{8}$

23. $\frac{8}{9} \times \frac{3}{4}$

24. $\frac{7}{9} \times \frac{3}{14}$

25. $\frac{9}{10} \times \frac{2}{3}$

26. $\frac{3}{7} \times \frac{14}{15}$

27. $\frac{3}{5} \times \frac{25}{30}$

28. $\frac{7}{12} \times \frac{4}{21}$

29. $\frac{9}{10} \times \frac{18}{35}$

30. $\frac{5}{16} \times \frac{4}{25}$

31. $\frac{18}{36} \times \frac{24}{45}$

32. $\frac{6}{35} \times \frac{10}{18}$

33. $\frac{9}{14} \times \frac{12}{45}$

34. $\frac{20}{30} \times \frac{15}{16}$

Multiply. Use canceling where possible.

35. $\frac{1}{2} \times \frac{3}{8} \times \frac{3}{5}$ **36.** $\frac{2}{7} \times \frac{3}{5} \times \frac{1}{5}$ **37.** $\frac{3}{8} \times \frac{9}{10} \times \frac{3}{4}$ **38.** $\frac{2}{3} \times \frac{4}{5} \times \frac{1}{3}$

39. $\frac{5}{6} \times \frac{2}{5} \times \frac{1}{4}$ **40.** $\frac{4}{5} \times \frac{2}{3} \times \frac{7}{10}$ **41.** $\frac{3}{8} \times \frac{5}{9} \times \frac{1}{10}$ **42.** $\frac{2}{3} \times \frac{3}{4} \times \frac{2}{5}$

43. $\frac{7}{9} \times \frac{5}{11} \times \frac{6}{35}$ **44.** $\frac{3}{16} \times \frac{5}{12} \times \frac{4}{5}$ **45.** $\frac{2}{3} \times \frac{3}{10} \times \frac{5}{8}$ **46.** $\frac{7}{12} \times \frac{16}{21} \times \frac{5}{8}$

47. $\frac{2}{3} \times \frac{4}{5} \times \frac{1}{2}$ **48.** $\frac{2}{5} \times \frac{3}{4} \times \frac{15}{16}$ **49.** $\frac{5}{14} \times \frac{2}{7} \times \frac{3}{10}$ **50.** $\frac{11}{15} \times \frac{5}{22} \times \frac{2}{7}$

Multiply.

51. $\frac{24}{90} \times \frac{75}{120}$ **52.** $\frac{24}{60} \times \frac{35}{54}$ **53.** $\frac{45}{135} \times \frac{120}{270}$ **54.** $\frac{45}{100} \times \frac{30}{63}$

The answers to review and odd-numbered exercises are in the back of the book.

4–6 DIVISION OF FRACTIONS

Proper and Improper Fractions

In section 4–1, we discussed fractions that name a whole or more than one whole.

■ **Example 1**

Write a fractional number for the shaded part.

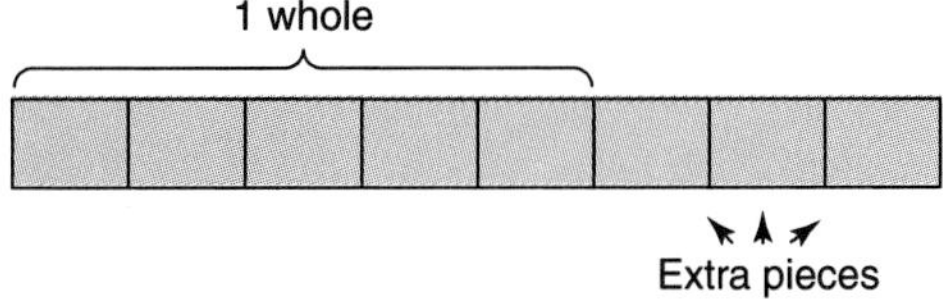

Figure 4–30

Solution Since the whole is divided into 5 equal pieces, each piece is $\frac{1}{5}$ of the whole. We have eight $\frac{1}{5}$s or $\frac{8}{5}$ (eight-fifths).

Notice that the numerator is larger than the denominator.

Rule

A fractional number that has the numerator equal to or larger than the denominator is called an improper fraction.

All of these fractions are improper fractions:

$$\frac{3}{2} \quad \frac{5}{3} \quad \frac{26}{13} \quad \frac{7}{7} \quad \frac{53}{24}$$

These fractional numbers name a number greater than or equal to 1. If $\frac{18}{7}$ is an improper fraction, what is a proper fraction? You guessed it!

Rule

A fractional number that has the numerator smaller than the denominator is a proper fraction.

All of the fractions listed here are proper fractions. They name 0 or a number between 0 and 1.

$$\frac{1}{2} \quad \frac{6}{7} \quad \frac{29}{46} \quad \frac{119}{263} \quad \frac{0}{8}$$

Practice check: Do exercises 129–134 in the margin.

State whether the following fractions are proper or improper.

129. $\frac{5}{6}$ **130.** $\frac{16}{1}$ **131.** $\frac{0}{5}$

132. $\frac{7}{8}$ **133.** $\frac{33}{8}$ **134.** $\frac{56}{56}$

The answers are in the margin on page 166.

Reciprocals

To divide fractions we will need to know about the reciprocal of a fraction.

The reciprocal of $\frac{3}{4}$ is $\frac{4}{3}$.
The reciprocal of $\frac{1}{6}$ is $\frac{6}{1}$.
The reciprocal of $\frac{9}{7}$ is $\frac{7}{9}$.
The reciprocal of 23 is $\frac{1}{23}$ (*Note:* $23 = \frac{23}{1}$.)

All we need to do to form a reciprocal of a fraction is to exchange the numerator and denominator.

Be careful on this one: $\frac{0}{8}$ does not have a reciprocal, since $\frac{8}{0}$ is not a number.

Rule

Zero does not have a reciprocal!

Practice check: Do exercises 135–141 in the margin.

State the reciprocal of the following fractions.

135. $\frac{3}{7}$ **136.** $\frac{1}{11}$ **137.** $\frac{0}{15}$

138. $\frac{17}{8}$ **139.** 34 **140.** $\frac{136}{7}$

141. $\frac{28}{157}$

The answers are in the margin on page 166.

Meaning of Division

Consider $6 \div 2$.
↗ Dividend ↖ Divisor

This asks the question, "How many 2s are there in 6?" We know that there are *three* 2s in 6.

Now, suppose we have $\frac{3}{4} \div \frac{1}{4}$. Using the same principle, this asks the question, "How many $\frac{1}{4}$s are there in $\frac{3}{4}$?" To see how many, we can use a picture.

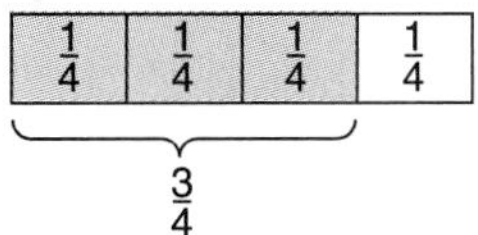

Figure 4–31

The shaded part represents $\frac{3}{4}$ of the whole object.
Since each piece is $\frac{1}{4}$ of the whole, we can see that there are *three* $\frac{1}{4}$s in $\frac{3}{4}$. Thus,

$$\frac{3}{4} \div \frac{1}{4} = 3$$

It is rather impractical to do division of fractions by drawing pictures. We can get the answer by using multiplication. If we multiply $\frac{3}{4}$ by the reciprocal of $\frac{1}{4}$, the result will be 3.

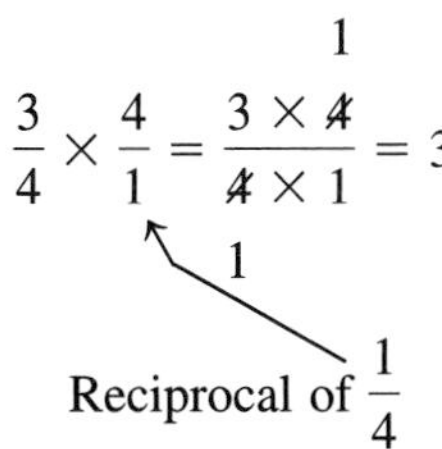

$$\frac{3}{4} \times \frac{4}{1} = \frac{3 \times \overset{1}{\cancel{4}}}{\underset{1}{\cancel{4}} \times 1} = 3$$

Rule

To divide two fractions, multiply the first fraction by the reciprocal of the second fraction.

Memorize this rule.

■ Example 2

Divide.

$$\frac{2}{3} \div \frac{3}{4}$$

Solution $\frac{2}{3} \div \frac{3}{4} = \frac{2}{3} \times \frac{4}{3} = \frac{8}{9}$

Keep the first fraction the same.

Change to multiplication.

Multiply by the reciprocal of $\frac{3}{4}$.

Practice check: Do exercises 142–145 in the margin.

Divide.

142. $\frac{2}{3} \div \frac{5}{4}$

143. $\frac{3}{8} \div \frac{5}{7}$

144. $\frac{2}{7} \div \frac{3}{10}$

145. $\frac{4}{7} \div \frac{3}{5}$

The answers are in the margin on page 167.

Answers to exercises 129–134

129. proper **130.** improper
131. proper **132.** proper
133. improper **134.** improper

Answers to exercises 135–141

135. $\frac{7}{3}$ **136.** $\frac{11}{1}$ or 11 **137.** does not have a reciprocal

138. $\frac{8}{17}$ **139.** $\frac{1}{34}$ **140.** $\frac{7}{136}$

141. $\frac{157}{28}$

Divide.

146. $\frac{17}{10} \div \frac{1}{2}$ **147.** $\frac{7}{8} \div \frac{3}{4}$

148. $\frac{5}{6} \div \frac{5}{8}$ **149.** $\frac{9}{10} \div \frac{48}{14}$

The answers are in the margin on page 169.

You should cancel whenever possible.

■ Example 3

Divide.

$$\frac{5}{12} \div \frac{1}{3}$$

Solution $\frac{5}{12} \div \frac{1}{3} = \frac{5}{12} \times \frac{3}{1} = \frac{5 \times \overset{1}{\cancel{3}}}{\underset{4}{\cancel{12}} \times 1} = \frac{5}{4}$

↑ Reciprocal of $\frac{1}{3}$ (pointing to $\frac{3}{1}$)

Practice check: Do exercises 146–149 in the margin.

■ Looking Back

You should now be able to

1. Recognize proper and improper fractions.
2. State the reciprocal of a fraction.
3. Divide a fraction by another fraction.

■ Problems 4–6

Review

1. Write a fractional number that tells what portion of the group are box-shaped.

2. Name the point on the number line.

33 34

3. Find the factors of 16.

4. Find the missing numerator: $\frac{3}{8} = \frac{?}{48}$.

State whether the following fractions are proper or improper.

5. $\frac{2}{3}$ 6. $\frac{7}{7}$ 7. $\frac{9}{8}$ 8. $\frac{0}{17}$

9. $\frac{13}{15}$ 10. $\frac{23}{23}$ 11. $\frac{29}{63}$ 12. $\frac{116}{116}$

13. $\frac{36}{29}$ 14. $\frac{129}{356}$ 15. $\frac{1081}{756}$ 16. $\frac{0}{231}$

17. $\frac{3468}{3468}$

Answers to exercises 142–145

142. $\frac{8}{15}$ **143.** $\frac{21}{40}$ **144.** $\frac{20}{21}$ **145.** $\frac{20}{21}$

State the reciprocal of the following fractions.

18. $\frac{4}{5}$ 19. $\frac{7}{9}$ 20. $\frac{1}{8}$ 21. $\frac{0}{31}$

22. $\frac{20}{3}$ 23. 56 24. $\frac{4}{11}$ 25. 117

26. $\frac{0}{18}$ 27. $\frac{1}{67}$ 28. $\frac{25}{2}$ 29. 2106

30. $\frac{5}{4}$ 31. 21 32. $\frac{8}{3}$ 33. $\frac{109}{656}$

34. 689

Divide.

35. $\frac{7}{8} \div \frac{4}{5}$ 36. $\frac{5}{6} \div \frac{3}{7}$ 37. $\frac{4}{5} \div \frac{7}{8}$ 38. $\frac{7}{9} \div \frac{4}{5}$

39. $\frac{5}{6} \div \frac{2}{3}$ **40.** $\frac{7}{8} \div \frac{7}{4}$ **41.** $\frac{4}{9} \div \frac{2}{3}$ **42.** $\frac{7}{8} \div \frac{5}{12}$

43. $\frac{6}{7} \div \frac{9}{10}$ **44.** $\frac{7}{8} \div \frac{7}{6}$ **45.** $\frac{2}{3} \div \frac{10}{3}$ **46.** $\frac{2}{3} \div \frac{2}{9}$

47. $\frac{8}{9} \div \frac{3}{4}$ **48.** $\frac{11}{15} \div \frac{2}{3}$ **49.** $\frac{4}{5} \div \frac{2}{3}$ **50.** $\frac{4}{5} \div \frac{20}{35}$

51. $\frac{15}{16} \div \frac{5}{14}$ **52.** $\frac{7}{8} \div \frac{21}{24}$

Divide.

53. $\frac{45}{4} \div \frac{10}{8}$ **54.** $\frac{12}{25} \div \frac{36}{75}$ **55.** $\frac{18}{7} \div \frac{12}{42}$ **56.** $\frac{15}{63} \div \frac{20}{21}$

The answers to review and odd-numbered exercises are in the back of the book.

4–7 DIVIDING FRACTIONS IN COMPLEX FORM

Let's again consider 6 divided by 2. Here are three ways to express 6 divided by 2:

$$6 \div 2 \qquad 2\overline{)6} \qquad \frac{6}{2}$$

↑ The last one is a fractional number.

We know that in $\frac{6}{2}$ the numerator is 6, and the denominator is 2. A fractional number indicates division: It tells us to divide the numerator by the denominator.

State the meaning of the following fractions.

150. $\frac{15}{3}$ **151.** $\frac{34}{17}$

152. $\frac{88}{4}$ **153.** $\frac{105}{5}$

The answers are in the margin on page 171.

■ **Examples**

1. $\frac{10}{5}$ means $10 \div 5$. **2.** $\frac{21}{7}$ means $21 \div 7$. **3.** $\frac{24}{12}$ means $24 \div 12$.

Practice check: Do exercises 150–153 in the margin.

Consider $\frac{\frac{5}{8}}{\frac{4}{5}}$, or $\frac{\frac{7}{16}}{21}$, or $\frac{15}{\frac{9}{17}}$. These are called complex fractional numbers because one or more of the numerators or denominators are fractions.

$\frac{\frac{5}{8}}{\frac{4}{5}}$ ← Numerator ← Denominator

Thus, $\frac{\frac{5}{8}}{\frac{4}{5}}$ means $\frac{5}{8} \div \frac{4}{5}$.

To simplify $\frac{\frac{5}{8}}{\frac{4}{5}}$ means to complete the division.

Answers to exercises 146–149

146. $\frac{17}{5}$ **147.** $\frac{7}{6}$ **148.** $\frac{4}{3}$

149. $\frac{21}{80}$

Example 4

Simplify.

$$\frac{\frac{5}{8}}{\frac{4}{5}}$$

Solution

$$\frac{\frac{5}{8}}{\frac{4}{5}} = \frac{5}{8} \div \frac{4}{5} = \frac{5}{8} \times \frac{5}{4} = \frac{25}{32}$$

↑ Reciprocal of $\frac{4}{5}$ (points to $\frac{5}{4}$)

Practice check: Do exercises 154–157 in the margin.

Simplify.

154. $\frac{\frac{21}{8}}{\frac{6}{35}}$ **155.** $\frac{\frac{1}{2}}{\frac{6}{8}}$

156. $\frac{\frac{4}{5}}{\frac{7}{9}}$ **157.** $\frac{\frac{21}{6}}{\frac{7}{9}}$

The answers are in the margin on page 171.

Looking Back

Given a complex fraction you should now be able to simplify it.

Problems 4–7

Review

1. Is 57 prime, composite, or neither?

2. Find the missing numerator. $\frac{7}{9} = \frac{?}{27}$

3. Reduce $\frac{18}{27}$ to lowest terms.

4. Is $\frac{66}{66}$ a proper or improper fraction?

Multiply or divide as indicated.

5. $\frac{2}{3} \times \frac{7}{5}$

6. $\frac{5}{5} \times \frac{7}{8}$

7. $\frac{4}{5} \div \frac{7}{8}$

8. $\frac{6}{8} \times \frac{12}{3}$

9. $\frac{3}{16} \div \frac{5}{12}$

10. $\frac{9}{10} \div \frac{15}{8}$

11. $\frac{3}{5} \div \frac{2}{7}$

12. $\frac{5}{9} \times \frac{3}{10}$

13. $\frac{16}{21} \times \frac{3}{4}$

14. $\frac{36}{49} \div \frac{12}{7}$

15. $\frac{12}{9} \div \frac{4}{15}$

16. $\frac{1}{6} \times \frac{5}{3}$

Simplify.

17. $\frac{\frac{6}{8}}{\frac{3}{4}}$

18. $\frac{\frac{36}{75}}{\frac{12}{25}}$

19. $\frac{\frac{8}{10}}{\frac{26}{15}}$

20. $\frac{\frac{3}{10}}{\frac{5}{7}}$

21. $\frac{\frac{6}{8}}{\frac{9}{12}}$

22. $\frac{\frac{9}{20}}{\frac{15}{8}}$

Simplify.

23. $\frac{\frac{7}{10} \times \frac{4}{9}}{\frac{2}{3}}$

24. $\frac{\frac{3}{10} \times \frac{5}{14}}{\frac{3}{7}}$

25. $\frac{\frac{5}{16}}{\frac{7}{8} \times \frac{12}{21}}$

26. $\frac{\frac{10}{15}}{\frac{9}{16} \times \frac{5}{6}}$

27. $\frac{\frac{5}{6} \div \frac{7}{12}}{\frac{3}{8} \times \frac{4}{5}}$

28. $\frac{\frac{3}{16} \times \frac{6}{7}}{\frac{6}{35} \div \frac{8}{15}}$

The answers to review and odd-numbered exercises are in the back of the book.

4–8 APPLIED PROBLEMS

■ Example 1

If a pie is cut into five equal pieces, how much of the pie is $\frac{1}{3}$ of a piece?

Solution The pie was divided into 5 pieces.

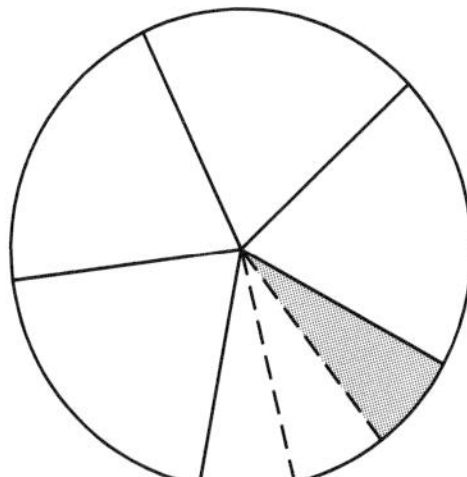

Figure 4–32

If we divide one of these 5 pieces into 3 pieces, how much of the pie is one of these pieces?

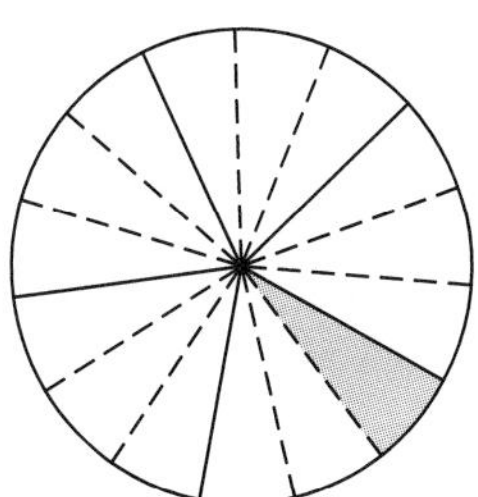

Figure 4–33

To find the answer, we divide each of the other four pieces into thirds. Now the pie is divided into 15 pieces. Thus, $\frac{1}{3}$ of the original piece of pie is $\frac{1}{15}$ of the whole pie.

We can get the answer directly by multiplication.

$$\frac{1}{3} \times \frac{1}{5} = \frac{1}{15} \quad \text{of the whole pie}$$

Answers to exercises 150–153

150. $\frac{15}{3}$ means $15 \div 3$.

151. $\frac{34}{17}$ means $34 \div 17$.

152. $\frac{88}{4}$ means $88 \div 4$.

153. $\frac{105}{5}$ means $105 \div 5$.

Answers to exercises 154–157

154. $\frac{245}{16}$ **155.** $\frac{2}{3}$ **156.** $\frac{36}{35}$

157. $\frac{9}{2}$

■ Problems 4–8

Solve.

1. **Generic.** During one winter quarter at Haysack Technical College, $\frac{3}{5}$ of the student population took an English class. If $\frac{1}{8}$ of those who took an English class were enrolled in English 89, what part of the student body took English 89?

2. **Generic.** What part of an hour is $\frac{2}{3}$ of $\frac{1}{2}$ an hour?

3. **Clerk.** Joan's duty in a pet store is to feed the Cakaw $\frac{1}{16}$ cup of seeds each day. How many Cakaws can be fed if each package of seeds contains $\frac{3}{8}$ cup?

4. **Machinist.** John Higgins, a machinist, was given $\frac{1}{4}$ of a job. At the end of the work day he completed $\frac{2}{3}$ of what he was given. How much of the job did he complete?

5. **Generic.** A kilometer is $\frac{3}{5}$ of a mile. How many kilometers are there in $\frac{1}{2}$ mile?

6. **Painters.** Each door requires $\frac{1}{10}$ gallon of varnish. How many doors can be varnished with $\frac{4}{5}$ gallon of varnish.

7. **Police Science.** In a certain city, $\frac{2}{5}$ of calls made by police officers led to arrests and $\frac{2}{3}$ of all the arrests led to convictions. What fraction of the calls actually led to convictions?

8. **Seamstress.** A seamstress was commissioned by a customer to design and make a bathing suit. She bought $\frac{7}{8}$ of a yard of cloth but used only $\frac{1}{4}$ of what she bought. What fraction of a yard did she use?

9. **Nursery.** Poole's Garden and Nursery's policy for planting beets is as follows: $\frac{1}{8}$ cup of seed for a 40-foot row. How many rows can be planted from $\frac{3}{2}$ cup of seed?

10. **Statistician.** Of the total number of hazardous waste sights, California and Texas have $\frac{16}{180}$ and $\frac{4}{180}$ respectively. How many times more sights does California have than Texas.

    (*Source:* Environmental Protection Agency)

11. **Generic.** Find the area of one of the rectangular sides of the steel bar as shown.

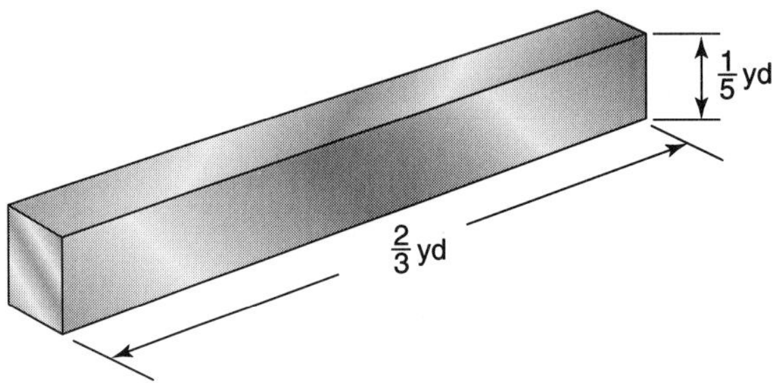

12. **Generic.** Find the volume of the aluminum box shown. The inside measurements are width, $\frac{2}{3}$ ft; length, $\frac{3}{4}$ ft; and height, $\frac{7}{8}$ ft. (Volume is equal to the product of length, width, and height.)

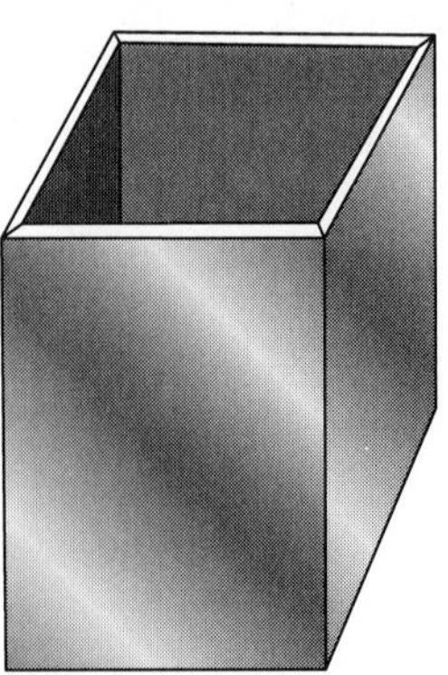

13. **Chef.** A recipe calls for $\frac{3}{4}$ cup of flour. If only a $\frac{1}{8}$-cup measure were available, how many $\frac{1}{8}$ cups would be needed to meet the recipe's requirement?

14. **Chef.** A recipe requires $\frac{7}{8}$ of a cup of chopped onions. A special dish only requires $\frac{1}{2}$ that amount. How many chopped onions should the chef use?

15. **Chef.** Each pasta dish requires $\frac{1}{12}$ pound of ground round. How many pasta dishes can be made from $\frac{1}{2}$ pound of ground round.

16. **Automobile Mechanic.** A metal container when full holds $\frac{7}{8}$ gallon of gasoline. Find the number of gallons if the container is half full of gasoline.

17. **Generic.** The United States and foreign countries have $\frac{3}{5}$ and $\frac{2}{5}$ of the endangered species in the world respectively. How many times more endangered species are there in the United States than in foreign countries?

18. **Generic.** At Jason Technical College $\frac{3}{5}$ of the welding class received a grade of B and $\frac{1}{3}$ of those who received a grade of B were women. What portion of the entire class were women who received a grade of B?

19. **Business.** Two thirds of the cordless telephones sold at a phone center are cellular phones. Of the cellular phones, $\frac{3}{5}$ operate with nickel cadmium batteries. What fraction of the cordless telephones sold used nickel cadmium batteries?

20. **Secretary.** Only $\frac{7}{8}$ of a ream of dual-purpose xerographic paper is left. If a secretary uses $\frac{1}{16}$ of the ream each day, how many days will the rest of the ream last?

The answers to odd-numbered exercises are in the back of the book.

CHAPTER 4 OVERVIEW

Summary

4–1 You learned to write fractional names for a part of a whole, for 1, for 0, for whole numbers, and for a combination of whole numbers and fractions.

4–2 You learned tests for divisibility of a natural number by 2, 3, or 5, and to find the factors of a natural number.

4–3 You learned to factor a whole number greater than 1 into its prime factorization.

4–4 You learned to raise a fraction to higher terms and to reduce a fraction to lowest terms.

4–5 You learned to multiply fractions using canceling.

4–6 You learned to recognize proper and improper fractions, and to divide a fraction by another fraction.

4–7 You learned to simplify complex fractions.

4–8 You learned to solve applied problems involving multiplication and fractions.

Terms To Remember

	Page		***Page***
Canceling	153	Divisibility	141
Complex Fractions	168	Divisor	164
Composite	147	Factors	144
Denominator	135	Numerator	135
Dividend	164	Prime Factorization	148
		Prime Number	147

Rules

- A fraction that has the same numerator and denominator is another name for 1 whole except when the numerator and denominator are 0.
- A fraction that has a numerator of zero and a whole number other than zero for a denominator is another name for 0.
- A fraction having a denominator of 1 is always equal to the numerator.
- A number is divisible by 2 if its ones digit is divisible by 2.
- A number is divisible by 3 if the sum of its digits is divisible by 3.
- A number is divisible by 5 if its ones digit is 0 or 5.
- The factors of a number are 1 and itself, and all the natural numbers between that evenly divide the number.
- All natural numbers that have only two different factors, 1 and the numbers itself, are called *prime numbers*.
- All natural numbers that have more than two different factors are called *composite numbers*.
- The first six prime numbers are 2, 3, 5, 7, 11, and 13.
- To find the prime factorization of a number

    1. Start with the smallest prime number that will divide the number.

    2. Divide each resulting quotient by a prime number.

    3. Stop when a resulting quotient is a prime number.

4. The factorization is the product of all the prime number divisors and the last prime number quotient.

- We can multiply the numerator and denominator of a fraction by any whole number greater than 0 to obtain another fraction that is equal to the first. The resulting fraction is in higher terms.
- To multiply fractional numbers, we multiply the numerators together and multiply the denominators together.
- A fractional number that has the numerator equal to or larger than the denominator is called an *improper fraction.*
- A fractional number that has the numerator smaller than the denominator is a *proper fraction.*
- Zero does not have a reciprocal!
- To divide two fractions, multiply the first fraction by the reciprocal of the second fraction.

Self-Test

The answers are in the back of the book.

4–1 **1.** Name a fraction that the shaded part of the object suggests.

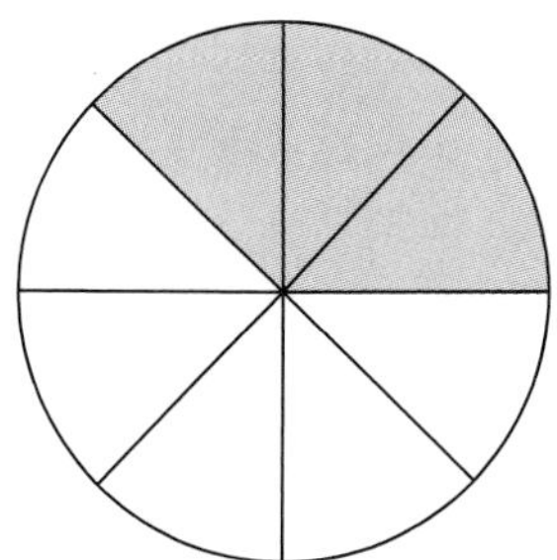

2. Draw a number line and locate the point $\frac{2}{7}$.

3. Name the numerator of $\frac{16}{21}$.

4. Simplify $\frac{562}{1}$.

5. Name the point on the number line.

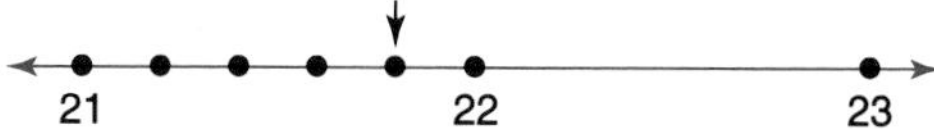

4–2 **6.** Is 114 divisible by 3 or by 5?

7. List the factors of 33.

4–3 **8.** Is 83 prime or composite?

9. State the prime factorization of 36.

4–4 **10.** Reduce $\frac{45}{60}$ to lowest terms.

4–5 **11.** Multiply. Use canceling when possible.

$$\frac{13}{51} \times \frac{17}{39}$$

4–6 **12.** Divide and simplify.

$$\frac{8}{9} \div \frac{4}{5}$$

4–7 **13.** Simplify.

$$\frac{\frac{6}{7}}{\frac{18}{21}}$$

4–8 **14.** A shop spent $\frac{1}{2}$ of its profits on rent, maintenance, and supplies. The owner pays $\frac{3}{7}$ of this amount for rent. What fraction of the profits goes for rent?

15. Suppose to make one complete revolution the distance a nut moves on a pulley is $\frac{2}{5}$ inch. How many turns are needed for that nut to move $\frac{3}{5}$ inch?

Chapter Test

Directions: This test will aid in your preparation for a possible chapter test given by your instructor. The answers are in the back of the book. Go back and review in the appropriate section(s) if you missed any test items.

4–1 **1.** Name a fraction that the shaded part of the object suggests.

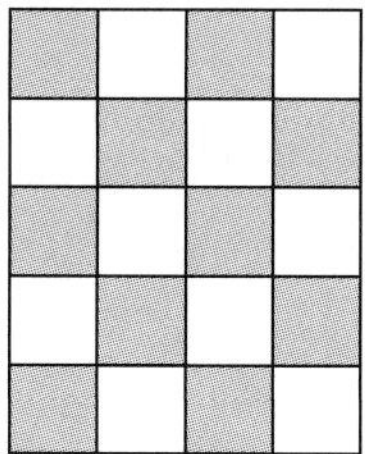

2. Shade in a portion of an object that is named by the fraction $\frac{6}{7}$.

3. Draw a number line and locate the point $\frac{3}{5}$.

4. Write a fractional number that tells what portion of the group are triangles.

5. Name the denominator of $\frac{15}{37}$.

Simplify.

6. $\dfrac{32}{32}$ **7.** $\dfrac{0}{18}$ **8.** $\dfrac{81}{0}$

Simplify.

9. $\frac{1}{1}$ **10.** $\frac{682}{1}$

11. Write a fractional number for the shaded part.

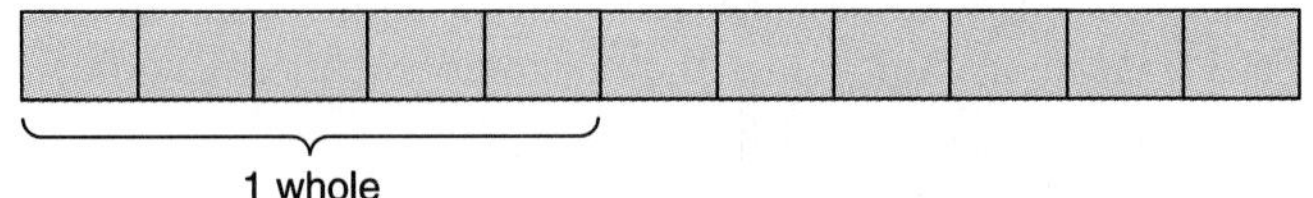

12. Find the number $29\frac{4}{5}$ on a number line.

13. Name the point on the number line.

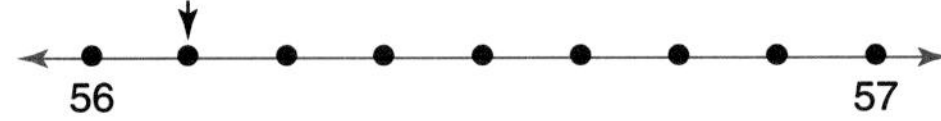

4–2 State whether the following numbers are divisible by 2, 3, or 5.

14. 50 **15.** 8811

16. Find the factors of 52.

4–3 Tell whether the following numbers are prime, composite, or neither.

17. 1 **18.** 53 **19.** 42

20. Find the prime factorization of 56.

4–4 **21.** Find the missing numerator: $\frac{2}{3} = \frac{?}{36}$.

22. Reduce $\frac{32}{48}$ to lowest terms.

4–5 Multiply. Use canceling where possible.

23. $\frac{2}{3} \times \frac{4}{5}$

24. $\frac{16}{30} \times \frac{18}{36}$

25. $\frac{11}{15} \times \frac{2}{7} \times \frac{5}{22}$

4–6 State whether the following fractions are proper or improper.

26. $\frac{5}{8}$

27. $\frac{0}{23}$

28. $\frac{15}{15}$

29. $\frac{29}{28}$

State the reciprocal of the following fractions.

30. $\frac{6}{8}$

31. $\frac{0}{18}$

32. 206

33. Divide and simplify: $\frac{6}{13} \div \frac{9}{11}$.

4–7 34. Divide and simplify: $\frac{\frac{8}{21}}{\frac{12}{7}}$.

4–8 35. The actual thickness of 1-inch lumber is $\frac{25}{32}$ inch. What would the thickness be if $\frac{3}{5}$ of that amount is shaved?

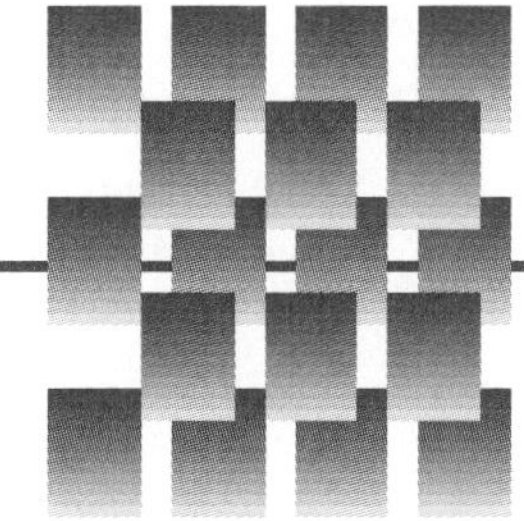

CHAPTER 5

Addition and Subtraction of Fractions

Choice: Either skip the Pretest and study all of Chapter 5 or take the Pretest and study only the sections you need to study.

Pretest

Directions: This Pretest will help you determine which sections to study in Chapter 5. If any of your answers are incorrect, or if you omitted any exercise, turn to the indicated section(s) and study the material; then take the Chapter Test.

5–1 **1.** Add and reduce: $\frac{4}{14} + \frac{7}{14}$.

5–2 **2.** Find the lowest common denominator of these two fractions: $\frac{3}{8}$ and $\frac{5}{6}$.

5–3 Add.

3. $\frac{7}{25} + \frac{3}{5}$ **4.** $\frac{1}{4} + \frac{5}{18}$

5–4 Subtract.

5. $\frac{5}{18} - \frac{1}{6}$ **6.** $\frac{5}{8} - \frac{7}{12}$

5–5 **7.** Decide which fraction is larger: $\frac{7}{9}$ or $\frac{5}{6}$?

8. A certain recipe calls for $\frac{3}{8}$ cup of milk and $\frac{1}{4}$ cup of syrup. How many cups would it make if the milk and syrup were mixed together?

9. The outside diameter of a pipe is $\frac{7}{8}$ inch, and the inside diameter is $\frac{3}{4}$ inch. How thick are the walls of the pipe?

The answers are in the margin on page 184.

5–1 ADDING FRACTIONS WITH LIKE DENOMINATORS

Adding fractions is something like adding filters.

2 filters + 3 filters + 1 filter = 6 filters

For the answer to make sense, the things we are adding must all be the same.

■ **Example 1**

Add $\frac{1}{5} + \frac{2}{5}$.

Solution We can say "one fifth + two fifths = three fifths." The result is fifths, just as the result of adding filters is filters. Thus,

$$\frac{1}{5} + \frac{2}{5} = \frac{3}{5}$$

■ **Example 2**

Add $\frac{2}{6}$ and $\frac{5}{6}$ using shaded objects.

Solution

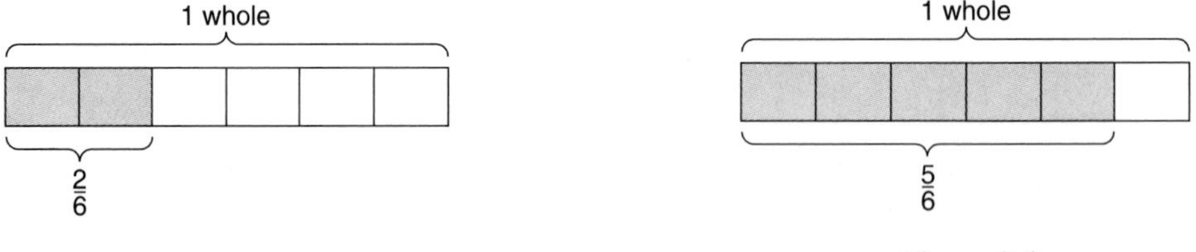

Figure 5–1 **Figure 5–2**

Each piece is $\frac{1}{6}$ of the whole, and together there are 7 shaded pieces, for a total of $\frac{7}{6}$.

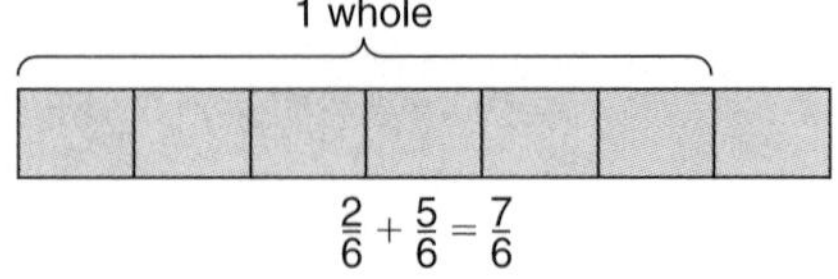

Figure 5–3

■ Example 3

Add $\frac{2}{4}$ and $\frac{3}{4}$ using a number line.

Solution **Step 1** We first divide the number line into fourths.

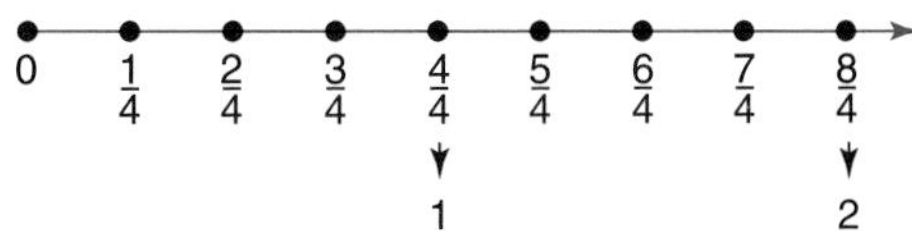

Figure 5–4

Step 2 We start at 0 and make 2 jumps of $\frac{1}{4}$ each to $\frac{2}{4}$.

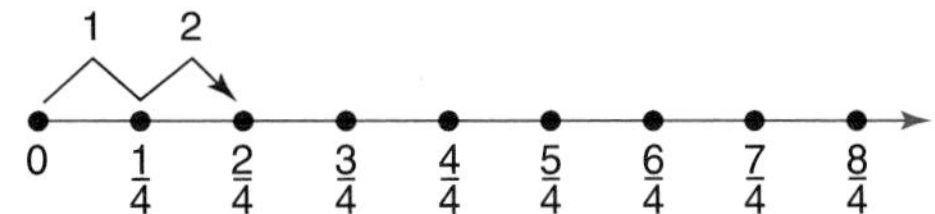

Figure 5–5

Step 3 From $\frac{2}{4}$ we make 3 jumps of $\frac{1}{4}$ each to $\frac{5}{4}$.

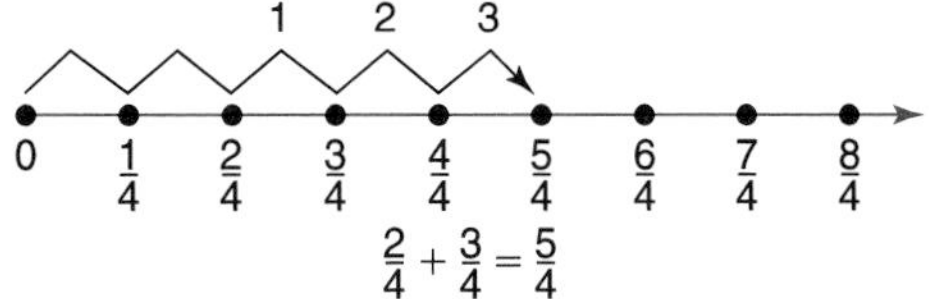

Figure 5–6

The results of Examples 1, 2, and 3 can be stated as follows:

$$\frac{1}{5}+\frac{2}{5}=\frac{3}{5}, \quad \frac{2}{6}+\frac{5}{6}=\frac{7}{6}, \quad \text{and} \quad \frac{2}{4}+\frac{3}{4}=\frac{5}{4}$$

Notice:

1. In each example the fractions being added have the same denominator.

2. In all three cases, to find the sum we added the numerators and kept the same denominator.

All fractions having the same denominator are called *like fractions.*

Rule

When adding like fractions, all we do is add the numerators and keep the common denominator.

■ Example 4

Add $\frac{2}{7}+\frac{1}{7}$.

Answers to Pretest

Circle the answers you missed or omitted. Study the corresponding section(s), then take the Chapter Test.

1. $\frac{6}{7}$ **2.** 24 **3.** $\frac{22}{25}$ **4.** $\frac{19}{36}$ **5.** $\frac{1}{9}$

6. $\frac{1}{24}$ **7.** $\frac{5}{6}$ **8.** $\frac{5}{8}$ cup **9.** $\frac{1}{16}$ in.

Add.

1. $\frac{3}{8}+\frac{4}{8}$ **2.** $\frac{8}{12}+\frac{3}{12}$

The answers are in the margin on page 186.

Add.

3. $\frac{2}{5}+\frac{2}{5}$ **4.** $\frac{6}{25}+\frac{9}{25}$

5. $\frac{3}{4}+\frac{2}{4}$ **6.** $\frac{3}{7}+\frac{4}{7}$

The answers are in the margin on page 186.

Add.

7. $\frac{1}{3}+\frac{3}{3}+\frac{4}{3}$ **8.** $\frac{3}{15}+\frac{5}{15}+\frac{1}{15}$

The answers are in the margin on page 186.

Solution Add numerators

$$\frac{2}{7}+\frac{1}{7}=\frac{2+1}{7}=\frac{3}{7}$$

Keep common denominator.

Practice check: Do exercises 1 and 2 in the margin.

After we find the sum, we must reduce the fraction to its simplest form whenever possible.

■ Example 5

Add $\frac{3}{8}+\frac{1}{8}$

Solution $\frac{3}{8}+\frac{1}{8}=\frac{4}{8}$ ← This must be reduced.

$$\frac{4}{8}=\frac{\overset{1}{\cancel{\overset{2}{\cancel{4}}}}}{\underset{2}{\cancel{\underset{4}{\cancel{8}}}}}=\frac{1}{2}$$

$$\frac{3}{8}+\frac{1}{8}=\frac{4}{8}=\frac{1}{2}$$

Practice check: Do exercises 3–6 in the margin.

When adding more than two fractions the same methods are used.

■ Example 6

Add $\frac{7}{8}+\frac{1}{8}+\frac{3}{8}$.

Solution In this case we still find the sum of all the numerators and keep the common denominator.

$$\frac{7}{8}+\frac{1}{8}+\frac{3}{8}=\frac{7+1+3}{8}=\frac{11}{8}$$

Practice check: Do exercises 7 and 8 in the margin.

■ Looking Back

You should now be able to add like fractions and reduce the sum when neccessary.

■ Problems 5–1

Review

Multiply or divide as indicated. Use canceling.

1. $\frac{7}{8} \times \frac{2}{3}$ **2.** $\frac{11}{12} \div \frac{5}{6}$ **3.** $\frac{3}{5} \div \frac{9}{10}$ **4.** $\frac{2}{9} \times \frac{6}{25} \times \frac{15}{18}$

Add and reduce when necessary.

5. $\frac{3}{6} + \frac{2}{6}$ **6.** $\frac{1}{6} + \frac{3}{6}$ **7.** $\frac{3}{4} + \frac{1}{4}$ **8.** $\frac{3}{7} + \frac{2}{7}$

9. $\frac{5}{8} + \frac{2}{8}$ **10.** $\frac{5}{8} + \frac{3}{8}$ **11.** $\frac{3}{10} + \frac{9}{10}$ **12.** $\frac{1}{15} + \frac{9}{15}$

13. $\frac{3}{8} + \frac{1}{8}$ **14.** $\frac{4}{27} + \frac{5}{27}$ **15.** $\frac{5}{10} + \frac{9}{10}$ **16.** $\frac{5}{17} + \frac{6}{17}$

17. $\frac{11}{20} + \frac{9}{20}$ **18.** $\frac{11}{18} + \frac{5}{18}$ **19.** $\frac{1}{2} + \frac{3}{2} + \frac{9}{2}$ **20.** $\frac{1}{2} + \frac{3}{2} + \frac{5}{2}$

21. $\frac{7}{16} + \frac{2}{16} + \frac{5}{16}$ **22.** $\frac{1}{12} + \frac{5}{12} + \frac{9}{12}$ **23.** $\frac{72}{91} + \frac{7}{91}$ **24.** $\frac{18}{105} + \frac{22}{105}$

25. $\frac{35}{80} + \frac{27}{80} + \frac{16}{80}$ **26.** $\frac{29}{45} + \frac{16}{45} + \frac{3}{45}$

The answers to review and odd-numbered exercises are in the back of the book.

5–2 FINDING THE LOWEST COMMON DENOMINATOR (LCD)

Before we begin this section, go back to sections 4–2, 4–3, and 4–4. Take an hour or two to review as many as you need of the practice exercises in these sections.

Answers to exercises 1 and 2

1. $\frac{7}{8}$ 2. $\frac{11}{12}$

Answers to exercises 3–6

3. $\frac{4}{5}$ 4. $\frac{15}{25}$ or $\frac{3}{5}$ 5. $\frac{5}{4}$ 6. $\frac{7}{7}$ or 1

Answers to exercises 7 and 8

7. $\frac{7}{3}$ 8. $\frac{9}{15}$ or $\frac{3}{5}$

Consider the sum $\frac{2}{3} + \frac{5}{8}$. Since one fraction is in thirds and the other in eighths, we cannot add these fractions in their present form. We must make both fractions have the same denominator. We can apply what we have learned in sections 4–2 to 4–4.

For instance, we could change $\frac{2}{3}$ to $\frac{16}{24}$ and $\frac{5}{8}$ to $\frac{15}{24}$. Now both fractions have the same denominator (24) and can be added. The common denominator used will be the smallest number that is divisible by each of the denominators. We call this number the *lowest common denominator* or the *LCD.*

■ Examples

1. The LCD of $\frac{1}{6}$ and $\frac{5}{12}$ is 12, since 12 is divisible by both 6 and 12, and 12 is the smallest such number.
2. The LCD of $\frac{2}{4}$ and $\frac{5}{18}$ is 36, since 36 is divisible by both 4 and 18, and 36 is the smallest such number.

There are a number of ways to find the LCD, but we shall examine only one.

■ Example 3

Find the LCD for $\frac{1}{6}$ and $\frac{5}{16}$.

Solution **Step 1** Write the denominators in a horizontal line.

$$\underline{)6 \quad 16}$$

Step 2 Divide by the smallest prime number that will evenly divide at least two of the numbers.

$$\begin{array}{r|ll} 2 & 6 & 16 \\ \hline & 3 & 8 \end{array}$$

If we cannot find another prime number that will divide at least two of the numbers, we are done. In this case we are done, because no prime number will divide both 3 and 8.

Step 3 Find the LCD.

$$\begin{array}{r|ll} 2 & 6 & 16 \\ \hline & 3 & 8 \end{array}$$

Multiply together all of the outside numbers. The LCD is $2 \times 3 \times 8 = 48$.

■ Example 4

Find the LCD for $\frac{2}{3}$, $\frac{3}{8}$, and $\frac{5}{6}$.

Solution **Step 1** Write the denominators. $\underline{)3 \quad 8 \quad 6}$

Step 2 Divide by the smallest prime number that will divide at least two of the numbers.

$$\begin{array}{r|lll} 2 & 3 & 8 & 6 \\ \hline & 3 & 4 & 3 \end{array}$$

Bring down any number that cannot be divided evenly by 2.

Step 3 Repeat Step 2. Divide by 3 this time.

```
2)3  8  6
3)3  4  3
  1  4  1
     ↖ Bring down the 4.
```

We are done, because no prime number will divide at least two of the bottom numbers.

The LCD is $2 \times 3 \times 1 \times 4 \times 1 = 24$.

Practice check: Do exercises 9–13 in the margin.

To find the LCD for more than two fractions, we can use the same process.

Find the LCD of each group of fractions.

9. $\frac{11}{12}$ and $\frac{3}{4}$ **10.** $\frac{17}{20}$ and $\frac{4}{5}$

11. $\frac{5}{6}$ and $\frac{17}{18}$ **12.** $\frac{3}{10}, \frac{1}{6},$ and $\frac{4}{14}$

13. $\frac{2}{6}, \frac{4}{9},$ and $\frac{2}{3}$

The answers are in the margin on page 188.

■ Example 5

Find the LCD of $\frac{4}{5}$, $\frac{1}{4}$, and $\frac{2}{3}$.

Solution)5 4 3

There is no prime number that will divide at least two of the three numbers, so the LCD is

$$5 \times 4 \times 3 = 60.$$

■ Example 6

Find the LCD of $\frac{1}{6}$, $\frac{13}{18}$, $\frac{4}{9}$, and $\frac{3}{4}$.

Solution For each step we must find a prime number that will divide at least two of the numbers.

Step 1 Divide each by 2.

```
2)6  18  9  4
  3   9  9  2  ← Bring down.
```

Step 2

```
2)6  18  9  4    We divided 3, 9 and 9 by 3
3)3   9  9  2    and brought down the 2.
  1   3  3  2
```

Step 3

```
2)6  18  9  4
3)3   9  9  2    We divided both 3s by 3
3)1   3  3  2    and brought down the 1
  1   1  1  2    and 2.
```

We are done!

$$\text{LCD} = 2 \times 3 \times 3 \times 1 \times 1 \times 1 \times 2 = 36$$

Find the LCD of the following fractions without looking at the steps above:

a. $\frac{4}{5}, \frac{1}{4},$ and $\frac{2}{3}$ **b.** $\frac{1}{6}, \frac{13}{18}, \frac{4}{9},$ and $\frac{3}{4}$.

Answers to exercises 9–13

9. 12 **10.** 20 **11.** 18 **12.** 210 **13.** 18

Here is a summary of the steps of a method for finding the LCD.

Rule

How to find the LCD:

Step 1 Write the denominators in a horizontal line.

Step 2 Divide by the smallest prime number that will evenly divide at least two of the numbers. Bring down any number that cannot be divided evenly.

Step 3 Repeat Steps 1 and 2 until no other prime number will evenly divide at least two of the numbers.

Step 4 The LCD is the product of all the outside numbers.

Practice check: Do exercises 14–17 in the margin.

Find the LCD of each group of fractions.

14. $\frac{1}{6}$ and $\frac{4}{7}$ **15.** $\frac{7}{9}$ and $\frac{1}{4}$

16. $\frac{7}{8}, \frac{5}{6}, \frac{3}{4}$

17. $\frac{4}{5}, \frac{3}{8}, \frac{7}{10}, \frac{1}{4}, \frac{7}{20}$

The answers are in the margin on page 191.

Forming Equal Fractions

Before reading further, review section 4–4, and redo any of exercises 5–16 in Problems 4–4 that you need for your review.

■ Example 7

Find the equal fractions having the LCD as their denominators. Do not find the sum.

$$\frac{1}{6} \text{ and } \frac{3}{8}$$

Solution **Step 1** Find the LCD.

$$\begin{array}{r|ll} 2 & 6 & 8 \\ \hline & 3 & 4 \end{array} \qquad \text{LCD} = 2 \times 3 \times 4 = 24$$

Step 2 Form equal fractions with 24 as the LCD. Do each fraction one at a time.

a. $\frac{1}{6} = \frac{?}{24}$

$$\frac{1}{6} = \frac{1}{6} \times \frac{4}{4} = \frac{4}{24}$$

To get 24 in the denominator, we multiplied $\frac{1}{6}$ by $\frac{4}{4}$.

b. $\frac{3}{8} = \frac{?}{24}$

$$\frac{3}{8} = \frac{3}{8} \times \frac{3}{3} = \frac{9}{24}$$

We multiplied $\frac{3}{8}$ by $\frac{3}{3}$ to get 24 in the denominator.

The fraction equal to $\frac{1}{6}$ is $\frac{4}{24}$. The fraction equal to $\frac{3}{8}$ is $\frac{9}{24}$.

Practice check: Do exercises 18–21 in the margin.

Find equal fractions having the LCD as their denominators.

18. $\frac{11}{12}$ and $\frac{9}{10}$ **19.** $\frac{3}{6}$ and $\frac{7}{18}$

20. $\frac{3}{4}$ and $\frac{3}{10}$ **21.** $\frac{11}{16}$ and $\frac{5}{12}$

The answers are in the margin on page 191.

■ Example 8

Find equal fractions having the LCD as their denominators.

$$\frac{3}{12}, \quad \frac{5}{6}, \quad \text{and} \quad \frac{4}{9}$$

Solution **Step 1** Find the LCD.

$$\begin{array}{r|rrr} 2 & 12 & 6 & 9 \\ 3 & 6 & 3 & 9 \\ \hline & 2 & 1 & 3 \end{array}$$

$$\text{LCD} = 2 \times 3 \times 2 \times 1 \times 3 = 36$$

Step 2 Form equal fractions with 36 as the LCD.

a. $\frac{3}{12} = \frac{?}{36}$ **b.** $\frac{5}{6} = \frac{?}{36}$ **c.** $\frac{4}{9} = \frac{?}{36}$

$\frac{3}{12} \times \frac{3}{3} = \frac{9}{36}$ $\frac{5}{6} \times \frac{6}{6} = \frac{30}{36}$ $\frac{4}{9} \times \frac{4}{4} = \frac{16}{36}$

The equal fractions are $\frac{9}{36}$, $\frac{30}{36}$, and $\frac{16}{36}$.

Practice check: Do exercises 22–26 in the margin.

Find equal fractions having the LCD as their denominators.

22. $\frac{2}{3}$ and $\frac{3}{5}$ **23.** $\frac{1}{6}$ and $\frac{3}{7}$

24. $\frac{3}{14}$ and $\frac{3}{4}$ **25.** $\frac{5}{8}, \frac{1}{6}$, and $\frac{3}{4}$

26. $\frac{2}{5}, \frac{3}{16}$, and $\frac{7}{20}$

The answers are in the margin on page 191.

■ Looking Back

You should now be able to find the LCD of two or more fractions and form equal fractions having that LCD as the denominators.

■ Problems 5–2

Review

Solve. Use canceling.

1. $\frac{\frac{5}{8}}{\frac{3}{4}}$ **2.** $\frac{7}{10} \times \frac{5}{14}$ **3.** $\frac{9}{10} \div \frac{1}{3}$

Add and reduce when necessary.

4. $\frac{3}{8} + \frac{4}{8}$ **5.** $\frac{2}{10} + \frac{3}{10}$ **6.** $\frac{3}{5} + \frac{2}{5}$ **7.** $\frac{2}{25} + \frac{8}{25}$ **8.** $\frac{6}{13} + \frac{5}{13}$

Just find the LCD of each of the groups of fractions.

9. $\frac{4}{9}$ and $\frac{1}{6}$

10. $\frac{2}{3}$ and $\frac{1}{6}$

11. $\frac{3}{5}$ and $\frac{17}{20}$

12. $\frac{5}{8}$ and $\frac{7}{6}$

13. $\frac{7}{8}$ and $\frac{5}{24}$

14. $\frac{7}{9}$ and $\frac{3}{4}$

15. $\frac{1}{6}$ and $\frac{3}{5}$

16. $\frac{2}{9}$ and $\frac{7}{15}$

17. $\frac{8}{9}$ and $\frac{5}{4}$

18. $\frac{2}{15}$ and $\frac{11}{12}$

19. $\frac{5}{14}$ and $\frac{7}{20}$

20. $\frac{7}{35}$ and $\frac{13}{21}$

21. $\frac{5}{8}$, $\frac{3}{10}$, and $\frac{11}{12}$

22. $\frac{7}{10}$, $\frac{1}{2}$, and $\frac{2}{5}$

23. $\frac{7}{16}$, $\frac{5}{8}$, and $\frac{11}{24}$

24. $\frac{3}{12}$, $\frac{7}{16}$, and $\frac{5}{21}$

25. $\frac{13}{12}$ and $\frac{29}{32}$

26. $\frac{7}{65}$ and $\frac{5}{39}$

27. $\frac{3}{42}$, $\frac{18}{50}$, and $\frac{21}{63}$

28. $\frac{23}{50}$, $\frac{31}{75}$, and $\frac{18}{25}$

Find equal fractions having the LCD as their denominators. Do not find the sum.

29. $\frac{4}{9}$ and $\frac{1}{6}$

30. $\frac{5}{8}$ and $\frac{3}{4}$

31. $\frac{2}{3}$ and $\frac{1}{4}$

32. $\frac{4}{7}$ and $\frac{3}{5}$

33. $\frac{4}{15}$ and $\frac{5}{12}$

34. $\frac{11}{12}$ and $\frac{13}{18}$

35. $\frac{2}{5}$ and $\frac{3}{4}$

36. $\frac{7}{12}$ and $\frac{5}{9}$

37. $\frac{7}{21}$ and $\frac{8}{14}$

38. $\frac{11}{15}$ and $\frac{17}{25}$

39. $\frac{13}{16}$ and $\frac{5}{18}$

40. $\frac{5}{14}$ and $\frac{7}{6}$

41. $\frac{5}{8}$, $\frac{3}{6}$, and $\frac{7}{4}$

42. $\frac{3}{5}$, $\frac{1}{6}$, and $\frac{7}{9}$

43. $\frac{3}{12}$, $\frac{5}{16}$, and $\frac{9}{20}$

44. $\frac{1}{5}$, $\frac{7}{10}$, and $\frac{23}{6}$

45. $\frac{29}{54}$ and $\frac{13}{48}$ **46.** $\frac{123}{240}$ and $\frac{53}{270}$ **47.** $\frac{15}{96}, \frac{13}{64}$, and $\frac{39}{80}$ **48.** $\frac{5}{11}, \frac{19}{30}$, and $\frac{27}{33}$

The answers to review and odd-numbered exercises are in the back of the book.

Answers to exercises 14–17

14. 42 **15.** 36 **16.** 24 **17.** 40

Answers to exercises 18–21

18. LCD = 60; equal fractions are $\frac{55}{60}$ and $\frac{54}{60}$.

19. LCD = 18; equal fractions are $\frac{9}{18}$ and $\frac{7}{18}$.

20. LCD = 20; equal fractions are $\frac{15}{20}$ and $\frac{6}{20}$.

21. LCD = 48; equal fractions are $\frac{33}{48}$ and $\frac{20}{48}$.

Answers to exercises 22–26

22. $\frac{10}{15}$ and $\frac{9}{15}$ **23.** $\frac{7}{42}$ and $\frac{18}{42}$

24. $\frac{6}{28}$ and $\frac{21}{28}$ **25.** $\frac{15}{24}, \frac{4}{24}$, and $\frac{18}{24}$

26. $\frac{32}{80}, \frac{15}{80}$, and $\frac{28}{80}$

5–3 ADDING FRACTIONS WITH UNLIKE DENOMINATORS

Rule

To add fractions with unlike denominators, we must

Step 1 Find the LCD.

Step 2 Convert the unlike fractions to like fractions using the LCD.

Step 3 Add the like fractions.

■ Example 1

Add $\frac{2}{5} + \frac{3}{7}$.

Solution **Step 1** Find the LCD. There is no one prime number that will divide both 5 and 7, so the LCD = 5 × 7 or 35.

Step 2 Form equal fractions.

$$\frac{2}{5} = \frac{2}{5} \times \frac{7}{7} = \frac{14}{35} \qquad \frac{3}{7} = \frac{3}{7} \times \frac{5}{5} = \frac{15}{35}$$

Step 3 Add the like fractions.

$$\frac{14}{35} + \frac{15}{35} = \frac{29}{35}$$

Practice check: Do exercises 27–30 in the margin.

Add.

27. $\frac{1}{3} + \frac{3}{4}$ **28.** $\frac{3}{8} + \frac{2}{7}$

29. $\frac{3}{5} + \frac{1}{4}$ **30.** $\frac{7}{9} + \frac{1}{2}$

The answers are in the margin on page 193.

■ Example 2

Add $\frac{5}{7} + \frac{9}{14}$.

Solution **Step 1** Find the LCD.

$$\begin{array}{r|rr} 7 & 7 & 14 \\ \hline & 1 & 2 \end{array} \qquad \text{LCD} = 7 \times 1 \times 2 = 14$$

Step 2 Form equal fractions.

$$\frac{5}{7} = \frac{5}{7} \times \frac{2}{2} = \frac{10}{14} \qquad \frac{9}{14} = \frac{9}{14} = \frac{9}{14}$$

Step 3 Add the like fractions.

$$\frac{10}{14} + \frac{9}{14} = \frac{19}{14}$$

Example 3

Add $\frac{7}{9} + \frac{5}{12}$.

Solution **Step 1** Find the LCD.

$$3\overline{)9 \quad 12} \qquad \text{LCD} = 3 \times 3 \times 4 = 36$$
$$3 \quad 4$$

Step 2 Form equal fractions.

$$\frac{7}{9} = \frac{7}{9} \times \frac{4}{4} = \frac{28}{36} \qquad \frac{5}{12} = \frac{5}{12} \times \frac{3}{3} = \frac{15}{36}$$

Step 3 Add the like fractions.

$$\frac{28}{36} + \frac{15}{36} = \frac{43}{36}$$

Practice check: Do exercises 31–35 in the margin.

Add.

31. $\frac{1}{2} + \frac{5}{10}$ 32. $\frac{15}{16} + \frac{5}{8}$

33. $\frac{2}{9} + \frac{5}{6}$ 34. $\frac{3}{18} + \frac{1}{12}$

35. $\frac{9}{16} + \frac{7}{24}$

The answers are in the margin on page 194.

Example 4

Add $\frac{2}{3} + \frac{1}{16} + \frac{6}{7}$.

Solution **Step 1** Find the LCD.

$$3\overline{)3 \quad 6 \quad 7} \qquad \text{LCD} = 3 \times 1 \times 2 \times 7 = 42$$
$$1 \quad 2 \quad 7$$

Step 2 Form equal fractions.

$$\frac{2}{3} = \frac{2}{3} \times \frac{14}{14} = \frac{28}{42} \qquad \frac{1}{6} = \frac{1}{6} \times \frac{7}{7} = \frac{7}{42} \qquad \frac{6}{7} = \frac{6}{7} \times \frac{6}{6} = \frac{36}{42}$$

Step 3 Add the like fractions.

$$\frac{28}{42} + \frac{7}{42} + \frac{36}{42} = \frac{28 + 7 + 36}{42} = \frac{71}{42}$$

Practice check: Do exercises 36–39 in the margin.

Add.

36. $\frac{4}{7} + \frac{1}{2} + \frac{2}{3}$ 37. $\frac{7}{8} + \frac{3}{4} + \frac{1}{2}$

38. $\frac{7}{10} + \frac{3}{5} + \frac{5}{4}$ 39. $\frac{5}{24} + \frac{6}{8} + \frac{1}{3}$

The answers are in the margin on page 194.

Looking Back

You should now be able to add unlike fractions using the LCD.

■ Problems 5–3

Perform the indicated operations.

Answers to exercises 27–30

27. $\frac{13}{12}$ 28. $\frac{37}{56}$ 29. $\frac{17}{20}$ 30. $\frac{23}{18}$

Rule

We do not need to find the LCD when multiplying or dividing fractions.

Review

1. $\frac{7}{15} + \frac{3}{15}$
2. $\frac{3}{4} \times \frac{2}{7}$
3. $\frac{3}{10} \div \frac{5}{8}$
4. $\frac{3}{8} + \frac{5}{8}$
5. $\frac{3}{4} \times \frac{1}{7} \times \frac{28}{33}$
6. $\frac{24}{45} \div \frac{10}{9}$

Add.

7. $\frac{2}{3} + \frac{1}{4}$
8. $\frac{3}{5} + \frac{1}{4}$
9. $\frac{5}{10} + \frac{3}{5}$
10. $\frac{3}{8} + \frac{3}{4}$
11. $\frac{2}{5} + \frac{5}{6}$
12. $\frac{7}{9} + \frac{4}{7}$
13. $\frac{1}{8} + \frac{5}{6}$
14. $\frac{5}{9} + \frac{3}{6}$
15. $\frac{7}{12} + \frac{1}{6}$
16. $\frac{3}{7} + \frac{9}{14}$
17. $\frac{3}{8} + \frac{2}{5}$
18. $\frac{4}{7} + \frac{3}{5}$
19. $\frac{13}{16} + \frac{11}{10}$
20. $\frac{3}{10} + \frac{7}{12}$
21. $\frac{15}{32} + \frac{1}{4}$
22. $\frac{7}{6} + \frac{19}{24}$
23. $\frac{1}{4} + \frac{5}{6} + \frac{1}{3}$
24. $\frac{3}{5} + \frac{1}{3} + \frac{9}{10}$
25. $\frac{3}{8} + \frac{1}{4} + \frac{5}{16}$
26. $\frac{5}{12} + \frac{3}{8} + \frac{1}{4}$
27. $\frac{15}{24} + \frac{7}{16}$
28. $\frac{5}{18} + \frac{11}{16}$
29. $\frac{5}{16} + \frac{7}{12} + \frac{15}{24}$
30. $\frac{7}{18} + \frac{13}{36} + \frac{5}{24}$

The answers to review and odd-numbered exercises are in the back of the book.

Answers to exercises 31–35

31. 1 **32.** $\frac{25}{16}$ **33.** $\frac{19}{18}$ **34.** $\frac{1}{4}$

35. $\frac{41}{48}$

Answers to exercises 36–39

36. $\frac{73}{42}$ **37.** $\frac{17}{8}$ **38.** $\frac{51}{20}$ **39.** $\frac{31}{24}$

5–4 SUBTRACTION OF FRACTIONS

Subtraction of Like Fractions

■ **Example 1**

Show $\frac{3}{4} - \frac{1}{4}$ using a shaded object.

Solution

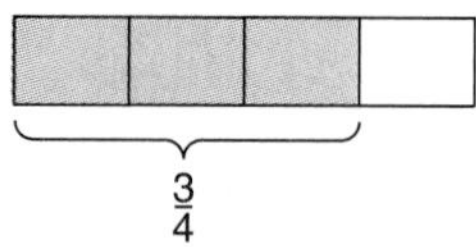

Figure 5–7

Each shaded piece is $\frac{1}{4}$ of the whole, and the problem tells us to take one of these pieces (or $\frac{1}{4}$) away from the $\frac{3}{4}$.

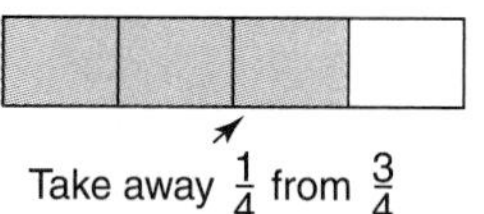

Figure 5–8

This is what we have left:

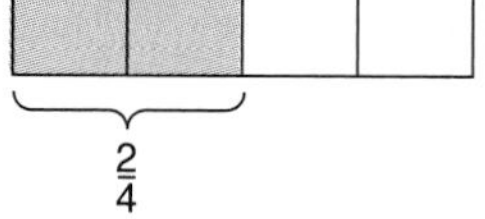

Figure 5–9

We have $\frac{2}{4}$ left. (Notice that $\frac{2}{4}$ is the same as $\frac{1}{2}$.)

■ **Example 2**

Show $\frac{3}{4} - \frac{1}{4}$ using a number line.

Solution

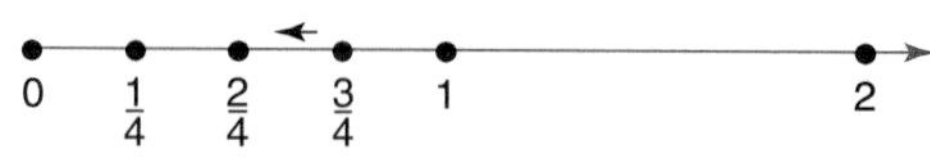

Figure 5–10

We begin at $\frac{3}{4}$ and draw an arrow $\frac{1}{4}$ of a unit to the left. The difference is $\frac{2}{4}$ or $\frac{1}{2}$.

Notice that, if we subtract the numerators and keep the common denominator, we get the same results.

$$\frac{3}{4} - \frac{1}{4} = \frac{3-1}{4} = \frac{2}{4} \text{ or } \frac{1}{2}$$

Rule

To find the difference between like fractions we subtract the numerators and keep the common denominator.

■ Example 3

Subtract $\frac{17}{35} - \frac{7}{35}$.

Solution

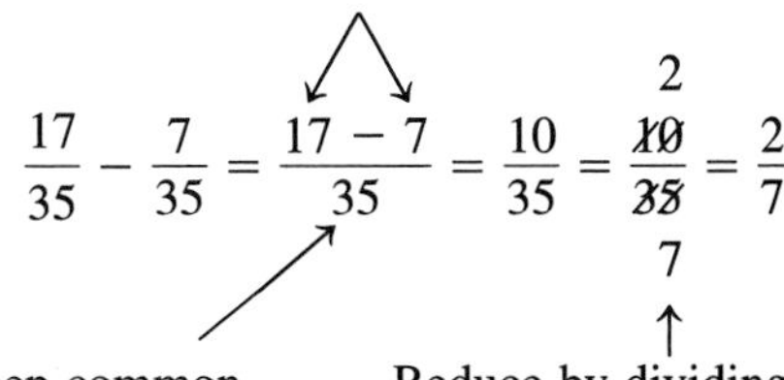

$$\frac{17}{35} - \frac{7}{35} = \frac{17-7}{35} = \frac{10}{35} = \frac{\cancel{10}^{2}}{\cancel{35}_{7}} = \frac{2}{7}$$

Practice check: Do exercises 40–45 in the margin.

Subtract and reduce when possible.

40. $\frac{5}{8} - \frac{1}{8}$ 41. $\frac{9}{17} - \frac{3}{17}$

42. $\frac{19}{20} - \frac{9}{20}$ 43. $\frac{9}{16} - \frac{3}{16}$

44. $\frac{28}{32} - \frac{17}{32}$ 45. $\frac{55}{64} - \frac{27}{64}$

The answers are in the margin on page 197.

Subtraction of Unlike Fractions

The method for finding the difference between two unlike fractions is similar to the method for adding unlike fractions.

Rule

To subtract fractions with unlike denominators, we must

Step 1 Find the LCD.
Step 2 Form equal fractions using the LCD.
Step 3 Subtract the numerators and keep the common denominator.

■ Example 4

Find the difference: $\frac{3}{4} - \frac{2}{3}$.

Solution **Step 1** Find the LCD.

$$\text{LCD} = 4 \times 3 = 12$$

(Note that if no prime number will divide both denominators, the LCD is the product of the denominators.)

Step 2 Form equal fractions and subtract.

$$\frac{3}{4} = \frac{3}{4} \times \frac{3}{3} = \frac{9}{12}$$
$$-\frac{2}{3} = \frac{2}{3} \times \frac{4}{4} = \frac{8}{12}$$
$$\frac{1}{12} \quad (9 - 8 = 1)$$

■ Example 5

Subtract $\frac{3}{4} - \frac{5}{8}$.

Solution **Step 1** Find the LCD.

$$\begin{array}{r|rr} 2 & 4 & 8 \\ 2 & 2 & 4 \\ \hline & 1 & 2 \end{array} \qquad \text{LCD} = 2 \times 2 \times 1 \times 2 = 8$$

(Note that if one denominator divides the other, the LCD is always the larger denominator.)

Step 2 Form equal fractions and subtract.

$$\frac{3}{4} = \frac{3}{4} \times \frac{2}{2} = \frac{6}{8}$$
$$-\frac{5}{8} = \qquad \frac{5}{8}$$
$$\frac{1}{8} \quad (6 - 5 = 1)$$

■ Example 6

Subtract $\frac{7}{18} - \frac{3}{20}$.

Solution LCD = 180.

$$\frac{7}{18} = \frac{7}{18} \times \frac{10}{10} = \frac{70}{180}$$
$$-\frac{3}{20} = \frac{3}{20} \times \frac{9}{9} = \frac{27}{180}$$
$$\frac{43}{180} \quad (70 - 27 = 43)$$

Practice check: Do exercises 46–51 in the margin.

Subtract.

46. $\frac{5}{8} - \frac{2}{5}$ **47.** $\frac{6}{11} - \frac{5}{13}$

48. $\frac{5}{9} - \frac{7}{18}$ **49.** $\frac{13}{16} - \frac{19}{32}$

50. $\frac{7}{15} - \frac{5}{12}$ **51.** $\frac{9}{14} - \frac{15}{42}$

The answers are in the margin on page 199.

■ Looking Back

You should now be able to find the difference between two unlike fractions.

■ Problems 5–4

Rule

We do not need to find the LCD when multiplying or dividing fractions.

Answers to exercises 40–45

40. $\frac{4}{8}$ or $\frac{1}{2}$ **41.** $\frac{6}{17}$ **42.** $\frac{1}{2}$

43. $\frac{6}{16}$ or $\frac{3}{8}$ **44.** $\frac{11}{32}$ **45.** $\frac{28}{64}$ or $\frac{7}{16}$

Perform the indicated operations.

Review

1. $\frac{7}{8} \times \frac{16}{21}$ **2.** $\frac{5}{16} + \frac{7}{8}$ **3.** $\frac{8}{9} - \frac{3}{9}$ **4.** $\frac{3}{16} \div \frac{5}{12}$

5. $\frac{2}{3} + \frac{5}{11}$ **6.** $\frac{3}{8} + \frac{5}{8}$ **7.** $\frac{15}{16} - \frac{5}{16}$ **8.** $\frac{4}{27} \div \frac{2}{9}$

Subtract.

9. $\frac{3}{4} - \frac{1}{3}$ **10.** $\frac{2}{3} - \frac{1}{4}$ **11.** $\frac{17}{20} - \frac{2}{5}$ **12.** $\frac{15}{16} - \frac{3}{4}$

13. $\frac{3}{5} - \frac{5}{12}$ **14.** $\frac{4}{7} - \frac{5}{12}$ **15.** $\frac{5}{12} - \frac{3}{16}$ **16.** $\frac{3}{4} - \frac{5}{18}$

17. $\frac{27}{32} - \frac{3}{4}$ **18.** $\frac{3}{5} - \frac{13}{25}$ **19.** $\frac{13}{15} - \frac{3}{5}$ **20.** $\frac{13}{4} - \frac{17}{24}$

21. $\frac{7}{11} - \frac{1}{6}$ **22.** $\frac{7}{6} - \frac{4}{7}$ **23.** $\frac{13}{12} - \frac{11}{18}$ **24.** $\frac{11}{16} - \frac{7}{12}$

25. $\frac{3}{6} - \frac{5}{16}$ **26.** $\frac{5}{8} - \frac{3}{12}$ **27.** $\frac{31}{35} - \frac{15}{28}$ **28.** $\frac{27}{35} - \frac{9}{20}$

29. $\frac{5}{18} - \frac{7}{63}$ **30.** $\frac{47}{72} - \frac{53}{84}$ **31.** $\frac{25}{80} - \frac{15}{48}$ **32.** $\frac{23}{30} - \frac{7}{54}$

The answers to review and odd-numbered exercises are in the back of the book.

5–5 COMPARING FRACTIONS

When we solve word problems using fractions, we sometimes need to know whether one fraction is larger than another, especially when we need to subtract one fraction from another.

■ Example 1

Which is more, $\frac{3}{4}$ or $\frac{2}{4}$?

Solution

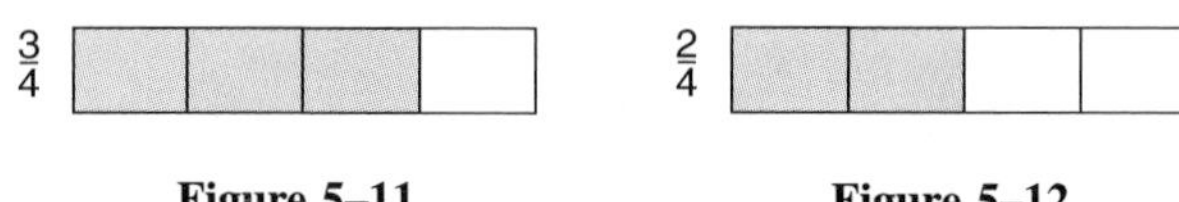

Figure 5–11 **Figure 5–12**

It is obvious by comparing the two shaded figures that $\frac{3}{4}$ is larger than $\frac{2}{4}$.

The number line shows the same comparison.

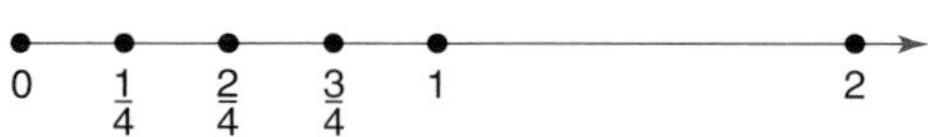

Figure 5–13

Since $\frac{3}{4}$ is to the right of $\frac{2}{4}$ on the number line, $\frac{3}{4}$ is greater than $\frac{2}{4}$.

Note: From chapter 1, you learned that the $>$ symbol means "greater than" and $<$ means "less than." Recall that the symbol points to the smaller number. Without having to draw shaded figures, we can tell which fraction is larger by comparing the numerators when the denominators are the same. Look at $\frac{3}{4}$ and $\frac{2}{4}$ again. Since the 3 is larger than the 2, we can conclude that $\frac{3}{4} > \frac{2}{4}$.

■ Example 2

Which fraction is larger, $\frac{7}{16}$ or $\frac{9}{16}$? Compare the numerators.

Solution Since 9 is larger than 7, then $\frac{9}{16}$ is larger than $\frac{7}{16}$, or $\frac{9}{16} > \frac{7}{16}$.

Practice check: Do exercises 52–56 in the margin.

Which fraction is larger? Compare the numerators.

52. $\frac{7}{8}$ or $\frac{3}{8}$ **53.** $\frac{5}{2}$ or $\frac{2}{2}$

54. $\frac{17}{27}$ or $\frac{19}{27}$ **55.** $\frac{31}{52}$ or $\frac{8}{52}$

56. $\frac{104}{103}$ or $\frac{102}{103}$

The answers are in the margin on page 201.

Suppose we must compare two unlike fractions. We change the two fractions to like fractions using the LCD and then compare.

■ Example 3

Which is greater, $\frac{3}{8}$ or $\frac{4}{9}$?

Solution **Step 1** Change to like fractions.

$$\text{LCD} = 72 \qquad \frac{3}{8} = \frac{27}{72} \qquad \frac{4}{9} = \frac{32}{72}$$

Step 2 Compare. Since $\frac{32}{72}$ is greater than $\frac{27}{72}$, then $\frac{4}{9} > \frac{3}{8}$.

■ Example 4

Which is greater, $\frac{5}{27}$ or $\frac{1}{6}$?

Solution **Step 1** Change to like fractions.

$$\text{LCD} = 54 \qquad \frac{5}{27} = \frac{10}{54} \qquad \frac{1}{6} = \frac{9}{54}$$

Step 2 Compare. Since $\frac{10}{54}$ is greater than $\frac{9}{54}$, then $\frac{5}{27} > \frac{1}{6}$.

Practice check: Do exercises 57–62 in the margin.

Answers to exercises 46–51

46. $\frac{9}{40}$ **47.** $\frac{23}{143}$ **48.** $\frac{3}{18}$ or $\frac{1}{6}$

49. $\frac{7}{32}$ **50.** $\frac{1}{20}$ **51.** $\frac{2}{7}$

Decide which fraction is larger.

57. $\frac{4}{5}$ or $\frac{3}{4}$ **58.** $\frac{3}{8}$ or $\frac{3}{10}$

59. $\frac{5}{8}$ or $\frac{7}{12}$ **60.** $\frac{5}{3}$ or $\frac{8}{5}$

61. $\frac{13}{16}$ or $\frac{7}{6}$ **62.** $\frac{16}{19}$ or $\frac{15}{21}$

The answers are in the margin on page 201.

■ Looking Back

Given two fractions you should now be able to decide which one is larger.

■ Problems 5–5

Perform the indicated operations. Remember, we do not need to find the LCD when multiplying or dividing fractions.

Review

1. $\frac{7}{12} \times \frac{5}{8} \times \frac{16}{21}$ **2.** $\frac{5}{7} - \frac{3}{8}$ **3.** $\frac{\frac{7}{12}}{\frac{8}{9}}$ **4.** $\frac{4}{3} + \frac{1}{6}$

5. $\frac{2}{3} - \frac{4}{9}$ **6.** $\frac{5}{9} \times \frac{7}{12}$ **7.** $\frac{4}{9} \times \frac{3}{2}$ **8.** $\frac{5}{8} - \frac{6}{10}$

9. $\frac{4}{7} \div \frac{2}{5}$ **10.** $\frac{3}{4} + \frac{7}{8} + \frac{1}{6}$

State which fraction is greater.

11. $\frac{5}{8}$ or $\frac{7}{8}$ **12.** $\frac{18}{15}$ or $\frac{28}{15}$ **13.** $\frac{106}{1000}$ or $\frac{1005}{1000}$ **14.** $\frac{350}{900}$ or $\frac{305}{900}$

15. $\frac{7}{9}$ or $\frac{3}{5}$ **16.** $\frac{6}{7}$ or $\frac{7}{14}$ **17.** $\frac{9}{7}$ or $\frac{9}{12}$ **18.** $\frac{1}{2}$ or $\frac{8}{10}$

19. $\frac{6}{3}$ or $\frac{7}{2}$

20. $\frac{2}{9}$ or $\frac{7}{12}$

21. $\frac{2}{15}$ or $\frac{7}{10}$

22. $\frac{3}{14}$ or $\frac{5}{7}$

23. $\frac{3}{4}$ or $\frac{9}{10}$

24. $\frac{4}{45}$ or $\frac{11}{20}$

The answers to review and odd-numbered exercises are in the back of the book.

5–6 APPLIED PROBLEMS

This section contains word problems with fractions that can be solved by addition or subtraction. Remember to write a statement for your answer.

■ **Example 1**

Jane used $\frac{1}{4}$ of a tank of gasoline to drive 123 miles. She then drove 151 miles and used $\frac{3}{8}$ of a tank of gasoline. What part of a tank of gasoline did she use altogether?

Solution To find the answer we add the two portions.

$$\frac{1}{4} = \frac{1}{4} \times \frac{2}{2} = \frac{2}{8} \qquad \frac{3}{8} = \frac{3}{8}$$

Thus, $\frac{2}{8} + \frac{3}{8} = \frac{5}{8}$. Jane used $\frac{5}{8}$ of a tank of gasoline to drive 274 miles.

■ **Example 2**

Fred Lander bought two pieces of property as an investment. It cost him \$8500 for $\frac{3}{8}$ acre and \$6200 for $\frac{7}{9}$ acre. What was the difference in acreage between the two pieces of property?

Solution We are being asked to determine the difference between the acreage, not the price difference. We must first determine which piece of property is the larger, for we must subtract the smaller amount from the larger.

Step 1 Change to like fractions.

$$\text{LCD} = 72 \qquad \frac{3}{8} = \frac{27}{72} \qquad \frac{7}{9} = \frac{56}{72}$$

This shows that the larger piece of land is $\frac{7}{9}$ acre.

Step 2 Find the difference: $\frac{7}{9} - \frac{3}{8}$.

$$\begin{array}{rl} \frac{7}{9} = & \frac{56}{72} \\ -\frac{3}{8} = & \underline{\frac{27}{72}} \\ & \frac{29}{72} \end{array}$$

The difference in acreage between the two pieces of property was $\frac{29}{72}$ acre.

Answers to exercises 52–56

52. $\frac{7}{8}$ 53. $\frac{5}{2}$ 54. $\frac{19}{27}$ 55. $\frac{31}{52}$

56. $\frac{104}{103}$

Answers to exercises 57–62

57. $\frac{4}{5}$ 58. $\frac{3}{8}$ 59. $\frac{5}{8}$ 60. $\frac{5}{3}$

61. $\frac{7}{6}$ 62. $\frac{16}{19}$

■ Example 3

Find the length of this pin.

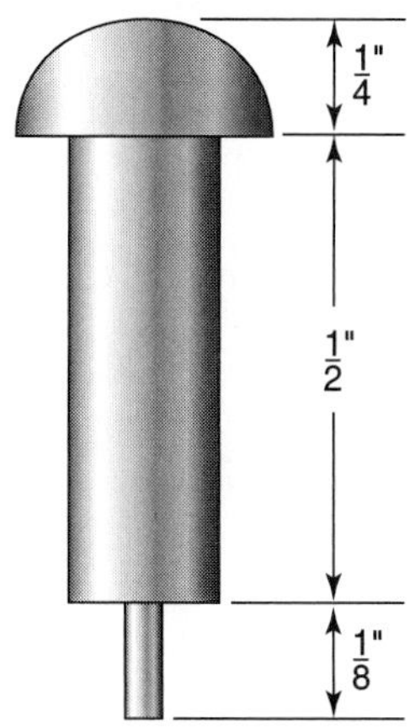

Figure 5–14

Solution To find the answer we add the fractions.
LCD = 8

$$\frac{1}{4} = \frac{1}{4} \times \frac{2}{2} = \frac{2}{8} \qquad \frac{1}{2} = \frac{1}{2} \times \frac{4}{4} = \frac{4}{8} \qquad \frac{1}{8} = \frac{1}{8}$$

Thus, $\frac{2}{8} + \frac{4}{8} + \frac{1}{8} = \frac{7}{8}$. The pin is $\frac{7}{8}$ inch long.

■ Example 4

The outside diameter of a pipe is $\frac{3}{4}$ inch, and the inside diameter is $\frac{1}{2}$ inch. How thick are the walls of the pipe?

Solution We first draw a picture.

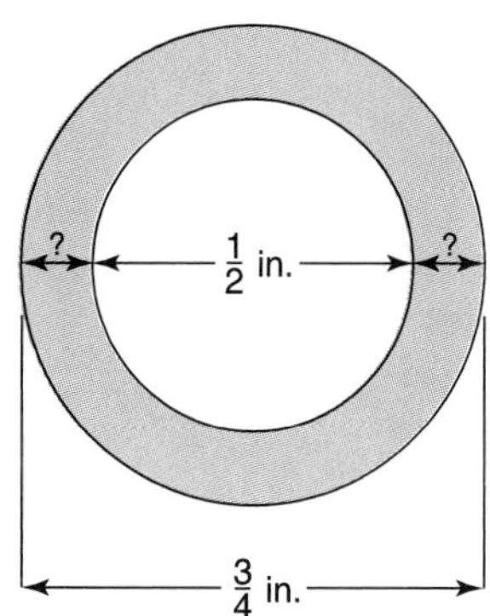

Figure 5–15

By subtracting the inside diameter from the outside diameter we get double the thickness of the walls.

$$\frac{3}{4} - \frac{1}{2} = \frac{3}{4} - \frac{2}{4} = \frac{1}{4} \text{ inch}$$

In order to get the thickness of the walls, we must divide this difference by 2.

$$\frac{1}{4} \div 2 = \frac{1}{4} \times \frac{1}{2} = \frac{1}{8} \text{ inch}$$

Practice check: Do exercises 63–66 in the margin.

Solve.

63. A certain recipe requires $\frac{1}{8}$ cup honey, $\frac{1}{3}$ cup lime juice, and $\frac{1}{2}$ cup of hot water. How many cups of ingredients were used in this recipe?

64. Juan bought $\frac{3}{4}$ yard of cotton fabric and $\frac{5}{8}$ yard of wool fabric. How much more cotton than wool did he buy?

65. Three pieces of wood were glued together. Two of the pieces are $\frac{3}{16}$ inch thick and the other is $\frac{1}{8}$ inch thick. Find the thickness of the final piece of wood.

66. A recipe calls for $\frac{3}{5}$ cup of flour. Dick has only $\frac{1}{2}$ cup on hand. How much more flour is needed?

The answers are in the margin on page 205.

■ Looking Back

You should now be able to solve applied problems using the operations of addition and subtraction.

■ Problems 5–6

Solve. Write a statement for your answer.

1. **Generic.** A certain recipe requires $\frac{1}{2}$ cup of raisins and $\frac{1}{4}$ cup of sunflower seeds. How many cups would it make if the raisins and sunflower seeds were mixed together?

2. **Machinist.** Find the total thickness of the stack of steel bars in the figure.

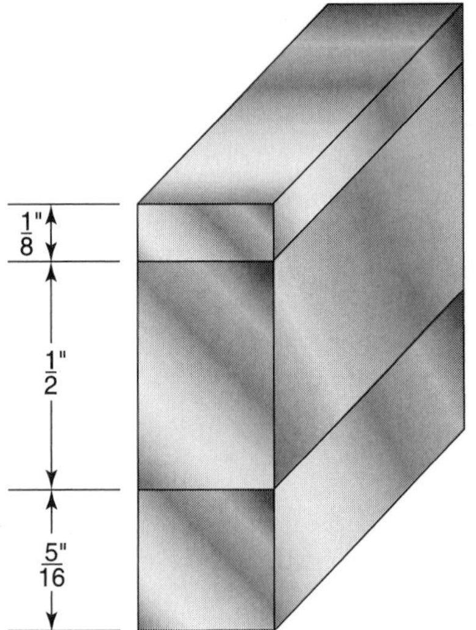

3. **Generic.** Phil blended $\frac{2}{3}$ lb of tobacco with $\frac{1}{6}$ lb of another brand of tobacco to come up with an enjoyable blend. How many pounds of the blend did Phil have?

4. **Generic.** A certain recipe calls for $\frac{5}{8}$ cup of flour. Henry found he had only $\frac{1}{3}$ cup left in his bin. How much did he have to borrow from his neighbor?

5. **Generic.** Jim ate $\frac{3}{5}$ of a raspberry ring and Sue ate $\frac{3}{10}$ of it. What fraction more did Jim eat than Sue?

6. **Machinist.** A bolt and nut with two washers are needed to fasten two metal strips together in the figure. What is the minimum length of the bolt needed?

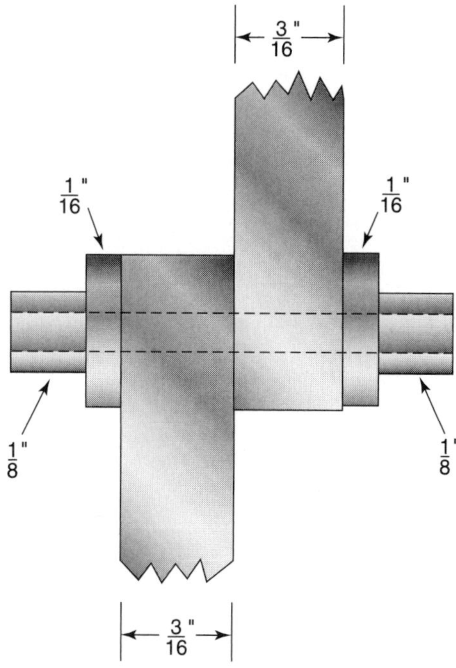

7. **Painter.** John painted $\frac{7}{8}$ of a house green and $\frac{1}{12}$ of it white. What fraction more of the house was painted green than white?

8. **Generic.** Sue Beacham spends $\frac{1}{4}$ of the day sleeping, $\frac{5}{12}$ of the day working, and $\frac{1}{12}$ of the day eating. What portion of the day does she spend doing these activities?

9. **Butcher.** A meat cutter trimmed $\frac{2}{9}$ pound of fat from an $\frac{11}{12}$-pound steak. What did the steak weigh after the fat was trimmed?

10. **Carpenter.** A kitchen counter has a $\frac{7}{8}$-in. plywood base that is overlaid with a $\frac{1}{16}$-in. Formica top. How thick is the counter?

11. Carpenter. Jo Anne has a piece of wood that is $\frac{7}{8}$ foot long. She needs to fit it into a space that is $\frac{9}{16}$ foot long. How much must she cut from the piece of wood to make it fit?

12. Plumber. What is the outside diameter of a pipe if the inside diameter is $\frac{5}{8}$ in. and the pipe wall is $\frac{3}{16}$ in.?

13. Machinist. Find the missing dimension in the figure.

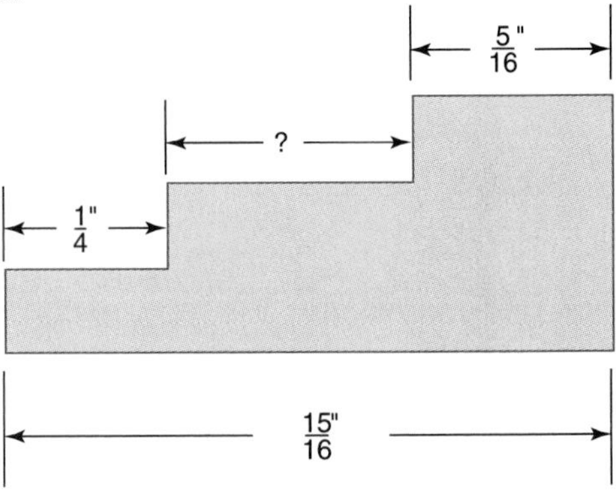

14. Plumber. The outside diameter of a pipe is $\frac{5}{6}$ in., and the inside diameter is $\frac{3}{5}$ in. How thick are the walls of the pipe?

15. Generic. In an election Larry Yetter received $\frac{2}{9}$ of the votes cast, and Betty Ricky received $\frac{2}{27}$ of the votes cast. What fractional part of the total votes did both candidates receive together?

16. Machinist. Jim has a pin that is $\frac{5}{8}$ in. in length. By measuring he found that it is $\frac{3}{16}$ in. too long. What length of pin does he need?

17. Finance. American Concrete Pipe stock went up $\frac{5}{8}$ of a point in one day and fell $\frac{7}{16}$ of a point the next day. Did the stock register a gain or a loss at the end of the second day?

18. Machinist. What is the length of the screw shown in the figure?

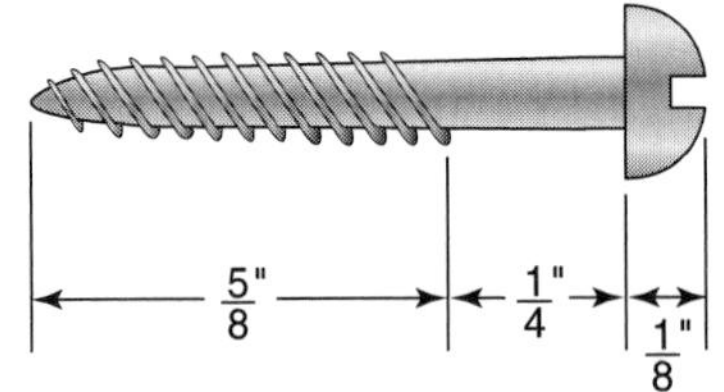

19. Auto Mechanic. A certain truck brake lining new is $\frac{3}{4}$ in. thick. A brake check after 20,000 miles showed that $\frac{7}{16}$ in. of the lining remained. How many inches were worn in 20,000 miles?

The answers to the odd-numbered exercises are in the back of the book.

CHAPTER 5 OVERVIEW

Summary

5–1 You learned to add like fractions

5–2 You learned to find the LCD of two or more fractions and form equivalent fractions having the LCD as the denominators.

5–3 You learned to add unlike fractions using the LCD.

5–4 You learned to subtract like and unlike fractions.

5–5 You learned to decide which of two fractions is larger.

5–6 You learned to solve application problems involving addition and subtraction of fractions.

Answers to exercises 63–66

63. $\frac{23}{24}$ cup **64.** $\frac{1}{8}$ yd **65.** $\frac{1}{2}$ in.

66. $\frac{1}{10}$ cup

Terms To Remember

	Page
Like fractions	183
Lowest common denominator	185

Rules

- When adding like fractions, all we do is add the numerators and keep the common denominator.
- How to find the LCD:

  Step 1 Write the denominators in a horizontal line.

  Step 2 Divide by the smallest prime number that will evenly divide at least two of the numbers. Bring down any number that cannot be divided evenly.

  Step 3 Repeat Steps 1 and 2 until no other prime number will evenly divide at least two of the numbers.

  Step 4 The LCD is the product of all the outside numbers.
- To add fractions with unlike denominators, we must

  Step 1 Find the LCD.

  Step 2 Convert the unlike fractions to like fractions using the LCD.

  Step 3 Add the like fractions.
- To find the difference between like fractions we subtract the numerators and keep the common denominator.
- To subtract fractions with unlike denominators, we must

  Step 1 Find the LCD.

  Step 2 Form equal fractions using the LCD.

  Step 3 Subtract the numerators and keep the common denominator.
- We do not need to find the LCD when multiplying or dividing fractions.

Self-Test

The answers are in the back of the book.

5–1 Add and reduce where possible.

1. $\frac{5}{9} + \frac{3}{9}$ **2.** $\frac{5}{6} + \frac{1}{6}$ **3.** $\frac{7}{12} + \frac{2}{12}$

5–2 Find the lowest common denominator of the following groups of fractions.

4. $\frac{1}{5}$ and $\frac{1}{8}$ **5.** $\frac{7}{6}$ and $\frac{2}{3}$ **6.** $\frac{5}{12}$ and $\frac{7}{18}$.

5–3 Add.

7. $\frac{5}{6} + \frac{7}{9}$ **8.** $\frac{5}{16} + \frac{7}{8}$ **9.** $\frac{16}{15} + \frac{3}{10}$ **10.** $\frac{22}{3} + \frac{13}{7}$

5–4 Subtract and reduce where possible.

11. $\frac{7}{10} - \frac{3}{10}$ **12.** $\frac{4}{9} - \frac{2}{9}$ **13.** $\frac{17}{27} - \frac{8}{27}$

Subtract.

14. $\frac{3}{4} - \frac{2}{7}$ **15.** $\frac{7}{6} - \frac{11}{18}$ **16.** $\frac{5}{12} - \frac{3}{8}$ **17.** $\frac{23}{16} - \frac{13}{12}$

5–5 Decide which fraction is larger.

18. $\frac{1}{2}$ or $\frac{5}{9}$ **19.** $\frac{5}{16}$ or $\frac{3}{8}$ **20.** $\frac{15}{27}$ or $\frac{7}{18}$

5–5 Solve.

21. A lathe reduces the diameter of a shaft by $\frac{3}{100}$ in. If the original diameter was $\frac{760}{1000}$ in., what is the new diameter of the shaft?

22. What is the thickness of 3 pieces of steel whose thickness are $\frac{3}{8}$ in., $\frac{1}{4}$ in., and $\frac{5}{16}$ in.?

23. Will an electric cable with diameter $\frac{13}{16}$ in. fit inside a PVC pipe with an inside diameter of $\frac{7}{8}$ in.? If not, by how much?

24. The diameters of the large and small ends of a tapered shaft are $\frac{11}{16}$ in. and $\frac{5}{32}$ in. respectively. Find the amount of taper by finding the difference.

Chapter Test

Directions: This test will aid in your preparation for a possible chapter test given by your instructor. The answers are in the back of the book. Go back and review in the appropriate section(s) if you missed any items.

Add or subtract and reduce where possible.

1. $\frac{5}{8} + \frac{1}{8}$ **2.** $\frac{7}{12} - \frac{5}{12}$ **3.** $\frac{5}{6} - \frac{1}{6}$ **4.** $\frac{5}{16} + \frac{9}{16}$

5. $\frac{15}{27} + \frac{9}{27}$ **6.** $\frac{7}{9} - \frac{5}{9}$ **7.** $\frac{16}{15} - \frac{7}{15}$ **8.** $\frac{7}{10} + \frac{3}{10}$

9. $\frac{5}{7} + \frac{1}{7}$

Find the lowest common denominator of the following groups of fractions.

10. $\frac{2}{3}$ and $\frac{3}{5}$ **11.** $\frac{5}{9}$ and $\frac{1}{12}$ **12.** $\frac{5}{21}$ and $\frac{9}{14}$ **13.** $\frac{3}{10}$ and $\frac{7}{12}$

14. $\frac{1}{6}, \frac{1}{9},$ and $\frac{1}{2}$ **15.** $\frac{5}{6}, \frac{1}{3},$ and $\frac{3}{10}$

Add or subtract.

16. $\frac{5}{7} - \frac{3}{5}$

17. $\frac{2}{7} + \frac{3}{14}$

18. $\frac{5}{12} + \frac{7}{15}$

19. $\frac{7}{12} - \frac{5}{9}$

20. $\frac{13}{24} + \frac{7}{16}$

21. $\frac{5}{14} - \frac{5}{21}$

Decide which fraction is larger.

22. $\frac{4}{5}$ or $\frac{3}{4}$

23. $\frac{7}{10}$ or $\frac{5}{6}$

24. $\frac{1}{16}$ or $\frac{1}{12}$

Solve.

25. A $\frac{1}{16}$-lb slice of baloney was cut from a $\frac{7}{8}$-lb piece of baloney. How much was left after the cut?

26. A flour bin had only $\frac{3}{8}$ cup of flour remaining. A recipe called for $\frac{1}{3}$ cup more. How many cups did the recipe require?

27. Robert walks to his college but stops at the donut shop each morning before reaching school. The distance from his home to the donut shop is $\frac{2}{3}$ mile and from the donut shop to the college is $\frac{3}{5}$ mile. How far is the walk from his home to the college?

28. How much thicker is a $\frac{3}{8}$-inch shim than a $\frac{3}{16}$-inch shim?

29. James completed his job in $\frac{4}{5}$ of an hour. William completed the same job in $\frac{5}{6}$ of an hour. Who was the faster worker? By how much?

CHAPTER 6

Operations with Mixed Numbers

Choice: Either skip the Pretest and study all of Chapter 6 or take the Pretest and study only the sections you need to study.

Pretest

Directions: This Pretest will help you determine which sections to study in Chapter 6. If any of your answers are incorrect, or if you omitted any exercise, turn to the indicated section(s) and study the material; then take the Chapter Test.

6–1 **1.** Change $\frac{5}{3}$ to a mixed number.

2. Change $4\frac{3}{4}$ to an improper fraction.

6–2 Multiply. Leave the answers as mixed numbers when possible.

3. $2\frac{1}{6} \times 12$

4. $3\frac{3}{4} \times 3\frac{3}{5}$

6–3 Divide. Leave the answers as mixed numbers when possible.

5. $4 \div 4\frac{2}{3}$

6. $2\frac{4}{9} \div 3\frac{1}{5}$

6–4 Add and simplify.

7. $7 + 9\frac{5}{7}$ **8.** $23\frac{5}{8} + 12\frac{7}{12}$

6–5 Subtract and simplify.

9. $43\frac{7}{12} - 14\frac{5}{9}$ **10.** $6 - 3\frac{7}{8}$

6–6 **11.** Speed Saxon jogged $11\frac{3}{8}$ miles one day and $12\frac{5}{12}$ miles the next. How many total miles did Speed jog those two days?

12. A case of 24 cans of oil weighs $15\frac{3}{8}$ pounds. How much does each can of oil weigh?

The answers are in the margin on page 213.

6–1 MIXED NUMBERS AND CONVERSIONS

In this section we will be discussing numbers and fractional numbers that are equal to or greater than 1. Examples of these kinds of numbers are $\frac{6}{6}$, $4\frac{1}{2}$, $\frac{15}{8}$, and $107\frac{3}{4}$. A number like $4\frac{1}{2}$ has a special name.

Rule

Any number that is a combination of a whole number and a proper fractional number is called a mixed number.

$5\frac{3}{4}$, $23\frac{8}{9}$, $106\frac{7}{16}$, and $7\frac{2}{3}$ are all examples of mixed numbers.

■ **Example 1**

Write the mixed number that is represented by the shaded object.

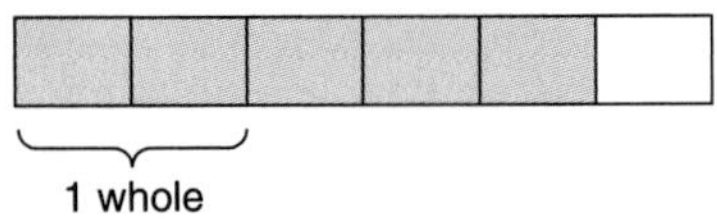

Figure 6–1

Solution $\frac{5}{3} = 1\frac{2}{3}$

Practice check: Do practice exercises 1–3 in the margin.

Write the mixed number that is represented either by the shaded object or the number line.

1. $\frac{12}{5} = ?$

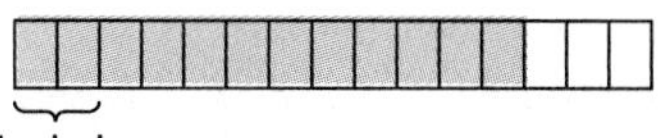

2. $\frac{7}{4} = ?$

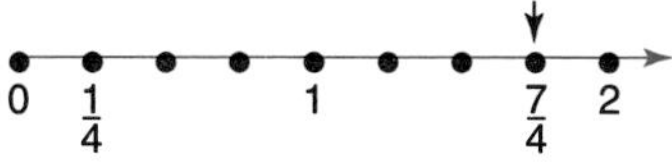

3. $\frac{7}{3} = ?$

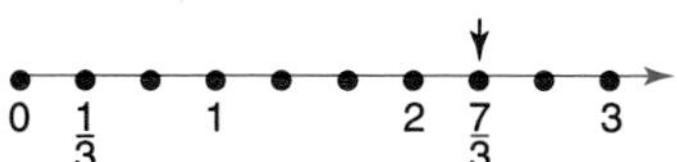

The answers are in the margin on page 213.

Changing Improper Fractions to Mixed Numbers

The value of an improper fraction is sometimes more meaningful when it is changed to a mixed number. Example 2 shows two methods for changing an improper fraction to a mixed number.

■ Example 2

Change $\frac{9}{5}$ to a mixed number.

Solution 1 Expanded method:

Step 1 Change 9 to the sum of the two addends.

a. The first addend is the largest number that can be divided evenly by the denominator.

b. The second addend must be smaller than the denominator.

$$\frac{9}{5} = \frac{5+4}{5}$$

Step 2 Expand to the sum of two fractions by reversing the procedure used in addition.

$$\frac{9}{5} = \frac{5+4}{5} = \frac{5}{5} + \frac{4}{5}$$

Step 3 Simplify.

$$\frac{9}{5} = \frac{5+4}{5} = \frac{5}{5} + \frac{4}{5}$$

$$= 1 + \frac{4}{5}$$

$$= 1\frac{4}{5}$$

Solution 2 Short method:

Rule

A fraction always indicates division of the numerator by the denominator.

From the above definition we know that $\frac{9}{5}$ means $9 \div 5$.

Step 1 Divide.

$$\begin{array}{r} 1 \\ \text{Divisor} \rightarrow 5\overline{)9} \\ \underline{5} \\ 4 \end{array} \leftarrow \text{Remainder}$$

Step 2 Use the remainder as the numerator and the divisor as the denominator.

$$\begin{array}{r} 1\tfrac{4}{5} \\ 5\overline{)9} \\ \underline{5} \\ 4 \end{array} \begin{array}{l} \leftarrow \text{Remainder} \\ \leftarrow \text{Divisor} \end{array} \qquad \frac{9}{5} = 1\frac{4}{5}$$

The short method is usually used because it is simple and quick.

■ **Example 3**

Change $\frac{38}{3}$ to a mixed number using the short method.

Solution **Step 1** Divide.

$$\begin{array}{r} 12 \\ \text{Divisor} \nearrow 3\overline{)38} \\ \underline{3} \\ 8 \\ \underline{6} \\ 2 \end{array} \leftarrow \text{Remainder}$$

Step 2 Write the whole number and fraction.

$$12\frac{2}{3} \begin{array}{l} \leftarrow \text{Remainder} \\ \leftarrow \text{Divisor} \end{array}$$

Rule

A short method for changing an improper fraction to a mixed number:

Step 1 Divide the numerator by the denominator.

Step 2 Write the mixed number. The quotient is the whole number part. The remainder is the numerator, and the divisor is the denominator of the fraction part. The result of Step 1 is a whole number only if the remainder is 0.

Practice check: Do exercises 4–9 in the margin.

Change to mixed numbers or whole numbers using the short method.

4. $\frac{7}{5}$ **5.** $\frac{13}{3}$ **6.** $\frac{105}{5}$

7. $\frac{43}{9}$ **8.** $\frac{57}{4}$ **9.** $\frac{132}{13}$

The answers are in the margin on page 215.

Changing a Mixed Number to an Improper Fraction

In order to do multiplication and division with mixed numbers, you will need to know how to change a mixed number to an improper fraction.

■ **Example 4**

Change $2\frac{3}{5}$ to an improper fraction.

Solution 1 Expanded method:

Step 1 Change 2 to a fraction with a denominator of 5.

$$2\frac{3}{5} = 2 + \frac{3}{5}$$
$$\downarrow$$
$$= \frac{10}{5} + \frac{3}{5}$$

Step 2 Add.

$$\frac{10}{5} + \frac{3}{5} = \frac{13}{5}$$

Solution 2 Short method: Do the following steps mentally.

Step 1 Multiply the whole number by the denominator.

$$2\frac{3}{5} \quad \text{(Think: } 2 \times 5 = 10\text{)}$$

Step 2 Add this product to the numerator.

$$2\frac{3}{5} \quad \text{(Think: } 10 + 3 = 13\text{)}$$

Product Numerator

Step 3 Use this result as the numerator of the improper fraction while keeping the same denominator.

$$2\frac{3}{5} = \frac{(2 \times 5) + 3}{5} = \frac{10 + 3}{5} = \frac{13}{5}$$

You do these steps mentally.

■ **Example 5**

Change $13\frac{2}{3}$ to an improper fraction using the short method.

Whole number Numerator Denominator

$$13\frac{2}{3} = \frac{(13 \times 3) + 2}{3} = \frac{39 + 2}{3} = \frac{41}{3}$$

Rule

A short method for changing a mixed number to an improper fraction:

Step 1 Multiply the whole number part by the denominator of the fraction part, and to this result add the numerator of the fraction part.

Step 2 Write the improper fraction. Use the result of Step 1 for the numerator. The denominator remains the same.

Practice check: Do practice exercises 10–14.

■ Looking Back

You should now be able to

1. Change an improper fraction to a mixed number.
2. Change a mixed number to an improper fraction.

Answers to Pretest

Circle the answers you missed or omitted. Cross them off after you've studied the corresponding section(s), then take the Chapter Test.

1. $1\frac{2}{3}$ **2.** $\frac{19}{4}$ **3.** 26

4. $13\frac{1}{2}$ **5.** $\frac{6}{7}$ **6.** $\frac{55}{72}$

7. $16\frac{5}{7}$ **8.** $36\frac{5}{24}$ **9.** $29\frac{1}{36}$

10. $2\frac{1}{8}$ **11.** $23\frac{19}{24}$ **12.** $\frac{41}{64}$ pound

Answers to exercises 1–3

1. $2\frac{2}{5}$ **2.** $1\frac{3}{4}$ **3.** $2\frac{1}{3}$

Change the following mixed numbers to improper fractions using the short method.

10. $1\frac{3}{5}$ **11.** $4\frac{1}{3}$ **12.** $10\frac{3}{8}$

13. $21\frac{7}{8}$ **14.** $17\frac{2}{15}$

The answers are in the margin on page 215.

Problems 6–1

Review

Perform the indicated operations.

1. $\frac{3}{5} \div \frac{3}{7}$
2. $\frac{7}{4} + \frac{3}{4}$
3. $\frac{33}{10} \times \frac{6}{11}$
4. $\frac{25}{8} - \frac{11}{4}$
5. $\frac{4}{9} + \frac{3}{12} + \frac{7}{10}$
6. $\frac{15}{12} - \frac{5}{18}$

Change the following improper fractions to mixed numbers or whole numbers.

7. $\frac{13}{5}$
8. $\frac{25}{5}$
9. $\frac{59}{9}$
10. $\frac{43}{12}$
11. $\frac{35}{2}$
12. $\frac{73}{5}$
13. $\frac{55}{5}$
14. $\frac{48}{3}$
15. $\frac{213}{5}$
16. $\frac{189}{18}$
17. $\frac{132}{5}$
18. $\frac{119}{53}$
19. $\frac{63}{10}$
20. $\frac{76}{37}$
21. $\frac{5320}{51}$
22. $\frac{6070}{231}$

Change the following mixed numbers or whole numbers to improper fractions.

23. $2\frac{2}{3}$
24. $1\frac{8}{9}$
25. $15\frac{2}{7}$
26. 7

27. $15\frac{4}{9}$ **28.** $4\frac{7}{10}$ **29.** $3\frac{9}{16}$ **30.** 18

31. $5\frac{7}{13}$ **32.** $21\frac{8}{9}$ **33.** $1\frac{9}{16}$ **34.** $39\frac{3}{7}$

35. $50\frac{5}{8}$ **36.** $10\frac{7}{100}$ **37.** $60\frac{13}{1000}$

38. $23\frac{17}{35}$ **39.** $15\frac{29}{133}$ **40.** $321\frac{139}{432}$

Answers to exercises 4–9

4. $1\frac{2}{5}$ **5.** $4\frac{1}{3}$ **6.** 21 **7.** $4\frac{7}{9}$
8. $14\frac{1}{4}$ **9.** $10\frac{2}{13}$

Answers to exercises 10–14

10. $\frac{8}{5}$ **11.** $\frac{13}{3}$ **12.** $\frac{83}{8}$ **13.** $\frac{175}{8}$
14. $\frac{275}{15}$

The answers to review and odd-numbered exercises are in the back of the book.

6–2 MULTIPLICATION WITH WHOLE NUMBERS AND MIXED NUMBERS

Multiplication by One or Zero

As with whole numbers, multiplication by 1 or 0 gives the same results.

■ **Example 1**

Multiply $1 \times \frac{9}{15}$.

Solution

$$1 \times \frac{9}{15}$$
$$\downarrow$$
$$= \frac{1}{1} \times \frac{9}{15}$$ We know that 1 is the same as $\frac{1}{1}$.

$$= \frac{1 \times 9}{1 \times 15} = \frac{9}{15}$$

Rule

The product of one and a fraction is that fraction.

■ **Example 2**

Multiply $\frac{7}{13} \times 0$.

Solution

$$\frac{7}{13} \times 0$$

$$= \frac{7}{13} \times \frac{0}{1} \quad \text{0 is the same as } \frac{0}{1}.$$

$$= \frac{7 \times 0}{13 \times 1}$$

$$= \frac{0}{13} \quad \text{Zero divided by any whole number greater than zero is equal to zero.}$$

$$= 0$$

Rule

The product of zero and a fraction is zero.

Practice check: Do exercises 15–20 in the margin.

Multiply.

15. $\frac{11}{84} \times 0$ **16.** $0 \times \frac{3}{10}$

17. $1 \times \frac{7}{12}$ **18.** $0 \times \frac{17}{56}$

19. $\frac{35}{16} \times 1$ **20.** $\frac{8}{15} \times 0$

The answers are in the margin on page 219.

Multiplying Fractions and Whole Numbers Greater than One

To multiply with a whole number, we first change the whole number to an improper fraction.

■ **Example 3**

Multiply $24 \times \frac{9}{16}$.

Solution

$$24 \times \frac{9}{16}$$

$$= \frac{24}{1} \times \frac{9}{16} \quad \text{Changing 24 to } \frac{24}{1}$$

$$= \frac{\overset{3}{\cancel{24}}}{1} \times \frac{9}{\underset{2}{\cancel{16}}} \quad \text{Canceling by dividing both 24 and 16 by 8}$$

$$= \frac{27}{2} \quad \text{or} \quad 13\frac{1}{2}$$

Practice check: Do exercises 21–27 in the margin.

Multiply.

21. $\frac{3}{4} \times 4$ **22.** $8 \times \frac{7}{18}$

23. $0 \times \frac{5}{2}$ **24.** $\frac{4}{27} \times 12$

25. $\frac{9}{10} \times 1$ **26.** $\frac{14}{3} \times 0$

27. $34 \times \frac{9}{17}$

The answers are in the margin on page 219.

Multiplying Fractions and Mixed Numbers

To multiply with mixed numbers, we first change the mixed numbers to improper fractions.

■ **Example 4**

Multiply $12\frac{2}{3} \times \frac{3}{14}$.

Solution

$$12\frac{2}{3} \times \frac{3}{14}$$

$$= \frac{38}{3} \times \frac{3}{14} \quad \text{Changing } 12\tfrac{2}{3} \text{ to } \tfrac{38}{3}$$

$$= \frac{\overset{19}{\cancel{38}}}{\underset{1}{\cancel{3}}} \times \frac{\overset{1}{\cancel{3}}}{\underset{7}{\cancel{14}}} \quad \text{Canceling}$$

$$= \frac{19}{7}, \text{ or } 2\frac{5}{7}$$

Practice check: Do exercises 28–33 in the margin.

Multiply.

28. $5\frac{2}{3} \times \frac{3}{4}$ **29.** $\frac{4}{5} \times 5\frac{7}{8}$

30. $0 \times \frac{21}{8}$ **31.** $2\frac{1}{2} \times 1$

32. $3\frac{1}{7} \times \frac{7}{8}$ **33.** $8\frac{3}{4} \times 0$

The answers are in the margin on page 219.

Multiplying a Mixed Number by a Mixed Number

■ **Example 5**

Multiply $3\frac{3}{4} \times 1\frac{1}{5}$.

Solution

$$3\frac{3}{4} \times 1\frac{1}{5}$$

$$= \frac{15}{4} \times \frac{6}{5} \quad \text{Changing to improper fractions}$$

$$= \frac{\overset{3}{\cancel{15}}}{\underset{2}{\cancel{4}}} \times \frac{\overset{3}{\cancel{6}}}{\underset{1}{\cancel{5}}} \quad \text{Canceling}$$

$$= \frac{9}{2}, \text{ or } 4\frac{1}{2}$$

Rule

When multiplying with whole numbers and mixed numbers, first change them to improper fractions, then find the product.

Practice check: Do exercises 34–39 in the margin.

Multiply.

34. $1\frac{3}{5} \times 1\frac{1}{4}$ **35.** $3\frac{3}{8} \times 1\frac{7}{9}$

36. $4 \times 7\frac{3}{4}$ **37.** $1\frac{4}{7} \times 2\frac{1}{2}$

38. $0 \times \frac{6}{7}$ **39.** $6\frac{3}{4} \times \frac{8}{15}$

The answers are in the margin on page 219.

■ Looking Back

You should now be able to do multiplications involving proper fractions, one, zero, improper fractions, and mixed numbers.

■ Problems 6–2

Review

1. $\frac{3}{8} + \frac{1}{8}$

2. $\frac{\frac{7}{8}}{\frac{21}{22}}$

3. $\frac{5}{6} - \frac{3}{18}$

4. $\frac{7}{5} + \frac{2}{3}$

5. $\frac{5}{12} - \frac{5}{16}$

6. $\frac{7}{16} \div \frac{1}{2}$

Multiply and write the answers as a mixed number where possible.

7. $1 \times \frac{2}{3}$

8. $\frac{7}{9} \times 9$

9. $\frac{5}{7} \times 7$

10. $\frac{15}{16} \times 0$

11. $\frac{21}{4} \times 0$

12. $\frac{5}{16} \times 32$

13. $26 \times \frac{7}{13}$

14. $1 \times 3\frac{1}{2}$

15. $2\frac{2}{3} \times 9$

16. $5\frac{1}{7} \times 21$

17. $8\frac{1}{2} \times 3\frac{1}{9}$

18. $18 \times 2\frac{5}{8}$

19. $\frac{5}{6} \times 2\frac{2}{5}$

20. $2\frac{5}{8} \times 1\frac{2}{7}$

21. $6\frac{7}{8} \times 1$

22. $3\frac{2}{3} \times 5\frac{2}{5}$

23. $1\frac{4}{5} \times 4\frac{2}{3}$

24. $5\frac{5}{6} \times \frac{2}{7}$

25. $5\frac{2}{4} \times 3\frac{1}{11}$

26. $\frac{32}{47} \times 1$

27. $\frac{7}{15} \times 25$ **28.** $10\frac{5}{8} \times \frac{4}{5}$ **29.** $0 \times 3\frac{1}{8}$

30. $2\frac{6}{7} \times 3\frac{2}{4}$ **31.** $5\frac{1}{2} \times \frac{7}{16} \times \frac{14}{33}$ **32.** $1\frac{1}{8} \times 2\frac{1}{4} \times 3\frac{5}{5}$

33. $6\frac{2}{3} \times 18 \times 3\frac{1}{4}$ **34.** $5\frac{1}{6} \times \frac{8}{9} \times 1\frac{3}{10}$

The answers to review and odd-numbered exercises are in the back of the book.

Answers to exercises 15–20

15. 0 **16.** 0 **17.** $\frac{7}{12}$ **18.** 0

19. $\frac{35}{16}$ **20.** 0

Answers to exercises 21–27

21. 3 **22.** $3\frac{1}{9}$ **23.** 0 **24.** $1\frac{7}{9}$

25. $\frac{9}{10}$ **26.** 0 **27.** 18

Answers to exercises 28–33

28. $\frac{17}{4}$, or $4\frac{1}{4}$ **29.** $\frac{47}{10}$, or $4\frac{7}{10}$

30. 0 **31.** $2\frac{1}{2}$ **32.** $\frac{11}{4}$, or $2\frac{3}{4}$

33. 0

Answers to exercises 34–39

34. 2 **35.** 6 **36.** 31

37. $\frac{55}{14}$, or $3\frac{13}{14}$ **38.** 0

39. $\frac{18}{5}$, or $3\frac{3}{5}$

6–3 DIVISION WITH WHOLE NUMBERS AND MIXED NUMBERS

A Review of Division of Fractions

■ **Example 1**

Divide $\frac{2}{3} \div \frac{5}{16}$.

Solution

$$\frac{2}{3} \div \frac{5}{16} = \frac{2}{3} \times \frac{16}{5}$$

Keep the same. Change to multiplication. Change to reciprocal.

$$= \frac{2}{3} \times \frac{16}{5}$$

$$= \frac{32}{15}, \quad \text{or} \quad 2\frac{2}{15}$$

Practice check: Do exercises 40–42 in the margin.

Divide.

40. $\frac{9}{10} \div \frac{3}{5}$ **41.** $\frac{3}{4} \div \frac{1}{3}$ **42.** $\frac{2}{3} \div \frac{1}{2}$

The answers are in the margin on page 221.

Rule ______________________________

When we divide a number by 1, the quotient is always the other number.

■ **Examples**

2. $5 \div 1 = 5$

3. $\frac{2}{3} \div 1 = \frac{2}{3}$

4. $3\frac{1}{2} \div 1 = 3\frac{1}{2}$

Practice check: Do exercises 43–46 in the margin.

Divide.

43. $\frac{7}{7} \div 1$ **44.** $5\frac{2}{3} \div 1$

45. $105\frac{3}{8} \div 1$ **46.** $8 \div 1$

The answers are in the margin on page 223.

Division and Zero

Rule

When we divide zero by any number other than zero, the quotient is always zero.

■ **Examples**

5. $0 \div 5 = 0$

6. $0 \div \frac{3}{2} = 0$

7. $0 \div 3\frac{1}{8} = 0$

Rule

Remember: Never divide by zero! It is impossible.

Practice check: Do exercises 47–52 in the margin.

Divide.

47. $0 \div \frac{4}{3}$ **48.** $0 \div 9$

49. $3\frac{1}{2} \div 0$ **50.** $0 \div 33\frac{5}{8}$

51. $0 \div 1050\frac{2}{3}$ **52.** $\frac{8}{3} \div 0$

The answers are in the margin on page 223.

Division Involving Counting Numbers and Mixed Numbers

■ **Example 8**

Divide $3\frac{3}{8} \div 9$.

Solution

$$3\frac{3}{8} \div 9$$

$$= \frac{27}{8} \div \frac{9}{1} \quad \text{Changing to fractions}$$

$$= \frac{27}{8} \times \frac{1}{9} \quad \text{Multiplying by reciprocal}$$

$$= \frac{\overset{3}{\cancel{27}}}{8} \times \frac{1}{\underset{1}{\cancel{9}}} \quad \text{Canceling}$$

$$= \frac{3}{8}$$

Practice check: Do exercises 53–57 in the margin.

Divide.

53. $\frac{5}{8} \div 5$ **54.** $5\frac{3}{5} \div 14$

55. $8 \div 1\frac{1}{2}$ **56.** $1 \div 2\frac{5}{8}$

57. $6\frac{3}{4} \div 18$

The answers are in the margin on page 223.

Division Involving Mixed Numbers and Fractions

Answers to exercises 40–42

40. $1\frac{1}{2}$ **41.** $2\frac{1}{4}$ **42.** $1\frac{1}{3}$

■ **Example 9**

Divide $2\frac{1}{3} \div \frac{14}{27}$.

Solution

$$2\frac{1}{3} \div \frac{14}{27}$$

$$= \frac{7}{3} \div \frac{14}{27} \quad \text{Changing to improper fraction}$$

$$= \frac{7}{3} \times \frac{27}{14} \quad \text{Multiplying by reciprocal}$$

$$= \frac{\overset{1}{\cancel{7}}}{\underset{1}{\cancel{3}}} \times \frac{\overset{9}{\cancel{27}}}{\underset{2}{\cancel{14}}} \quad \text{Canceling}$$

$$= \frac{9}{2}, \text{ or } 4\frac{1}{2}$$

■ **Example 10**

Divide $1\frac{3}{5} \div 8\frac{2}{3}$.

Solution

$$1\frac{3}{5} \div 8\frac{2}{3}$$

$$= \frac{8}{5} \div \frac{26}{3} \quad \text{Changing to improper fractions}$$

$$= \frac{8}{5} \times \frac{3}{26} \quad \text{Multiplying by reciprocal}$$

$$= \frac{\overset{4}{\cancel{8}}}{5} \times \frac{3}{\underset{13}{\cancel{26}}} \quad \text{Canceling}$$

$$= \frac{12}{65}$$

Rule

When division involves whole numbers and mixed numbers, first change them to improper fractions, then find the quotient.

Practice check: Do exercises 58–61 in the margin.

Divide.

58. $\frac{3}{8} \div 3\frac{3}{4}$ **59.** $1\frac{1}{4} \div 3\frac{3}{7}$

60. $2\frac{2}{5} \div \frac{4}{5}$ **61.** $3\frac{3}{4} \div 3\frac{1}{8}$

The answers are in the margin on page 223.

■ Looking Back

You should now be able to divide with proper fractions, whole numbers, improper fractions, and mixed numbers.

■ Problems 6–3

Review

Reduce and change to a mixed number when possible.

1. $\frac{6}{5} \times \frac{7}{8}$

2. $\frac{1}{3} + \frac{2}{5}$

3. $5\frac{1}{4} \times \frac{3}{14}$

4. $\frac{12}{16} - \frac{5}{12}$

5. $\frac{7}{10} + \frac{3}{5}$

6. $3\frac{1}{8} \times 1\frac{3}{5}$

Divide. Change answers to mixed numbers when possible.

7. $\frac{3}{2} \div 1$

8. $0 \div 8\frac{2}{3}$

9. $\frac{3}{5} \div 3\frac{3}{4}$

10. $\frac{5}{8} \div 35$

11. $4\frac{2}{3} \div 21$

12. $\frac{5}{8} \div 1$

13. $0 \div \frac{3}{7}$

14. $6 \div 4\frac{1}{2}$

15. $2\frac{7}{8} \div 5\frac{1}{2}$

16. $4\frac{3}{4} \div \frac{7}{16}$

17. $8 \div \frac{1}{4}$

18. $3\frac{1}{2} \div 4\frac{1}{5}$

19. $2\frac{4}{9} \div 11$

20. $23\frac{9}{10} \div 0$

21. $5\frac{1}{8} \div 0$

22. $6\frac{4}{9} \div 29$

23. $2\frac{1}{4} \div 8\frac{4}{5}$

24. $1\frac{2}{5} \div 3\frac{8}{9}$

25. $1\frac{1}{9} \div \frac{15}{27}$

26. $106\frac{3}{8} \div 1$

27. $3\frac{7}{16} \div 5$

28. $1 \div 14\frac{1}{8}$

29. $9\frac{3}{5} \div 2\frac{5}{6}$

30. $\frac{5}{6} \div 5\frac{5}{8}$

31. $6\frac{5}{8} \div 1$ **32.** $27 \div 8\frac{2}{5}$ **33.** $1 \div 3\frac{4}{9}$ **34.** $3\frac{3}{8} \div 5\frac{1}{3}$

35. $29\frac{3}{8} \div 100$ **36.** $2\frac{1}{14} \div 4\frac{2}{49}$ **37.** $21\frac{1}{8} \div 1\frac{3}{10}$ **38.** $17\frac{1}{2} \div 40\frac{25}{26}$

The answers to review and odd-numbered exercises are in the back of the book.

Answers to exercises 43–46

43. $\frac{6}{7}$ **44.** $5\frac{2}{3}$ **45.** $105\frac{3}{8}$ **46.** 8

Answers to exercises 47–52

47. 0 **48.** 0 **49.** impossible!
50. 0 **51.** 0 **52.** impossible!

Answers to exercises 53–57

53. $\frac{1}{8}$ **54.** $\frac{2}{5}$ **55.** $5\frac{1}{3}$ **56.** $\frac{8}{21}$

57. $\frac{3}{8}$

Answers to exercises 58–61

58. $\frac{1}{10}$ **59.** $\frac{35}{96}$ **60.** 3 **61.** $1\frac{1}{5}$

6–4 ADDITION WITH MIXED NUMBERS

In this section you will learn how to add mixed numbers and whole numbers using a short method.

Adding Mixed Numbers and Whole Numbers

■ **Example 1**

Add $5 + 7\frac{2}{5}$.

Solution Expanded method:

$$5 + 7\frac{2}{5} = 5 + \left(7 + \frac{2}{5}\right)$$

$$= (5 + 7) + \frac{2}{5} \quad \text{Using the associative property}$$

$$= 12 + \frac{2}{5}, \quad \text{or} \quad 12\frac{2}{5}$$

Solution Short method:

Step 1 Place one number over the other.

$$\begin{array}{r} 5 \\ +7\frac{2}{5} \\ \hline \end{array}$$

Step 2 Add the whole numbers.

$$\begin{array}{r} 5 \\ +7\frac{2}{5} \\ \hline 12 \end{array}$$

Step 3 Bring down the fraction.

$$\begin{array}{r} 5 \\ +7\frac{2}{5} \\ \hline 12\frac{2}{5} \end{array}$$

Since it is easier to add mixed numbers vertically, we shall use the short method.

Practice check: Do exercises 62–65 in the margin.

Add using the short method.

62. $8 + 7\frac{5}{11}$ **63.** $18\frac{2}{9} + 6$

64. $1 + 9\frac{1}{2}$ **65.** $25\frac{7}{15} + 33$

The answers are in the margin on page 227.

Adding Mixed Numbers with Like Denominators

■ **Example 2**

Add $3\frac{3}{5} + 5\frac{1}{5}$.

Solution **Step 1** Place one number over the other.

$$\begin{array}{r} 3\frac{3}{5} \\ +5\frac{1}{5} \\ \hline \end{array}$$

Step 2 Add the whole numbers.

$$\begin{array}{r} 3\frac{3}{5} \\ +5\frac{1}{5} \\ \hline 8 \end{array}$$

Step 3 Add the fractions.

$$\begin{array}{r} 3\frac{3}{5} \\ +5\frac{1}{5} \\ \hline 8\frac{4}{5} \end{array}$$

■ **Example 3**

Add $2\frac{5}{8} + 4\frac{7}{8}$.

Solution

$$\begin{array}{r} 2\frac{5}{8} \\ +4\frac{7}{8} \\ \hline 6\frac{12}{8} \end{array}$$

$$6\frac{12}{8} \text{ means } 6 + \frac{12}{8} \leftarrow$$ This is an improper fraction. We can't keep the answer in this form.

$$= 6 + \frac{3}{2}$$ Reducing $\frac{12}{8}$

$$= 6 + 1\frac{1}{2}$$ Changing to a mixed number

$$= 7\frac{1}{2}$$

■ **Example 4**

Add $7\frac{2}{4} + 2\frac{3}{4}$.

Solution

$$\begin{array}{r} 7\frac{2}{4} \\ +2\frac{3}{4} \\ \hline 9\frac{5}{4} \end{array}, \text{ or } 10\frac{1}{4}$$

Note: $9\frac{5}{4} = 9 + \frac{5}{4}$

$= 9 + 1\frac{1}{4}$ Changing to a mixed number

$= 10\frac{1}{4}$

Practice check: Do exercises 66–69 in the margin.

Add and simplify.

66. $3\frac{5}{12} + 7\frac{3}{12}$ **67.** $11\frac{7}{8} + 6\frac{7}{8}$

68. $4\frac{5}{11} + 8\frac{6}{11}$ **69.** $31\frac{11}{12} + 13\frac{7}{12}$

The answers are in the margin on page 227.

Adding Mixed Numbers with Unlike Denominators

■ **Example 5**

Add $4\frac{5}{6} + 2\frac{7}{15}$.

Solution **Step 1** Place one number over the other and write in equal signs.

$$\begin{array}{r} 4\frac{5}{6} = \\ +2\frac{7}{15} = \\ \hline \end{array}$$

Step 2 Carry over the whole numbers.

$$\begin{array}{r} 4\frac{5}{6} = 4 \\ +2\frac{7}{15} = 2 \\ \hline \end{array}$$

Step 3 Find the LCD and write in the equal fractions.

$$\text{LCD} = 30 \qquad \begin{array}{r} 4\frac{5}{6} = 4\frac{25}{30} \\ \underline{+2\frac{7}{15}} = \underline{2\frac{14}{30}} \end{array}$$

Step 4 Add and simplify.

$$\begin{array}{r} 4\frac{5}{6} = 4\frac{25}{30} \\ \underline{+2\frac{7}{15}} = \underline{2\frac{14}{30}} \\ 6\frac{39}{30}, \end{array} \quad \text{or} \quad 7\frac{3}{10}$$

Note: $6\frac{39}{30} = 6\frac{13}{10}$

$= 6 + \frac{13}{10}$

$= 6 + 1\frac{3}{10}$

$= 7\frac{3}{10}$

Practice check: Do exercises 70–72 in the margin.

Add and simplify.

70. $4\frac{3}{8} + 3\frac{1}{7}$ **71.** $6\frac{3}{8} + 2\frac{1}{4}$

72. $13\frac{7}{12} + 9\frac{7}{8}$

The answers are in the margin on page 229.

Example 6 shows how to add a mixed number and a fraction.

■ Example 6

Add $8\frac{3}{4} + \frac{5}{8}$.

Solution **Step 1** Place one number over the other and write in equal signs.

$$\begin{array}{r} 8\frac{3}{4} = \\ \underline{+\frac{5}{8} =} \end{array}$$

Step 2 Carry over the whole number.

$$\begin{array}{r} 8\frac{3}{4} = 8 \\ \underline{+\frac{5}{8}} = \end{array}$$

Step 3 Find the LCD and write in equal fractions.

$$\begin{array}{r} 8\frac{3}{4} = 8\frac{6}{8} \\ \underline{+\frac{5}{8}} = \underline{\frac{5}{8}} \end{array}$$

Step 4 Add and simplify.

$$\begin{array}{r} 8\frac{3}{4} = 8\frac{6}{8} \\ +\frac{5}{8} = \frac{5}{8} \\ \hline 8\frac{11}{8}, \text{ or } 9\frac{3}{8} \end{array}$$

Rule

A short method for adding mixed numbers with unlike denominators:

Step 1 Place one number over the other and write in equal signs.
Step 2 Carry over the whole numbers.
Step 3 Find the LCD and write in the equal fractions.
Step 4 Add and simplify.

Practice check: Do exercises 73–76 in the margin.

Answers to exercises 62–65

62. $15\frac{5}{11}$ **63.** $24\frac{2}{9}$ **64.** $10\frac{1}{2}$

65. $58\frac{7}{15}$

Answers to exercises 66–69

66. $10\frac{2}{3}$ **67.** $18\frac{3}{4}$ **68.** 13

69. $45\frac{1}{2}$

Add and simplify.

73. $9\frac{3}{5} + \frac{1}{3}$ **74.** $7\frac{1}{6} + \frac{3}{8}$

75. $13\frac{2}{3} + 5\frac{1}{65}$ **76.** $34\frac{7}{12} + \frac{5}{9}$

The answers are in the margin on page 229.

■ Looking Back

You should now be able to add mixed numbers, whole numbers, and proper fractions.

■ Problems 6–4

Review

Perform the operations as indicated.

1. $\frac{7}{8} - \frac{5}{16}$ **2.** $7\frac{3}{8} \times \frac{4}{9}$ **3.** $3\frac{3}{8} \div 1$ **4.** $3\frac{1}{5} \times 3\frac{3}{4}$

5. $2\frac{1}{4} \div 5\frac{1}{4}$ **6.** $0 \div \frac{4}{7}$ **7.** $1\frac{3}{5} \div 18$ **8.** $\frac{5}{8} + \frac{3}{16}$

Add and simplify.

9. $8\frac{3}{8} + 7\frac{1}{8}$ **10.** $13\frac{5}{11} + 18\frac{7}{11}$ **11.** $13 + 7\frac{5}{7}$ **12.** $29 + 35\frac{7}{16}$

13. $23\frac{1}{6} + 9\frac{2}{3}$ **14.** $3\frac{3}{10} + 9\frac{2}{5}$ **15.** $56\frac{3}{4} + 13\frac{7}{18}$ **16.** $31\frac{17}{24} + 5\frac{13}{16}$

17. $\frac{9}{16} + 32\frac{3}{8}$ **18.** $\frac{7}{8} + 13\frac{1}{4}$ **19.** $73\frac{3}{7} + 31\frac{1}{7}$ **20.** $56\frac{2}{9} + 81\frac{7}{9}$

21. $39\frac{9}{10} + 88$ **22.** $56\frac{7}{13} + 111$ **23.** $105\frac{7}{15} + 81\frac{3}{10}$ **24.** $72\frac{13}{16} + 120\frac{7}{12}$

25. $217\frac{1}{2} + 155\frac{4}{5}$ **26.** $318\frac{7}{8} + 132\frac{2}{5}$ **27.** $17\frac{7}{10} + \frac{5}{8}$ **28.** $\frac{1}{6} + 56\frac{9}{10}$

29. $7\frac{1}{6} + 10\frac{5}{12} + 5\frac{1}{3}$ **30.** $8\frac{3}{5} + 30 + 15\frac{7}{10}$

31. $58\frac{5}{12} + 108 + 39\frac{5}{8}$ **32.** $69\frac{3}{4} + \frac{11}{20} + 24\frac{37}{40}$

33. $1468\frac{7}{12} + 406\frac{9}{16} + 31\frac{5}{8}$ **34.** $31\frac{3}{8} + 18\frac{5}{6} + 52\frac{5}{12}$

The answers to review and odd-numbered exercises are in the back of the book.

6–5 SUBTRACTION WITH MIXED NUMBERS

We use the short method when we subtract with mixed numbers. The problems are written in the same manner as with addition.

Subtraction with Mixed Numbers Without "Borrowing"

■ **Example 1**

Subtract $15\frac{7}{12} - 9\frac{5}{12}$.

Solution **Step 1** Place the first number over the second one.

$$\begin{array}{r} 15\frac{7}{12} \\ -9\frac{5}{12} \\ \hline \end{array}$$

Step 2 Subtract the whole numbers.

$$\begin{array}{r} 15\frac{7}{12} \\ -9\frac{5}{12} \\ \hline 6 \end{array}$$

Step 3 Subtract the fractions and simplify.

$$\begin{array}{r} 15\frac{7}{12} \\ -9\frac{5}{12} \\ \hline 6\frac{2}{12}, \text{ or } 6\frac{1}{6} \end{array}$$

Practice check: Do exercises 77–79 in the margin.

■ Example 2

Subtract $7\frac{3}{11} - 4\frac{3}{11}$.

Solution

$$\begin{array}{r} 7\frac{3}{11} \\ -4\frac{3}{11} \\ \hline 3 \end{array}$$

$\left(\text{We know that } \frac{3}{11} - \frac{3}{11} = \frac{0}{11} = 0, \text{ so we don't show 0.}\right)$

■ Example 3

Subtract $39\frac{8}{13} - 16$.

Solution

$$\begin{array}{rl} 39\frac{8}{13} & \\ -16 & \leftarrow \text{We imagine a } \frac{0}{13} \text{ here.} \\ \hline 23\frac{8}{13} & \leftarrow \text{We bring down the } \frac{8}{13}. \end{array}$$

Practice check: Do exercises 80–84 in the margin.

Answers to exercises 70–72

70. $7\frac{29}{56}$ **71.** $8\frac{5}{8}$ **72.** $23\frac{11}{24}$

Answers to exercises 73–76

73. $9\frac{14}{15}$ **74.** $7\frac{13}{24}$ **75.** $18\frac{5}{6}$

76. $35\frac{5}{36}$

Subtract and simplify.

77. $\begin{array}{r} 8\frac{7}{8} \\ -4\frac{5}{8} \\ \hline \end{array}$ **78.** $\begin{array}{r} 18\frac{13}{16} \\ -15\frac{7}{16} \\ \hline \end{array}$

79. $52\frac{11}{20} - 23\frac{3}{20}$

The answers are in the margin on page 231.

Subtract and simplify.

80. $\begin{array}{r} 29\frac{9}{13} \\ -15\frac{5}{13} \\ \hline \end{array}$ **81.** $\begin{array}{r} 13\frac{5}{8} \\ -9\frac{5}{8} \\ \hline \end{array}$

82. $68\frac{2}{3} - 39\frac{2}{3}$ **83.** $\begin{array}{r} 7\frac{5}{9} \\ -3 \\ \hline \end{array}$

84. $53\frac{5}{21} - 19$

The answers are in the margin on page 231.

■ **Example 4**

Subtract $15\frac{3}{4} - 8\frac{1}{8}$.

Solution **Step 1** Place the first number over the second and write equal signs, because the fractions are not like fractions.

$$\begin{array}{r} 15\frac{3}{4} = \\ -8\frac{1}{8} = \\ \hline \end{array}$$

Step 2 Bring over the whole numbers.

$$\begin{array}{rcr} 15\frac{3}{4} & = & 15 \\ -8\frac{1}{8} & = & 8 \\ \hline \end{array}$$

Step 3 Find the LCD and write equal fractions.

$$\text{LCD} = 8 \qquad \begin{array}{rcr} 15\frac{3}{4} & = & 15\frac{6}{8} \\ -8\frac{1}{8} & = & 8\frac{1}{8} \\ \hline \end{array}$$

Step 4 Subtract.

$$\begin{array}{rcr} 15\frac{3}{4} & = & 15\frac{6}{8} \\ -8\frac{1}{8} & = & 8\frac{1}{8} \\ \hline & & 7\frac{5}{8} \end{array}$$

Practice check: Do exercises 85–89 in the margin.

Subtract.

85. $\begin{array}{r} 8\frac{17}{21} \\ -5\frac{8}{21} \\ \hline \end{array}$ **86.** $\begin{array}{r} 5\frac{5}{13} \\ -1 \\ \hline \end{array}$

87. $\begin{array}{r} 31\frac{7}{16} \\ -29\frac{7}{16} \\ \hline \end{array}$ **88.** $15\frac{1}{3} - 4\frac{1}{5}$

89. $32\frac{3}{4} - 17\frac{7}{10}$

The answers are in the margin on page 233.

Subtracting Mixed Numbers with "Borrowing"

Subtract $6\frac{2}{5} - 3\frac{4}{5}$.

$6\frac{2}{5}$ ← The fractions are like fractions

$-3\frac{4}{5}$ but we can't subtract $\frac{4}{5}$ from $\frac{2}{5}$.

To complete this example, we will have to learn how to "borrow." This is different than place value "borrowing," as you will learn in Examples 5 through 8.

■ **Example 5**

Change $6\frac{2}{5}$ by "borrowing" 1 from the 6.

Solution

$$6\frac{2}{5} = 6 + \frac{2}{5}$$

$$= (5 + 1) + \frac{2}{5} \qquad \text{Changing 6 to } (5 + 1)$$

$$= 5 + \frac{5}{5} + \frac{2}{5} \qquad \text{Changing 1 to } \tfrac{5}{5}$$

We chose $\frac{5}{5}$ in order to have a denominator the same as in $\frac{2}{5}$.

$$= 5 + \frac{7}{5} \qquad \text{Adding } \tfrac{5}{5} \text{ and } \tfrac{2}{5}$$

$$= 5\frac{7}{5}$$

So, $6\frac{2}{5} = 5\frac{7}{5}$

■ **Example 6**

Change $17\frac{5}{11}$ by "borrowing" 1 from 17.

Solution

$$17\frac{5}{11} = 17 + \frac{5}{11}$$

$$= (16 + 1) + \frac{5}{11} \qquad \text{Changing 17 to } (16 + 1)$$

$$= 16 + \frac{11}{11} + \frac{5}{11} \qquad \text{Changing 1 to } \tfrac{11}{11}$$

We choose $\frac{11}{11}$ in order to have a denominator the same as in $\frac{5}{11}$.

$$= 16 + \frac{16}{11} \qquad \text{Adding } \tfrac{11}{11} \text{ and } \tfrac{5}{11}$$

$$= 16\frac{16}{11}$$

So, $17\frac{5}{11} = 16\frac{16}{11}$

Practice check: Do exercises 90–93 in the margin.

Let's do the same "borrowing" but with a short method.

■ **Example 7**

"Borrow" $7\frac{5}{8}$ using the short method.

Solution **Step 1** Take 1 from the 7 and write down 6: $7\frac{5}{8} = 6$.

Answers to exercises 77–79

77. $4\frac{2}{8}$, or $4\frac{1}{4}$ **78.** $3\frac{6}{16}$, or $3\frac{3}{8}$

79. $29\frac{8}{20}$, or $29\frac{2}{5}$

Answers to exercises 80–84

80. $14\frac{4}{13}$ **81.** 4 **82.** 29

83. $4\frac{5}{9}$ **84.** $34\frac{5}{21}$

Change the following mixed numbers by "borrowing" 1.

90. $10\frac{4}{5}$ **91.** $23\frac{5}{7}$

92. $5\frac{6}{19}$ **93.** $64\frac{13}{25}$

The answers are in the margin on page 233.

Step 2 *Mentally* take the 1 that you "borrowed," change it to $\frac{8}{8}$, and add it to $\frac{5}{8}$. This gives you $\frac{13}{8}$.

Step 3 Place $\frac{13}{8}$ next to the 6: $7\frac{5}{8} = 6\frac{13}{8}$.

Practice doing this example a few times.

■ Example 8

"Borrow" $31\frac{17}{23}$ using the short method.

Solution **Step 1** Take 1 from the 31 and write down 30: $31\frac{17}{23} = 30$.

Step 2 *Mentally* take the 1 that you "borrowed," change it to $\frac{23}{23}$, and add it to $\frac{17}{23}$. This gives you $\frac{40}{23}$.

Step 3 Place $\frac{40}{23}$ next to 30: $31\frac{17}{23} = 30\frac{40}{23}$.

Rule

A short method for "borrowing" when subtracting with mixed numbers:

Step 1 Take 1 from the whole number part and write the new whole number.

Step 2 Mentally change the 1 that was "borrowed" to a fractional name for 1, using the denominator of the mixed number.

Step 3 Mentally add the fractional name for one to the fraction part of the mixed number.

Step 4 Place the result of Step 3 next to the new whole number.

"Borrow" using the short method.

94. $12\frac{5}{6}$ **95.** $17\frac{23}{32}$

96. $29\frac{9}{16}$

The answers are in the margin on page 234.

Practice check: Do exercises 94–96 in the margin.

Now that you know how to "borrow," we can complete a subtraction.

■ Example 9

Subtract $6\frac{2}{5} - 3\frac{4}{5}$.

Solution

$$\begin{array}{r} 6\frac{2}{5} \\ -3\frac{4}{5} \\ \hline \end{array}$$

The fractions are like fractions. We can't subtract $\frac{5}{5}$ from $\frac{2}{5}$.

Step 1 Change $6\frac{2}{5}$ to $5\frac{7}{5}$, and bring over the $3\frac{4}{5}$.

$$\begin{array}{rcr} 6\frac{2}{5} & = & 5\frac{7}{5} \\ -3\frac{4}{5} & = & 3\frac{4}{5} \\ \hline \end{array}$$

Step 2 Subtract.

$$\begin{array}{rcr} 6\frac{2}{5} & = & 5\frac{7}{5} \\ -3\frac{4}{5} & = & 3\frac{4}{5} \\ \hline & & 2\frac{3}{5} \end{array}$$

■ Example 10

Subtract $4\frac{3}{7} - 1\frac{5}{7}$.

Solution

$$\begin{array}{r} 4\frac{3}{7} \\ -1\frac{5}{7} \\ \hline \end{array}$$

The fractions are like fractions. We can't subtract $\frac{5}{7}$ from $\frac{3}{7}$.

Step 1 Change $4\frac{3}{7}$ to $3\frac{10}{7}$, and bring over the $1\frac{5}{7}$.

$$\begin{array}{rcr} 4\frac{3}{7} & = & 3\frac{10}{7} \\ -1\frac{5}{7} & = & 1\frac{5}{7} \\ \hline & & 2\frac{5}{7} \end{array}$$

Practice check: Do exercises 97–102 in the margin.

■ Example 11

Subtract $39 - 21\frac{2}{7}$.

Solution

$$\begin{array}{r} 39 \\ -21\frac{2}{7} \\ \hline \end{array}$$

← The fraction here is assumed to be $\frac{0}{7}$ or 0. We can't subtract $\frac{5}{7}$ from $\frac{0}{7}$.

Step 1 Change 39 to $38\frac{7}{7}$ and bring over the $21\frac{2}{7}$.

$$\begin{array}{rcr} 39 & = & 38\frac{7}{7} \\ -21\frac{2}{7} & = & 21\frac{2}{7} \\ \hline \end{array}$$

Note: $39 = 38 + 1 = 38 + \frac{7}{7} = 38\frac{7}{7}$

Step 2 Subtract.

$$\begin{array}{rcr} 39 & = & 38\frac{7}{7} \\ -21\frac{2}{7} & = & 21\frac{2}{7} \\ \hline & & 17\frac{5}{7} \end{array}$$

Practice check: Do exercises 103–107 in the margin.

When mixed numbers have unlike fractions, you must first find the LCD.

■ Example 12

Subtract $13\frac{7}{10} - 8\frac{3}{4}$.

Answers to exercises 85–89

85. $3\frac{3}{7}$ **86.** $4\frac{5}{13}$ **87.** 2
88. $11\frac{2}{15}$ **89.** $15\frac{1}{20}$

Answers to exercises 90–93

90. $9\frac{9}{5}$ **91.** $22\frac{12}{7}$ **92.** $4\frac{25}{19}$
93. $63\frac{38}{25}$

Subtract.

97. $15\frac{1}{10} - 9\frac{7}{10}$

98. $32\frac{2}{5} - 23\frac{4}{5}$

99. $8\frac{7}{32} - 4\frac{18}{32}$

100. $63\frac{5}{12} - 13\frac{9}{12}$

101. $4\frac{1}{3} - 2$

102. $19\frac{2}{7} - 10\frac{2}{7}$

The answers are in the margin on page 234.

Subtract.

103. $21\frac{3}{5} - 16\frac{7}{15}$

104. $7\frac{5}{10} - 6\frac{9}{10}$

105. $56 - 30\frac{3}{8}$

106. $9 - \frac{6}{19}$

107. $23 - 19\frac{21}{33}$

The answers are in the margin on page 234.

Answers to exercises 94–96

94. $11\frac{11}{6}$ **95.** $16\frac{55}{32}$ **96.** $28\frac{25}{16}$

Answers to exercises 97–102

97. $5\frac{2}{5}$ **98.** $8\frac{3}{5}$ **99.** $3\frac{21}{32}$

100. $49\frac{2}{3}$ **101.** $2\frac{1}{3}$ **102.** 9

Answers to exercises 103–107

103. $5\frac{2}{15}$ **104.** $\frac{3}{5}$ **105.** $25\frac{5}{8}$

106. $8\frac{13}{19}$ **107.** $3\frac{12}{33}$

Solution

$$\begin{array}{r} 13\frac{7}{10} \\ -8\frac{3}{4} \\ \hline \end{array} \quad \text{Unlike fractions}$$

Step 1 Find the LCD and form equal numbers.

$$\text{LCD} = 20 \qquad \begin{array}{rcr} 13\frac{7}{10} & = & 13\frac{14}{20} \\ -8\frac{3}{4} & = & 8\frac{15}{20} \\ \hline \end{array}$$

We can't subtract $\frac{15}{20}$ from $\frac{14}{20}$.

Step 2 "Borrow" and then subtract.

$$\begin{array}{rcrcr} 13\frac{7}{10} & = & 13\frac{14}{20} & = & 12\frac{34}{20} \\ -8\frac{3}{4} & = & 8\frac{15}{20} & = & 8\frac{15}{20} \\ \hline & & & & 4\frac{19}{20} \end{array}$$

■ **Example 13**

Subtract $21\frac{5}{7} - 3\frac{4}{5}$.

Solution

$$\begin{array}{r} 21\frac{5}{7} \\ -3\frac{4}{5} \\ \hline \end{array} \quad \text{Unlike fractions}$$

Step 1 Find the LCD and form equal numbers.

$$\text{LCD} = 35 \qquad \begin{array}{rcr} 21\frac{5}{7} & = & 21\frac{25}{35} \\ -3\frac{4}{5} & = & 3\frac{28}{35} \\ \hline \end{array}$$

Unable to subtract

Step 2 "Borrow" and then subtract.

$$\begin{array}{rcrcr} 21\frac{5}{7} & = & 21\frac{25}{35} & = & 20\frac{60}{35} \\ -3\frac{4}{5} & = & 3\frac{28}{35} & = & 3\frac{28}{35} \\ \hline & & & & 17\frac{32}{35} \end{array}$$

Rule

A short method for subtracting with mixed numbers involving "borrowing":

Step 1 Place one number over the other.

Step 2 If the numbers to be subtracted have unlike fractions, find the LCD and write equal mixed numbers.

Step 3 If the bottom fraction is larger than the top fraction, "borrow," and write another equal top mixed number.

Step 4 Place the unchanged bottom number directly below the top number from Step 3 and subtract.

Practice check: Do exercises 108–112 in the margin.

Subtract.

108. $15 - \frac{3}{8}$

109. $23 - 5\frac{4}{19}$

110. $12\frac{3}{8} - 4\frac{9}{16}$

111. $22\frac{1}{2} - 7\frac{3}{5}$

112. $56\frac{3}{8} - 33\frac{5}{12}$

The answers are in the margin on page 237.

■ Looking Back

You should now be able to subtract mixed numbers with or without "borrowing."

■ Problems 6–5

Review

1. $5\frac{7}{8} \times \frac{4}{7}$

2. $3\frac{1}{8} \div 6\frac{3}{4}$

3. $6\frac{2}{9} \div 12$

4. $6\frac{3}{10} \times 3\frac{7}{15}$

5. $9\frac{4}{7} + 16\frac{1}{7}$

6. $15 + 9\frac{2}{3}$

7. $17\frac{3}{10} + 5\frac{2}{5}$

8. $\frac{5}{8} + 15\frac{7}{9}$

Subtract and simplify.

9. $7\frac{5}{9} - 4\frac{2}{9}$

10. $5\frac{7}{12} - 4\frac{1}{12}$

11. $13\frac{2}{3} - 10\frac{2}{3}$

12. $18\frac{3}{8} - 9\frac{3}{8}$

13. $39\frac{4}{13} - 23$

14. $43\frac{5}{16} - 39$

15. $10\frac{5}{12} - 6\frac{3}{16}$

16. $16\frac{7}{12} - 10\frac{5}{18}$

17. $13\frac{13}{28} - 9\frac{17}{28}$

18. $105\frac{23}{40} - 68\frac{31}{40}$

19. $49 - 15\frac{3}{7}$

20. $98 - 73\frac{1}{6}$

21. $18 - \frac{7}{18}$

22. $24 - \frac{5}{6}$

23. $39\frac{17}{20} - 22\frac{7}{8}$

24. $56\frac{5}{8} - 38\frac{19}{24}$

25. $21\frac{14}{31} - 18\frac{17}{31}$

26. $121\frac{9}{19} - 106\frac{15}{19}$

27. $44\frac{6}{19} - 39\frac{6}{19}$

28. $72\frac{7}{8} - 56\frac{7}{8}$

29. $2 - \frac{11}{12}$

30. $72 - \frac{15}{37}$

31. $33\frac{19}{27} - 9\frac{8}{27}$

32. $256\frac{21}{37} - 198\frac{8}{37}$

33. $56 - 13\frac{5}{17}$

34. $78 - 43\frac{13}{51}$

35. $50\frac{5}{18} - 17$

36. $18\frac{7}{13} - 5$

37. $46\frac{3}{7} - 16\frac{2}{5}$

38. $53\frac{4}{25} - 29\frac{2}{15}$

39. $15\frac{5}{18} - 7\frac{11}{12}$

40. $117\frac{5}{12} - 89\frac{7}{9}$

41. $5\frac{19}{64} - 4\frac{31}{48}$

42. $39\frac{3}{26} - 18\frac{5}{38}$

43. $31\frac{7}{27} - 21\frac{11}{24}$

44. $309\frac{11}{76} - 156\frac{7}{32}$

The answers to review and odd-numbered exercises are in the back of the book.

6–6 APPLIED PROBLEMS

■ Example 1

A rectangular plot of land measures $1\frac{5}{9}$ rods by $5\frac{1}{4}$ rods. What is the area of the land in square rods?

Solution To find the area, we multiply the length by the width.

$$1\frac{5}{9} \times 5\frac{1}{4}$$

$$= \frac{\overset{7}{\cancel{14}}}{\underset{3}{\cancel{9}}} \times \frac{\overset{7}{\cancel{21}}}{\underset{2}{\cancel{4}}}$$

$$= \frac{49}{6}, \quad \text{or} \quad 8\frac{1}{6}$$

The area is $8\frac{1}{6}$ square rods.

■ Example 2

Sam Alley found that a crate of 4 bowling balls weighs $38\frac{1}{2}$ pounds. What is the average weight of each bowling ball?

Solution Sometimes the method to use in solving the problem becomes clear when we substitute whole numbers for the original numbers. Suppose instead that Sam Alley found that a crate of 4 bowling balls weighs 20 pounds, and we had to find the average weight of each ball. If we had the weight of each of the bowling balls we would *multiply* the weight of a ball by 4 to get 20 pounds. Since we don't have the weight of each ball, we *divide:* $20 \div 4 = 5$. Thus, the average weight is 5 pounds.

Now back to the original problem—to find the weight of each ball, we *divide.*

$$38\frac{1}{2} \div 4$$

$$= \frac{77}{2} \div \frac{4}{1}$$

$$= \frac{77}{2} \times \frac{1}{4}$$

$$= \frac{77}{8}, \quad \text{or} \quad 9\frac{5}{8}$$

The average weight of each bowling ball is $9\frac{5}{8}$ pounds.

■ Example 3

Alice drove her "semi" truck $53\frac{5}{6}$ miles to Junction City, then drove another $79\frac{1}{4}$ miles to Coulee. How many miles did she drive?

Solution We need to find the *total* miles driven, so this is an addition problem.

$$\begin{array}{rcl} 56\frac{5}{6} & = & 56\frac{10}{12} \\ +79\frac{1}{4} & = & 79\frac{3}{12} \\ \hline & & 135\frac{13}{12}, \quad \text{or} \quad 136\frac{1}{12} \end{array}$$

Alice drove $136\frac{1}{12}$ miles.

Answers to exercises 108–112

108. $14\frac{5}{8}$ **109.** $17\frac{15}{19}$ **110.** $7\frac{13}{16}$

111. $14\frac{9}{10}$ **112.** $22\frac{23}{24}$

Example 4

Before Sam began to jog, he weighed $172\frac{3}{16}$ pounds. After jogging 10 miles, Sam weighed $163\frac{5}{8}$ pounds. How many pounds did he lose?

Solution We must find the difference on this one.

$$\begin{array}{rcrcr} 172\frac{3}{16} & = & 172\frac{3}{16} & = & 171\frac{19}{16} \\ -163\frac{5}{8} & = & 163\frac{10}{16} & = & 163\frac{10}{16} \\ \hline & & & & 8\frac{9}{16} \end{array}$$

Sam lost $8\frac{9}{16}$ pounds.

Practice check: Do exercises 113 and 114 in the margin.

Solve.

113. How many pieces of pipe each $1\frac{1}{2}$ inches long can be cut from a pipe $22\frac{1}{8}$ inches long?

114. Find the miles per gallon of a Ace Delivery van that has traveled 512 miles on $25\frac{3}{5}$ gallons of gasoline.

The answers are in the margin on page 241.

Looking Back

You should now be able to solve applied problems using mixed numbers and the four basic operations.

Problems 6–6

Solve. Write a statement for your answer.

1. **Confectioner.** How many pounds of candy are there in 5 boxes if each box contains candy that weighs $2\frac{1}{4}$ pounds?

2. **Generic.** A car traveled for $5\frac{3}{8}$ hours. If it averaged 50 miles per hour, how many miles did it travel?

3. **Transportation.** Sandra drove 68 miles in $3\frac{2}{5}$ hours. What was her average speed?

4. **Construction.** If $18\frac{2}{3}$ yards of concrete are needed for the foundation of one house and $12\frac{3}{8}$ yards are needed for another house, how many yards of concrete are needed altogether?

5. **Auto Mechanics.** A fuel oil barrel has a capacity of 50 gallons. If it took $34\frac{5}{8}$ gallons to fill the tank, how many gallons were in the tank before it was filled?

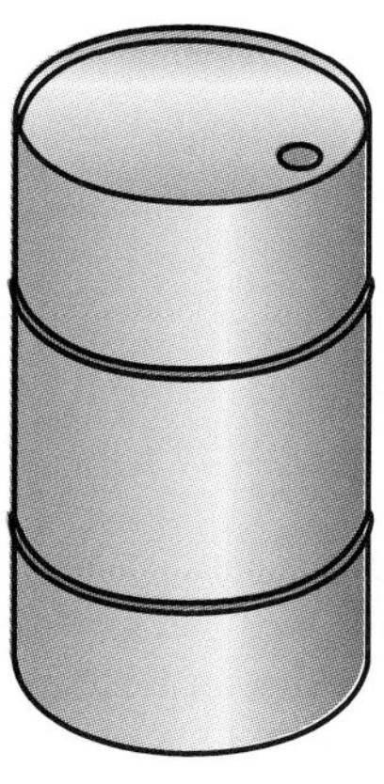

6. **Librarian.** John Counts, a part-time librarian, worked $3\frac{1}{5}$ hours on Monday, $4\frac{3}{4}$ hours each on Tuesday and Wednesday, 4 hours on Thursday, and $3\frac{3}{5}$ hours on Friday. How many hours did he work that week?

7. **Interior Decorator.** Figuring a bedroom carpeting job for his company, Jack Fort found the measurements to be $5\frac{5}{8}$ yards by $4\frac{7}{12}$ yards. How many square yards of carpeting will he need to carpet the bedroom?

8. **Chauffeur.** Maria Santini, a chauffeur, used $7\frac{7}{10}$ gallons of gasoline in the morning and $8\frac{1}{2}$ gallons in the afternoon. How many gallons did she use that day?

9. **Janitorial.** A drum of liquid wax weights $46\frac{2}{9}$ pounds. If the drum weighs $11\frac{11}{12}$ pounds, how much does the liquid wax weigh?

10. **Carpentry.** How many pieces of board measuring $8\frac{3}{4}$ inches can be cut from a board measuring $52\frac{1}{2}$ inches?

11. **Generic.** Sam Jenkins can walk an average of $3\frac{3}{8}$ miles per hour. How long would it take him to walk $7\frac{5}{9}$ miles?

12. **Transportation.** By traveling a different route, a bus can now make a certain trip in $5\frac{5}{6}$ hours. This saves $1\frac{3}{8}$ hours over the old route. How many hours did the bus take using the old route?

13. **Landscape.** Bill can dig $4\frac{1}{2}$ post holes in one hour. How many post holes can he dig in $5\frac{1}{3}$ hours?

14. **Landscape.** One case of 15 identical statues weighs $35\frac{5}{7}$ pounds. What is the weight of each statue?

15. **Auto Mechanic.** Rita Grubb can give her car a complete tune-up in $3\frac{1}{4}$ hours. A qualified mechanic can do the same job in $1\frac{7}{8}$ hours. How much time is saved by having the mechanic do the tune-up?

16. **Meteorologist.** In 1997, it rained $18\frac{5}{6}$ inches in Watson County. If the record rainfall is $35\frac{5}{9}$ inches, by how many inches did the rainfall in 1997 fall short of the record?

17. **Carpentry.** If the total thickness of 18 identical books is $23\frac{1}{4}$ inches, what is the thickness of each book?

18. **Chef.** Two recipes required different amounts of flour; one called for $5\frac{1}{3}$ cups, and the other for $3\frac{1}{4}$ cups. How many cups of flour did the two recipes call for?

19. **Generic.** Find the distance around the sides of a triangle whose sides measure $3\frac{2}{3}$ inches, $4\frac{1}{6}$ inches, and $3\frac{7}{12}$ inches.

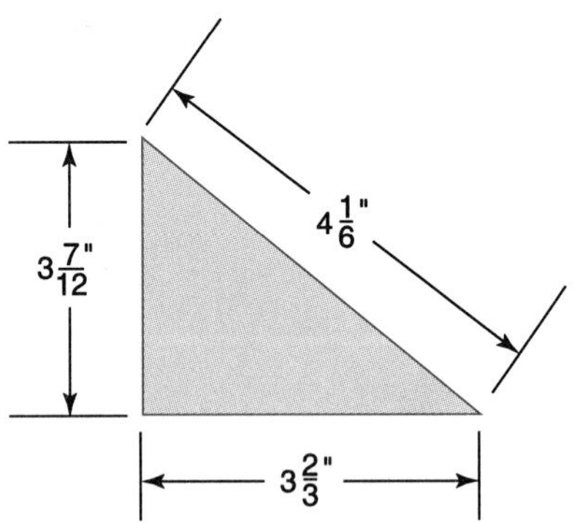

20. **Surveyor.** A certain city lot contained $1\frac{5}{6}$ acres. If the house covered $\frac{2}{9}$ of the lot, how many acres did the house cover?

21. **Finance.** American Consolidated opened at $78\frac{5}{16}$ on the stock exchange and closed at $78\frac{1}{8}$. Find the net change in the stock.

22. **Generic.** Bonnie Jackson sold $16\frac{5}{6}$ tons of scrap metal from a total of $54\frac{5}{8}$ tons. How many tons did she have left to sell?

The answers to odd-numbered exercises are in the back of the book.

CHAPTER 6 OVERVIEW

Summary

6–1 You learned to change an improper fraction to a mixed number, and a mixed number to an improper fraction.

6–2 You learned to multiply numbers in the form of proper fractions, one, zero, improper fractions, and mixed numbers.

6–3 You learned to divide numbers in the form of proper fractions, whole numbers, improper fractions, and mixed numbers.

6–4 You learned to add numbers in the form of mixed numbers, whole numbers, and proper fractions.

6–5 You learned to subtract numbers in the form of mixed numbers, whole numbers, and proper fractions with or without "borrowing."

6–6 You learned to solve application problems involving mixed numbers using addition, subtraction, multiplication, or division.

Terms To Remember

	Page		*Page*
Mixed numbers	210	"Borrowing"	230

Rules

- Any number that is a combination of a whole number and a proper fractional number is called a *mixed number.*
- A fraction always indicates division of the numerator by the denominator.
- A short method for changing an improper fraction to a mixed number:

  Step 1 Divide the numerator by the denominator.

  Step 2 Write the mixed number. The quotient is the whole number part. The remainder is the numerator, and the divisor is the denominator of the fraction part. The result of Step 1 is a whole number only if the remainder is 0.

- A short method for changing a mixed number to an improper fraction:

  Step 1 Multiply the whole number part by the denominator of the fraction part, and to this result add the numerator of the fraction part.

  Step 2 Write the improper fraction. Use the result of Step 1 for the numerator. The denominator remains the same.

- The product of one and a fraction is that fraction.
- The product of zero and a fraction is zero.
- When multiplying with whole numbers and mixed numbers, first change them to improper fractions, then find the product.
- When we divide a number by 1, the quotient is always the other number.
- When we divide zero by any number other than zero, the quotient is always zero.
- Remember: Never divide by zero! It is impossible.
- When division involves whole numbers and mixed numbers, first change them to improper fractions, then find the quotient.

Answers to exercises 113–114

113. The number of pieces that can be cut is 14.

114. The van traveled 20 miles per gallon.

- A short method for adding mixed numbers with unlike denominators:

  Step 1 Place one number over the other and write in equal signs.

  Step 2 Carry over the whole numbers.

  Step 3 Find the LCD and write in the equal fractions.

  Step 4 Add and simplify.

- A short method for "borrowing" when subtracting with mixed numbers:

  Step 1 Take 1 from the whole number part and write the new whole number.

  Step 2 Mentally change the 1 that was "borrowed" to a fractional name for 1, using the denominator of the mixed number.

  Step 3 Mentally add the fractional name for one to the fraction part of the mixed number.

  Step 4 Place the result of Step 3 next to the new whole number.

- A short method for subtracting with mixed numbers involving "borrowing":

  Step 1 Place one number over the other.

  Step 2 If the numbers to be subtracted have unlike fractions, find the LCD and write equal mixed numbers.

  Step 3 If the bottom fraction is larger than the top fraction, "borrow," and write another equal top mixed number.

  Step 4 Place the unchanged bottom number directly below the top number from Step 3 and subtract.

Self-Test

The answers are in the back of the book.

6–1 Change the following fractions to mixed numbers.

1. $\frac{8}{3}$

2. $\frac{41}{19}$

Change the following mixed numbers to improper fractions.

3. $6\frac{3}{7}$

4. $31\frac{4}{9}$

6–2 Multiply. Leave the answers as mixed numbers when possible.

5. $1 \times \frac{5}{8}$

6. $0 \times 5\frac{1}{2}$

7. $2\frac{3}{4} \times 4$

8. $8\frac{1}{2} \times 2\frac{2}{3}$

6–3 Divide. Leave the answers as mixed numbers when possible.

9. $6\frac{7}{8} \div 1$ **10.** $4 \div 2\frac{3}{4}$ **11.** $0 \div 12\frac{1}{4}$ **12.** $4\frac{4}{15} \div 6\frac{2}{5}$

6–4 Add and simplify.

13. $23\frac{5}{12} + 11\frac{7}{12}$ **14.** $11 + 5\frac{3}{8}$ **15.** $23\frac{7}{15} + 56\frac{2}{5}$ **16.** $21\frac{5}{18} + 50\frac{7}{12}$

6–5 Subtract and simplify.

17. $\begin{array}{r} 27\frac{5}{9} \\ -22\frac{2}{9} \\ \hline \end{array}$ **18.** $\begin{array}{r} 31\frac{23}{24} \\ -25\frac{5}{8} \\ \hline \end{array}$ **19.** $\begin{array}{r} 44 \\ -16\frac{2}{3} \\ \hline \end{array}$ **20.** $\begin{array}{r} 69\frac{7}{12} \\ -52\frac{13}{18} \\ \hline \end{array}$

21. $41\frac{2}{7} - 30$

6–6 Solve.

22. How many $8\frac{1}{2}$-in. lengths of conduit can be cut from a 72-in. length of conduit.

23. Janis needs to mail 35 copies of a book each weighing $1\frac{1}{4}$ pounds. What is the weight of all the books?

24. Find the total distance (perimeter) around the triangle below.

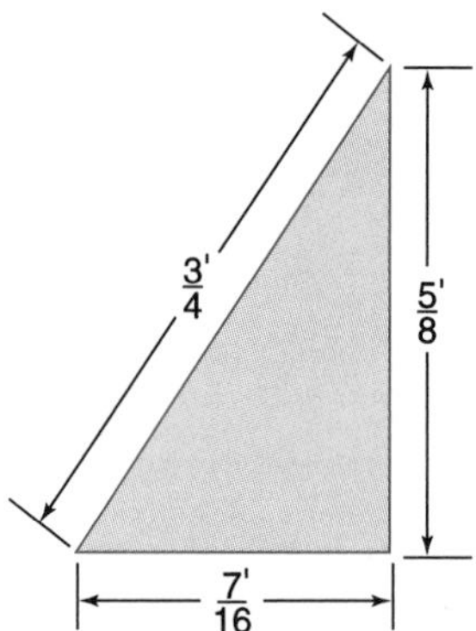

25. One end of a tapered shaft has a diameter of $3\frac{5}{16}$ inches while the smaller end is $1\frac{5}{32}$ inches in diameter. Find the difference in the diameters.

Chapter Test

Directions: This test will aid in your preparation for a possible chapter test given by your instructor. The answers are in the back of the book. Go back and review in the appropriate section(s) if you missed any test items.

6–1 Change the following fractions to mixed numbers.

1. $\frac{9}{2}$

2. $\frac{56}{17}$

Change the following mixed numbers to improper fractions.

3. $7\frac{5}{8}$

4. $29\frac{3}{7}$

6–2 Multiply. Leave the answers as mixed numbers when possible.

5. $\frac{5}{7} \times 1$ **6.** $\frac{8}{9} \times 30$ **7.** $2\frac{1}{8} \times 0$ **8.** $7\frac{3}{5} \times 18$

9. $\frac{5}{8} \times 2\frac{2}{7}$ **10.** $3\frac{3}{4} \times 7\frac{1}{3}$

6–3 Divide. Leave the answers as mixed numbers when possible.

11. $5\frac{3}{8} \div 1$ **12.** $6 \div 7\frac{1}{7}$ **13.** $0 \div 26\frac{5}{7}$ **14.** $15\frac{3}{4} \div 13\frac{1}{2}$

15. $\frac{7}{12} \div 1\frac{17}{18}$ **16.** $3\frac{5}{13} \div 0$

6–4 Add and simplify.

17. $15\frac{2}{9} + 23\frac{4}{9}$ **18.** $5 + 8\frac{7}{16}$ **19.** $56\frac{3}{8} + 9\frac{3}{4}$ **20.** $19\frac{5}{16} + 24\frac{7}{24}$

21. $\frac{9}{14} + 15\frac{8}{21}$

6–5 Subtract and simplify.

22. $\begin{array}{r} 31\frac{4}{21} \\ -19\frac{4}{21} \\ \hline \end{array}$ **23.** $\begin{array}{r} 51\frac{17}{24} \\ -38\frac{3}{8} \\ \hline \end{array}$ **24.** $45 - 8\frac{5}{9}$ **25.** $8\frac{7}{9} - 1\frac{5}{12}$

26. $\begin{array}{r} 18\frac{13}{33} \\ -14\frac{9}{33} \\ \hline \end{array}$ **27.** $106\frac{4}{7} - 99$ **28.** $\begin{array}{r} 58\frac{29}{41} \\ -39\frac{33}{41} \\ \hline \end{array}$ **29.** $5 - \frac{7}{13}$

6–6 **30.** If lettuce sells for 12¢ a pound, how much would $3\frac{5}{6}$ pounds cost?

31. One crate of hammers weighed $77\frac{2}{3}$ pounds, and another crate weighed $69\frac{4}{9}$ pounds. How many pounds did the two crates weigh altogether?

32. If $52\frac{1}{5}$ pounds of bone meal cost $17\frac{2}{5}$ dollars, what is the cost per pound?

33. A $20\frac{9}{16}$-in. casting shrinks $\frac{3}{8}$ in. when cooled. Find the dimension of the casting after cooling.

Test Your Memory

These problems review Chapters 1–6.

1. Name the digits in 3104.

2. Write an expanded number for 20,503.

3. Write a word name for 351,068,009.

4. Give a number meaning to the digit 7 in 568,070,566.

Place < or > symbols between the following numbers to make a true statement:

5. 0   3

6. 667   666

7. 9381   9383

Round to the nearest ten.

8. 25

9. 564

Round to the nearest hundred.

10. 738

11. 2889

Round to the nearest thousand.

12. 8461

13. 56,581

Add using the short method.

14.
$$\begin{array}{r} 28 \\ +37 \\ \hline \end{array}$$

15. 3992 + 5603

16. Add.
```
 3224
   68
  752
 3021
```

17. Estimate the sum to the nearest ten. $76 + 42 + 53 + 65$

Solve.

18. The populations of four cities in Alaska are Anchorage, 48,081; Fairbanks, 14,771; Fort Richardson, 10,751; and Seward, 18,089. Find the total population of all four cities.

Subtract using the short method.

19. $\begin{array}{r} 9665 \\ \underline{2431} \end{array}$

20. $\begin{array}{r} 737 \\ \underline{473} \end{array}$

21. $\begin{array}{r} 6430 \\ \underline{3745} \end{array}$

Multiply.

22. $\begin{array}{r} 431 \\ \underline{\times 6} \end{array}$

23. $\begin{array}{r} 5108 \\ \underline{\times 8} \end{array}$

Multiply by writing the answer directly.

24. 40×34

25. $\begin{array}{r} 304 \\ \underline{\times 6000} \end{array}$

Multiply.

26. $\begin{array}{r} 571 \\ \times 329 \\ \hline \end{array}$

27. $\begin{array}{r} 5007 \\ \times 803 \\ \hline \end{array}$

28. A company makes 31 cars a day. How many cars will the company make in a year? (Assume 252 working days in a year.)

Divide.

29. $5\overline{)370}$

30. $7\overline{)15{,}183}$

31. $54\overline{)27{,}462}$

32. $745\overline{)51{,}405}$

33. Jack drove 1798 miles in a 31-day month. How many miles per day did Jack drive?

34. Draw a number line and locate the point $\frac{3}{8}$.

35. Write a fractional number for the shaded part.

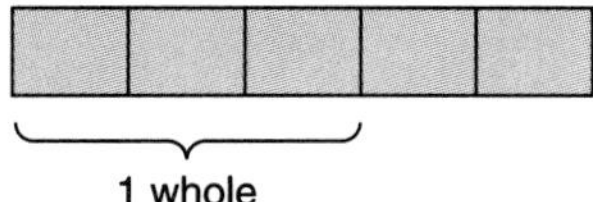

36. Find the prime factorization of 28.

Multiply. Use cancelation where possible.

37. $\frac{2}{3} \times \frac{21}{24}$

38. $\frac{2}{9} \times \frac{3}{7} \times \frac{14}{15}$

39. The distance around the track is $\frac{3}{4}$ of a mile. The first time around, Kristen jogged $\frac{2}{3}$ of the distance and stopped to rest. How far did she jog?

Divide.

40. $\frac{2}{3} \div \frac{3}{7}$

41. $\frac{5}{9} \div \frac{5}{6}$

42. Simplify. $\frac{\frac{5}{6}}{\frac{2}{3}}$

Add and reduce where possible.

43. $\frac{2}{8} + \frac{5}{8}$

44. $\frac{5}{6} + \frac{1}{6}$

45. $\frac{5}{7} + \frac{3}{5}$

46. $\frac{9}{10} + \frac{7}{12}$

47. In one evening, Reed M. Fast read $\frac{3}{7}$ of a book. The next evening, he read $\frac{1}{4}$ more of the book. What fraction of the book did he read in these two evenings?

Subtract.

48. $\dfrac{17}{27} - \dfrac{8}{27}$

49. $\dfrac{4}{5} - \dfrac{2}{7}$

50. $\dfrac{5}{8} - \dfrac{7}{12}$

51. Decide which fraction is larger, $\dfrac{7}{16}$ or $\dfrac{5}{8}$.

52. Jim takes $\frac{3}{5}$ of an hour to cut a lawn. Bill takes $\frac{1}{2}$ of an hour to cut the same lawn. How much more time does Jim take?

Change the following fractions to mixed numbers.

53. $\dfrac{7}{5}$

54. $\dfrac{15}{2}$

Change the following mixed numbers to improper fractions.

55. $3\dfrac{1}{8}$

56. $13\dfrac{2}{7}$

Multiply and simplify.

57. $\dfrac{7}{16} \times 12$

58. $4\dfrac{2}{3} \times 3\dfrac{3}{4}$

Divide and simplify.

59. $5 \div 6\frac{1}{4}$

60. $1\frac{1}{6} \div 9\frac{1}{3}$

Add and simplify.

61. $2\frac{5}{8} + 1\frac{2}{8}$

62. $6\frac{5}{12} + 21$

63. $2\frac{3}{8} + 1\frac{5}{12}$

64. $\frac{11}{18} + 5\frac{7}{9}$

Subtract and simplify.

65. $9\frac{5}{8} - 4\frac{7}{8}$

66. $36 - 18\frac{5}{16}$

67. $67\frac{7}{10} - 31\frac{3}{5}$

68. $11\frac{5}{16} - 5\frac{3}{4}$

Solve.

69. A car traveled for $4\frac{2}{5}$ hours. If it averaged 45 miles per hour, how many miles did it travel?

70. From a cedar post $9\frac{3}{4}$ ft long, a carpenter cut a piece $5\frac{7}{8}$ ft long. How much of the post remained?

The answers are in the back of the book.

PART THREE

Decimals, Ratio and Proportion, and Percent

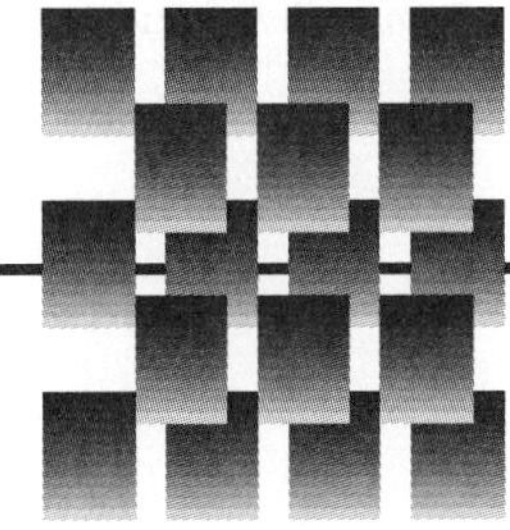

CHAPTER 7

Addition and Subtraction of Decimals

Choice: Either skip the Pretest and study all of Chapter 7 or take the Pretest and study only the sections you need to study.

Pretest

Directions: This Pretest will help you determine which sections to study in Chapter 7. If any of your answers are incorrect, or if you omitted any exercise, turn to the indicated section(s) and study the material; then take the Chapter Test.

7–1 **1.** Name the fifth place to the right of the decimal point.

2. Write an expanded number for 7.042.

7–2 **3.** Write a word name for 6.428.

7–3 **4.** Which is larger, 0.040 or 0.4?

7–4 Add.

5. $4.031 + 0.0018 + 2$

6. $0.08 + 8 + 0.8$

7–5 **7.** Estimate the sum to the nearest tenth: $53.301 + 13.652$.

Subtract.

7–6 **8.** 32.691 − 1.350

7–7 **9.** 53.461 − 15.18

10. 30 − 16.98

7–8 **11.** Sandra paid the following amounts for clothing: $43.16, $17.95, $152.23, and $5.07. How much did she spend on clothing?

12. Sam had a checkbook balance of $232.71. He wrote checks for $43.98 and $3.08. What is his new balance?

13. Ace Printers received a shipment of printing supplies from a wholesale company. They were as follows: fax paper, 46.4 lb; litho developer, 29.7 lb; toner, 17.6 lb; and litho paper, 126.2 lb. Find the total weight of the shipment.

14. Find the missing measurements on the figure.

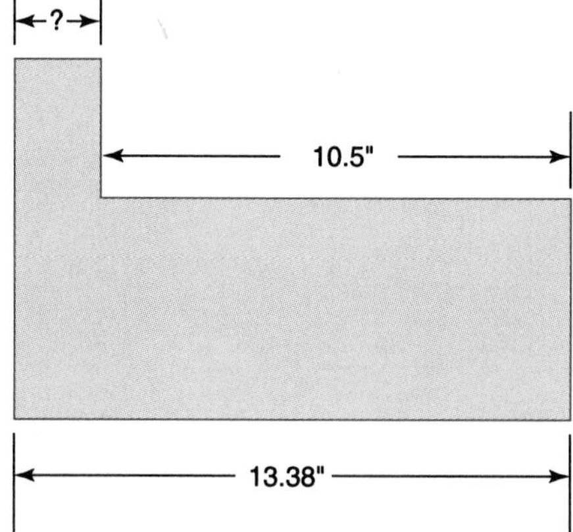

The answers are on page 258.

7–1 DECIMAL NUMBERS AND PLACE VALUE

From sections 1–2 and 1–3 we learned about place value of whole numbers. If each square below represents a digit, the place value of each digit would be as follows:

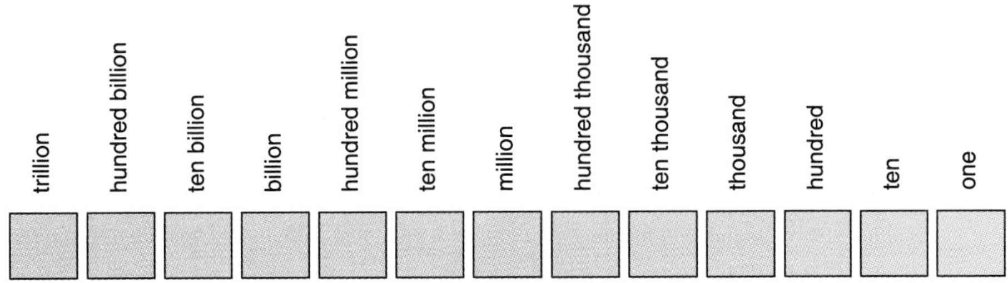

Figure 7–1

The word name for 6,406,349,170 is six billion, four hundred six million, three hundred forty-nine thousand, one hundred seventy.

Practice check: Do exercises 1 and 2 in the margin.

1. In the number 378,410,056 state the value of

   a. 8 **b.** 5 **c.** 4 **d.** 7

2. Write a word name for 5,006,987.

The answers are in the margin on page 258.

From Section 6–1 we learned that a number like $6\frac{3}{10}$ is called a *mixed number.* The 6 is a whole number and the $\frac{3}{10}$ is a proper fraction. By separating the whole number and the fraction with a *decimal point,* we can write this type of number with digits.

6.3
↑
Decimal
point

We call numbers like 6.3, 206.541, 0.6004, and 4,051.210 *decimal numbers,* or just plain *decimals.* Note that the decimal point is placed to the right of the ones place.

Here is the place value chart for decimal numbers.

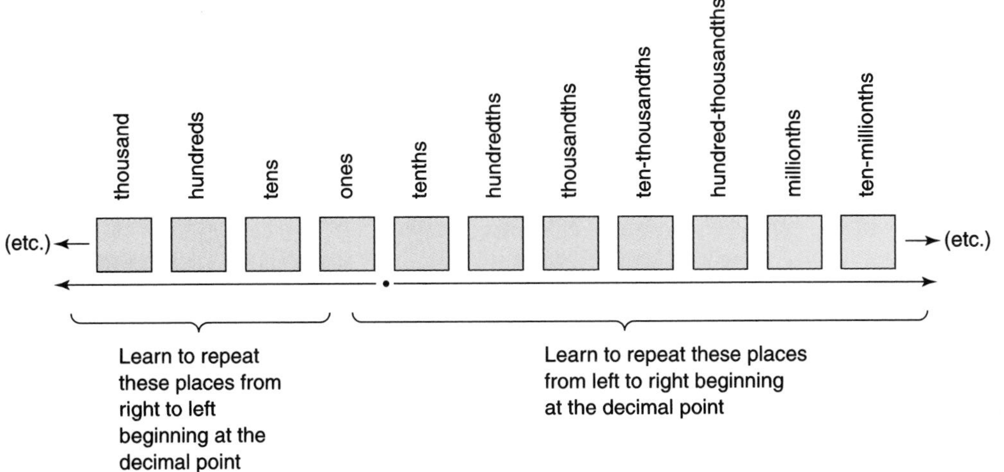

Figure 7–2

Note also that the decimal point separates the ones place and the tenths place.

Practice check: Do exercises 3–7 in the margin.

3. Name the fifth place to the right of the decimal point.
4. Name the eighth place to the right of the decimal point.
5. Name the sixth place to the left of the decimal point.
6. Name the sixth place to the right of the decimal point.
7. Name the ninth place to the right of the decimal point.

The answers are in the margin on page 258.

Answers to Pretest

Circle the answers you missed or omitted. Cross them off after you've studied the corresponding section(s); then take the Chapter Test.

1. hundred thousandths

2. $7 + \frac{0}{10} + \frac{4}{100} + \frac{2}{1000}$

3. six and four hundred twenty-eight thousandths.
4. 0.4 **5.** 6.0328 **6.** 8.88
7. 67.0 **8.** 31.341 **9.** 38.281
10. 13.02 **11.** \$218.41
12. \$185.65 **13.** 219.9 lb
14. 2.88″

Answers to exercises 1 and 2

1. a. 8 million **b.** 5 tens
c. 4 hundred thousand
d. 70 million
2. Five *million,* six *thousand,* nine hundred eighty seven.

Answers to exercises 3–7

3. hundred thousandths
4. hundred millionths
5. hundred thousandths
6. millionths **7.** billionths

Expanding Decimals

Just as we expanded whole numbers in section 1–2, we can expand decimals.

■ Example 1

Write an expanded number for 0.205. A zero is usually placed in the ones place to show that there are no ones.

Solution

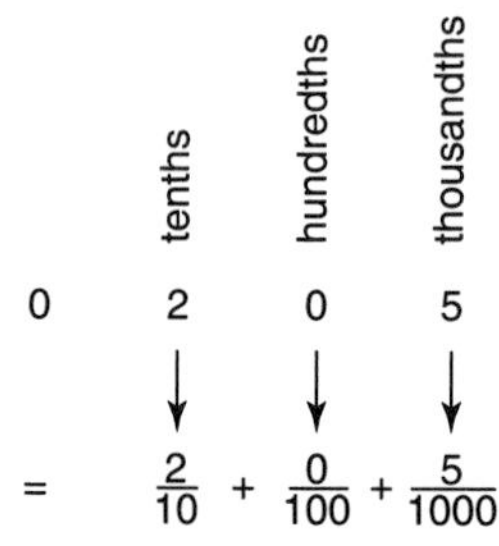

Figure 7–3

■ Example 2

Write an expanded number for 542.342.

Solution

hundreds	tens	ones	tenths	hundredths	thousandths
5	4	2	3	4	2

$$= 500 + 40 + 2 + \frac{3}{10} + \frac{4}{100} + \frac{2}{1000}$$

Figure 7–4

Practice check: Do exercises 8–13 in the margin.

Write an expanded number for

8. 0.678 **9.** 0.6103
10. 21.18 **11.** 436.03
12. 70.0046 **13.** 7046.5904

The answers are in the margin on page 261.

We can rewrite an expanded number as a decimal number.

■ Examples

Expanded Number	*Decimal Number*
3. $\frac{5}{10}$	0.5
4. $5 + \frac{2}{10} + \frac{7}{100}$	5.27
5. $20 + 6 + \frac{0}{10} + \frac{5}{100} + \frac{3}{1000}$	26.053
6. $200 + 7 + \frac{7}{10} + \frac{5}{1000}$	207.705

In Example 6, the zeros in the tens and hundredths place are *placeholders* showing that there are no tens and hundredths.

Practice check: Do exercises 14–18 in the margin.

Write as decimal number.

14. $\frac{7}{10}$ **15.** $7 + \frac{6}{10} + \frac{9}{100}$

16. $50 + \frac{3}{10} + \frac{5}{100}$

17. $100 + 50 + \frac{2}{100} + \frac{4}{1000}$

18. $3000 + 60 + \frac{2}{10} + \frac{8}{10{,}000}$

The answers are in the margin on page 261.

■ Example 7

Write a decimal number for $\frac{2}{100}$ (two hundredths).

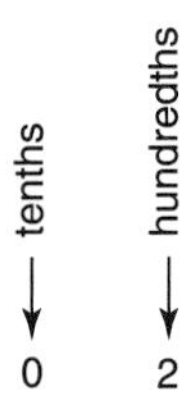

Figure 7–5

The 2 must be placed in the hundredths place. We write a zero in the tenths place as a placeholder to show that there are no tenths.

$$\frac{2}{100} = 0.02$$

■ Example 8

Write a decimal number for $23\frac{9}{1000}$.

(Don't confuse this end of sentence period with a decimal point.)

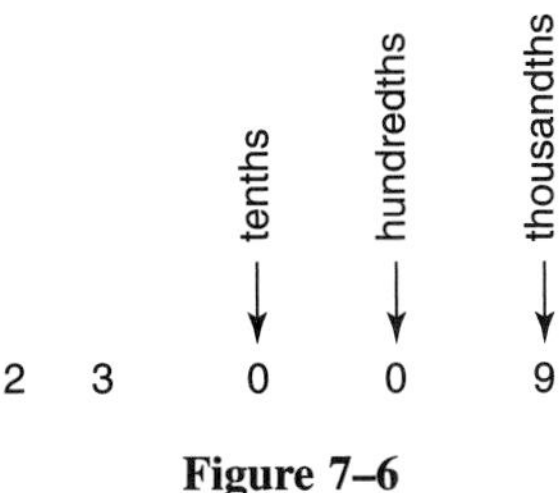

Figure 7–6

Practice check: Do exercises 19–23 in the margin.

Write as decimal numbers.

19. $\frac{5}{10}$ **20.** $\frac{8}{1000}$ **21.** $17\frac{6}{10}$

22. $\frac{1}{1{,}000{,}000}$ **23.** $560\frac{3}{10{,}000}$

The answers are in the margin on page 261.

Examples 9 and 10 show how to rewrite decimals containing zeros as placeholders as fractional numbers.

■ Example 9

Write a fractional number for 0.03.

Solution 0.03
↗
The 3 is in the hundredths place, so $0.03 = \frac{3}{100}$.

■ Example 10

Write a mixed number for 3.0009.

Solution 3.0009

The 9 is in the ten-thousandths place, so $3.0009 = 3\frac{9}{10,000}$.

Practice check: Do exercises 24–28 in the margin.

Write as a fractional numbers or as a mixed number.

24. 0.06 **25.** 19.7

26. 0.0004 **27.** 106.009

28. 0.00001

The answers are in the margin on page 262.

■ Looking Back

You should now be able to

1. Name the place values in the decimal number system.
2. Write an expanded number from a decimal, or a decimal from an expanded number.
3. Convert a fraction like $\frac{3}{100}$ to decimals, and convert a decimal to a fraction or a mixed number.

■ Problems 7–1

1. Name the second place to the right of the decimal point.
2. Name the sixth place to the right of the decimal point.
3. Name the eighth place to the left of the decimal point.
4. Name the fourth place to the right of the decimal point.
5. Name the sixth place to the left of the decimal point.

Write an expanded number for

6. 0.4 **7.** 24.46 **8.** 0.04 **9.** 9.6

10. 4.704 **11.** 0.3004 **12.** 206.507 **13.** 3090.5

14. 80.07 **15.** 5002.007 **16.** 0.6914 **17.** 10,501.0009

Write a decimal number for

18. $\frac{2}{10}$ **19.** $\frac{3}{10} + \frac{5}{100}$ **20.** $15 + \frac{7}{10} + \frac{7}{100}$ **21.** $700 + 30 + 5 + \frac{3}{100}$

22. $30 + \frac{5}{10}$ **23.** $9 + \frac{8}{10} + \frac{2}{1000}$ **24.** $\frac{5}{1000} + \frac{1}{10,000}$

25. $800 + 7 + \frac{6}{10} + \frac{9}{1000}$ **26.** $4000 + 80 + \frac{6}{100}$

27. $5000 + 1 + \frac{1}{10} + \frac{7}{10,000}$

Write a decimal number for

28. $\frac{8}{100}$ **29.** $\frac{5}{10}$ **30.** $\frac{6}{10,000}$ **31.** $5\frac{7}{100}$

32. $\frac{7}{1,000,000}$ **33.** $31\frac{1}{10,000,000}$ **34.** $\frac{2}{1000}$ **35.** $356\frac{4}{100,000}$

36. $\frac{3}{10}$ **37.** $19\frac{9}{1000}$ **38.** $6504\frac{5}{10,000}$

Write a fractional number or a mixed number for

39. 0.007 **40.** 0.01 **41.** 0.00002 **42.** 8.4

43. 0.0001 **44.** 0.005 **45.** 68.000009 **46.** 0.6

47. 579.0004 **48.** 1009.08

The answers to odd-numbered exercises are in the back of the book.

Answers to exercises 8–13

8. $\frac{6}{10} + \frac{7}{100} + \frac{8}{1000}$

9. $\frac{6}{10} + \frac{1}{100} + \frac{0}{1000} + \frac{2}{10,000}$

10. $20 + 1 + \frac{1}{10} + \frac{8}{100}$

11. $400 + 30 + 6 + \frac{0}{10} + \frac{3}{100}$

12. $70 + \frac{0}{10} + \frac{0}{100} + \frac{4}{1000} + \frac{6}{10,000}$

13. $7000 + 40 + 6 + \frac{5}{10} + \frac{9}{100} + \frac{0}{1000} + \frac{4}{10,000}$

Answers to exercises 14–18

14. 0.7 **15.** 7.69 **16.** 50.35
17. 150.024 (*Note*: There are no tenths, so we write a zero as a placeholder in the tenths place.)
18. 3060.2008

Answers to exercises 19–23

19. 0.5 **20.** 0.008 **21.** 17.6
22. 0.000001 **23.** 560.0003

Answers to exercises 24–28

24. $0.06 = \frac{6}{100}$ **25.** $19.7 = 19\frac{7}{10}$

26. $0.0004 = \frac{4}{10{,}000}$

27. $106.009 = 106\frac{9}{1000}$ (thousandths)

28. $0.00001 = \frac{1}{100{,}000}$ (hundred thousandths)

7–2 DECIMAL WORD NAMES

When we say or write decimal word names, we must carefully note place value.

Writing Word Names from Decimals

■ **Example 1**

Write the word name for 0.68

Solution **Step 1** To say or write word names for decimals between 0 and 1, we first determine the place value of the last number

0.68 ← Hundredths

Step 2 Next we say or write the number as if the decimal were not there.

Sixty-eight

Step 3 Then, we say or write out the place value of the number on the far right, placing it after the number in Step 2.

Sixty-eight hundredths

■ **Example 2**

Write the word name for 0.201.

Solution **Step 1** Determine the place value of the last number.

0.201 Thousandths

Step 2 Say or write the number as if the decimal were not there.

Two hundred one

Step 3 End with thousandths.

Two hundred one thousandths

Practice check: Do exercises 29–35 in the margin.

Write word names for

29. 0.2 **30.** 0.05 **31.** 0.32

32. 0.364 **33.** 0.509

34. 0.7075 **35.** 0.00941

The answers are in the margin on page 265.

Next we will learn to say or write word names for decimals that are greater than 1.

■ **Example 3**

Write a word name for 46.14.

Solution

Hundredths

46 . 14

Forty-six and fourteen hundredths

Figure 7–7

(When we come to the decimal point, we read "and.")

■ **Example 4**

Write a word name for 105.074.

Solution

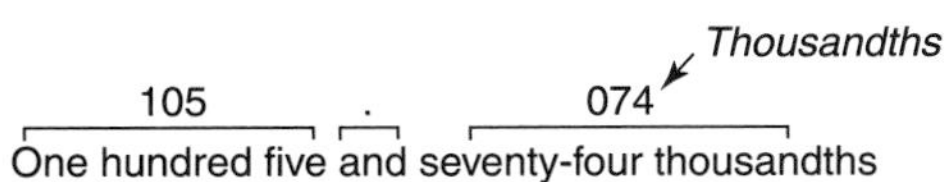

Figure 7–8

■ Example 5

Write a word name for $6\frac{15}{100}$.

Solution Six *and* fifteen hundredths.

Practice check: Do practice exercises 36–42 in the margin.

Write word names for

36. 3.56 **37.** 0.047

38. 12.345 **39.** 439.0017

40. 2.4013 **41.** $29\frac{36}{100}$

42. $538\frac{49}{1000}$

The answers are in the margin on page 265.

Writing Decimals from Decimal Word Names

0.6	The tenths are one place to the right of the decimal point.
0.19	The hundredths are two places to the right of the decimal point.
0.056	The thousandths are three places to the right of the decimal point.
0.6041	The ten-thousandths are four places to the right of the decimal point.

■ Example 6

Write a decimal number for sixty-two hundredths.

Solution **Step 1** Write the number sixty-two.

62

Step 2 Place the decimal point. We need to have the last number in the hundredths place, or two places to the right of the decimal point, so the decimal point is placed in front of the 6.

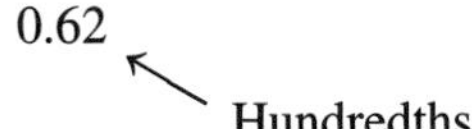

■ Example 7

Write a decimal number for twenty-eight thousandths.

Solution **Step 1** Write the number twenty-eight.

28

Step 2 Place the decimal point. We need to have the last number in the thousandths place, or three places to the right of the decimal point, so we need to place a zero in front of the 2 and the decimal point in front of the zero.

0.028
↗
Thousandths

■ Example 8

Write a decimal number for three hundred nineteen and fifty-three ten-thousandths.

Solution

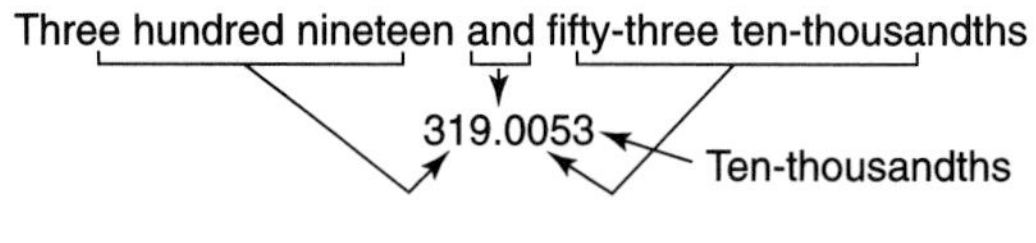

Figure 7–9

Write a decimal number for

43. five hundred fifteen and ninety-nine hundredths

44. twenty and two thousandths

45. six thousand four and thirty-eight-hundred-thousandths

The answers are in the margin on page 266.

Practice check: Do exercises 43–45 in the margin.

■ Looking Back

You should now be able to say or write word names for decimal numbers, and, conversely, write a decimal from a word name.

■ Problems 7–2

Review

1. Name the fourth place to the right of the decimal point.

2. Name the fourth place to the left of the decimal point.

3. Write a decimal number for $\frac{9}{10,000}$.

4. Write a fractional number for .007

Write an expanded number for

5. 0.503

6. 504.609

Write a decimal number for

7. $\frac{7}{10} + \frac{9}{10,000}$

8. $600 + 7 + \frac{5}{100} + \frac{1}{10,000}$

Write word names for

9. 0.8

10. 0.07

11. 0.91

12. 38.5

13. 248.19

14. 0.005

15. 0.063

16. 0.253

17. 197.649 **18.** 0.4960 **19.** 3.004 **20.** 15.00007

21. $22\frac{17}{100}$ **22.** $105\frac{23}{1000}$ **23.** $68\frac{204}{10,000}$ **24.** $1053\frac{2}{100,000}$

Write decimal numbers for

25. eight tenths

26. six and one tenth

27. six hundredths

28. four hundred seven and twenty-five hundredths

29. three hundred seven thousandths

30. nineteen and three thousand five hundred sixty ten-thousandths

31. eight hundred and thirty-five thousandths

32. seventy-four and fourteen hundredths

33. eighty-six and six hundred seventy-three hundred-thousandths

Answers to exercises 29–35

29. two tenths
30. five hundredths
31. thirty-two hundredths
32. three hundred sixty-four thousandths
33. five hundred nine thousandths
34. seven thousand seventy-five ten-thousandths
35. nine hundred forty-one hundred-thousandths

Answers to exercises 36–42

36. three and fifty-six hundredths
37. forty-seven thousandths
38. twelve and three hundred forty-five thousandths
39. four hundred thirty-nine and seventeen ten-thousandths
40. two and four thousand thirteen ten-thousandths
41. twenty-nine and thirty-six hundredths
42. five hundred thirty-eight and forty-nine thousandths

The answers to review and odd-numbered exercises are in the back of the book.

7–3 COMPARING DECIMALS

Sometimes we will need to know whether one decimal is larger or smaller than another, but first we must show that writing any number of zeros after the farthest right digit does not change its value.

■ Examples

1. $0.3 = 0.30$, since $0.3 = \frac{3}{10} = \frac{3}{10} \times \frac{10}{10} = \frac{30}{100} = 0.30$.

2. $0.41 = 0.4100$, since $0.41 = \frac{41}{100} = \frac{41}{100} \times \frac{100}{100} = \frac{4100}{10,000} = 0.4100$.

3. $0.308 = 0.3080$, since $0.308 = \frac{308}{1000} = \frac{308}{1000} \times \frac{10}{10} = \frac{3080}{10,000} = 0.3080$

Answers to exercises 43–45

43. 515.99 **44.** 20.002
45. 6004.00038

Rule

To compare decimal numbers we

1. Write as many extra zeros as necessary so that both decimals have the same number of places to the right of the decimal point.
2. Mentally drop the decimal point and compare them as you do whole numbers.

■ **Example 6**

Which number is larger, 0.8 or 0.67?

Solution **Step 1** Make the same number of places to the right of the decimal point in both numbers.

0.80 ↗ 2 places   0.67 ↗ 2 places

(We made two places to the right of the decimal point by writing one zero.)

Step 2 Compare the two numbers without decimal points.

Think: $80 > 67$, so $0.8 > 0.67$.

■ **Example 7**

Which number is larger, 0.5678 or 0.06587?

Solution **Step 1** Make the same number of places to the right of the decimal point in both numbers.

0.56780 ↗ 5 places   0.06587 ↗ 5 places

(We made five places to the right of the decimal point by writing one zero.)

Step 2 Compare the two numbers without decimal points.

Think: $56{,}780 > 6587$, so $0.56780 > 0.06587$.

■ **Example 8**

Which number is larger, 9.206 or 9.2?

Solution *Write:* 9.206   9.200

Think: $9206 > 9200$, so $9206 > 9.2$.

■ **Example 9**

Which number is larger, 0.09 or 0.107?

Solution *Write:* 0.090   0.107

Think: $90 < 107$, so $0.09 < 0.107$.

Practice check: Do exercises 46–48 in the margin.

Which of the numbers in the following pairs is larger?

46. 0.4 or 0.39

47. 1.784 or 0.784

48. 5.406 or 5.4036

The answers are in the margin on page 269.

■ Looking Back

You should now be able to state which of two decimal numbers is larger.

■ Problems 7–3

Do as many exercises as suggested by your instructor.

Review

1. Name the third place to the right of the decimal point.

2. Name the ninth place to the right of the decimal point.

Write an expanded number for

3. 0.074

4. 7050.6097

Write a decimal number for

5. $\frac{5}{100} + \frac{8}{10{,}000}$

6. $50 + 5 + \frac{6}{10} + \frac{8}{1000}$

Write word names for

7. 0.942

8. 300.0056

Write decimal numbers for

9. eight hundred and twenty-nine thousandths

10. three hundred fifty and four hundred six ten-thousandths

Which of the numbers in the following pairs is larger?

11. 0.04 or 0.004

12. 0.8 or 0.083

13. 0.74 or 5.6

14. 3.9 or 0.898

15. 0.79 or 0.8

16. 0.40 or 0.04

17. 0.0041 or 0.005

18. 5.1 or 4.010

19. 1.749 or 0.874

20. 3.0549 or 3.05495

21. 0.3 or 0.31

22. 6.605 or 6.6045

The answers to review and odd-numbered exercises are in the back of the book.

7–4 ADDITION OF DECIMALS

Our money system (dollars and cents) is based on decimals. Thus, if a small crescent wrench for $5.35 and a dozen washers for $0.51 are purchased at Shelton's Tool and Die, what is the total cost without tax?

Solution

$$\begin{array}{r} \$5.35 \\ +0.51 \\ \hline \$5.86 \end{array} \quad \text{The total cost is \$5.86.}$$

We can show the meaning of adding decimals by using fractions.

■ **Example 1**

Add 0.4 + 0.2 using the fraction method.

Solution

$$\begin{aligned} 0.4 &= \frac{4}{10} \\ +0.2 &= \frac{2}{10} \\ &\quad \frac{6}{10} \end{aligned} \qquad \text{So, } 0.4 + 0.2 = 0.6$$

■ **Example 2**

Add 2.14 + 23.456 using the fraction method.

Solution **Step 1** Write in enough zeros so that there are the same number of decimal places to the right of the decimal point in both numbers.

$$\begin{array}{r} 2.140 \\ 23.456 \end{array} \leftarrow \text{We wrote in one extra zero.}$$

Step 2 Add using fractions.

$$\begin{aligned} 2.140 &= 2\frac{140}{1000} \\ 23.456 &= 23\frac{456}{1000} \\ &\quad 25\frac{596}{1000} \end{aligned} \qquad \text{So, } 2.14 + 23.456 = 25.596$$

Practice check: Do exercises 49–54 in the margin.

Add using the fraction method.

49. 0.2 + 0.6

50. 0.245 + 0.102

51. 0.6 + 8.19

52. 7.053 + 32.81

53. 132.3 + 40.398

54. 7.5218 + 102.201

The answers are in the margin on page 271.

It is easy to observe from Examples 1 and 2 that adding decimals is similar to adding whole numbers. When adding decimals, we have to make sure that the decimal points are lined up in a straight vertical line. This ensures that all the digits are in their correct places according to their place values.

Answers to exercises 46–48

46. 0.4 > 0.39
47. 1.784 > 0.784
48. 5.406 > 5.4036

■ Example 3

Add 3.45 and 0.032

Solution **Step 1** Line up the decimal points.

$$\begin{array}{r} \downarrow \\ 3.450 \\ +0.032 \\ \hline . \end{array}$$

Write in a zero here. ← Bring down the decimal point first.

Step 2 Add the same as you would whole numbers.

$$\begin{array}{r} 3.450 \\ +0.032 \\ \hline 3.482 \end{array}$$

When we add whole numbers and decimal numbers, we have to remember that all whole numbers have a decimal point after the ones place.

■ Examples

4. 7 = 7.0

5. 180 = 180.0

6. 42 = 42.0

■ Example 7

Add 6, 32.46, and 121.0346.

Solution Line up the decimal points.

$$\begin{array}{r} 6.0000 \\ 32.4600 \\ +121.0346 \\ \hline 159.4946 \end{array}$$

Place zeros here.

Practice check: Do exercises 55–61 in the margin.

Add.

55. 0.34 + 0.53

56. 0.28 + 0.01

57. 35.4 + 3.4

58. 1.02 + 21.34

59. 0.1043 + 81.5 + 6

60. 12 + 0.005 + 31.602

61. 8.043 + 0.0051 + 11

The answers are in the margin on page 271.

When we add decimals, we can "carry" just as we do with whole numbers.

■ Example 8

Add 0.584 + 0.268.

Solution **Step 1** Add thousandths.

1 ← "Carry" $\frac{1}{100}$.

$$\begin{array}{r} 0.584 \\ +0.268 \\ \hline .2 \end{array}$$

(So you won't forget, bring down the decimal point immediately.)

Step 2 Add hundredths.

```
   11    "Carry" 1/10.
 0.584
+0.268
 ------
  . 52
```

Step 3 Add tenths.

```
   11
 0.584
+0.268
 ------
 0.852
```

■ Example 9

Add 3.417 and 5.639.

Solution Be sure the decimal points are in a column.

```
 1. 1
 3.417
+5.639
 ------
 9.056
```

■ Example 10

Add 17 + 0.256 + 3.24 + 0.3689.

Solution Make sure that the decimal points are in a column.

```
17.0000
 0.2560   Place zeros here
 3.2400
 0.3689
-------
20.8649
```

To check addition of decimals, we add in reverse order.

Rule

A short method for adding decimal numbers.

Step 1 Place the numbers over one another. Make sure the decimal points are in a column.

Step 2 Place zeros so that each number has the same number of places to the right of the decimal point.

Step 3 Bring down the answer decimal point.

Step 4 Add as you would whole numbers.

Practice check: Do exercises 62–67 in the margin.

Add and check.

62. 5.98 + 9.23

63. 0.4841 + 0.1832

64. 426.328 + 8.946

65. 0.53 + 1.7 + 8.39

66. 0.94 + 8 + 5.54 + 0.056

67. 0.14 + 0.5468 + 5.007

The answers are in the margin on page 273.

■ Looking Back

You should now be able to add decimal numerals with or without "carrying."

■ Problems 7–4

Review

1. Name the seventh place to the right of the decimal point.

2. Write an expanded number for 26.57.

Write a decimal number for

3. $31 + \frac{8}{10} + \frac{5}{100}$

4. $15\frac{16}{1000}$

5. Write a fractional number for 0.0031.

6. Write a word name for 32.042.

Which of the numbers in the following pairs is larger?

7. 0.9 or 0.094

8. 9.704 or 9.7004

Add.

9. 1.5 + 0.03

10. 0.26 + 4.319

11. 62.2 + 7.1

12. 81.005 + 0.972

13. 0.45 + 3.32

14. 0.18 + 0.7

15. 1.5 + 5 + 32.041

16. 23 + 5.17 + 0.2205 + 0.001

17. 0.04 + 1.3 + 32 + 3.526 + 0.013

18. 4.3 + 0.04 + 8 + 1.63 + 0.0128

19. 4.837 + 3.67

20. 4.06 + 2.876

21. 7.126 + 2.897

Answers to exercises 49–54

49. $\frac{8}{10}$ or 0.8 **50.** $\frac{347}{1000}$ or 0.347

51. $8\frac{79}{100}$, or 8.79

52. $39\frac{863}{1000}$, or 39.863

53. $172\frac{698}{1000}$, or 172.698

54. $109\frac{7228}{10{,}000}$, or 109.7228

Answers to exercises 55–61

55. 0.87 **56.** 0.29 **57.** 38.8
58. 22.36 **59.** 87.6043
60. 43.607 **61.** 19.0481

---

22. 4.76 + 4.897

23. 4.287 + 3.7 + 7.69

24. 5.741 + 5.36 + 1.9

25. 5.38 + 7.817 + 6.7

26. 5.8 + 7.318 + 7.05

27. 18.389 + 13.57 + 19.6

28. 17.76 + 21.587 + 13.5

29. 4.19 + 3.0749 + 5.87

30. 5.4 + 3.9841 + 3.87

31. 52.6941 + 13.871 + 34.89

32. 73.99 + 14.8647 + 19.591

33. 18.734 + 7 + 13.76

34. 31.5176 + 71.841 + 19

35. 3.814 + 7 + 19.61 + 31.9

36. 58.491 + 15 + 22.5546 + 11

37. 15.416 + 32.61 + 10.9 + 18

38. 42 + 2.0043 + 41.2 + 6

The answers to review and odd-numbered exercises are in the back of the book.

7–5 ROUNDING AND ESTIMATING SUMS

Sometimes information must be in a decimal form with fewer places than you already have. This is easily done by rounding, just as we did with whole numbers.

Rounding to the Nearest tenth

■ **Example 1**

Round 0.24 to the nearest tenth.

Solution **Step 1** Decide which digit is in the tenths place.

0.24 ↖ Tenths

Step 2 Look at the number to the right of the number in the tenths place to decide if it is less than 5, or it is 5 or greater.

Tenths ↘ 0.24 ← Less than 5

Step 3 This tells us to drop the 4 hundredths and keep the 2 tenths.

0.24 ← Hundredths digit is less than 5.
↑
0.2 Keeping the same tenths digit; dropping the 4

After rounding, no digits should remain on the right of the tenths place.

Answers to exercises 62–67

62. 15.21 **63.** 0.6673
64. 435.274 **65.** 10.62
66. 14.536 **67.** 5.6938

■ **Example 2**

Round 341.653 to the nearest tenth.

Solution **Step 1** Decide which digit is in the tenths place.

341.653 ↖ Tenths

Step 2 Look at the number to the right of 6 tenths.

Tenths 341.653

5 or more tells us to round up.

Step 3 We round up to 7 tenths and drop the rest.

341.653
↓
341.7 Rounded to the nearest tenth

Practice check: Do exercises 68–72 in the margin.

Round to the nearest tenth.

68. 4.546 **69.** 1.5097
70. 7.6804 **71.** 2.0098
72. 8.95

The answers are in the margin on page 275.

Rounding to the Nearest Hundredth

■ **Example 3**

Round 17.4684 to the nearest hundredth.

Solution **Step 1** Decide which digit is in the hundredths place.

17.4684 Hundredths

Step 2 Look at the number to the right of the number in the hundredths place to decide if it is less than 5 or is 5 or greater.

17.4684 Hundredths
↗
5 or greater

Step 3 This tells us to drop the 8 and 4, and raise 6 hundredths to 7 hundredths.

17.4684 ← Thousandths digit is 5 or greater.
↑
17.47
Raising 6 hundredths to 7 hundredths; dropping the 8 and 4

After rounding to the nearest hundredth, no digits should remain on the right of the hundredths place.

■ **Example 4**

Round 431.053 to the nearest hundredths.

Solution Hundredths
↓
431.053 ⟵ Less than 5
↑
431.05 Keep 5 hundredths; drop the rest.

Practice check: Do exercises 73–77 in the margin.

Round to the nearest hundredth.

73. 0.018504 **74.** 7.3047

75. 3.0042 **76.** 82.59522

77. 9.995

The answers are in the margin on page 277.

Rounding to the Nearest Thousandth

■ **Example 5**

Round 5.53951 to the nearest thousandth.

Solution **Step 1** Decide which digit is in the thousandths place.

5.5395 ← Thousandths

Step 2 Look at the number to the right of the number in the thousandths place to decide if it is less than 5 or is 5 or greater.

Thousandths → 5.53951 ← 5 or greater

Step 3 This tells us to drop the 5 and 1, and raise 9 thousandths to 10 thousandths.

5 or greater → 5 53951 → 5.540 Raising 9 thousandths to 10 thousandths

Note:
$$\begin{array}{r} 5.539 \\ +.001 \\ \hline 5.540 \end{array}$$
Raising 1 thousandth

After rounding to the nearest thousandth, no digits should remain on the right of the thousandths place.

Practice check: Do exercises 78–82 in the margin.

Round to the nearest thousandth.

78. 0.43182 **79.** 3.40752

80. 82.02047 **81.** 292.04956

82. 9.9995

The answers are in the margin on page 277.

Rule

To round decimals:

1. Locate the digit of the place value to be rounded.
2. Look at the digit just to the right of the digit in Step 1.
   a. If this digit is 5 or more, round up.
   b. If this digit is less than 5, round down.

   No digit should remain on the right of the digit in Step 1.

Answers to exercises 68–72

68. 4.5 **69.** 1.5 **70.** 7.7
71. 2.0 **72.** 9.0

Estimating Sums

We can also check our addition by estimating. If the estimate is not close to the exact answer, we should add again.

■ Example 6

Find the exact sum of 0.45 + 4.6 + 0.236 and then check by estimating.

Solution We estimate in this case by rounding each number to the nearest tenth and then adding.

	Nearest tenth
0.45	0.5
4.6	4.6
0.236	0.2
5.286	5.3
↗ Exact sum	↗ Notice that the estimate is close to the exact sum, so we can conclude that the exact sum is probably correct.

Practice check: Do exercises 83–85 in the margin.

Find the exact sums and then check by estimating to the nearest tenth.

83. 7.48 + 1.847

84. 22.787 + 13.9

85. 0.046 + 5 + 3.534 + 18.362

The answers are in the margin on page 277.

■ Looking Back

You should now be able to round decimals and estimate sums by rounding.

■ Problems 7–5

Review

1. Write a decimal number for $800 + 6 + \frac{4}{100}$.

2. Write a mixed number for 17.05.

3. Write a decimal number for twenty-four and sixty-nine hundredths.

4. Which number is larger, 0.81 or 0.801?

Add.

5. 5.604 + 21.72

6. 10.04 + 15 + 0.612

Round to the nearest tenth, hundredth, and thousandth.

7. 0.089476

8. 5.07843

9. 6.408697

10. 8.16739

11. 53.0308747

12. 0.003846

Find the exact sum and check by estimating to the nearest tenth.

13. 1.625 + 7.13 + 0.007 + 5.627

14. 1.49 + 18.39 + 8.36 + 0.73

15. 62.7942 + 1.6 + 4.318 + 17.68

16. 0.409 + 18 + 6.047 + 23.5

Find the exact sum and check by estimating to the nearest hundredth.

17. 9.4308 + 0.0047 + 1.252 + 7.3718

18. 28.291 + 6.84 + 13.3587 + 56

19. 13.714 + 38.8953 + 7.69

20. 504.005 + 0.1044 + 16 + 29.1496

The answers to review and odd-numbered exercises are in the back of the book.

7–6 SUBTRACTION OF DECIMALS

Subtracting decimals is similar to subtracting dollars and cents. For instance, an electrical outlet for $1.66 and an electrical outlet cover for $0.32 were purchased at Jay Hardware. Find the difference between the two items.

Solution

$$\begin{array}{r} \$1.66 \\ -0.32 \\ \hline \$1.34 \end{array}$$

The difference is $1.34.

As with addition of decimals, we must be careful to line up the decimal points in a vertical column.

■ Example 1

Subtract 23.6845 − 2.551 by using the short method.

Solution **Step 1** Place the first number over the second and line up the decimal points.

```
 23.6845
 −2.551
 -------
    .   ⟵ Bring down the decimal point.
```

Step 2 Place zeros so that the numbers have the same number of decimal places to the right of the decimal point.

```
 23.6845
 −2.5510 ⟵ Write in one extra zero.
```

Step 3 Subtract from right to left, begining with the ten-thousandths place.

```
 23.6845  ↖ Begin here
 −2.5510
 -------
 21.1335
```

Answers to exercises 73–77

73. 0.02 **74.** 7.3
75. 3.00 **76.** 82.60
77. 10.00

Answers to exercises 78–82

78. 0.432 **79.** 3.408
80. 82.020 **81.** 392.050
82. 10.000

Answers to exercises 83–85

83. 9.327, 9.3 **84.** 36.687, 36.7
85. 26.942, 26.9

Practice check: Do exercises 86–89 in the margin.

Subtract.

86. 0.74 − 0.43 **87.** 5.873 − 0.61

88. 59.8415 − 18.0304

89. 578.60719 − 254.4030

The answers are in the margin on page 278.

Example 2 shows how to "borrow" when subtracting decimals.

■ Example 2

Subtract 0.3621 − 0.1737 using the short method.

Solution Place the numbers vertically, aligning the decimal points, and subtract as if they were whole numbers.

Step 1 To subtract ten-thousandths, we must "borrow" from the thousandths.

```
        1 11
  0 . 3 6 2 1
 −0 . 1 7 3 7
 -------------
    .       4
```

Step 2 To subtract thousandths, we must "borrow" from the hundredths.

```
         11
        5 1 11
  0 . 3 6 2 1
 −0 . 1 7 3 7
 -------------
    .     8 4
```

Step 3 To subtract hundredths, we must "borrow" from the tenths.

```
      1 5 1 1
      2 5 1 11
  0 . 3 6 2 1
 −0 . 1 7 3 7
 -------------
    .   8 8 4
```

Answers to exercises 86–89

86. 0.31 **87.** 5.263
88. 41.8111 **89.** 324.20418

Step 4 Subtract tenths.

```
   1511
   25111
 0.3621
-0.1737
 0.1884
```

To check, we add the difference to the subtrahend.

```
 0.1737   Subtrahend
+0.1884   Difference
 0.3621   Minuend
```

■ Example 3

Subtract 87.416 − 6.39.

Solution Place the numbers vertically and subtract as if they were whole numbers.

```
   311
87.416                     Check:   6.390
-6.390 ← Place a zero              +81.026
81.026                              87.416
```

Practice check: Do exercises 90–93 in the margin.

Subtract and check.

90. 0.82 − 0.74

91. 0.75438 − 0.74847

92. 8.36 − 2.59

93. 81.649 − 39.85

The answers are in the margin on page 281.

■ Looking Back

You should now be able to subtract decimals with or without "borrowing."

■ Problems 7–6

Review

1. Write a decimal number for $53\,\frac{3}{100{,}000}$.

2. Write a word name for 23.00416.

3. Which of the numbers is larger, 5.62 or 5.602?

4. Add 15.461 + 8 + 0.0046.

5. Find the exact sum and check by estimating to the nearest tenth: 2.69 + 15.413 + 0.061 + 20.

6. Find the sum of the following numbers: fifty-six and forty-one hundredths, six tenths, eighteen and one and five hundredths.

Subtract.

7. 0.57 − 0.26

8. 5.68 − 3.43

9. 8.719 − 5.516

10.
$$\begin{array}{r} 37.4826 \\ -16.0312 \\ \hline \end{array}$$

11. 398.6042 − 76.2011

12. 29.42 − 9.3

13.
$$\begin{array}{r} 53.437 \\ -31.42 \\ \hline \end{array}$$

14. 681.3057 − 70.004

15. 757.5679 − 355.204

16. 6528.07418 − 304.0611

Subtract and check.

17. 4.879 − 3.244

18. 0.53 − 0.38

19. 0.86 − 0.79

20. 0.5621 − 0.4839

21. 0.7642 − 0.1388

22. 24.46 − 6.47

23. 38.712 − 19.635

24. 10.846 − 3.98

25. 156.427 − 79.56

26. 291.3184 − 156.342

The answers to review and odd-numbered exercises are in the back of the book.

7–7 "BORROWING" AND ZEROS

Subtracting decimals containing zeros that involves "borrowing" is similar to the same operation with whole numbers.

■ **Example 1**

Subtract 0.507 − 0.37.

Solution **Step 1** Subtract thousandths.

$$\begin{array}{r} 0.507 \\ \underline{-0.370} \leftarrow \text{Place a zero.} \\ .\;\;7 \end{array}$$

Step 2 To subtract hundredths we "borrow" from the tenths.

$$\begin{array}{r} 4\,10\;\; \\ 0.\cancel{5}\,\cancel{0}\,7 \\ \underline{-0.3\,7\,0} \\ .\;3\,7 \end{array}$$

Step 3 Subtract tenths.

$$\begin{array}{r} 4\,10\;\; \\ 0.\cancel{5}\,\cancel{0}\,7 \\ \underline{-0.3\,7\,0} \\ 0.1\,3\,7 \end{array}$$

Check:
$$\begin{array}{r} 0.370 \\ \underline{+0.137} \\ 0.507 \end{array}$$

■ **Example 2**

Subtract 30.07 − 9.68.

Solution **Step 1** To subtract hundredths, we have to start "borrowing" from the tens.

$$\begin{array}{r} 9\;\;9\;\;\;\;\; \\ 2\;\cancel{10}\;\cancel{10}\;17 \\ \cancel{3}\;\;\cancel{0}\,.\;\cancel{0}\;\;\cancel{7} \\ \underline{-9\,.\;6\;\;8} \end{array}$$

Step 2 Complete the subtraction.

$$\begin{array}{r} 9\;\;9\;\;\;\;\; \\ 2\;\cancel{10}\;\cancel{10}\;17 \\ \cancel{3}\;\;\cancel{0}\,.\;\cancel{0}\;\;7 \\ \underline{-9\,.\;6\;\;8} \\ 2\;\;0\,.\;3\;\;9 \end{array}$$

Check:
$$\begin{array}{r} 9.68 \\ \underline{+20.39} \\ 30.07 \end{array}$$

■ Example 3

Find the difference: 83.0407 − 29.6028.

```
    12     9
   7 2 10 3 10 17
   8 3 . 0 4  0  7
  −2 9 . 6 0  2  8
   5 3 . 4 3  7  9
```

Check: 29.6028
+53.4379
83.0407

Practice check: Do exercises 94–97 in the margin.

■ Example 4

Subtract 9.4 − 5.68.

Solution **Step 1** Write in enough zeros so there are the same number of places to the right of the decimal point.

```
 9.40 ← Place a zero
−5.68
```

Step 2 Subtract hundredths.

```
     3 10
  9 . 4 0
 −5 . 6 8
    .   2
```

Step 3 Subtract tenths and ones.

```
    13
  8  3 10
  9 . 4 0
 −5 . 6 8
  3 . 7 2
```

Check: 5.68
+3.72
9.40

■ Example 5

Find the difference: 7 − 0.576.

Solution **Step 1** Place zeros.

```
 7.000
−0.576
```

Recall that the decimal point is just to the right of the whole number.

Answers to exercises 90–93

90. 0.08 **91.** 0.00591
92. 5.77 **93.** 41.799

Subtract and check.

94. 0.706 − 0.539

95. 7.0405 − 2.7875

96. 8.3004 − 5.4878

97. 40.508 − 17.849

The answers are in the margin on page 283.

Step 2 Subtract.

$$\begin{array}{r} \quad 9\ \ 9 \\ 6\ \cancel{10}\ \cancel{10}\ 10 \\ \cancel{7}.\ \cancel{0}\ \ \cancel{0}\ \ \cancel{0} \\ -0.\ 5\ \ 7\ \ 6 \\ \hline 6.\ 4\ \ 2\ \ 4 \end{array}$$

Practice check: Do exercises 98–101 in the margin.

Subtract.

98. 4.8 − 0.38

99. 13.64 − 2.047

100. 21.02 − 0.0041

101. 3 − 1.46

The answers are in the margin on page 285.

■ Looking Back

You should now be able to subtract decimals involving zeros and "borrowing."

■ Problems 7–7

Review

1. Write an expanded number for 470.061.

2. Write a decimal number for $21\frac{6}{1000}$.

3. Write a word name for 58.609.

4. Which of the numbers is larger, 0.0021 or 0.003?

Subtract.

5. 3.0204 − 2.5843

6. 7.0304 − 3.8674

7. 9.0205 − 6.7856

8. 4.7002 − 2.8926

9. 6 − 3.7296

10. 3.9 − 2.8475

11. 4.06 − 2.8163

12. 15 − 9.5374

13. 13.7 − 8.7276

14. 5.06 − 3.7261

Answers to exercises 94–97

94. 0.167 **95.** 4.2530
96. 2.8126 **97.** 22.659

Mixed Practice

15. 7.46 + 2.795

16. 5.328 + 5.897

17. 5.482 − 4.03

18. 5.158 + 3.7 + 4.68

19. 7.618 − 1.387

20. 22.483 − 3.587

21. 17.78 + 31.585 + 11.9

22. 32.0046 − 29.6987

23. 5.8 + 7.318 + 6.03

24. 7 − 2.8148

The answers to review and odd-numbered exercises are in the back of the book.

7–8 APPLIED PROBLEMS

■ Example 1

The Smyths had investments in four companies. Their stock holdings on a certain day were worth \$175.50, \$6347.75, \$909.75, and \$1919.25 respectively. What was the total worth of the stocks on that day.

Solution In order to solve we simply add the worth of each stock.

$$\begin{array}{r} \$\ 175.50 \\ 6347.75 \\ 909.75 \\ +1919.25 \\ \hline \$9352.25 \end{array}$$

The stocks were worth \$9352.25.

■ Example 2

The figure shows the dimensions of an odd-shaped city lot. The owner wanted to fence the lot completely. How much fencing is required? Round to the next larger foot.

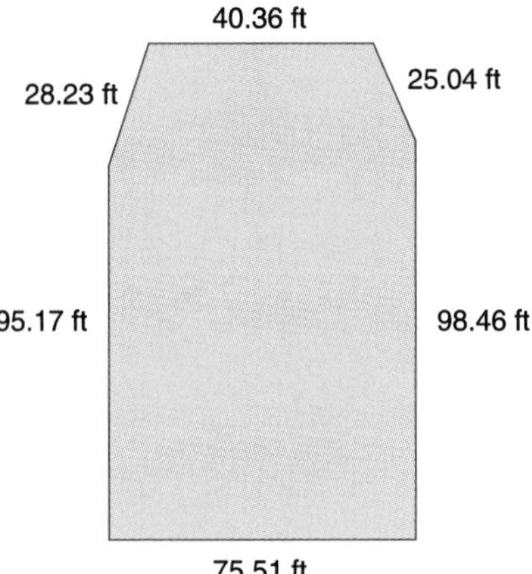

Figure 7–10

Solution Finding the required amount of fencing is the same as finding the perimeter of the lot.

We do this by finding the sum of the measurements of all the sides of the lot.

$$\begin{array}{r} 28.32 \\ 40.36 \\ 25.04 \\ 98.46 \\ 75.51 \\ +95.17 \\ \hline 362.86 \end{array}$$

The owner needs 362.86 feet of fencing, or 363 feet to the next larger foot.

■ Example 3

Stan bought a new suit for \$146.29, shoes for \$44.03, a shirt for \$17.95, and a tie for \$4.55. He paid for all of these clothes with a personal check. If Stan had \$256.38 in his checking account, what was his balance after he wrote the check?

Solution We must first find the amount of the check. We do this by finding the total bill.

$$\begin{array}{r} \$146.29 \\ 44.03 \\ 17.95 \\ +4.55 \\ \hline \$212.82 \end{array}$$ He paid \$212.82 for the clothes.

Next we subtract the amount of the bill from the amount in the checking account to get the balance.

$$\begin{array}{r} \$256.38 \\ -212.82 \\ \hline \$\ \ 43.56 \end{array}$$ Stan had \$43.56 left in his checking account.

Practice check: Do exercise 102 in the margin.

102. The outside diameter of a certain plastic tubing is 3.285 in., and its thickness is 0.195 in. What is the inside diameter?

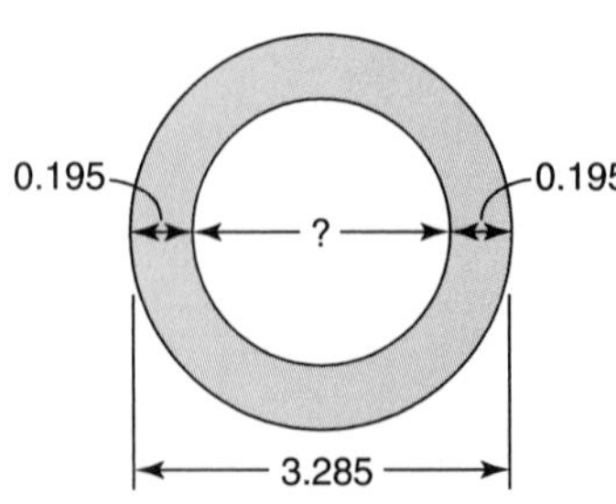

The answer is in the margin on page 287.

■ Problems 7–8

Solve. Write a statement for the answer.

1. **Automotive.** Bill, a do-it-yourself mechanic, needed to tune up his car. He paid $15.48 for a tune-up kit, $17.53 for a set of spark plugs, $27.09 for a set of ignition wires, $2.35 for a gas filter, $15.19 for a distributor cap, and $5.15 for an air filter. How much did the parts cost Bill?

2. **Contractor.** A contractor submitted a bill for a remodeling job that cost $4563.35 for labor, $6320.56 for material, and $653.75 for overhead. What was the total bill for the job?

3. **Plumber.** A piece of PCV pipe 4.25 feet long is shortened by 0.9 feet. How long is the remaining piece? Disregard the width of the cut.

4. **Generic.** Sue had a checkbook balance of $198.03. She wrote checks for $32.95, $5.68, $29.53, and $.89. What was her new balance?

5. **Generic.** Car 56 of the Green Baum Taxi Company traveled 382.4 miles on Monday, 246.6 miles on Tuesday, 198.1 miles on Wednesday, 304 miles on Thursday, and 215.3 miles on Friday. How many miles did Car 56 travel during those five days?

6. **Plumber.** The outside diameter of a piece of copper tubing is 1.567 inches, and the wall thickness is 0.009 inches. What is the inside diameter of the tubing?

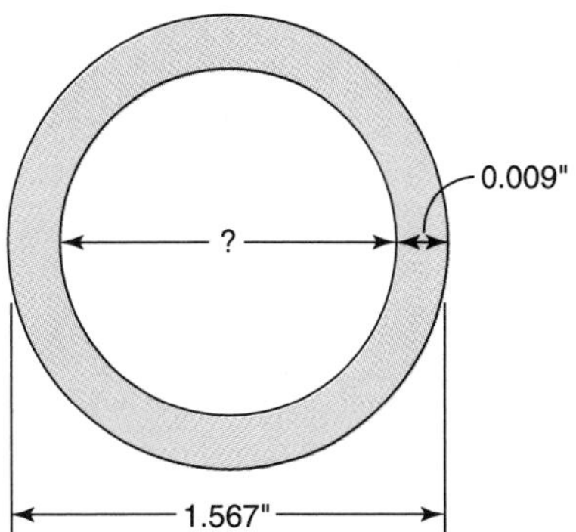

7. **Generic.** From one hundred fifty four and twenty-three thousandths subtract one hundred seventeen and five thousand eighty-nine ten-thousandths.

8. **Generic.** Find the sum of the following numbers: twenty-eight and forty-five hundredths, three and nine tenths, eighteen and twenty-seven thousandths, fifty-seven hundredths, and one hundred six and nine ten-thousandths.

Answers to exercises 98–101

98. 4.42 **99.** 11.593
100. 21.0159 **101.** 1.54

9. **Seamstress.** Carol has to make five dresses for a wedding party. The material needed for each dress is as follows: 2.6 yards, 3 yards, 2.8 yards, 2.6 yards, and 2.9 yards. How many yards of cloth does she have to buy?

10. **Machinist.** An aluminum block 10.936 cm thick was milled with a cut of 0.116 cm. Find the new thickness of the block.

11. **Welding.** Find the missing dimension.

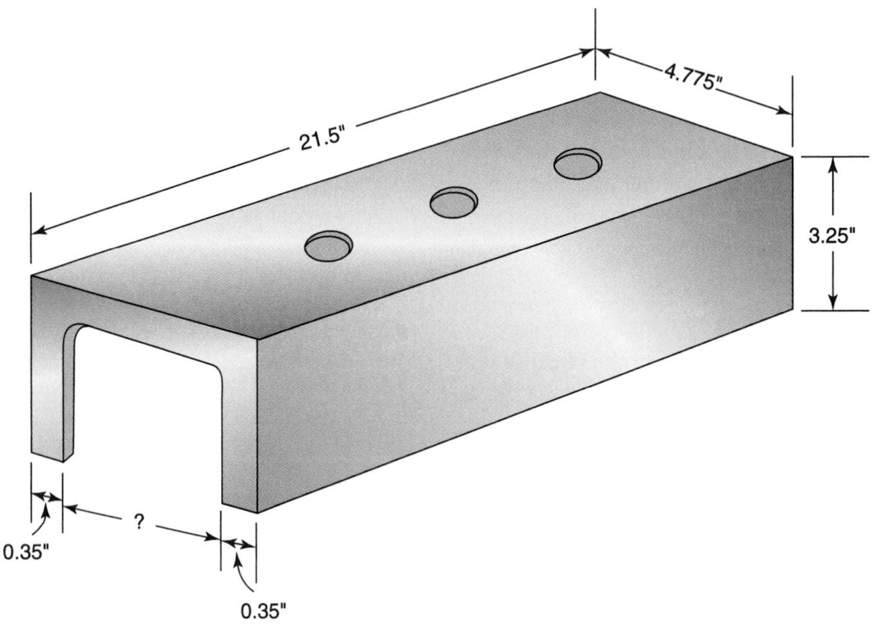

12. **Trucker.** During a five-day delivery trip, Samuel filled his truck's diesel tank each morning. The amounts were 28.9 gal, 32.6 gal, 27.1 gal, 25.7 gal, and 17.5 gal. How much diesel did he buy during his trip?

13. **Mason.** To make ends meet, an apprentice mason worked in three different jobs. During one week he spent 5.5 hours on the first job, 9.25 hours on the second job, and 10.75 hours on the third job. How many hours did he work that week?

14. **Generic.** How much change should you receive if you bought a set of golf balls for $8.56 and you gave the clerk $10.00?

15. **Auto Mechanic.** A piston when cold measures 4.625 in. When hot it measures 4.638 in. How many inches does it expand when heated?

16. **Fire Science.** A Kelly tool has claw length 1.63 in., has handle length 30.38 in., and has head thickness 3.68 in. What is the length of the tool?

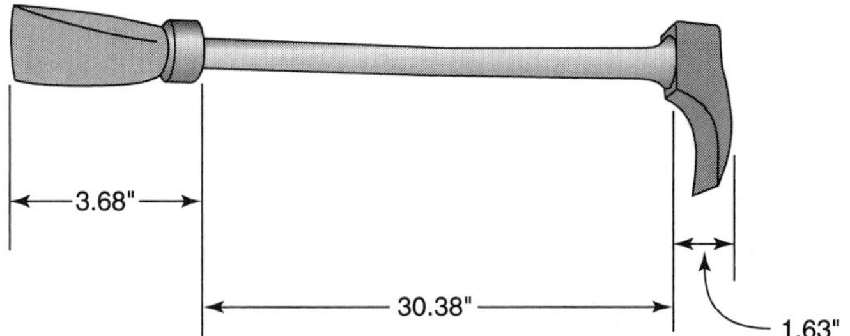

Answer to exercise 102

The inside diameter is 2.895 inches.

17. **Agriculture.** A farmer needs to replace a fence for a field whose four sides measure 377.5 ft, 496.4 ft, 298.2 ft, and 703.7 ft. How much fencing is required? Round to the next larger foot.

18. **Machinist.** If one box of 1-inch bolts weighs 63.5 pounds and another weighs 47.9 lb, what is the difference between the weight of the two boxes of 1-inch bolts?

19. **Electrician.** Sam calculated the electrical loads for a house as follows:

	Kilowatts
Air-conditioning	9.2
Oven	4.0
Kitchen stove	6.1
Water heater	5.2
Dishwasher	1.5
Compactor	0.8
Washer/dryer	5.3

What is the total electrical load?

20. **Utilities.** During a 1-year period, the following amounts of natural gas were consumed in a home.

Month	Therm Units	Month	Therm Units
January	220.84	July	11.74
February	174.10	August	25.15
March	142.66	September	13.77
April	119.45	October	49.16
May	73.84	November	76.27
June	30.88	December	96.43

How many more therm units were used during the first 6 months than during the last 6 months of the year?

CHAPTER 7 OVERVIEW

Summary

7–1 You learned to name the place values in the decimal number system, expand numbers, convert fractions like $\frac{23}{1000}$ to decimals, and convert a decimal to a fraction or a mixed number.

7–2 You learned to say or write word names for decimal numbers.

7–3 You learned to state which of two decimals is larger.

7–4 You learned to add decimals with or without "borrowing."

7–5 You learned to round decimals and estimate sums by rounding.

7–6 You learned to subtract decimals with or without "borrowing."

7–7 You learned to subtract decimals involving zeros and "borrowing."

7–8 You learned to solve applied problems involving addition and subtraction of decimals.

Terms To Remember

	Page
Decimal point	257
Decimal numbers	256
Place holders	259

Rules

- To compare decimal numbers we

  1. Write as many extra zeros as necessary so that both decimals have the same number of places to the right of the decimal point.
  2. Mentally drop the decimal point and compare the numbers as you do whole numbers.

- A short method for adding decimal numbers:

  Step 1 Place the numbers over one another. Make sure the decimal points are in a column.

  Step 2 Place zeros so that each number has the same number of places to the right of the decimal point.

  Step 3 Bring down the answer decimal point.

  Step 4 Add as you would whole numbers.

- To round decimals,

  1. Locate the digit of the place value to be rounded.
  2. Look at the digit just to the right of the digit in Step 1.
     a. If this digit is 5 or more, round up.
     b. If this digit is less than 5, round down.

No digit should remain on the right of the digit in Step 1.

Self-Test

7–1 **1.** Name the seventh place to the left of the decimal point.

2. Write a decimal number for $23\frac{17}{1000}$.

3. Write a mixed number for 9.0005.

4. Write a decimal number for $900 + 6 + \frac{5}{1000} + \frac{7}{10,000}$.

7–2 **5.** Write a word name for 56.709.

7–3 **6.** Which number is larger, 0.7199 or 0.072?

Add.

7–4 **7.** $96.5 + 7.3$

8. $47.04 + 6.849$

9. $0.86 + 2 + 0.7$

10. $4.06 + 0.007 + 10.43 + 14$

7–5 **11.** Round 83.30972 to the nearest thousandth.

Subtract.

7–6 **12.** $58.498 - 36.18$

13. $8.59 - 0.3$

7–7 **14.** 29.5 − 18.29 **15.** 32 − 3.601

7–8 **16.** To seat a new exhaust valve properly in an automobile engine, the mechanic must grind off 0.004 in. If the new valve measures 1.825 in., what is the size of the valve after it has been ground?

17. A shim is required to close up an opening between two parts that measure 0.0045 in. If an available shim is only 0.00125 in., how much more is needed to close the opening?

18. Find the length of the valve.

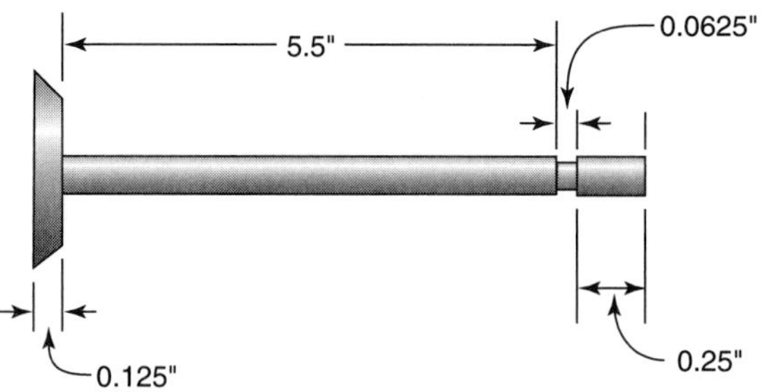

19. A dealer at Art's Auto Parts had the following sales figures for the first week in May: Monday, $2349.46; Tuesday, $1921.32; Wednesday, $2510.11; Thursday, $2004.23; and Friday, $844.02. Find the total sales for that week.

The answers are in the back of the book.

Chapter Test

Directions: This test will aid in your preparation for a possible chapter test given by your instructor. The answers are in the back of the book. Review the appropriate section(s) if you miss any test items.

7–1 **1.** Name the sixth place to the right of the decimal point.

2. Name the sixth place to the left of the decimal point.

3. Write a decimal number for $15\frac{8}{1000}$.

4. Write a mixed number for 13.0015.

5. Write an expanded number for 6.804.

6. Write a decimal number for $4000 + 90 + \frac{8}{10} + \frac{5}{10{,}000}$.

7–2 **7.** Write a word name for 41.605.

8. Write a decimal number for eight and twenty-one thousandths.

7–3 **9.** Which number is larger, 0.680 or 0.6799?

7–4 Add.

10. 23.4 + 6.2

11. 7.052 + 0.0017 + 12

12. 76.03 + 5.989

13. 0.77 + 8 + 0.6

14. 3.89 + 0.006 + 7.43 + 13

7–5 **15.** Round to the nearest hundredth: 746.05674.

16. Estimate the sum to the nearest tenth: 78.85 + 15.421.

Subtract.

7–6 **17.** 76.398 − 46.17

18. 0.5896 − 0.5837

19. 9.34 − 0.7

20. 98.246 − 39.78

7–7 **21.** 31.8 − 22.37

22. 0.75 − 0.683

23. 13 − 6.502

7–8 **24.** Find the length of the part below.

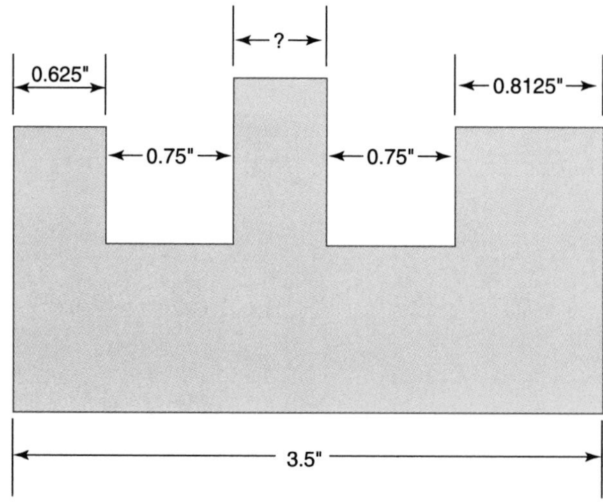

25. Sandy is writing checks for $3.80, $15, $39.66, and $0.56. If her checkbook balance was $89.15, what is her new balance?

26. A tapered shaft has a small end diameter of 1.045 in. and a large end diameter of 2.410 in. What is the difference in the diameters?

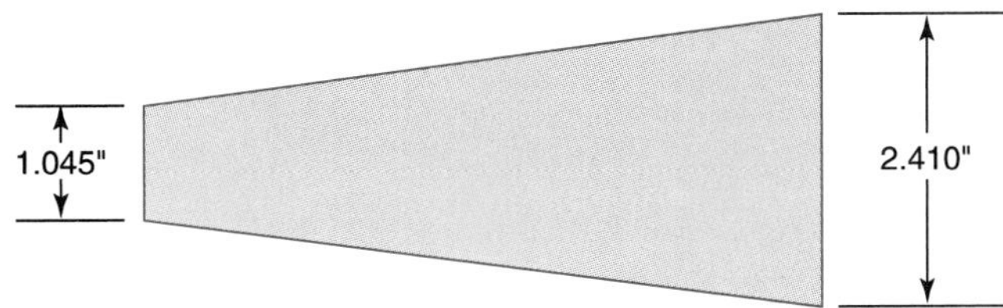

27. Margaret wrote checks for the following amounts: $4.52, $23.82, $19.06, $55.30, and $8.19. She had $120.15 in her checking account. Did she have enough to cover all of her checks?

28. Jack sawed a piece of wood 12.68 inches long from a piece 43.07 inches long. How much of the piece was left?

29. Mary bought some jewelry for $18.39 and gave the clerk a $20 bill. How much was her change?

CHAPTER 8

Multiplication and Division of Decimals

Choice: Either skip the Pretest and study all of Chapter 8 or take the Pretest and study only the sections you need to study.

Pretest

Directions: This Pretest will help you determine which sections to study in Chapter 8. If any of your answers are incorrect, or if you omitted any exercise, turn to the indicated section(s) and study the material; then take the Chapter Test.

8–1 **1.** Multiply: 0.4606×32.

8–2 **2.** 0.8×3.9 **3.** 42.6×3.402

8–3 **4.** Multiply by writing the answer directly: 4.89×1000.

8–4 **5.** Find the exact answer, then check by estimating: 3.52×6.1.

8–5 **6.** Divide: $7\overline{)43.4}$.

7. Write the quotient directly: $35.7 \div 1000$.

8–6 **8.** Divide: $3.41\overline{)1.02982}$.

8–7 **9.** Round the quotient to the nearest tenth: $0.21\overline{)4.618}$.

8–8 **10.** Express $\frac{3}{4}$ as a decimal.

11. Express 0.04 as a reduced common fraction.

8–9 **12.** A car traveled 435.6 miles in 9.4 hours. What was its average speed? (Round to the nearest tenth.)

13. Find the total tax on 43.6 gallons of gasoline if the tax is 9.7 cents per gallon. (Round to the nearest cent.)

The answers are on page 299.

8–1 MULTIPLICATION OF WHOLE NUMBERS WITH DECIMALS

The shortest method of multiplication of decimals requires us to know the number of places to the right of the decimal point in decimal numbers. Let's practice.

■ **Example 1**

How many places to the right of the decimal point are there in 6312.004?

Solution 6 3 1 2 . 0 0 4
↑ ↑ ↑
1 2 3

There are 3 places to the right of the decimal point.

Practice check: Do exercises 1–6 in the margin.

How many places are there to the right of the decimal point?

1. 0.67 **2.** 239.0006

3. 46.700 **4.** 0.3468

5. 40.4 **6.** 55

The answers are in the margin on page 299.

Example 2 shows the fraction method and short method of multiplying decimals.

■ **Example 2**

Multiply 5×0.3.

Solution Fraction method:

$$5 \times 0.3 = \frac{5}{1} \times \frac{3}{10} = \frac{5 \times 3}{1 \times 10} = \frac{15}{10} = 1\frac{5}{10} = 1.5$$

↑ Improper fraction ↖ Mixed number

$$5 \times 0.3 = 1.5$$

Solution Short method:

Step 1 Place the numbers vertically and multiply as if they were whole numbers.

$$\begin{array}{r} 5 \\ \times 0.3 \\ \hline 1\,5 \end{array}$$

Step 2 Total the number of places to the right of the decimal point in the factors.

$$\left.\begin{array}{r} 5 \longrightarrow 0 \\ \times 0.3 \longrightarrow 1 \end{array}\right\} \text{Total: 1 decimal place}$$
$$1\,5$$

Step 3 Place the decimal point in the product so that the number of places to the right of the decimal point is the same as the sum of the number of places in the factors.

$$\begin{array}{r} 5 \\ \times 0.3 \\ \hline 1.\underline{5} \end{array}$$

↑ 1 decimal place

When multiplying decimals, it is easier and quicker to use the short method.

■ Example 3

Find the product of 3.18 × 5 using the short method.

Solution **Step 1** Multiply as if they were whole numbers.

$$\begin{array}{r} 3.18 \\ \times 5 \\ \hline 1590 \end{array}$$

Step 2 Total the number of places to the right of the decimal point.

$$\left.\begin{array}{r} 3.18 \longrightarrow 2 \\ \times 5 \longrightarrow 0 \end{array}\right\} \text{Total: 2 decimal place}$$
$$1\,5\,9\,0$$

Step 3 Place the decimal point in the product.

$$\begin{array}{r} 3.18 \\ \times 5 \\ \hline 15.90 \end{array}$$

↑ 2 decimal places

Practice check: Do exercises 7–12 in the margin.

Multiply by using the short method.

7. 4×0.1 **8.** 48×0.01

9. 0.7×8 **10.** 34×0.5

11. 0.46×3 **12.** 19×4.05

The answers are in the margin on page 301.

■ Example 4

Multiply 29×4.5106.

Solution **Step 1** Multiply as if they were whole numbers.

$$\begin{array}{r} 4.5106 \\ \times 29 \\ \hline 405954 \\ 90212 \\ \hline 1308074 \end{array}$$

Step 2 Total the number of places to the right of the decimal points.

$$\begin{array}{r} 4.5106 \longrightarrow 4 \\ \times 29 \longrightarrow 0 \\ \hline 405954 \\ 90212 \\ \hline 1308074 \end{array}$$

Total: 4 places

Step 3 Place the decimal point in the product.

$$\begin{array}{r} 4.5106 \\ \times 29 \\ \hline 405954 \\ 90212 \\ \hline 130.8074 \end{array}$$

↑ 4 decimal places

■ Example 5

Multiply 4×0.002.

Solution

$$\begin{array}{r} 4 \longrightarrow 0 \\ \times 0.002 \longrightarrow 3 \\ \hline 0.008 \end{array}$$

Total: 3 places

↑ 3 places

We must fill in 2 zeros to make 3 places to the right of the decimal point.

Practice check: Do exercises 13–17 in the margin.

Multiply.

13. 7×0.57

14. 0.378×34

15. 46×3.505

16. 87×13.1406

17. 0.0007×9

The answers are in the margin on page 301.

Example 6

A farmer needs to ship 450 pounds of fertilizer. Determine his freight charge at $0.078 per pound.

Solution

$$\begin{array}{r} 450 \\ \times 0.078 \\ \hline 3\,600 \\ 31\,50 \\ \hline 35.100, \end{array} \quad \text{or} \quad \$35.10$$

Looking Back

You should now be able to multiply whole numbers with decimals.

Answers to Pretest

1. 14.7392 **2.** 3.12 **3.** 144.9252 **4.** 4890 **5.** 21.472, 24 **6.** 6.2 **7.** 0.0357 **8.** 0.302 **9.** 22.0 **10.** 0.75 **11.** $\frac{1}{25}$ **12.** 46.3 mph **13.** 423¢, or $4.23

Answers to exercises 1–6

1. 2 places **2.** 4 places **3.** 3 places **4.** 4 places **5.** 1 place **6.** 0 places

Problems 8–1

Multiply.

1. 6 × 0.5 **2.** 8 × 0.7 **3.** 28 × 0.51 **4.** 0.137 × 6

5. 5.547 × 24 **6.** 3 × 2.508 **7.** 36 × 0.6043 **8.** 6435 × 1.3104

9. 35 × 0.6 **10.** 48 × 0.06 **11.** 0.04 × 4 **12.** 31 × 0.007

13. 8 × 0.001 **14.** 5 × 0.005 **15.** 77.15 × 83 **16.** 38 × 4.8

17. 370 × 7.045 **18.** 5.1816 × 52 **19.** 17 × 0.0001 **20.** 8 × 0.000005

21. Printer. How thick are 9000 sheets of bond paper if each sheet is 0.009 inch thick?

22. Photographer. Find the cost of 42 boxes of litho film if the cost per box is $33.45.

23. **Welder.** A welder makes 25 tee-irons each weighing 3.185 pounds. Find the total weight.

24. **Sheet Metal.** Five sheet metal workers were on a job for 7.5 hours. Determine the total of man hours required for the job.

25. **Job Marketing.** The R.C. Jake Screw Corporation purchased 85 boxes of screws at a cost of $3.375 per box and 36 boxes of washers at $1.25 per box. Find the total purchase cost to the nearest cent.

26. **Forestry.** Determine the cost of excavating 31,468 cu yd of forest land at $1.25 per cu yd.

27. **Machinist.** How long would it take to machine 60 steel shafts if each shaft takes 20.5 minutes.

28. **Machinist.** A machinist's helper gets $15.37 per hour. How much will she earn in a week of 40 hours.

29. **Nursing.** Each of 56 student nurses is to receive 8.4 mℓ of a test solution. How many mℓ must be measured out?

30. **Fire Science.** A fire truck from Engine 5 has a water tank capacity of 552 gallons. How much weight will be added when the tank is filled if water weighs 8.323 pounds per gallon?

The answers to odd-numbered exercises are in the back of the book.

8–2 MULTIPLICATION OF TWO DECIMALS

Next you will learn how to multiply two decimals using the short method.

■ **Example 1**

Multiply.

$$\begin{array}{r} 92.6 \\ \times 0.523 \\ \hline \end{array}$$

Answers to exercises 7–12

7. 0.04 **8.** 0.48 **9.** 5.6
10. 17.0 **11.** 1.38 **12.** 76.95

Answers to exercises 13–17

13. 3.99 **14.** 12.852
15. 161.184 **16.** 1143.2322
17. 0.0063

Solution **Step 1** We multiply as if we were multiplying whole numbers.

$$\begin{array}{r} 92.6 \\ \underline{\times 0.523} \\ 2778 \\ 18520 \\ \underline{464000} \\ 484298 \end{array}$$

Step 2 Total the numbers of places to the right of the decimal point in the factors.

$$\left.\begin{array}{r} 92.6 \longrightarrow 1 \\ \underline{\times 0.523} \longrightarrow 3 \end{array}\right\} \text{Total: 4 decimal places}$$

Step 3 Place the decimal point in the product so that the number of places to the right of the decimal point is the same as the sum of the number of places in the factors.

$$\begin{array}{r} 92.6 \\ \underline{\times 0.523} \\ 2778 \\ 18520 \\ \underline{464000} \\ 48.\underline{4298} \end{array}$$

↑ 4 places

■ **Example 2**

Multiply.

$$\begin{array}{r} 5.41 \\ \underline{\times 0.038} \end{array}$$

Solution

$$\left.\begin{array}{r} 5.41 \longrightarrow 2 \\ \underline{\times\ 0.038} \longrightarrow 3 \end{array}\right\} \text{Total: 5 places}$$

$$\begin{array}{r} 4328 \\ \underline{16230} \\ 0.\underline{20558} \end{array}$$

↑ 5 places

■ **Example 3**

A printer finds that offset paper costs \$0.556 per pound. Find the cost of 2.125 pounds of the paper.

Solution If one pound costs \$0.556, then 2.125 pounds would cost \$1.18.

```
     0.556
    ×2.125
      2780
     1112
     556
  1 112
  1.181500,   or   $1.18
```

Rule

A short method for multiplying decimals:

Step 1 *Place the numbers vertically and multiply as if they were whole numbers.*

Step 2 *Total the number of places to the right of the decimal point in the factors.*

Step 3 *Place the decimal point in the product so that the number of places to the right of the decimal point is the same as the sum of the number of places in the factors.*

Multiply.

18. 0.5 × 0.8 **19.** 8.2 × 0.7
20. 23.4 × 2.09
21. 30.64 × 0.746

The answers are in the margin on page 305.

Practice check: Do exercises 18–21 in the margin.

■ Looking Back

You should now be able to multiply two decimals.

■ Problems 8–2

Multiply.

1. 0.7 × 0.9 **2.** 4.5 × 0.6 **3.** 50 × 7.3 **4.** 7.9 × 3.3

5. 0.84 × 0.7 **6.** 2.16 × 7.2 **7.** 1.217 × 32 **8.** 0.39 × 0.54

9. 316.49 × 0.07 **10.** 0.07 × 0.04 **11.** 0.8 × 0.004 **12.** 28.19 × 0.717

13. 531.60 × 3.481 **14.** 0.05 × 3.115 **15.** 3.6842 × 98 **16.** 0.00561 × 0.302

17. 19×0.06

18. 0.03147×56.2

19. 600×7.047

20. 9000×0.0001

21. Interior Decorator. If carpet cost $6.83 per yard, what is the cost of 22.57 yards?

22. Police Science. A steel cap weighs 146.4 lb. If aluminum weighs 0.507 as much as steel, find the weight of an aluminum cap.

23. Sheet Metal. The width of a rectangular piece of sheet metal is 0.354 times its length. Find the width if the length is 7.75 feet.

24. Mason. A 50-gallon drum is full with grout. Find the weight of the drum if one gallon of grout weighs 8.75 pounds and the drum weights 35.4 pounds.

25. Carpenter. Find the floor area of a master bedroom 20.375 ft by 35.625 ft. Round to the nearest hundredth of a foot.

26. Marketing. A buyer bought $23\frac{5}{8}$ yards of satin priced at $3.25 per yard. What did the buyer pay for the satin?

27. Machinist. The dimensions of a piece of metal are 14.7 in. by 26.25 in. by 156 in. Determine the number of cubic inches in the piece.

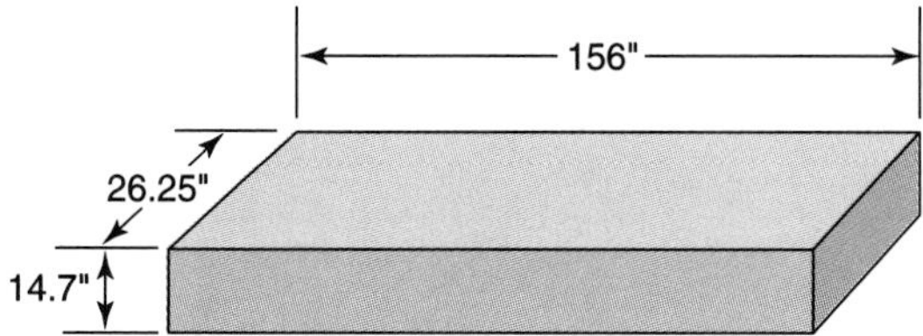

28. Welding. If a welder receives an hourly wage of $17.28 with time and a half for hours worked over 40, find his gross pay if he worked 53 hours.

29. Printer. A pressman works 7.1 hours on Monday, 6.2 hours on Thursday, and 5.6 hours on Friday. If he earns $21.17 per hour, determine the pressman's pay for that week.

30. Private Pilot. Ace Flying Service finds that one of its Cessnas uses 7.5 gallons of gasoline per hour. How much gas will it use on a flight lasting 5.8 hours?

The answers to odd-numbered exercises are in the back of the book.

8–3 MULTIPLICATION BY POWERS OF 10

In this section you will learn a method for writing the product directly when multiplying by powers of 10.

■ Example 1

Multiply 0.06745 × 1000.

Solution

$$\begin{array}{rl} 0.06745 & \rightarrow 5 \text{ places} \\ \underline{\times 1000} & \rightarrow 0 \text{ places} \\ 67.45000 & \rightarrow 5 \text{ places} \end{array}$$

We can drop the 3 zeros on the end and use 67.45 as our product.

Notice: To write the product 67.45 directly from 0.06745 × 1000, the decimal point was moved 3 places to the right. Also notice that 1000 has 3 zeros.

Example 1 leads us to a short-cut rule for multiplying decimals by powers of 10.

Rule

To get the product directly when multiplying a decimal by a power of 10, move the decimal point in the decimal number as many places to the right as the number of zeros in the power of 10.

■ Examples 2–4

Write the product directly.

2. 31.76 × 10 = 317.6 1 place to right

3. 4.0741 × 1000 = 4074.1 3 places to right

4. 0.67 × 100 = 67.0 2 places to right

■ Example 5

Multiply 0.7 × 100.

Solution To move the decimal point 2 places, we must place a zero to the right of 7 to make the second place.

0.70 × 100 = 70.

2 places to the right

Example 6

If a tank of oil holds 1000 gallons, what is the volume of the tank in cubic inches? (A gallon of oil has a volume of 231 cubic inches.)

Solution Multiply the number of gallons by the number of cubic inches in one gallon.

$$231 \times 1000 = 231{,}000 \text{ cu in.}$$

Practice check: Do exercises 22–30 in the margin.

Answers to exercises 18–21

18. 0.40 **19.** 5.74 **20.** 48.906 **21.** 22.88728

Write the product directly.

22. 460.471 × 100

23. 0.0642 × 10

24. 7.6842 × 1000

25. 0.31476 × 10,000

26. 0.00005 × 1000

27. 0.8 × 1000

28. 0.04 × 10,000

29. 38.7 × 100

30. 16 × 100

The answers are in the margin on page 307.

Looking Back

You should now be able to multiply decimals by powers of 10.

Problems 8–3

Write the product directly.

1. 4.86 × 10 **2.** 1000 × 0.074 **3.** 10 × 0.318 **4.** 1.7 × 1000

5. 0.17 × 1000 **6.** 10 × 0.876 **7.** 1000 × 7.187 **8.** 31.77 × 10,000

9. 0.0207 × 10 **10.** 348 × 1000

Multiply.

11. 78 × 0.61 **12.** 0.008 × 0.7 **13.** 76.48 × 1.217 **14.** 100 × 0.146

15. 0.1704 × 105 **16.** 0.001 × 8 **17.** 0.0318 × 1000 **18.** 347.69 × 0.04

19. 6.4872 × 1.41 **20.** 600 × 3.09

21. **Printer.** How high are a stack of 1000 sheets of dual-purpose reprographic paper if each sheet is 0.009 inch thick?

22. **Welder.** A steel plate used to make a brace weighs 3.486 pounds. What is the total weight if the job order calls for 100 braces?

23. **Masons.** Find the weight of 1000 square feet of ceramic tile if the average weight per square foot is 5.25 pounds.

24. **Sheet Metal.** Find the total height of 100 sheets of sheet metal each 0.0185 inch thick.

25. **Electrician.** Milliamperes of current are found by multiplying the number of amperes by 1000. Find the number of milliamperes in (a) 0.75 amperes, (b) 7.5 amperes, and (c) 0.075 amperes.

The answers to odd-numbered exercises are in the back of the book.

8–4 ROUNDING AND ESTIMATING

We can check the reasonableness of an exact product by estimating.

■ Example 1

Multiply 6.4 × 4.

Solution Estimate by rounding.

$$\begin{array}{rcr} 6.4 & \longrightarrow & 6 \\ \underline{\times 4} & \longrightarrow & \underline{\times 4} \\ 25.6 & & 24 \end{array}$$

↑ This is a reasonable answer.

Suppose instead we multiplied and mistakenly misplaced the decimal point.

$$\begin{array}{r} 6.4 \\ \underline{\times 4} \\ 2.56 \end{array}$$

The estimate of 24 would show us that we made a mistake.

■ Example 2

Multiply 3.8 × 0.74 and check by estimating.

Solution

```
   3.8       4
 ×0.74    ×0.7
   152     2.8
  2660
 2.812
   ↑
   This is a reasonable answer.
```

Answers to exercises 22–30

22. 46047.1 **23.** 0.642
24. 7684.2 **25.** 3147.6 **26.** 0.05
27. 800 **28.** 400 **29.** 3870
30. 1600

■ **Example 3**

How many pounds of water are there in a tank that contains 198.67 cu ft? (One cubic foot of water weighs 62.5 lb.)

Solution

```
    198.67        200
    ×62.5         ×60
    99 335     12,000
   397 34
 11 920 2
 12,416.875 <———— This is a reasonable answer.
```

We round each factor to an easily multiplied number. Sometimes this means that we'll round up for one factor and down for another.

Practice check: Do exercises 31–33 in the margin.

Find the exact product, and then check by estimating.

31. 2.86×3.7

32. 5.316×0.25

33. 0.86×9.601

The answers are in the margin on page 309.

■ **Looking Back**

You should now be able to check multiplication of decimals by estimating.

■ **Problems 8–4**

Find the exact product, and then check by estimating.

1. 3.6×0.8 **2.** 5.7×3.2 **3.** 30×0.25 **4.** 0.37×0.22

5. 0.385×1.3 **6.** 32.44×0.631 **7.** 7.964×1.49 **8.** 56×0.5032

9. 17.3×0.046 **10.** 0.0221×1.7

11. **Marketing.** Find the total value of 150 shares of Wimar stock at 20.375.

12. **Forestry.** Determine the cost of 11.25 miles of logging road at the rate of $550 per mile.

13. **Machinist.** A steel plate weighs 346.5 pounds. A piece of cast iron weighs 0.918 times as much as steel. How much would a piece of cast iron weigh?

14. **Wood Products.** Woodson Developers bought a parcel of land for a hardboard plant that measures 2460.5 ft by 653.75 ft. Find the square feet of the acquisition.

15. **Printer.** A printer receives an hourly wage of $18.30 with time and a half for any hours over 40 during a given week. Find her gross pay for a week in which she works 54 hours.

The answers to odd-numbered exercises are in the back of the book.

8–5 DIVIDING A DECIMAL BY A WHOLE NUMBER OR A POWER OF 10

The process used in dividing decimals is the same as the one used in dividing whole numbers. Example 1 reviews this process.

■ **Example 1**

Divide $152\overline{)31801}$.

Solution **Step 1**

$$\begin{array}{r} 2 \\ \mathbf{152}\overline{)\mathbf{318}01} \\ 304 \\ \hline 14 \end{array}$$

Divide 318 by 152.
Multiply 2 by 152.
Subtract.

Step 2

$$\begin{array}{r} 2 \\ 152\overline{)31801} \\ 304\downarrow \\ \hline 140 \end{array}$$

Bring down.

Step 3

$$\begin{array}{r} 20 \\ \mathbf{152}\overline{)31801} \\ 304 \\ \hline \mathbf{140} \\ 0 \\ \hline 140 \end{array}$$

140 divided by 152 is 0.

Multiply 0 by 152.
Subtract.

Step 4
```
       20
152)31801
    304
     140
       0↓
    1401    Bring down.
```

Step 5
```
       209
152)31801    Divide 1401 by 152.
    304
     140
       0
    1401
    1368    Multiply 9 by 152.
      33    Subtract.
```

Since 33 is smaller than 152 and there is nothing further to bring down, the division is complete.

209 R 33

Practice check: Do exercises 34–38 in the margin.

■ Example 2

Divide 7.2 ÷ 3 using the fraction method.

Solution

$$7.2 \div 3 = 7\frac{2}{10} \div 3$$

$$= \frac{72}{10} \div \frac{3}{1}$$

$$= \frac{\overset{24}{\cancel{72}}}{10} \times \frac{1}{\underset{1}{\cancel{3}}}$$

$$= \frac{24}{10} = 2\frac{4}{10}, \quad \text{or} \quad 2.4$$

■ Example 3

Divide 7.2 ÷ 3 using the short division method.

Solution **Step 1** $3\overline{)7.2}$ (with a dot placed above the decimal point) Place the quotient decimal point above the dividend decimal point.

Step 2 Divide as if you were dividing whole numbers; in other words, divide as if the decimal points were not there.

```
   2.4
3)7.2
  6
  12
  12
```

Answers to exercises 31–33

31. 10.582; one estimate is 12.
32. 1.32900; one estimate is 1.5.
33. 8.25686; one estimate is 9.0.

Divide.

34. $6\overline{)23{,}004}$ **35.** $18\overline{)6347}$

32. $768\overline{)522{,}490}$ **37.** $68\overline{)20{,}470}$

38. $589\overline{)417{,}012}$

The answers are in the margin on page 311.

When dividing decimals, it is easier and quicker to use the short method.

■ Example 4

Divide $156\overline{)17.9556}$.

Solution **Step 1** $156\overline{)17.9556}$ — Write in the decimal point.

Step 2 Divide.

$$\begin{array}{r} .1151 \\ 156\overline{)17.9556} \\ \underline{156} \\ 235 \\ \underline{156} \\ 795 \\ \underline{780} \\ 156 \\ \underline{156} \end{array} \qquad \text{Check: } \begin{array}{r} 0.1151 \\ \underline{\times 156} \\ 6906 \\ 57550 \\ \underline{115100} \\ 17.9556 \end{array}$$

The answer is 0.1151.

Practice check: Do exercises 39–41 in the margin.

Divide.

39. $6\overline{)53.4}$ **40.** $19\overline{)76.38}$

41. $142\overline{)29.1668}$

The answers are in the margin on page 313.

■ Example 5

Divide $60\overline{)0.1980}$.

Solution **Step 1** $60\overline{)0.1980}$ — Write in the decimal point.

Step 2 Divide.

Place zeros here as placeholders.

$$\begin{array}{r} .0033 \\ 60\overline{)0.1980} \\ \underline{180} \\ 180 \\ \underline{180} \end{array} \qquad \text{Check: } \begin{array}{r} 0.0033 \\ \underline{\times 60} \\ 0.1980 \end{array}$$

■ Example 6

A steel sheet 12.125 inches wide is to be cut into 5 strips of equal length. How wide will each strip be?

Solution

$$\begin{array}{r} 2.425 \\ 5\overline{)12.125} \\ \underline{10} \\ 21 \\ \underline{20} \\ 12 \\ \underline{10} \\ 25 \end{array} \qquad \text{Check: } \begin{array}{r} 2.425 \\ \underline{\times 5} \\ 12.125 \end{array}$$

Each strip will be 2.425 inches.

Practice check: Do exercises 42–46 in the margin.

Dividing by a Power of 10

In Section 8–3 we learned a process for writing the answer directly when multiplying by powers of 10.

Rule

To get the product directly when multiplying a decimal or a whole number by a power of 10, move the decimal point of the multiplicand as many places to the right as the number of zeros in the power of 10.

■ Examples

7. $15 \times 100 = 1500$
8. $18.7 \times 10 = 187$ (Notice that the resulting product is larger than the multiplicand.)
9. $0.0648 \times 1000 = 64.8$

Suppose, instead, we divide a decimal or whole number by a power of 10.

■ Example 10

Divide $10\overline{)4.68}$.

Solution

$$\begin{array}{r} 0.468 \\ 10\overline{)4.680} \\ \underline{40} \\ 68 \\ \underline{60} \\ 80 \\ \underline{80} \end{array}$$

Compare the dividend 4.68 and the quotient 0.468. *Notice:* To write the the quotient 0.468 directly from 4.68, the decimal point was moved 1 place to the left.

Rule

To get the quotient directly when dividing a decimal or whole number by a power of 10, move the decimal point of the dividend as many places to the left as the number of zeros in the power of 10.

■ Example 11

Write the quotient directly: $30 \div 10$.

Solution Move the decimal point 1 place to the left.

$$30 \div 10 = 3.0 = 3$$

Answers to exercises 35–38

34. 3834 **35.** 352 R 11
36. 680 R 250 **37.** 301 R 2
38. 708

Divide.

42. $3\overline{)0.0930}$

43. $76\overline{)0.54644}$

44. $502\overline{)1.02910}$

45. $8\overline{)24.48}$

46. $82\overline{)192.290}$

The answers are in the margin on page 313.

■ **Example 12**

Write the quotient directly: 0.8 ÷ 10.

Solution Move the decimal point 1 place to the left.

$$0.8 \div 10 = 0.08$$

Practice check: Do exercises 47–51 in the margin.

Write the quotient directly.

47. 700 ÷ 10

48. 89 ÷ 10

49. 9 ÷ 10

50. 0.07 ÷ 10

51. 235.16 ÷ 10

The answers are in the margin on page 315.

■ **Example 13**

Write the quotient directly: 4000 ÷ 100.

Solution Move the decimal point 2 places to the left.

40.00

The answer is 40.00 or 40.

■ **Example 14**

Write the quotient directly: 0.32 ÷ 100.

Solution Move the decimal point 2 places to the left.

0.0032 Write in 2 zeros.

The answer is 0.0032.

■ **Example 15**

Write the quotient directly: 85 ÷ 1000.

Solution Move the decimal point 3 places to the left.

The answer is 0.085.

■ **Example 16**

A farmer applies 100 gallons per acre of spray. If his discharge rate is 6.4 gallons per minute, how many acres will he cover per minute?

Solution Write the quotient directly.

$$6.4 \div 100 = 0.064$$ Move the decimal point 2 places to the left.

The answer is 0.064 acres per minute.

Practice check: Do exercises 52–62 in the margin.

Write the quotient directly.

52. 486 ÷ 100

53. 0.06 ÷ 100

54. 1700.6 ÷ 100

55. 5000 ÷ 1000

56. 3820 ÷ 1000

57. 6 ÷ 1000

58. 0.37 ÷ 1000

59. 392.6 ÷ 1000

60. 0.47 ÷ 10

61. 78.38 ÷ 100

62. 409 ÷ 10

The answers are in the margin on page 315.

■ Looking Back

You should now be able to divide decimals by whole numbers.

■ Problems 8–5

Divide and check.

1. $7\overline{)87.5}$ **2.** $6\overline{)11.04}$ **3.** $2\overline{)40.92}$

4. $6\overline{)57.06}$ **5.** $8\overline{)0.4328}$ **6.** $4\overline{)0.4140}$

7. $14\overline{)128.044}$ **8.** $30\overline{)108.12}$ **9.** $49\overline{)1.8424}$

10. $24\overline{)0.14520}$ **11.** $313\overline{)67.8584}$ **12.** $400\overline{)5237.60}$

13. $739\overline{)6.660607}$ **14.** $119\overline{)357.6188}$

For exercises 15–18, write the quotient directly.

15. 30 ÷ 10 **16.** 739 ÷ 100 **17.** 7 ÷ 10

18. 315 ÷ 1000 **19.** 69 ÷ 100 **20.** 0.8 ÷ 100

21. 0.987 ÷ 10 **22.** 38.47 ÷ 10 **23.** 3986.76 ÷ 1000

Answers to exercises 39–41

39. 8.9 **40.** 4.02 **41.** 0.2054

Answers to exercises 42–46

42. 0.0310 **43.** 0.00719
44. 0.00205 **45.** 3.06 **46.** 2.345

24. $4 \div 100$

25. $39{,}000 \div 1000$

26. $0.046 \div 1000$

27. $6000 \div 100$

28. $0.815 \div 100$

29. Aeronautical Maintenance. An airplane propeller 3.752 inches thick has 4 layers of uniformly thick laminating material. Find the thickness of one layer.

30. Auto Mechanics. After five treatments of rust proofing to an automobile, the coating on the metal surface is 3.54 mm. About how much coating was applied each time?

31. Machinist. A stack of 65 sheets of brass shims is 4.2575 inches high. What is the thickness of each sheet?

32. Contractor. The cost for a 19 cu yd concrete slab floor was $648.47. Find the cost per cu yd.

33. Nursing. The medication issued was 0.747 grains. If the medication is to be divided into 9 equal parts, how many grains should be in each part?

34. Trucking. A trucking company has a fleet of 100 trucks. During a working week of 5 days, the trucks traveled a total of 152,600 miles. What was the average distance traveled by each truck per day?

35. Water Management. A gasoline engine water pump uses 10 gallons of gasoline per hour. Find the number of hours the engine can operate on 122.6 gallons.

The answers to odd-numbered exercises are in the back of the book.

8–6 DIVIDING DECIMALS BY DECIMALS

When we divide decimals, it is best to divide by a whole number, but suppose the divisor is not a whole number? We know from section 4–4 that the *value* of a fraction does not change when the numerator and denominator are multiplied by the same number.

■ **Example 1**

Rewrite $\frac{4}{0.5}$ so the denominator is a whole number.

Solution The denominator has one decimal place, so we multiply.

$$\frac{4}{0.5} \text{ by } \frac{10}{10}$$

$$\frac{4}{0.5} \times \frac{10}{10} = \frac{4 \times 10}{0.5 \times 10} = \frac{40}{5}$$

Answers to exercises 47–51

47. 70 **48.** 8.9 **49.** 0.9
50. 0.007 **51.** 23.516

Answers to 52–62

52. 4.86 **53.** 0.0006 **54.** 17.006
55. 5 **56.** 3.82 **57.** 0.006
58. 0.00037 **59.** 0.3926
60. 0.047 **61.** 0.7838 **62.** 40.9

■ **Example 2**

Rewrite $\frac{0.418}{2.73}$ so the denominator is a whole number.

Solution The denominator has two decimal places, so we multiply $\frac{0.418}{2.73}$ by $\frac{100}{100}$.

$$\frac{0.418}{2.73} \times \frac{100}{100} = \frac{0.418 \times 100}{2.73 \times 100} = \frac{41.8}{273}$$

We moved the decimal point 2 places to the right in both the numerator and the denominator.

Practice check: Do exercises 63–67 in the margin.

Rewrite the fractions so the denominator is a whole number.

63. $\frac{5.76}{0.8}$ **64.** $\frac{0.0469}{5.68}$

65. $\frac{23.9}{0.047}$ **66.** $\frac{89.341}{0.0018}$

67. $\frac{3041.6147}{391.104}$

The answers are in the margin on page 317.

■ **Example 3**

Divide $0.6\overline{)2.586}$.

Solution The divisor is not a whole number. We can put the division example in the form of a fraction. Now

$$\frac{2.586}{0.6} = \frac{2.586 \times 10}{0.6 \times 10} = \frac{25.86}{6}$$

and becomes $6\overline{)25.86}$. Now we divide.

```
   4.31
6)25.86      Check  4.31
  24                ×0.6    Original divisor
   18              2.586    Original dividend
   18
    6
    6
    0
```

The quotient is 4.31. Notice that the decimal point was moved 1 place to the right in both the divisor and dividend.

■ **Example 4**

Divide $4.03\overline{)41.509}$

Solution $4.03\overline{)41.509}$

↑ Not a whole number

Step 1 Mentally multiply the dividend and divisor by 100 by moving the decimal point two places to the right.

$4.03\overline{)41.509}$ becomes $403\overline{)4150.9}$

2 places 2 places

Step 2 Divide.

```
     10.3        Check:  10.3
403)4150.9               4.03    Original divisor
    403                   309
    120                 41200
      0                 41.509   Original dividend
    120 9
    120 9
```

The quotient is 10.3.

Practice check: Do exercises 68 and 69 in the margin.

Divide.

68. $2.9\overline{)0.0348}$ **69.** $0.15\overline{)4.6575}$

The answers are in the margin on page 319.

■ Example 5

Divide $0.0037\overline{)0.08473}$.

Solution **Step 1** Change the divisor to a whole number.

$0.0037\overline{)0.08473}$ becomes $37\overline{)847.3}$

4 places 4 places

Step 2 Divide.

```
    22.9      Check:   22.9
37)847.3             ×0.0037   Original divisor
   74                  1603

   107
    74               0.08473   Original dividend
    33 3
    33 3
```

The quotient is 22.9.

Rule

A short method for dividing a decimal by a decimal:

Step 1 Move the decimal point in the divisor as many places to the right as needed to make it a whole number.

Step 2 Move the decimal point in the dividend the same number of places to the right as you did in the divisor.

Step 3 Place a decimal point directly above the dividend decimal point in the quotient place.

Step 4 Divide as you would with whole numbers.

Practice check: Do exercises 70 and 71 in the margin.

■ Looking Back

You should now be able to divide a decimal by a decimal without a remainder.

■ Problems 8–6

Divide and check.

1. $0.4\overline{)344.8}$ **2.** $0.5\overline{)3.315}$ **3.** $0.8\overline{)16.0}$

4. $0.9\overline{)0.0072}$ **5.** $0.3\overline{)96}$ **6.** $2.5\overline{)46.5}$

7. $7.6\overline{)19.684}$ **8.** $0.03\overline{)368.61}$ **9.** $0.06\overline{)0.49806}$

10. $0.14\overline{)56}$ **11.** $0.89\overline{)5.696}$ **12.** $0.05\overline{)0.00025}$

13. $0.49\overline{)375.34}$ **14.** $0.52\overline{)3.5464}$ **15.** $1.34\overline{)175.272}$

16. $0.35\overline{)70}$ **17.** $0.007\overline{)82.488}$ **18.** $0.122\overline{)95.16}$

19. $0.021\overline{)0.6552}$ **20.** $3.415\overline{)12.2940}$ **21.** $0.038\overline{)70.11}$

22. $0.018\overline{)222.93}$

Answers to exercises 63–67

63. $\frac{57.6}{8}$ **64.** $\frac{4.69}{568}$ **65.** $\frac{23900}{47}$

66. $\frac{893410}{18}$ **67.** $\frac{3041614.7}{391104}$

Divide.

70. $0.04\overline{)0.020592}$

71. $0.182\overline{)39.13}$

The answers are in the margin on page 319.

23. **Fire Science.** The discharge of a water pump is 0.1685 gallons per revolution. How many revolutions must the pump make to discharge 684.11 gallons?

24. **Waste Water Management.** A particular sludge weighs 9.75 lb per gallon. If the total weight of the tank full of sludge is 12,441 lb, find the capacity of the tank.

25. **Finance.** If the annual interest charge is $352.50 and the rate is 0.075 per year, what is the amount of money owed?

26. **Interior Decorator.** Upholstery fabric costs $10.95 a yard. How many yards would $65.70 buy?

27. **Carpentry.** A carpenter's hourly wage was $12.25. Find how many hours he worked if he earned $85.75.

28. **Nursing.** How many magnesium strips, each weighing 2.36 grams, can be cut from a strip weighing 33.04 grams?

29. **Catering.** Irma's Catering Service sells chicken dinners at $10.95 per serving. If the total change to an organization was $613.20, how many dinners were served at the function?

30. **Metal Trade.** A milling machine removes 0.013 in. of metal per cut. How many cuts are needed to shave off 1.326 in. of metal?

The answers to odd-numbered exercises are in the back of the book.

8–7 NONTERMINATING DIVISION

Thus far our division exercises all have had zero remainders. We call this type of division of decimals *terminating*. The division problems that do not come out even, or never have a zero remainder, are called *nonterminating*.

■ **Example 1**

Divide $3.1\overline{)23.62}$.

Answers to exercises 68 and 69

68. 0.012 **69.** 31.05

Answers to exercises 70 and 71

70. 0.5148 **71.** 215

Solution

```
        76.193
 3.1)23.6 2000   ←— Write in zeros as place holders.
     21 7
      1 9 2
      1 8 6
          60
          31
          290
          279
           110
            93
            17
```

We can keep writing in zeros and divide forever, but we will never have a zero remainder. This quotient is *nonterminating.*

We can express this type of quotient in short form as a decimal and fraction, or as a rounded decimal.

■ **Example 2**

Divide 0.27)1.6. Express the quotient as a two-place decimal and a fraction.

Solution 0.27)1.60 becomes 27)160.

Step 1 We want a two-place decimal, so we write in two zeros to the right of the decimal point.

27)160.00 2 places

Step 2 Divide.

```
       5.92 16/27
 27)160.00        We use a fraction to shorten the quotient. The fraction
    135           is obtained by placing the remainder over the divisor.
     250
     243
       70
       54
       16
```

The quotient is read "Five and ninety-two and sixteen twenty-sevenths hundredths." That's a mouthfull!

■ **Example 3**

Divide 0.007)0.0314. Express the quotient as a three-place decimal and a fraction.

Solution 0.007)0.0314 becomes 7)31.4.

Step 1 We want a three-place decimal, so we write in two zeros to the right of the decimal point.

7)31.400 3 places

Step 2

$$\begin{array}{r} 4.485\tfrac{5}{7} \\ 7\overline{)31.400} \\ \underline{28} \\ 34 \\ \underline{28} \\ 60 \\ \underline{56} \\ 40 \\ \underline{35} \\ 5 \end{array}$$

Rule

To express a quotient in nonterminating division as a decimal fraction:

Step 1 Move the decimal points in the divisor and dividend the same number of places needed to make the divisor a whole number.

Step 2 In the dividend, write in as many zeros as necessary to make the number of places required.

Step 3 Divide.

Step 4 Form the fraction by placing the remainder over the divisor.

Practice check: Do exercises 72–74 in the margin.

Divide. Express the quotient as a two-place decimal.

72. $0.106\overline{)0.00839}$ **73.** $0.74\overline{)80}$

74. $2.15\overline{)7.1639}$

The answers are in the margin on page 323.

■ Example 4

Divide $0.557\overline{)0.06004}$. Round the quotient to the nearest tenth.

Solution In order to do this, we must carry our division to one place beyond the tenths place. So we must have 2 places in our dividend.

$0.557\overline{)0.06004}$ becomes $557\overline{)60.04}$ 2 places

We divide.

$$\begin{array}{r} .10 \\ 557\overline{)60.04} \\ \underline{557} \\ 434 \\ \underline{0} \\ 434 \end{array}$$

The quotient rounded to the nearest tenth is 0.1.

■ Example 5

Divide $0.067\overline{)1.86}$. Round the quotient to the nearest hundredth.

Solution To do this, we must carry our division to one place beyond the hundredths place. We must have 3 places in our dividend, so we write in 3 zeros.

$0.067\overline{)1.860}$ becomes $67\overline{)1860.000}$ 3 places

We divide.

$$\begin{array}{r} 27.761 \\ 67\overline{)1860.000} \\ \underline{134} \\ 520 \\ \underline{469} \\ 510 \\ \underline{469} \\ 410 \\ \underline{402} \\ 80 \\ \underline{67} \\ 13 \end{array}$$

27.761 rounds to 27.76 to the nearest hundredth.

If necessary, we can shorten the quotient on terminating division problems, as well as on nonterminating.

Rule

To express a quotient in a nonterminating division as a rounded decimal:

Step 1 Move the decimal points in the divisor and dividend the same number of places needed to make the divisor a whole number.

Step 2 In the dividend, write in as many zeros as necessary to make one more place beyond the required rounding place.

Step 3 Divide.

Step 4 Round the quotient to the desired place value.

Practice check: Do exercises 75–78 in the margin.

Divide. Round to the nearest tenth.

75. $0.09\overline{)0.089}$ **76.** $0.448\overline{)1.6076}$

77. $0.79\overline{)5.29}$ **78.** $0.039\overline{)4}$

The answers are in the margin on page 323.

■ Looking Back

You should now be able to

1. Divide decimals with nonterminating quotients.
2. Express the quotients either as decimals with fractions, or as rounded decimals to the nearest tenth or hundredth.

■ Problems 8–7

Divide. Express the quotient as a one-place decimal and a fraction.

1. $0.6\overline{)8.7}$ **2.** $0.4\overline{)19.68}$ **3.** $3.6\overline{)101.4}$ **4.** $0.07\overline{)4.6}$

Express the quotient as a two-place decimal and a fraction.

5. $0.046\overline{)89}$

6. $0.74\overline{)46.4}$

7. $0.004\overline{)0.091}$

8. $0.01\overline{)0.006}$

9. $1.6\overline{)0.71}$

10. $3.68\overline{)7}$

Round the quotient to the nearest tenth.

11. $0.7\overline{)1105.6}$

12. $3.8\overline{)15.987}$

13. $13.8\overline{)82.56}$

14. $0.08\overline{)0.645}$

Round the quotient to the nearest hundredth.

15. $0.046\overline{)70}$

16. $0.38\overline{)3.6}$

17. $3.31\overline{)14.77}$

18. $0.007\overline{)35}$

19. $0.014\overline{)76.8}$

20. $0.593\overline{)0.786}$

21. **Metal Trades.** A drill feeds into a copper piece at the rate of 0.7 in. per revolution. How many revolutions are needed to drill a hole 5.6 in. deep?

22. **Mason.** A stone mason contracts for laying 73.4 cu yd of stone at a price of $1630.54. Find the price per cu yd. Round to the nearest cent.

23. **Welder.** A welder cuts key stock into pieces 5.25 cm long. How many pieces can be made from a length of key stock 105.35 cm long? Round to the nearest whole piece.

24. Printer. If a one-pound can of ink costs \$7.83, how many one-pound cans may be purchased for \$129.08? Round to the nearest whole number.

25. Weatherman. During a torrential downpour, 10.35 in. of rain fell in 3.25 hours. What was the rate in inches per hour? Round to the nearest hundredth.

Answers to exercises 72–74

72. $0.07\frac{97}{106}$ **73.** $108.10\frac{30}{37}$

74. $3.33\frac{44}{215}$

Answers to exercises 75–78

75. 1.0 **76.** 3.6 **77.** 6.70
78. 102.56

The answers to odd-numbered exercises are in the back of the book.

8–8 CONVERTING FRACTIONS TO DECIMALS AND DECIMALS TO FRACTIONS

Your work with percentages in Chapters 10 and 11 will require you to know how to change fractions to decimals and decimals to fractions.

Converting Fractions to Decimals

■ Example 1

Convert $\frac{17}{100}$ to a decimal.

Solution $\frac{17}{100}$ means $17 \div 100$.
From section 8–5 we know that $17 \div 100 = 0.17$.

■ Example 2

Convert $\frac{42{,}905}{10{,}000}$ to a decimal.

Solution $\frac{42{,}905}{10{,}000}$ means $42{,}905 \div 10{,}000$.
There are 4 zeros, so we make the decimal a four-decimal place number.

$$\frac{42{,}905}{10{,}000} = 4.2905$$

↑ 4 zeros ↑ 4 places

Practice check: Do exercises 79–85 in the margin.

Convert to a decimal.

79. $\frac{8}{10}$ **80.** $\frac{58}{100}$

81. $\frac{571}{1000}$ **82.** $\frac{72}{1000}$

83. $\frac{38}{10{,}000}$ **84.** $\frac{5281}{100}$

85. $\frac{38{,}104}{1000}$

The answers are in the margin on page 325.

■ Example 3

Convert $\frac{5}{8}$ to a decimal.

Solution To convert we divide 5 by 8.

$$\begin{array}{r} .625 \\ 8\overline{)5.000} \\ \underline{4\,8} \\ 20 \\ \underline{16} \\ 40 \\ \underline{40} \end{array}$$

$$\frac{5}{8} = 0.625$$

■ Example 4

Convert $\frac{7}{9}$ to a decimal.

Solution To convert, we divide.

$$\begin{array}{r} .77 \\ 9\overline{)7.00} \\ \underline{6\,3} \\ 70 \\ \underline{63} \\ 7 \end{array}$$

This is a nonterminating quotient, so we can leave it as a two-place decimal with a fraction or round to the nearest tenth, hundredth, thousandth, and so forth. We choose to leave it as a two-place decimal with a fraction.

$$\begin{array}{r} 0.77\frac{7}{9} \\ 9\overline{)7.00}\phantom{\frac{7}{9}} \\ 6\,3\phantom{0\frac{7}{9}} \\ 70\phantom{\frac{7}{9}} \\ \underline{63}\phantom{\frac{7}{9}} \\ 7\phantom{\frac{7}{9}} \end{array}$$

$$\frac{7}{9} = 0.77\frac{7}{9}$$

■ Example 5

Convert $\frac{42}{18}$ to a decimal.

Solution **Step 1** Reduce. $\frac{42}{18} = \frac{7}{3}$

Step 2 Divide.

$$\begin{array}{r} 2.33 \\ 3\overline{)7.00} \\ \underline{6} \\ 10 \\ \underline{9} \\ 10 \\ \underline{9} \\ 1 \end{array}$$

$2.33\frac{1}{3}$, or 2.3 rounded to the nearest tenth.

Note: $\frac{7}{3}$ is exactly equal to $2.33\frac{1}{3}$, whereas 2.3 is approximately equal to $\frac{7}{3}$, since we rounded.

Rule

To convert a fraction to a decimal:

Step 1 Divide the numerator by the denominator.

Step 2 Leave the quotient as a decimal with a fraction or as a rounded decimal, if the division is nonterminating.

Practice check: Do exercises 86–92 in the margin.

■ Example 6

Convert $4\frac{3}{7}$ to a decimal.

Solution **Step 1** Separate the whole number and the fraction.

$$4 \quad \frac{3}{7}$$

Step 2 Find a decimal number for $\frac{3}{7}$.

$$\begin{array}{r} .428 \\ 7\overline{)3.000} \\ \underline{28} \\ 20 \\ \underline{14} \\ 60 \\ \underline{56} \\ 4 \end{array} \quad \text{or} \quad 0.43 \text{ to the nearest hundredth}$$

Step 3 Recombine.

$$4\frac{3}{7} \approx 4.43$$

Note: ≈ means "approximately equal to."

Answers to exercises 79–85

79. 0.8 **80.** 0.58 **81.** 0.571
82. 0.072 **83.** 0.0038 **84.** 52.81
85. 38.104

Convert the following fractions to decimals. Leave the quotient as a two-place decimal if it is nonterminating.

86. $\frac{1}{5}$ **87.** $\frac{1}{2}$

88. $\frac{3}{5}$ **89.** $\frac{2}{3}$

90. $\frac{7}{16}$ **91.** $\frac{14}{8}$

92. $\frac{23}{13}$

The answers are in the margin on page 327.

Rule

To convert mixed numbers to decimals:

Step 1 Separate the whole number from the fraction.
Step 2 Convert the fraction to a decimal.
Step 3 Recombine.

Practice check: Do exercises 93–96 in the margin.

Convert the following mixed numbers to decimals.

93. $2\frac{2}{5}$ **94.** $15\frac{7}{8}$

95. $106\frac{9}{20}$ **96.** $59\frac{13}{15}$

The answers are in the margin on page 329.

Converting Decimals to Fractions

■ Example 7

Convert 0.74 to a reduced common fraction.

Solution $0.74 = \frac{74}{100} = \frac{37}{50}$

■ Example 8

Convert 3.042 to a mixed number.

Solution $3.042 = 3\frac{42}{1000} = 3\frac{21}{500}$

Practice check: Do exercises 97–104 in the margin.

Convert the following decimals to reduced common fractions or mixed numbers.

97. 0.5 **98.** 0.54

99. 0.008 **100.** 15.73

101. 0.775 **102.** 3.864

103. 0.0014 **104.** 56.0625

The answers are in the margin on page 329.

■ Example 9

Convert $0.16\frac{2}{3}$ to a reduced common fraction.

Solution

Hundredths ↓

$$0.16\frac{2}{3} = \frac{16\frac{2}{3}}{100}$$

$$= 16\frac{2}{3} \div 100$$

$$= \frac{50}{3} \times \frac{1}{100}$$ We divide by multiplying by the reciprocal of 100.

$$= \frac{\overset{1}{\cancel{50}}}{3} \times \frac{1}{\underset{2}{\cancel{100}}}$$

$$= \frac{1}{6}$$

■ Example 10

Convert $0.3\frac{1}{4}$ to a reduced common fraction.

Solution

Tenths

$$0.3\frac{1}{4} = \frac{3\frac{1}{4}}{10}$$

$$= 3\frac{1}{4} \div 10$$

$$= \frac{13}{4} \times \frac{1}{10}$$

$$= \frac{13}{40}$$

Answers to exercises 86–92

86. 0.2 **87.** 0.5 **88.** 0.6

89. $0.66\frac{2}{3}$ **90.** 0.4375 **91.** 1.75

92. $1.76\frac{13}{12}$

Rule

To convert a decimal and fraction to a reduced common fraction:

Step 1 Change the decimal and fraction to a whole number and fraction over an appropriate power of 10.

Step 2 Complete the division and reduce if possible.

■ **Example 11**

Convert $7.33\frac{1}{3}$ to a mixed number.

Solution **Step 1** Separate

$$7 \quad 0.33\frac{1}{3}$$

Step 2 Convert the decimal fraction.

Hundredths

$$0.33\frac{1}{3} = \frac{33\frac{1}{3}}{100}$$

$$= 33\frac{1}{3} \div 100$$

$$= \frac{100}{3} \times \frac{1}{100}$$

$$= \frac{\overset{1}{\cancel{100}}}{3} \times \frac{1}{\underset{1}{\cancel{100}}}$$

$$= \frac{1}{3}$$

Convert to a reduced common fraction or a mixed number.

105. $15.16\frac{2}{3}$ **106.** $0.66\frac{2}{3}$

107. $31.83\frac{1}{3}$ **108.** $0.84\frac{1}{4}$

109. $7.008\frac{1}{3}$ **110.** $0.482\frac{3}{4}$

The answers are in the margin on page 331.

Step 3 Recombine.

$$7.33\frac{1}{3} = 7\frac{1}{3}$$

Practice check: Do exercises 105–110 in the margin.

■ Looking Back

You should now be able to change fractions or mixed numbers into decimals, and convert decimals into reduced common fractions or mixed numbers.

■ Problems 8–8

Review

Multiply.

1. $\begin{array}{r} 6.015 \\ \times 23.6 \\ \hline \end{array}$

2. $\begin{array}{r} 10.43 \\ \times 2.007 \\ \hline \end{array}$

3. 3.41×100

4. 0.0062×1000

Divide. Round to the nearest hundredth.

5. $0.32\overline{)14}$

6. $0.006\overline{)0.000416}$

7. $4.64\overline{)503.539}$

8. $4.004\overline{)5.66342}$

Convert the following fractions or mixed numbers to decimals. If the division is nonterminating, round the quotient as a two-place decimal.

9. $\frac{6}{100}$

10. $\frac{4}{5}$

11. $\frac{1}{3}$

12. $\frac{875}{1000}$

13. $17\frac{1}{2}$

14. $\frac{3}{5}$

15. $\frac{17}{20}$

16. $\frac{39}{10,000}$

17. $\frac{42}{50}$ **18.** $\frac{8461}{100}$ **19.** $37\frac{9}{13}$ **20.** $\frac{18}{7}$

21. $3\frac{1}{8}$ **22.** $\frac{1}{6}$ **23.** $132\frac{11}{12}$ **24.** $\frac{372}{20}$

25. $\frac{10}{4}$ **26.** $\frac{15}{40}$ **27.** $\frac{7}{8}$ **28.** $89\frac{5}{12}$

Convert to a reduced common fraction or a mixed number.

29. 0.2 **30.** 0.8 **31.** 0.35 **32.** 0.09

33. $6.33\frac{1}{3}$ **34.** 0.80 **35.** 0.25 **36.** 0.625

37. $0.66\frac{2}{3}$ **38.** 3.44 **39.** 16.875 **40.** 0.50

41. $3.83\frac{1}{3}$ **42.** 0.048 **43.** $0.41\frac{2}{3}$ **44.** 0.0575

Answers to exercises 93–96

93. 2.4 **94.** 15.875 **95.** 106.45 **96.** 59.87

Answers to exercises 97–104

97. $\frac{1}{2}$ **98.** $\frac{27}{50}$ **99.** $\frac{1}{125}$

100. $15\frac{73}{100}$ **101.** $\frac{31}{40}$ **102.** $3\frac{108}{125}$

103. $\frac{7}{5000}$ **104.** $56\frac{1}{16}$

---

45. 9.375 **46.** 1.125 **47.** $46.91\frac{2}{3}$ **48.** $7.145\frac{3}{4}$

The answers to review and odd-numbered exercises are in the back of the book.

8–9 APPLIED PROBLEMS

■ Example 1

If a cubic yard of garden bark costs \$7.43, how much would 5.5 cubic yards cost?

Solution To solve we multiply.

$$\begin{array}{r} 7.43 \\ \times 5.5 \\ \hline 3715 \\ 37150 \\ \hline 40.865 \end{array}$$

The price would be \$40.87.

■ Example 2

A jet traveled 1648.4 miles in 3.6 hours. What is its average ground speed? Round to the nearest tenth.

Solution In order to decide which operation to use, it sometimes helps to discuss the problem using whole numbers. Suppose the jet traveled 400 miles in 2 hours. It should be clear that in 1 hour it traveled 200 miles. To get 200 miles we divided 400 miles by 2 hours. To find the answer in the original problem, we divide 1648.4 miles by 3.6 hours.

$$\begin{array}{r} 457.88 \\ 3.6\overline{)16484.00} \\ 144 \\ \hline 208 \\ 180 \\ \hline 284 \\ 252 \\ \hline 320 \\ 288 \\ \hline 320 \\ 288 \\ \hline \end{array}$$

The jet's speed was 457.9 mph.

■ Example 3

On a certain flight, a plane consumed 53.8 gallons of fuel per hour.
How many gallons would the plane use in 7.25 hours? Round the answer to the nearest tenth.

Solution To solve we multiply.

$$\begin{array}{r} 53.9 \\ \times 7.25 \\ \hline 2695 \\ 10780 \\ 377300 \\ \hline 390.775 \end{array}$$

The airplane used 390.8 gallons of fuel.

■ Example 4

Susan paid $12.39 for a certain cut of meat, which sold for 2.84 a pound. How many pounds of meat did she buy? Round to the nearest hundredth of a pound.

Solution If Susan had paid $10.00 at $2 per pound, it is easy to see that she bought 5 pounds of meat, since $5 \times 2 = 10$. To get 5 lb we divided $10 by 2 lb. To find the answer in our original problem, we divide $12.39 by $2.84.

$$\begin{array}{r} 4.362 \\ 2.84\overline{)12.3900} \\ \underline{1136} \\ 1030 \\ \underline{852} \\ 1780 \\ \underline{1704} \\ 760 \\ \underline{568} \\ 192 \end{array}$$

Susan bought 4.36 pounds of meat.

Practice check: Do exercises 111–114 in the margin.

Answers to exercises 105–110

105. $15\frac{1}{6}$ **106.** $\frac{2}{3}$ **107.** $31\frac{5}{6}$

108. $\frac{337}{400}$ **109.** $7\frac{1}{120}$ **110.** $\frac{1931}{4000}$

Solve.

111. On a flight, a certain plane used 58.3 gallons of aviation fuel per hour. If the flight lasted 5.75 hours, how many gallons of fuel were consumed? Round to the nearest tenth.

112. A certain jet airplane burns 64.3 gallons of fuel per hour. Find the number of hours it can fly if its fuel tank holds 389.5 gallons. Round to the nearest tenth.

113. The per pound cost of tenderloin is $7.80. How much would 0.75 pound cost of this meat?

114. If a 6-ounce can of olives cost $1.29, find the cost per ounce. Round to the nearest tenth of a cent.

The answers are in the margin on page 333.

■ Problems 8–9

Solve.

1. **Mason.** An apprentice mason worked 38 hours at $7.82 per hour. What were his wages for that amount of time?

2. **Carpenter.** Sam cut a board into 3 equal pieces. If the board measured 15.72 feet, how long was each piece?

3. **Carpenter.** It takes an average of 3.7 hours to lay one floor square. Find the total number of hours to lay 13.25 squares.

4. **Secretarial.** Joe Housemann worked as a secretary and received $550.00 for a 35-hour week. What was his hourly wage? Round to the nearest cent.

5. **Butcher.** If New York steak costs $6.55 a pound, how much would 7.67 pounds cost? Round to the nearest cent.

6. **Apprentice.** Cindy Donaldson earned $354.16 during one week as an assembler. If she receives $9.82 per hour, how many hours did she work that week?

7. **Electrician.** The difference between two electric (kilowatt-hour) meter readings a month apart on the same meter is 352 kilowatt hours. Find the cost of electrical energy if the cost averages $0.045 per kilowatt hour (kWh).

8. **Baker.** A sack of flour weighing 100 pounds costs $24.95. What is the cost per pound? Round to the nearest cent.

9. **Water Management.** Water weighs 62.5 pounds per cubic foot. How much would 198.67 cubic feet of water weigh?

10. **Generic.** Sieg drove 529.4 miles using 25.2 gallons of gasoline. What was his average miles per gallon? Round to the nearest mile.

11. **Aeronautics.** A jet traveled an average of 594.74 miles per hour in 5.27 hours. How far did it travel in that time? Round to the nearest tenth of a mile.

12. **Machinist.** Janice's machinist's union negotiated an hourly wage of $14\frac{7}{8}$ dollars. What is the hourly wage to the nearest cent?

13. **Generic.** A certain car traveled 890 miles in one week. If it averages 27.3 miles per gallon, how many gallons did it use? Round to the nearest tenth of a gallon. If gasoline costs $1.41 per gallon, how much did the gasoline cost?

14. **Electrician.** Ms. Daniels paid $146.26 for 1000 light bulbs. How much did she pay per bulb? Round to the nearest cent.

15. **Aeronautics.** A certain jet uses 75.9 gallons of jet fuel each hour that it operates. If the jet has a capacity of 554.1 gallons, how many hours can this jet fly before running out of fuel?

16. Generic. In a ten-week period, Barbara Soloman earned \$4390.48. What were her earnings per week?

17. Publishing. A bookbinder had a pile of 37 books that measured 58.56 inches high. Find the average thickness of each book to the nearest thousandth.

Answers to exercises 111–114

111. 335.2 gallons **112.** 6.1 hr **113.** \$5.85 **114.** 21.5¢

The answers to odd-numbered exercises are in the back of the book.

CHAPTER 8 OVERVIEW

Summary

8–1 You learned to multiply whole numbers with decimals.

8–2 You learned to multiply a decimal by a decimal.

8–3 You learned to multiply decimals by powers of 10.

8–4 You learned to check multiplication by estimating.

8–5 You learned to divide decimals by whole numbers and powers of 10.

8–6 You learned to divide a decimal by a decimal without a remainder

8–7 You learned to divide decimals with nonterminating quotients.

8–8 You learned to solve applied problems involving decimals with multiplication and division.

Terms To Remember

	Page		*Page*
Powers of 10	304	Terminating decimal	318
Estimating	306	Nonterminating decimal	318
Rounding	306		

Rules

- A short method for multiplying decimals:

  Step 1 Place the numbers vertically and multiply as if they were whole numbers.

  Step 2 Total the number of places to the right of the decimal point in the factors.

  Step 3 Place the decimal point in the product so that the number of places to the right of the decimal point is the same as the sum of the number of places in the factors.

- To get the product directly when multiplying a decimal by a power of 10, move the decimal point in the decimal number as many places to the right as the number of zeros in the power of 10.

- To get the product directly when multiplying a decimal or a whole number by a power of 10, move the decimal point of the multiplicand as many places to the right as the number of zeros in the power of 10.
- To get the quotient directly when dividing a decimal or whole number by a power of 10, move the decimal point of the dividend as many places to the left as the number of zeros in the power of 10.
- A short method for dividing a decimal by a decimal:

  Step 1 Move the decimal point in the divisor as many places to the right as needed to make it a whole number.

  Step 2 Move the decimal point in the dividend the same number of places to the right as you did in the divisor.

  Step 3 Place a decimal point directly above the dividend decimal point in the quotient place.

  Step 4 Divide as you would with whole numbers.

- To express a quotient in nonterminating division as a decimal and fraction:

  Step 1 Move the decimal points in the divisor and dividend the same number of places needed to make the divisor a whole number.

  Step 2 In the dividend, write in as many zeros as necessary to make the number of places required.

  Step 3 Divide.

  Step 4 Form the fraction by placing the remainder over the divisor.

- To express a quotient in nonterminating division as a rounded decimal:

  Step 1 Move the decimal points in the divisor and dividend the same number of places needed to make the divisor a whole number.

  Step 2 In the dividend, write in as many zeros as necessary to make one more place beyond the required rounding place.

  Step 3 Divide.

  Step 4 Round the quotient to the desired place value.

- To convert a fraction to a decimal:

  Step 1 Divide the numerator by the denominator.

  Step 2 Leave the quotient as a decimal with a fraction or as a rounded decimal, if the division is nonterminating.

- To convert mixed numbers to decimals:

  Step 1 Separate the whole number from the fraction.

  Step 2 Convert the fraction to a decimal.

  Step 3 Recombine.

- To convert a decimal and a fraction to a reduced common fraction:

  Step 1 Change the decimal and fraction to a whole number and fraction over an appropriate power of 10.

  Step 2 Complete the division and reduce if possible.

Self-Test

Multiply.

8–1 **1.** 15 0.101 × 0.01 **2.** 4.09 × 9 **3.** 0.3609 × 51 **4.** 30 × 0.06

8–2 **5.** 0.7 × 2.8 **6.** 5.39 × 0.06 **7.** 20.4 × 1.204

8–3 Multiply by writing the answer directly.

8. 89.6 × 1000 **9.** 0.0006 × 10,000

8–4 Find the exact answer, then check by estimating.

10. 4.6 × 8 **11.** 8.9 × 0.64 **12.** 0.54 × 0.765

8–5 Divide.

13. $8\overline{)57.6}$ **14.** $406\overline{)2.2330}$

Write the quotient directly.

15. 246 ÷ 100 **16.** 6 ÷ 1000

8–6 Divide.

17. $0.6\overline{)4.188}$ **18.** $4.19\overline{)0.06285}$ **19.** $0.0312\overline{)56.628}$

8–7 Write the quotient as a two-place decimal and a fraction.

20. $4.9\overline{)78.05}$

Round the quotient to the nearest tenth.

21. $0.101\overline{)80.516}$

8–8 Change the following common fractions or mixed numbers to decimals. If the division is nonterminating, round the quotient to the nearest tenth.

22. $\frac{428}{10{,}000}$ **23.** $\frac{1}{4}$ **24.** $3\frac{5}{8}$

Express the following decimals as reduced common fractions or mixed numbers.

25. 0.008 **26.** $5.83\frac{1}{3}$ **27.** 39.375

8–9 **28.** Bill paid $1.46 per gallon for gasoline. If he used 45.6 gallons on a trip, what did it cost him for gasoline? Round to the nearest cent.

29. A certain piece of tagboard is 0.0084 inches thick. What would be the height of a stack of 130 pieces?

30. At $0.29 per pound, how many pounds of oranges can be purchased for $2.20? Round to the nearest tenth of a pound.

31. A baseball player was credited for 57 hits in 169 times at bat. What was his batting average? (*Hint:* Convert $\frac{57}{169}$ to a decimal fraction and round to the nearest thousandth.)

The answers are in the back of the book.

Chapter Test

Multiply.

1. 19×0.001
2. 0.41803×93
3. 0.6×3.9
4. 1.905×30.7

Multiply by writing the answer directly.

5. $0.0000072 \times 10{,}000$

Find the exact answer, then check by estimating.

6. 5.8×6
7. 0.846×0.46

Solve.

8. A grocery chain advertised pot roast at \$1.69 per pound. A customer bought 5.4 pounds. What was the cost to the customer? Round to the nearest cent.

9. A taxi driver filled his tank with 11.7 gallons of gasoline at \$1.049 per gallon. What did he pay for the gasoline? Round to the nearest cent.

Divide.

10. $508\overline{)1.4224}$ **11.** $104.5 \div 1{,}000$ **12.** $0.9\overline{)6.309}$ **13.** $0.0403\overline{)69.4772}$

Round to the nearest tenth.

14. $0.203\overline{)56.031}$

Change the following common fractions or mixed numbers to decimals. If division is non-terminating, round the quotient to the nearest tenth.

15. $\frac{3}{4}$ **16.** $23\frac{7}{8}$

Express the following decimals as reduced common fractions or mixed numbers.

17. $17.66\frac{2}{3}$ **18.** 0.225

19. A carton of 12 cans of a certain soft drink sold for $4.60. What was the cost per can? Round to the nearest cent.

20. At Jackson Brothers Complete Automobile Repair, 1.8 gallons of kerosene per day are sold from a 50-gallon drum. If the drum was full at the beginning, at this rate how many days will the drum last? (Round back to the nearest day.)

21. A car travels 382.7 miles in 7.5 hours. What was the average speed? Round to the nearest number of miles per hour.

22. A butcher at Central Meats and Sausage Company sold a cut of beef for $48.30 at $2.30 per pound. How many pounds were sold?

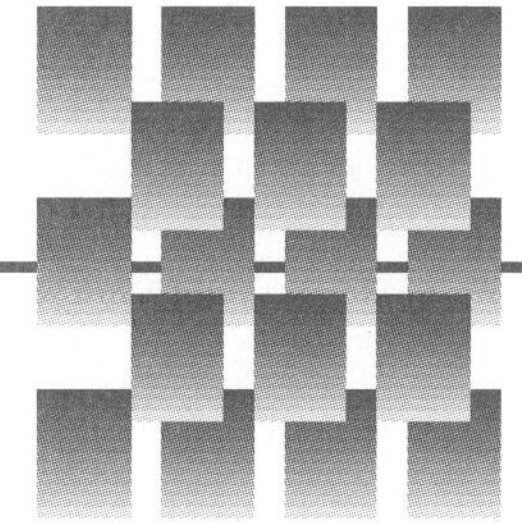

CHAPTER 9

Ratio and Proportion

Choice: Either skip the Pretest and study all of Chapter 9 or take the Pretest and study only the section(s) you need to study.

Pretest

Directions: This Pretest will help you determine which sections to study in Chapter 9. If any of your answers are incorrect, or if you omitted any exercise, turn to the indicated section(s) and study the material; then take the Chapter Test.

9–1 **1.** Express as a ratio in feet, in reduced form: 18 feet to 2 yards.

2. Are the following ratios equivalent?
4 cans to 6 bottles
10 cans to 15 bottles

9–2 **3.** State the following ratio in whole number terms: $4\frac{2}{3}$ to $1\frac{3}{7}$

9–3 **4.** Find the unit cost or ratio: $48.36 for 6.2 gallons.

9–4 **5.** Is this statement a true proportion? $\frac{9}{15} = \frac{6}{8}$

9–5 **6.** Show by the cross product method that the following statement is a true proportion: $\frac{0.6}{1.5} = \frac{3.4}{8.5}$

7. Solve for n: $\frac{24}{8} = \frac{21}{n}$

9–6 **8.** A map has a scale of 5 miles for every $\frac{3}{4}$ inch. How many miles are represented by 6 inches?

The answers are on page 345.

9–1 RATIO

Suppose a basketball team won 20 games and lost 4. We can compare by saying that for every 5 games won, the team lost 1 game.

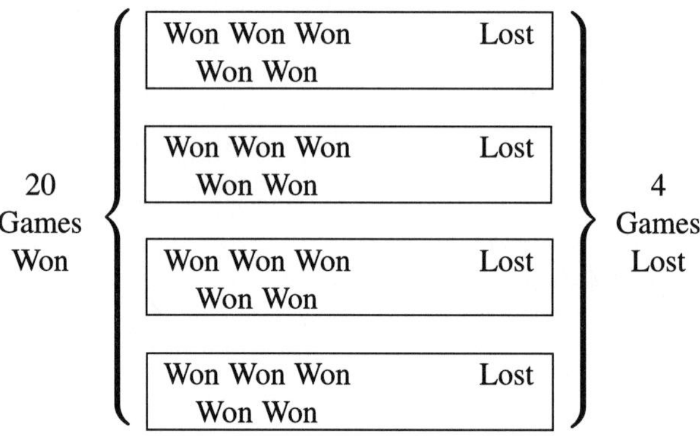

We call this kind of comparison a ratio. The *ratio* of games won to games lost is 20 to 4, or 5 to 1.

Rule

When two numbers or quantities are compared by division, the indicated division is called a ratio, *and the numbers or quantities are called* terms *of the ratio.*

Ratios may be stated as follows: $20 \div 4$, $\frac{20}{4}$, 20 to 4, or 20:4. For our purposes, we will use the second and third forms to express ratios.

■ **Example 1**

An automobile traveled 350 kilometers on 12 gallons of gasoline. Express kilometers to gallons as a ratio, then express gallons to kilometers as a ratio.

Solution $\frac{350}{12}$, or 350 kilometers to 12 gallons

$$\frac{12}{350}, \quad \text{or} \quad 12 \text{ gallons to } 350 \text{ kilometers}$$

Practice check: Do exercises 1–4 in the margin.

If possible, the numerator and denominator of a ratio should be expressed in the same units.

■ Example 2

Express the ratio of 6 inches to 4 feet in inches.

Solution **Step 1** Change 4 feet to inches.

$$1 \text{ foot} = 12 \text{ inches}$$
$$4 \text{ feet} = 4 \times 12 \text{ inches}$$
$$= 48 \text{ inches}$$

Step 2 Form the ratio in inches.

$$\frac{6 \text{ inches}}{4 \text{ feet}} = \frac{6 \text{ inches}}{48 \text{ inches}}, \quad \text{or} \quad \frac{6}{48}$$

Practice check: Do exercises 5–8 in the margin.

Ratios in Lowest or Highest Terms

Since a ratio can be expressed as a fraction, we can reduce the ratio to lowest terms or raise it to higher terms. Ordinarily a ratio is expressed in lowest terms.

■ Example 3

A class of 30 students had to share 18 minicalculators. Express the ratio of minicalculators to students in reduced form.

Solution $$\frac{18 \text{ minicalculators}}{30 \text{ students}} = \frac{18}{30} = \frac{3}{5}$$

The ratio is 3 minicalculators to 5 students.

Practice check: Do exercises 9–13 in the margin.

■ Example 4

A certain company learned that 7 out of 10 people preferred its brand over the competing one. Using this ratio, how many people preferred the brand if 60 people were polled?

Solution $$\frac{7}{10} = \frac{?}{60}$$

To raise $\frac{7}{10}$ to higher terms with a denominator of 60, we must multiply $\frac{7}{10}$ by $\frac{6}{6}$.

$$\frac{7}{10} = \frac{7}{10} \times \frac{6}{6} = \frac{42}{60}$$

42 people out of 60 people polled preferred their brand.

Express as a ratio in fraction form.

1. 38 TVs to 63 families
2. A map scale of 85 miles to 1 inch
3. 3 feet to 8 feet
4. 150 words to 4 minutes

The answers are in the margin on page 345.

Express as a ratio in the same units.

5. 5 minutes to 1 hour in minutes
6. 1 dollar to 60 cents in cents
7. 35 ounces to 2 pounds in ounces (1 pound = 16 ounces)
8. 3 yards to 8 feet in feet (1 yard = 3 feet)

The answers are in the margin on page 345.

Express each ratio in reduced form.

9. 6 cats to 14 dogs
10. 25 items right to 10 items wrong
11. 5 wins to 11 losses
12. A farmer planted 20 acres of wheat and 6 acres of corn. What is the ratio of corn to wheat?
13. A manufacturing firm found that for every 250 light bulbs, 8 were found to be defective. What is the ratio of defective bulbs to the total light bulbs manufactured?

The answers are in the margin on page 345.

14. If the ratio of the length to width of a picture frame is 5 to 3, what is the frame's length if the width is 15 inches?

15. A pro football team's ratio of losses to total games played is 1 to 5. If the team played 35 games, how many games did it lose?

The answers are in the margin on page 347.

Practice check: Do exercises 14 and 15 in the margin.

Equivalent Ratios

Two ratios are equivalent if they both can be reduced to the same ratio in lowest terms.

■ Example 5

Are the following ratios equivalent? 3 people to 6 chairs, and 7 people to 14 chairs.

Solution Reduce each ratio to lowest terms.

$$\frac{3}{6} = \frac{1}{2} \quad \text{and} \quad \frac{7}{14} = \frac{1}{2}$$

Since the ratios reduce to the same ratio in lowest terms, 3 to 6 and 7 to 14 are equivalent.

■ Example 6

One machine at Ace Manufacturing produces 28 widgets in 6 hours while another of the same type produces 70 widgets in 18 hours. Do the two machines produce at equivalent ratios?

Solution Reduce each ratio to lowest terms.

$$\frac{28}{6} = \frac{14}{3} \quad \text{and} \quad \frac{70}{18} = \frac{35}{9}$$

Since the ratios do not reduce to the same fraction, they are not equivalent.

Practice check: Do exercises 16–19 in the margin.

Are the following ratios equivalent?

16. 12 renters to 32 homeowners
18 renters to 48 homeowners

17. 8 motorcycles to 20 autos
32 motorcycles to 78 autos

18. 3 games won to 12 games lost
23 games won to 92 games lost

19. Are the following ratios of teachers to students equivalent in the following schools?

School A	School B
teachers 30	teachers 52
students 930	students 1716

The answers are in the margin on page 347.

■ Looking Back

You should now be able to

1. Compare two numbers or quantities as ratios.
2. Reduce a ratio to lowest terms.
3. Raise a ratio to higher terms.
4. Show that two ratios are equivalent.

■ Problems 9–1

Express the following as ratios in the same units where possible.

1. 50 males to 45 females
2. 108 chairs to 99 people
3. 2 dimes to 16 cents in cents
4. 3 hours to 35 minutes in minutes
5. 5 Fords to 7 Plymouths
6. 49 ounces to 3 pounds in ounces

7. 3 doorways to 15 windows

8. 33 watches to 18 clocks

9. 2 feet to 2 yards in feet

10. 18 gallons to 160 miles

11. Chef. A recipe calls for 2 cups of salt to 5 cups of sugar. What is the ratio of sugar to salt?

12. Machinist. A certain gear on a machine has 36 teeth while the meshing gear has 96 teeth. What is the ratio of the smaller toothed gear to the larger?

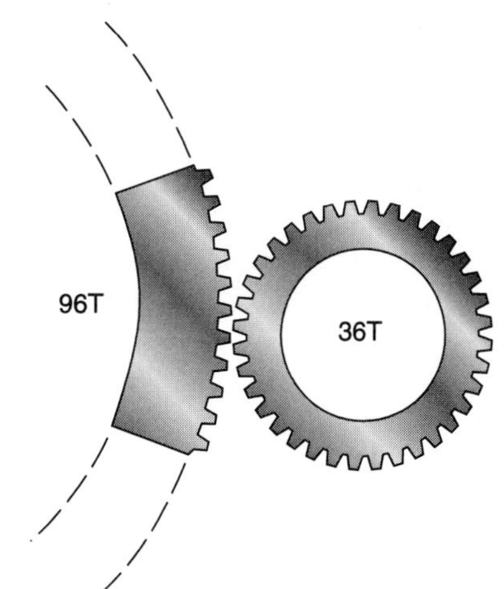

13. Generic. A soccer team lost 5 out of 14 games played. There were no tie games.

a. State the ratio of wins to losses.
b. State the ratio of wins to games played.
c. State the ratio of games played to losses.

14. Business. At the Quality Lion Hotels there are 250 managers and 650 assistant managers. What is the ratio of managers to assistant managers?

15. Agriculture. The Laddenburg Farms produced 224 bushels of carrots and 342 bushels of potatoes. Find the ratio of carrots to potatoes.

16. Generic. It takes two hours to walk to work and only 16 minutes by car. State the ratio of walking to riding in minutes.

Answers to Pretest

Circle the answers you missed or omitted. Cross them off after you've studied the corresponding section(s), then take the Chapter Test.

1. $\frac{3}{1}$ **2.** yes

3. $\frac{49}{15}$, or 49 to 15

4. $7.80 per gallon

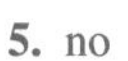

5. no **6.** $5.1 = 5.1$

7. $n = 7$ **8.** 40 miles

Answers to exercises 1–4

1. $\frac{38}{63}$ **2.** $\frac{85}{1}$ **3.** $\frac{3}{8}$ **4.** $\frac{150}{4}$

Answers to exercises 5–8

5. $\frac{5}{60}$ **6.** $\frac{100}{60}$ **7.** $\frac{35}{32}$ **8.** $\frac{9}{8}$

Answers to exercises 9–13

9. $\frac{3}{7}$ **10.** $\frac{5}{2}$ **11.** $\frac{5}{11}$ **12.** $\frac{3}{10}$

13. $\frac{4}{125}$

In exercises 17–24 express as ratios in reduced form.

17. 16 redheads to 52 blonds

18. 132 teachers to 22 administrators

19. 36 lace-leaf maples to 81 vine maples

20. 105 pens to 35 pencils

21. 39 bald men to 75 hair pieces

22. **Business.** Hunt and Peck Typewriter, Inc. manufactures electronic and manual typewriters at a ratio of 360 electronic to 12 manual. What is the ratio?

23. **Communications.** One newspaper uses a ratio of 18 columns of news to 12 columns of advertisement. What is the ratio?

24. **Business.** The city of Farnsworth found that 1001 homes were heated electrically and 231 were heated by gas. What is the ratio of homes heated by gas to those heated by electricity?

25. **Butcher.** Six pounds of pot roast will serve 10 persons. How many persons would be served with 18 pounds of pot roast?

26. **Machinist.** At Pfister Machine and Dye Company, a machinist produces 7 components in 30 minutes. How many components can the machinist produce in 7 hours?

27. **Generic.** The two triangles in the figure are in proportion. Find the missing measurement.

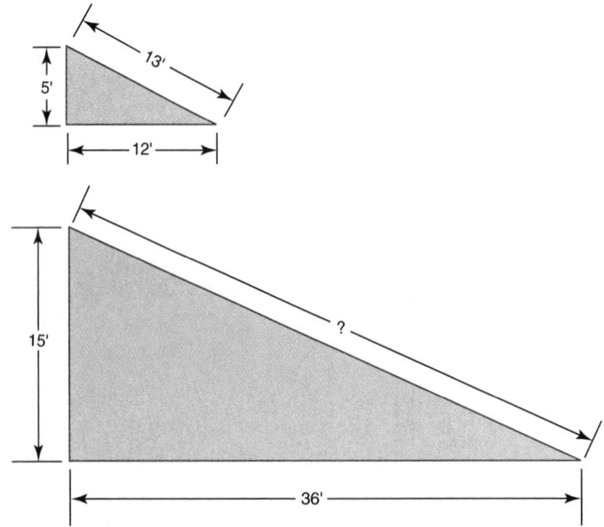

28. Generic. In a class election Bill received 36 votes. The ratio of votes cast to Bill's votes was 3 to 2. How many votes were cast?

29. Generic. Are the following ratios equivalent?

a. 4 men to 8 women and 22 men to 44 women

b. 105 candies to 3 cakes and 175 candies to 5 cakes

c. 32 pencils to 84 pens and 14 pencils to 42 pens

30. Business. Star Gear Company employed 25 minorities out of 150 workers. Jackson Iron Works' ratio of minorities to total workers is 82 to 492. Is this the same ratio?

The answers to odd-numbered exercises are in the back of the book.

Answers to exercises 14 and 15

14. $\frac{25}{15}$ The length is 25 in.

15. $\frac{7}{35}$ The team lost 7 games.

Answers to exercises 16–19

16. yes **17.** no **18.** yes
19. no

9–2 RATIOS WITH TERMS OTHER THAN WHOLE NUMBERS

Ratios with terms other than whole numbers are simplified in the same manner as the complex fractions found in section 4–8.

■ Example 1

Simplify, using whole number terms: $\frac{2}{3}$ to 5.

Solution $\frac{2}{3}$ to 5 means $\frac{\frac{2}{3}}{5}$, or $\frac{2}{3} \div 5$.

Completing the division, we get $\frac{2}{3} \div 5 = \frac{2}{3} \times \frac{1}{5} = \frac{2}{15}$.

The ratio is 2 to 15.

■ Example 2

Simplify to whole number terms: 3 hours to $5\frac{2}{5}$ minutes in minutes.

Solution $\frac{3 \text{ hours}}{5\frac{2}{5} \text{ minutes}} = \frac{180 \text{ minutes}}{5\frac{2}{5} \text{ minutes}}$ Changing hours to minutes

We complete the division.

$$180 \div 5\frac{2}{5} = \frac{180}{1} \div \frac{27}{5} = \frac{\overset{20}{\cancel{180}}}{1} \times \frac{5}{\underset{3}{\cancel{27}}} = \frac{100}{3}$$

The ratio is 100 minutes to 3 minutes.

Practice check: Do exercises 20–22 in the margin.

Simplify to whole number terms.

20. $3\frac{1}{2}$ cups of sugar to 14 cups of flour

21. $5\frac{1}{4}$ hours to $21\frac{3}{5}$ miles

22. $3\frac{1}{2}$ feet to 14 yards in feet

The answers are in the margin on page 349.

Example 3

Simplify using whole number terms: 3.6 ounces to 1.2 pounds in ounces.

Solution **Step 1** Change pounds to ounces (1 pound = 16 ounces)

$$\frac{3.6 \text{ ounces}}{1.2 \text{ pounds}} = \frac{3.6 \text{ ounces}}{19.2 \text{ ounces}}$$

Note: $1.2 \times 16 = 19.2$.

Step 2 Change the numerator and denominator to whole numbers and reduce.

$$\frac{3.6}{19.2} = \frac{3.6}{19.2} \times \frac{10}{10}$$
$$= \frac{36}{192} = \frac{3}{16}$$

The ratio is 3 ounces to 16 ounces, or 3 ounces to 1 pound.

Practice check: Do exercises 23–25 in the margin.

Simplify using whole number terms.

23. \$0.78 to \$4.24

24. 6.5 pounds to 15 adults

25. 7.5 yards to 67.05 feet in feet

The answers are in the margin on page 351.

Looking Back

Given a ratio with terms other than whole numbers, you should be able to find an equivalent ratio using whole numbers.

Problems 9–2

1. **Machinist.** When a machinist fabricated 8 automobile door panels, he wasted $1\frac{1}{2}$ pounds of steel. State the ratio using whole number terms.

2. **Construction.** To construct a retaining wall, $8\frac{1}{4}$ cu ft of sand are needed to make 24 cu ft of concrete. State the ratio using whole number terms.

3. **Machinist.** The cost of a can of cotter pins that weighs 24 ounces is \$0.88. State the ratio using whole number terms.

4. **Forestry.** A fire crew built 18 chains of fire line in the first $6\frac{3}{4}$ hours of their shift. State the ratio using whole number terms.

5. **Machinist.** A gear revolves $1\frac{1}{3}$ times for every $3\frac{1}{5}$ feet. Give the ratio of revolutions to feet traveled using whole number terms.

6. **Generic.** Frog A jumped $2\frac{1}{4}$ feet, while Frog B jumped 9.5 inches. State the ratio of A to B in inches using whole number terms.

7. **Machinist.** A pulley has a diameter 18 in. and another has a diameter 4.5 in. What is the ratio of the diameter of the first pulley to the diameter of the second pulley in whole number terms?

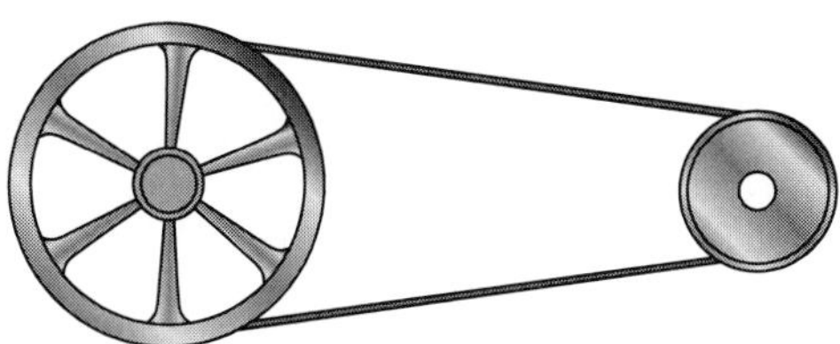

8. **Water Management.** A 7-in. water pipe under a fixed pressure discharges water at 20.3 gal per second. State the ratio of the diameter of the water pipe to the amount of discharge in whole number terms.

9. **Diesel Mechanics.** A value is changed by 0.006 in. with 3 clicks on an adjustment screw. State the ratio of clicks to the amount of change in whole number terms.

10. **Police Science.** A police department determined that their patrol cars made 18.4 miles per gallon at a constant speed of 60 mph. What is the ratio of miles per gallon to speed in whole number terms.

Answers to exercises 20–22

20. Ratio: 1 cup of sugar to 4 cups of flour
21. Ratio: 35 hours to 144 miles
22. Ratio: 1 foot to 12 feet, or 1 foot to 4 yards

9–3 RATES

■ Example 1

Jack Sanislo traveled 680 miles on 34 gallons of gasoline. What is the ratio of miles to gallons?

Solution $\frac{680}{34} = \frac{20}{1}$ Reducing

20 miles to 1 gallon

We could say that for every 20 miles, Jack used 1 gallon of gasoline, or that he got 20 miles per gallon.

When two different kinds of measures are compared, it is more meaningful to compare one of the measures to one unit of the other. The special kind of ratio is called a *rate.*

■ Example 2

Find the unit cost price or rate for a 10-ounce box of raisins costing 32¢.

Solution We need a ratio of cost to one ounce. To change the denominator to 1 ounce, we divide both numerator and denominator by 10.

$$\begin{matrix}\text{Cost}\rightarrow \\ \text{Weight}\rightarrow\end{matrix} \quad \frac{32}{10} = \frac{32 \div 10}{10 \div 10} = \frac{3.2}{1}$$

3.2¢ per ounce

Rule

To find a rate, *we simply divide the numerator of a ratio by the denominator.*

Thus, $\frac{32}{10} = 32 \div 10 = 3.2$, or 3.2¢ per ounce.

■ Example 3

Aaron can assemble $25\frac{1}{2}$ radio components in a 9-hour period. How many radio components per hour is this?

Solution

$$\begin{array}{r} \text{Radio components} \rightarrow \\ \text{Hours} \rightarrow \end{array} \frac{25\frac{1}{2}}{9} = 25\frac{1}{2} \div 9$$

$$= \frac{51}{2} \times \frac{1}{9}$$

$$= \frac{\overset{17}{\cancel{51}}}{2} \times \frac{1}{\underset{3}{\cancel{9}}}$$

$$= \frac{17}{6} = 2\frac{5}{6}$$

$2\frac{5}{6}$ radio components per hour.

Practice check: Do exercises 26–28 in the margin.

Find the rate.

26. A sump pump can pump 166.9 gallons in 1 hour. What is the rate per minute? (Round to the nearest tenth of a gallon.)

27. Janice can bake 24 dozen cookies in 3 hours. How many cookies is that per hour?

28. Bill drove 320.3 miles in 6.4 hours. What is the rate in miles per hour? (Round to the nearest mile.)

The answers are in the margin on page 352.

■ Looking Back

You should now be able to find the rate when comparing one measurement with another.

■ Problems 9–3

Find the unit cost or rate.

Cost	*Quantity*	
1. $1.62	3 pounds	
2. 45¢	10 ounces	
3. $2.14	10 cups	
4. $10.56	15 gallons	
5. $4.60	$3\frac{2}{3}$ pint	(Round to the nearest cent.)

Find the rate of distance per unit of time.

	Distance	*Time*
6.	175 miles	$5\frac{1}{2}$ hours (Round to the nearest mile.)
7.	100.6 kilometers	$2\frac{1}{4}$ hours
8.	320 feet	10 seconds
9.	940.4 miles	22 hours (Round to the nearest tenth of a mile.)
10.	5200 meters	$23\frac{3}{4}$ minutes (Round to the nearest tenth of a meter.)

11. Welding. A fabrication company had 16 female employees and 24 male employees. What is the rate of male employees to female employees? What is the rate of female employees to the total number of company employees?

12. Seamstress. Find the rate of cost to yards if 5 yards of cloth cost $9.90.

13. Welding. Aluminum weighs 160 lb per cu ft and steel weighs 490 lb per cu ft. What is the rate of the weight of steel to the weight of aluminum?

14. Water Management. In preparation for spring floods, East Stanwood found that one of their pumps can pump 1245 gallons of water in one hour. What is the rate of gallons per minute?

15. Generic. Find the unit price of cost per pound of 7.17 pounds of steak that cost $23.46.

The answers to odd-numbered exercises are in the back of the book.

Answers to exercises 23–25

23. $\frac{39}{212}$ **24.** $\frac{13}{30}$ **25.** $\frac{450}{1341}$

Answers to exercises 26–28

26. 2.8 gal per min.
27. 96 cookies per hour
28. 50 miles per hour

9–4 THE MEANING OF PROPORTION

In section 9–1, we learned that two ratios are equivalent if they both reduce to the same ratio in lowest terms. A statement that two ratios are equivalent is called a *proportion.* A proportion may be either true or false.

■ **Example 1** Which of the following statements are true proportions?

a. $\frac{8 \text{ girls}}{34 \text{ boys}} = \frac{16 \text{ girls}}{68 \text{ boys}}$

b. $\frac{2\frac{1}{2}}{3} = \frac{5}{\frac{3}{4}}$

Solution

a. $\frac{8 \text{ girls}}{34 \text{ boys}} = \frac{16 \text{ girls}}{68 \text{ boys}}$

This *is* a true proportion, since $\frac{8}{34} = \frac{4}{17}$, and $\frac{16}{18} = \frac{4}{17}$.

b. $\frac{2\frac{1}{2}}{3} = \frac{5}{\frac{3}{4}}$

This is *not* a true proportion, since

$$\frac{2\frac{1}{2}}{3} = 2\frac{1}{2} \div 3 = \frac{5}{2} \times \frac{1}{3} = \frac{5}{6}, \quad \text{and} \quad \frac{5}{\frac{3}{4}} = 5 \div \frac{3}{4} = \frac{5}{1} \times \frac{4}{3} = \frac{20}{3}.$$

Practice check: Do exercises 29–32 in the margin.

Which are true proportions and which are not?

29. $\frac{6}{8} = \frac{75}{100}$ **30.** $\frac{3\frac{1}{2}}{\frac{1}{4}} = \frac{7}{\frac{1}{2}}$

31. $\frac{8.4}{2.4} = \frac{\frac{7}{6}}{\frac{1}{2}}$

32. $\frac{9 \text{ inches}}{3 \text{ feet}} = \frac{7 \text{ inches}}{28 \text{ inches}}$

The answers are in the margin on page 355.

■ Looking Back

You should now be able to define proportion and determine if a certain statement of equality is a true proportion.

■ Problems 9–4

Review

Find the unit cost or rate.

Cost	*Quantity*
1. $5.34	6 pounds
2. $12.78	20 gallons

Find the rate of distance per unit of time

Distance	*Time*
3. 195 miles	7.8 hours
4. 470 feet	10 seconds

Determine if the following statements are true proportions.

5. $\frac{9}{15} = \frac{12}{20}$

6. $\frac{2.5}{1.5} = \frac{15}{9}$

7. $\frac{1\frac{1}{2}\text{ yr}}{4\text{ mo}} = \frac{27\text{ mo}}{6\text{ mo}}$

8. $\frac{21}{16} = \frac{20}{17}$

9. $\frac{7}{5} = \frac{12}{8}$

10. $\frac{6\text{ in.}}{1\text{ yd}} = \frac{3\text{ in.}}{18\text{ in.}}$

11. $\frac{6}{9} = \frac{49}{72}$

12. $\frac{1\frac{1}{6}}{\frac{1}{2}} = \frac{\frac{7}{9}}{\frac{1}{3}}$

13. $\frac{25\text{ cents}}{1\text{ can}} = \frac{\$6.25}{13\text{ cans}}$

14. $\frac{8\text{ pounds}}{\$2.40} = \frac{1\frac{1}{2}\text{ pounds}}{35\text{ cents}}$

The answers to review and odd-numbered exercises are in the back of the book.

9–5 CROSS PRODUCTS AND PROPORTION—SOLVING FOR THE UNKNOWN

We know that $\frac{6}{8}$ and $\frac{9}{12}$ are equal fractions because they both reduce to $\frac{3}{4}$. Notice this curious property with equal fractions:

a. Multiply the numerator of the first fraction by the denominator of the second fraction.

$$\frac{6}{8} \searrow \frac{9}{12} \qquad 6 \times 12 = 72$$

b. Multiply the denominator of the first fraction by the numerator of the second fraction.

$$\frac{6}{8} \nearrow \frac{9}{12} \qquad 8 \times 9 = 72$$

We find that the products are the same. The products 6×12 and 8×9 are called *cross products*.

Rule

A cross product *is the product of the numerator of one fraction and the denominator of another. With two fractions, there are two different cross products. These cross products are always equal when the fractions are equal.*

■ Example 1

Are $\frac{4}{7}$ and $\frac{12}{21}$ equal fractions?

Solution

$$\frac{4}{7} \times \frac{12}{21} \qquad \begin{array}{l} 7 \times 12 = 84 \\ 4 \times 21 = 84 \end{array}$$

$84 = 84$ The cross products are equal.
The fractions are equal.

■ **Example 2**

Are $\frac{2\frac{1}{3}}{4}$ and $\frac{\frac{3}{4}}{6}$ equal fractions?

Solution

$$\frac{2\frac{1}{3}}{4} \times \frac{\frac{3}{4}}{6} \qquad 4 \times \frac{3}{4} = \frac{4}{1} \times \frac{3}{4} = \frac{3}{1}$$

$$2\frac{1}{3} \times 6 = \frac{7}{3} \times \frac{6}{1} = \frac{14}{1}$$

$\frac{14}{1} \neq \frac{3}{1}$ The cross products are not equal.

The fractions are not equal.
Note: ≠ means not equal to.

Practice check: Do exercises 33–36 in the margin.

Decide by the cross product method whether the following pairs of fractions are either equal or not equal.

33. $\frac{16}{6}$ and $\frac{8}{3}$ **34.** $\frac{6}{9}$ and $\frac{4}{6}$

35. $\frac{63}{51}$ and $\frac{38}{30}$

36. $\frac{3\frac{1}{4}}{\frac{1}{5}}$ and $\frac{\frac{10}{13}}{2}$

The answers are in the margin on page 357.

Since ratios are expressed as a fraction, we can also use the cross product to decide if two ratios form a true proportion.

■ **Example 3**

Is the following proportion true or false?

$$\frac{15}{18} = \frac{20}{24}$$

Solution

$$\frac{15}{18} \times \frac{20}{24} \qquad 15 \times 24 = 18 \times 20$$

$$360 = 360$$

The cross products are equal, so the proportion is true.

Practice check: Do exercises 37–41 in the margin.

Decide by the cross product method whether or not the following statements are true proportions

37. $\frac{3}{8} = \frac{36}{96}$ **38.** $\frac{10}{25} = \frac{50}{125}$

39. $\frac{21}{7} = \frac{33}{11}$ **40.** $\frac{0.03}{0.27} = \frac{0.04}{0.36}$

41. $\frac{1\frac{2}{3}}{3} = \frac{2\frac{1}{3}}{5}$

The answers are in the margin on page 357.

Solving for an Unknown Term

Suppose we have to make $\frac{\square}{3} = \frac{8}{12}$ a true proportion. We must replace the box with a number that will cause it to be a true proportion. To find □, we begin by setting the cross products equal to each other.

$$\square \times 12 = 3 \times 8$$

$$\square \times 12 = 24$$

In words, this last statement reads, "What number multiplied by 12 is 24?" Of course, the number is 2. We can quickly find □ by dividing 24 by 12.

$$\square \times 12 = 24$$
$$\square = 24 \div 12$$
$$\square = 2$$

Check: $\frac{2}{3} = \frac{8}{12}$ is a true proportion, since $2 \times 12 = 3 \times 8$.

Answers to exercises 29–32

29. yes **30.** yes **31.** no
32. yes

Rule

Instead of a □, the unknown term of a proportion is usually represented by some letter of the alphabet.

■ Example 4

Find a number that will make $\frac{6}{9} = \frac{n}{15}$ a true proportion.

Solution **Step 1** Set the cross products equal to each other and multiply.

$$6 \times 15 = 9 \times n$$
$$90 = 9 \times n$$

In words, this last statement reads, "90 is 9 multiplied by what number?"

Step 2 To find the missing factor, we divide 90 by 9.

$$90 \div 9 = n$$
$$10 = n$$

Check: $\frac{6}{9} = \frac{10}{15}$ is a true proportion, since $6 \times 15 = 9 \times 10$.

Practice check: Do exercises 42–45 in the margin.

Find the number that will make the proportion true.

42. $\frac{n}{6} = \frac{8}{4}$ **43.** $\frac{4}{n} = \frac{8}{16}$

44. $\frac{18}{54} = \frac{n}{9}$ **45.** $\frac{3}{7} = \frac{9}{n}$

The answers are in the margin on page 357.

■ Example 5

Find a number that will make $\frac{1\frac{3}{8}}{6} = \frac{n}{3}$ a true proportion.

Solution **Step 1** Set the cross products equal to each other.

$$1\frac{1}{8} \times 3 = 6 \times n$$
$$\frac{11}{8} \times 3 = 6 \times n$$
$$\frac{33}{8} = 6 \times n$$

Step 2 To find the missing factor we divide $\frac{33}{8}$ by 6.

$$n = \frac{33}{8} \div 6$$
$$n = \frac{\overset{11}{\cancel{33}}}{8} \times \frac{1}{\underset{2}{\cancel{6}}}$$

$$n = \frac{11}{16}$$

Check: $\frac{1\frac{3}{8}}{6} = \frac{\frac{11}{16}}{3}$ is a true proportion, since $1\frac{3}{8} \times 3 = 6 \times \frac{11}{16}$.

Rule

To solve for a missing term in a proportion, set the cross products equal to each other, find the product of the number terms, then divide by the remaining number.

Practice check: Do exercises 46–49 in the margin.

Find the number that will make the proportion true.

46. $\frac{\frac{1}{2}}{4} = \frac{8}{n}$

47. $\frac{\frac{3}{4}}{\frac{3}{2}} = \frac{t}{\frac{5}{8}}$

48. $\frac{r}{1.7} = \frac{0.16}{1.36}$

49. $\frac{5.2}{n} = \frac{9.1}{4.2}$

The answers are in the margin on page 359.

■ Looking Back

You should now be able to solve for the unknown, using the cross product method, to make a true proportion.

■ Problems 9–5

Show by the cross product method which of the following statements are true proportions.

1. $\frac{1}{2} = \frac{3}{6}$

2. $\frac{3}{4} = \frac{36}{48}$

3. $\frac{8}{12} = \frac{16}{24}$

4. $\frac{9}{27} = \frac{3}{9}$

5. $\frac{27}{24} = \frac{15}{13}$

6. $\frac{22}{11} = \frac{16}{8}$

7. $\frac{3.08}{0.6} = \frac{0.77}{0.15}$

8. $\frac{0.008}{0.01} = \frac{5.6}{7}$

9. $\frac{3\frac{1}{3}}{\frac{1}{4}} = \frac{80}{6}$

10. $\frac{5\frac{2}{3}}{5} = \frac{15}{17}$

Find the number that will make the proportion true.

11. $\frac{3}{9} = \frac{2}{n}$

12. $\frac{5}{n} = \frac{4}{12}$

13. $\frac{10}{12} = \frac{t}{15}$

14. $\frac{n}{6} = \frac{18}{27}$

15. $\frac{8}{24} = \frac{18}{r}$

16. $\frac{40}{n} = \frac{32}{24}$

17. $\frac{54}{12} = \frac{n}{24}$

18. $\frac{40}{t} = \frac{25}{40}$

19. $\frac{7}{10} = \frac{r}{100}$

20. $\frac{r}{100} = \frac{13}{25}$

21. $\frac{\frac{5}{8}}{\frac{2}{5}} = \frac{\frac{5}{6}}{n}$

22. $\frac{1\frac{1}{3}}{t} = \frac{3}{\frac{4}{3}}$

23. $\frac{\frac{7}{8}}{6} = \frac{n}{\frac{1}{2}}$

24. $\frac{n}{\frac{6}{5}} = \frac{5}{\frac{2}{3}}$

25. $\frac{17}{\frac{3}{7}} = \frac{21}{t}$

26. $\frac{n}{0.42} = \frac{3}{0.14}$

27. $\frac{4.9}{9.1} = \frac{n}{16.9}$

28. $\frac{0.18}{10.8} = \frac{0.05}{t}$

29. $\frac{r}{100} = \frac{3.8}{5}$

30. $\frac{4.04}{20} = \frac{r}{100}$

The answers to odd-numbered exercises are in the back of the book.

9–6 APPLIED PROBLEMS

■ Example 1

If a man earns $36 in 6 hours, how many dollars can he earn in 28 hours?

Solution We make a proportion by using the number of dollars in the numerators and the number of hours in the denominators. The known ratio is $\frac{36}{6}$ and the unknown ratio is $d/28$.

$$\begin{matrix}\text{Dollars}\rightarrow \\ \text{Hours}\rightarrow\end{matrix}\frac{36}{6} = \frac{d}{28}\begin{matrix}\leftarrow\text{Dollars} \\ \leftarrow\text{Hours}\end{matrix}$$

$$36 \times 28 = 6 \times d$$

$$1008 = 6 \times d$$

$$168 = d \qquad \text{Dividing by 6.}$$

The man can earn $168 in 28 hours.

■ Example 2

In a certain milk product, the ratio of butterfat to skim milk is $\frac{3}{100}$. How many quarts of butterfat are there if the milk product has 900 quarts of skim milk?

Solution We make a proportion and solve. The known ratio is $\frac{3}{100}$ and the unknown ratio is $b/900$.

$$\begin{matrix}\text{Butterfat}\rightarrow \\ \text{Skim milk}\rightarrow\end{matrix}\frac{3}{100} = \frac{b}{900}\begin{matrix}\leftarrow\text{Butterfat} \\ \leftarrow\text{Skim milk}\end{matrix}$$

$$3 \times 900 = 100 \times b$$

$$2700 = 100 \times b$$

$$\frac{2700}{100} = b \qquad \text{Dividing by 100.}$$

$$27 = b$$

Answers to exercises 33–36

33. equal **34.** equal
35. not equal **36.** not equal

Answers to exercises 37–41

37. true **38.** true **39.** true
40. true **41.** false

Answers to exercises 42–45

42. $n = 12$ **43.** $n = 8$
44. $n = 3$ **45.** $n = 21$

There are 27 quarts of butterfat.

Note: The numerators of both ratios represent the same item (butterfat) and the denominators of both ratios represent the same item (skim milk). This needs to happen when using a proportion to solve a word problem.

Practice check: Do exercises 50 and 51 in the margin.

50. In one basketball game, a player made baskets at a ratio of 3 baskets to 7 attempts. At this ratio, how many attempts did the player make, if she made 15 baskets during the game?

51. If 25 tulip bulbs cost $1.10, how many bulbs can you buy for $5.50?

The answers are in the margin on page 361.

Example 3

A man sold $\frac{3}{16}$ of his stock for $1850. What is the value of $\frac{2}{5}$ of his stock at the same rate? (Round to the nearest dollar.)

Solution

$$\frac{\text{Stock} \rightarrow \frac{3}{16}}{\text{Dollars} \rightarrow 1850} = \frac{\frac{2}{5} \leftarrow \text{Stock}}{d \leftarrow \text{Dollars}}$$

$$\frac{3}{16} \times d = 1850 \times \frac{2}{5}$$

$$\frac{3}{16} \times d = \frac{\overset{370}{\cancel{1850}}}{1} \times \frac{2}{\underset{1}{\cancel{5}}}$$

$$\frac{3}{16} \times d = 740$$

$$d = 740 \div \frac{3}{16}$$

$$d = \frac{740}{1} \times \frac{16}{3}$$

$$d = \frac{740 \times 16}{3}$$

$$d = \frac{11{,}840}{3}$$

$d \approx \$3947$ to the nearest dollar

Practice check: Do exercises 52 and 53 in the margin.

52. If $\frac{7}{8}$ of a yard of cloth costs $2.80, how much will $12\frac{1}{2}$ yards cost?

53. A piece of iron $5\frac{1}{2}$ feet long weighs 11 pounds. What is the weight of a piece 17 ft long?

The answers are in the margin on page 361.

Looking Back

You should now be able to solve proportion application problems.

Problems 9–6

Solve. Write a statement for the answer.

1. **Banker.** If the interest on a note for 7 months is $34, what would be the interest on the same note for 15 months?

2. **Cartographer.** A map has a scale of 1 inch to 3 miles. How many miles would 7 inches represent?

3. **Marketing.** A party mixture of salted nuts has 5 pounds of peanuts to 3 pounds of cashews. At the same rate, how many pounds of cashews are needed if 20 pounds of peanuts are used in the mixture?

4. **Generic.** A worker can stack 40 boxes in 30 minutes. How long will it take to stack 100 boxes?

Answers to exercises 46–49

46. $n = 64$ **47.** $t = \frac{5}{16}$

48. $r = 0.2$ **49.** $n = 2.4$

5. **Landscaping.** A 10-ft-high fence post casts a shadow 8 ft long. What is the height of a tree casting a shadow 150 ft long at the same time of the day, all on level ground?

6. **Generic.** Light bulbs cost $1.40 for 5. At that rate how many light bulbs can be purchased for $16.80?

7. **Interior Decorator.** A gallon of paint covers 300 square feet of surface. How many gallons of paint will be needed to cover 1550 square feet?

8. **Sales.** For every 120 house contacts, a door-to-door sales rep makes 22 sales. How many sales can he expect to make with 300 house contacts?

9. **Plumber.** A 14-foot length of copper tubing weighs 8 pounds. How many pounds per foot does that copper tubing weigh?

10. **Cartographer.** A map has a scale of 10 miles for every $\frac{1}{2}$ inch. How many inches represent 162 miles?

11. **Catering.** A certain recipe provides $11\frac{1}{2}$ servings for 8 people. How many servings would be needed for 25 people? Round to the nearest whole serving.

12. **Generic.** Jill's pulse beats 70 times a minute. How many times will it beat in $3\frac{1}{4}$ hours?

13. Landscaping. If 15.5 yards of beauty bark cost $46.00, how much does 7.8 yards cost? Round to the nearest cent.

14. Generic. At an average rate of 50 mph a car uses 23.6 gallons in 788.2 miles. How many miles per gallon is this? Round to the nearest tenth of a mile.

15. Landscaping. If 50 pounds of fertilizer covers 2500 square feet of lawn, how many pounds would be needed to cover 6520 square feet? Round to the next greater whole pound.

The answers to odd-numbered exercises are in the back of the book.

CHAPTER 9 OVERVIEW

Summary

9–1 You learned to:

reduce a ratio to lowest terms,
raise a ratio to higher terms, and
show that two ratios are equivalent.

9–2 You learned to find a ratio involving numbers equivalent to a ratio containing terms other than whole numbers.

9–3 You learned to find the rate when comparing one measurement with another.

9–4 You learned to define proportion and determine if a certain statement of equality is a true proportion.

9–5 You learned to solve for the unknown using the cross product method to make a true proportion.

9–6 You learned to solve application problems that involve proportion.

Terms To Remember

	Page
Equivalency	344
Ratio	342
Terms	342
Rate	349
Proportion	352
Cross product	353

Rules

- When two numbers or quantities are compared by division, the indicated division is called a *ratio*, and the numbers or quantities are called *terms* of the ratio.

- To find a *rate*, we simply divide the numerator of a ratio by the denominator.
- A *cross product* is the product of the numerator of one fraction and the denominator of another. With two fractions, there are two different cross products. These cross products are always equal when the fractions are equal.
- Instead of a □, the unknown term of a proportion is usually represented by some letter of the alphabet.
- To solve for a missing term in a proportion, set the cross products equal to each other, find the product of the number terms, then divide by the remaining number.

Answers to exercises 50 and 51

50. 35 attempts **51.** 125 bulbs

Answers to exercises 52 and 53

52. $40 **53.** 34 lb

Self-Test

9–1 Express the following as reduced ratios in the same units where possible.

1. 7 people to 12 chairs **2.** 54 buttons to 18 bows

3. The low gear ratio in a truck is determined by the ratio of the number of teeth in the large gear and the small gear. Find the low gear ratio if the number of teeth in the large gear is 203 and the number of teeth in small gear is 21.

9–2 **4.** An electrician has a wire 845 feet long with a resistance of 3.5 ohms. Find the ratio in whole number terms.

9–3 **5.** Find the unit ratio.

Distance	***Quantity***
357 mi.	21 gal

6. A city sewer pipe measured a flow of 1260 gallons in 5 minutes. Give the rate of gallons per minute.

9–4 **7.** Is this statement a true proportion?

$$\frac{1.6}{2.2} = \frac{24}{32}$$

9–5 **8.** Show by the cross product method that the following statement is a true proportion.

$$\frac{0.004}{0.01} = \frac{2}{5}$$

9. Solve for *n*.

$$\frac{n}{21} = \frac{3}{9}$$

9–6 **10.** A manufacturer's direction for the correct pesticide mixture is 15 gallons of water to 2 pounds of pesticide per acre. How much pesticide should a farmer use if he has a 275-gallon spray tank?

The answers are in the back of the book.

Chapter Test

Directions: This test will aid in your preparation for a possible chapter test given by your instructor. The answers are in the back of the book. Go back and review in the appropriate section(s) if you missed any test items.

9–1 Express the following as ratios in the same units where possible.

1. 8 bottles to 7 cans

2. 3 nickels to 17 cents (in cents)

3. To make concrete, Joe mixed 4 buckets of cement and 13 buckets of sand. State the ratio of cement to sand.

Express each ratio in reduced form.

4. 18 bales of cotton to 15 bales of Dacron

5. 27 feet to 3 yards (in feet)

6. Are the following ratios equivalent? 5 mittens to 15 gloves and 18 mittens to 3 gloves.

9–2 7. A truck was driven 462.8 miles in 11.3 hours. State the ratio of miles to hours using whole numbers.

8. A certain drink calls for $2\frac{1}{2}$ oz gin to $3\frac{1}{3}$ oz mixer. Give the ratio of gin to mixer in whole numbers.

9–3 9. Find the unit cost or rate:

Cost	***Quantity***
\$3.84	5 pounds

10. Find the rate of distance per unit of time:

Distance	***Time***
368 miles	10 hours

11. Maria needs to save \$950 in 36 weeks for a set of tools. How much should she save per week?

9–4 Show that the following statements are true proportions. Use equivalency.

12. $\frac{6}{9} = \frac{8}{12}$

13. $\frac{7}{\frac{1}{5}} = \frac{525}{15}$

9–5 Show by the cross product method that the following statements are true.

14. $\frac{7}{8} = \frac{14}{16}$

15. $\frac{3.6}{0.8} = \frac{5.4}{1.2}$

16. $\frac{\frac{4}{5}}{6} = \frac{\frac{2}{3}}{5}$

Solve for the unknown term.

17. $\frac{n}{9} = \frac{24}{27}$

18. $\frac{\frac{2}{3}}{\frac{1}{15}} = \frac{r}{5}$

19. $\frac{0.7}{2.1} = \frac{5.6}{y}$

9–6 **20.** A buried cylindrical gasoline storage tank holds 1000 gallons when it is filled to a depth of 6 feet. How many gallons are there when it is filled to $8\frac{1}{2}$ feet?

21. A map has a scale of 25 miles for every inch. How many inches represent 378 miles?

22. If the ratio of pecans to cashews is 4 to 3, how many cups of cashews will there be with 16 cups of pecans?

Test Your Memory

These problems review Chapters 1–9.

1. Name the smallest whole number.

2. Name the third digit in 3,218,402.

3. Write an expanded number for 7109.

4. Write a word name for 18,601,492.

5. What is the meaning of 6 in the standard number 546,789,014?

6. Place the < or > symbol between 840 and 804 to make a true statement.

7. Round 8658 to the nearest hundred.

8. Add.

$$\begin{array}{r} 5870 \\ \underline{3349} \end{array}$$

9. Add.

$$\begin{array}{r} 7548 \\ \underline{5389} \end{array}$$

10. Add and check: 65 + 3903 + 89 + 458.

11. Estimate the sum to the nearest thousand, then find the exact sum and compare.

$$\begin{array}{r} 7052 \\ 4607 \\ 8493 \\ \underline{1850} \end{array}$$

12. The Jackson family used the following amounts of heating fuel oil during 1999: 189, 214, 233, 166, 179, and 105 gallons. How many gallons were used in 1999?

13. Subtract.

$$\begin{array}{r} 4893 \\ \underline{2701} \end{array}$$

14. Subtract.

$$\begin{array}{r} 3347 \\ \underline{1962} \end{array}$$

15. Subtract.

$$\begin{array}{r} 6002 \\ \underline{4328} \end{array}$$

16. A dress manufacturer has an order for 8617 dresses to be completed by May 31. On May 1 the company has completed a total of 6839 dresses. How many more dresses need to be completed by May 31?

Multiply.

17. $\begin{array}{r} 821 \\ \underline{\times 1000} \end{array}$

18. 6000×4

19. $\begin{array}{r} 4261 \\ \underline{\times 7} \end{array}$

20. $\begin{array}{r} 605 \\ \underline{\times 60} \end{array}$

21. $\begin{array}{r} 3204 \\ \underline{\times 54} \end{array}$

22. How far can a car travel on one tankful, if the tank holds 13 gallons of gasoline and the car averages 26 miles per gallon?

Divide.

23. $3\overline{)6612}$

24. $6\overline{)18{,}282}$

25. $364\overline{)158{,}607}$

26. At a cost of $25,674 a welder fabricated 33 aluminum water tanks. What is the cost of each tank?

27. Write a fractional number for the shaded part.

28. Find $25\frac{3}{4}$ on a number line.

29. Find the factors of 42.

30. Find the prime factorization of 42.

31. Reduce $\frac{32}{48}$ to lowest terms.

32. Multiply: $\dfrac{18}{25} \times \dfrac{1}{15} \times \dfrac{5}{6}$.

33. A block of steel has a volume of $\frac{5}{8}$ cu in. What is $\frac{1}{2}$ the volume?

34. State whether $\frac{29}{29}$ is a proper or improper fraction.

35. Divide and simplify: $\dfrac{\frac{10}{27}}{\frac{4}{9}}$.

36. Add: $\dfrac{17}{20} + \dfrac{5}{12}$.

37. Margaret bought $\frac{3}{4}$ yd of plain material and $\frac{5}{8}$ yd of print material. How much material did Margaret buy?

38. Subtract: $\dfrac{5}{12} - \dfrac{5}{18}$.

39. Decide which fraction is larger, $\frac{5}{8}$ or $\frac{17}{29}$.

40. A planer takes a $\frac{5}{32}$-in. cut from a piece of steel $\frac{5}{8}$ in. thick. What was the remaining thickness?

41. Change $\frac{25}{3}$ to a mixed number.

42. Multiply: $2\frac{1}{2} \times 1\frac{1}{4}$.

43. Divide: $2\frac{1}{3} \div 1\frac{5}{9}$.

44. Add: $5\frac{9}{10} + 1\frac{19}{25}$.

45. Subtract.

$$\begin{array}{r} 15\frac{1}{32} \\ 3\frac{9}{16} \\ \hline \end{array}$$

46. A foreman estimated a job to take $6\frac{1}{3}$ hours. If a worker only took $5\frac{3}{8}$ hours, by how much did the worker beat the foreman's estimate?

47. Write a mixed number for 32.046.

48. Write an expanded number for 8.618.

49. Write a word name for 7.057.

50. Which number is larger, 0.7020 or 0.703?

51. Add: $5.71 + 0.005 + 2.04 + 58$.

52. Estimate the sum to the nearest tenth: $67.518 + 8.079$.

53. Diamond Concrete delivered the following amounts of concrete to a construction site during the week: 21 cu yd, 33.4 cu yd, 16.2 cu yd, 38.0 cu yd, and 42.5 cu yd. How many cubic yards were delivered that week?

54. Subtract: $62.8 - 31.89$.

55. Subtract: $16 - 8.709$.

56. Jane Swift ran the 100-yard dash in 11.4 seconds on Wednesday. On Thursday, it took her 10.7 seconds. How much faster was she on Thursday?

57. Multiply: 4.76×0.04.

58. Write the product directly: $6.18 \times 10{,}000$.

59. A certain engine consumes 7.62 gallons of fuel per hour. How many gallons would be used if the engine ran for 72 hours?

60. Write the quotient directly: $48.6 \div 10{,}000$.

61. Divide: $6.21\overline{)0.19251}$.

62. Round to the nearest tenth: $18.6\overline{)81.54}$.

63. Express $0.33\frac{1}{3}$ as a common fraction.

64. A car travels 166.8 miles in 6.4 hours. What is its average speed per hour? Round to the nearest tenth of a mile.

65. Express as a ratio in minutes in reduced form: 20 minutes to 2 hours (in minutes).

66. Are the following ratios equivalent?

18 welders to 12 apprentices
9 welders to 6 apprentices

67. State the following ratio in whole number terms: 4 to $2\frac{2}{15}$.

68. Find the unit cost or ratio: \$432 for 30 yards of carpet.

69. Solve for n.

$$\frac{14}{5.25} = \frac{n}{24}$$

70. If a 232-lb casting cost \$580 what would a 500-lb casting cost?

The answers are in the back of the book.

PART FOUR

APPLICATIONS

CHAPTER 10

Percent and Percent Applications

Choice: Either skip the Pretest and study all of Chapter 10 or take the Pretest and study only the section(s) you need to study.

Pretest

Directions: This Pretest will help you determine which sections to study in Chapter 10. If any of your answers are incorrect, or if you omitted any exercise, turn to the indicated section(s) and study the material; then take the Chapter Test.

10–1 **1.** Change this fraction to a number with a percent symbol: $\frac{7.2}{100}$.

2. Change this percent number to a fraction: 13%.

10–2 **3.** Change this percent to a decimal: 5.9%.

4. Change this percent number to a common fraction in lowest terms: 8.4%.

10–3 **5.** Change this fraction to a percent: $\frac{3}{4}$.

6. Change this decimal to a percent: 0.406.

10–4 Identify B, P, and R, if B is the base, P is the percentage, and R is the rate in the following percent problems.

7. What is 56% of 89?

8. 32.5 is 7.7% of what number?

10–5 Solve.

9. What is 78% of 46?

10. 0.8% of what number is 6.4?

10–6 **11.** Sam spends 21% of his paycheck on food for his family. If his paycheck was $430, how much was spent on food?

10–7 **12.** A football team won 7 games out of 12 games played. What percent of the games played did the team win?

10–8 **13.** The sales tax on an article was $5.30. If the sales tax rate is 4%, what was the marked price of the article?

The answers are on page 378.

10–1 MEANING OF PERCENT

Percent means the ratio of some number to 100. $\frac{5}{100}, \frac{23}{100}, \frac{106}{100}, \frac{33.4}{100}$, and $\frac{\frac{2}{3}}{100}$ are all examples of percents.*

*The Latin word for *hundred* is *centum,* so percent means per *hundred.*

The rectangle represents one whole. It shows 16 shaded parts out of 100.

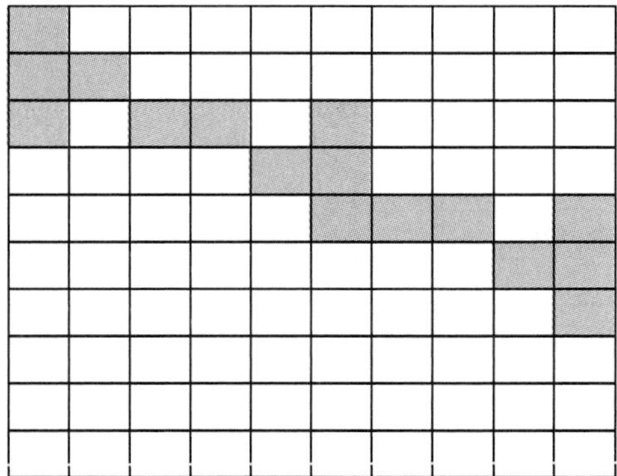

Figure 10–1

Each shaded part is $\frac{1}{100}$ of the whole. The fraction of the whole that is shaded is $\frac{16}{100}$ or 16 per 100.

Practice check: Do exercises 1 and 2 in the margin.

The symbol for *per 100* is %, called the *percent symbol.* Notice the slanted line between two zeros. This should remind you that percent is based on 100.

■ Examples

1. $\frac{29}{100}$ means 29 per 100, or 29% (read *29 percent*)

2. $\frac{6}{100}$ means 6 per 100, or 6%

3. $\frac{36.7}{100}$ means 36.7 per 100, or 36.7%

4. $\frac{\frac{4}{7}}{100}$ means $\frac{4}{7}$ per 100, or $\frac{4}{7}$ %.

Practice check: Do exercises 3–8 in the margin.

■ Example 5

Change 59% to a fraction.

Solution 59% means $\frac{59}{100}$.

Practice check: Do exercises 9–14 in the margin.

■ Looking Back

You should now be able to understand the meaning of percent.

What fraction of the whole is shaded?

1.

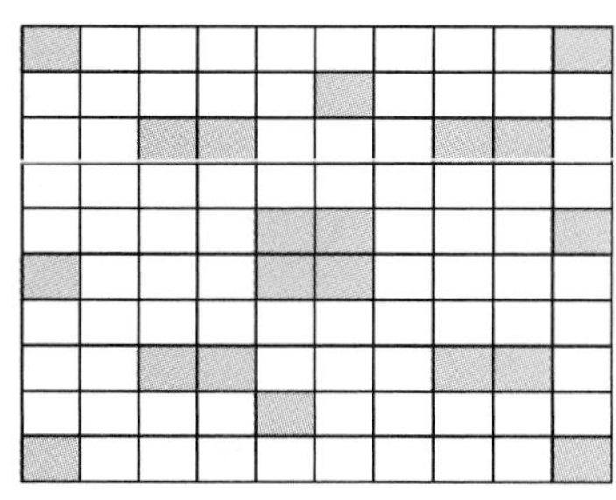

2.

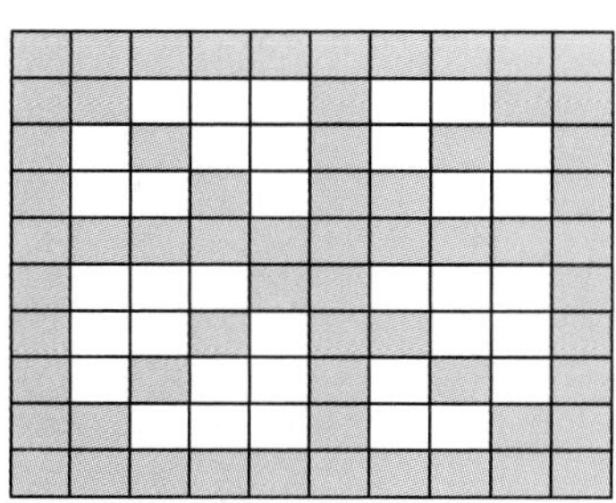

The answers are in the margin on page 379.

Change the fractions to a number with a percent symbol.

3. $\frac{31}{100}$ **4.** $\frac{93}{100}$ **5.** $\frac{150}{100}$

6. $\frac{45.2}{100}$ **7.** $\frac{\frac{2}{3}}{100}$ **8.** $\frac{33\frac{1}{3}}{100}$

The answers are in the margin on page 379.

Change the percent numbers to fractions.

9. 31% **10.** 83% **11.** 246%

12. 6.38% **13.** $12\frac{1}{2}$% **14.** $\frac{2}{3}$%

The answers are in the margin on page 379.

Answers to Pretest

Circle the answers you missed or omitted. Cross them off after you've studied the corresponding section(s); then take the Chapter Test.

1. 7.2% **2.** $\frac{13}{100}$ **3.** 0.059

4. $\frac{21}{250}$ **5.** 75% **6.** 40.6%

7. $B = 89$, $P =$ what?, $R = 56$

8. $B =$ what?, $P = 32.5$, $R = 7.7$

9. 35.88 **10.** 800 **11.** $90.30

12. $58\frac{1}{3}$ **13.** $132.50

■ Problems 10–1

What fraction of the whole is shaded?

1.

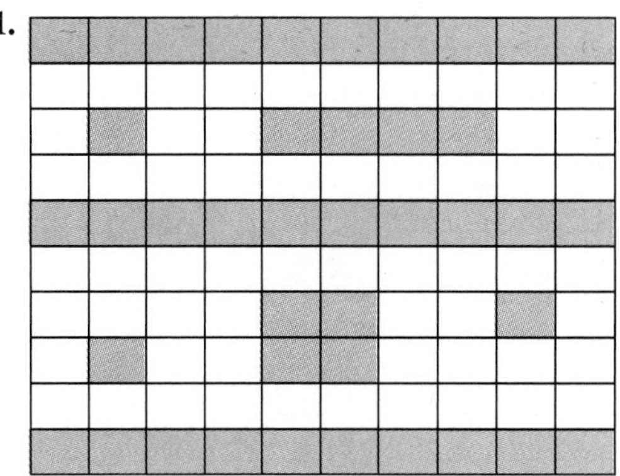

2.

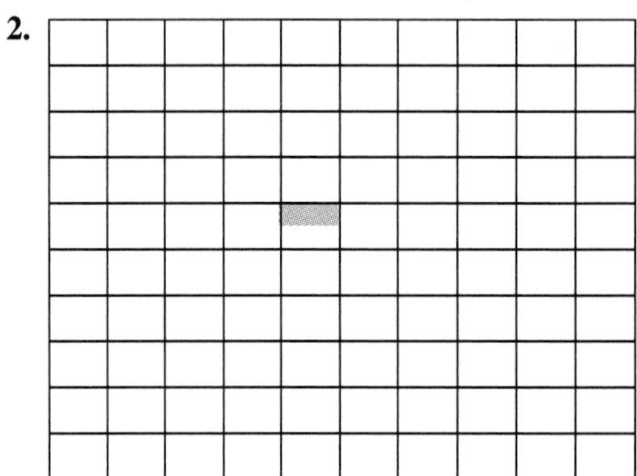

Construct a rectangle to represent the following percents.

3. $\frac{32}{100}$ **4.** $45\frac{1}{2}\,\%$

Change the fraction to a number with a percent symbol.

5. $\frac{62}{100}$ **6.** $\frac{8}{100}$ **7.** $\frac{56}{100}$ **8.** $\frac{9.8}{100}$ **9.** $\frac{156}{100}$ **10.** $\frac{66\frac{2}{3}}{100}$

11. $\frac{15.67}{100}$ **12.** $\frac{13}{100}$ **13.** $\frac{\frac{1}{2}}{100}$ **14.** $\frac{0.75}{100}$ **15.** $\frac{29}{100}$ **16.** $\frac{83\frac{1}{3}}{100}$

17. $\frac{360}{100}$ **18.** $\frac{0.18}{100}$ **19.** $\frac{2681}{100}$ **20.** $\frac{\frac{5}{8}}{100}$

Change the percent numbers to fractions.

21. 10% **22.** 39% **23.** 6.1% **24.** 182% **25.** $33\frac{1}{3}\%$ **26.** 13%

27. 57.84% **28.** 0.5% **29.** $62\frac{1}{2}\%$ **30.** $\frac{3}{8}\%$ **31.** 315% **32.** 78.04%

33. $83\frac{1}{3}\%$ **34.** 46% **35.** 0.004%

The answers to odd-numbered exercises are in the back of the book.

10–2 CHANGING PERCENTS TO DECIMALS AND FRACTIONS

In this section you will learn how to change percents to fractions in lowest terms.

Changing Percents to Decimals

■ **Example 1**

Change 67% to a decimal.

Solution **Step 1** Change the percent number to a fraction.

$$67\% = \frac{67}{100}$$

Step 2 Change to a decimal.

$$\frac{67}{100} = 0.67$$

Rule

When we divide by 100, we move the decimal point 2 places to the left.

■ **Example 2**

Change 7% to a decimal.

Solution

$$7\% = \frac{7}{100} = 0.07$$

Step 1 Step 2

We write a zero in front of the 7 in order to move the decimal point two places to the *left*.

Practice check: Do exercises 15–19 in the margin.

Answers to exercises 1 and 2

1. $\frac{20}{100}$ **2.** $\frac{64}{100}$

Answers to exercises 3–8

3. 31% **4.** 93% **5.** 150%

6. 45.2% **7.** $\frac{2}{3}\%$ **8.** $33\frac{1}{3}\%$

Answers to exercises 9–14

9. $\frac{31}{100}$ **10.** $\frac{83}{100}$ **11.** $\frac{246}{100}$

12. $\frac{6.38}{100}$ **13.** $\frac{12\frac{1}{2}}{100}$ **14.** $\frac{\frac{2}{3}}{100}$

Change the percents to decimals.

15. 79% **16.** 20% **17.** 5%

18. 6.9% **19.** 10%

The answers are in the margin on page 381.

■ **Example 3**

Change $83\frac{1}{3}\%$ to a decimal.

Solution

$$83\frac{1}{3}\% = \frac{83\frac{1}{3}}{100} = 0.83\frac{1}{3}$$

The decimal point is understood to be here. (arrow to after 83 in $83\frac{1}{3}\%$)

We divide by 100. (arrow to the 100)

Moving the decimal point 2 places to the left (arrow to $0.83\frac{1}{3}$)

Rule

To change a percent to a decimal, we move the decimal point two places to the left and delete the percent symbol.

Practice check: Do exercises 20–24 in the margin.

Change the percents to decimals.

20. $91\frac{2}{3}\%$ **21.** $8\frac{1}{3}\%$

22. $39.6\frac{1}{8}\%$ **23.** $41\frac{2}{3}\%$

24. $10\frac{1}{12}\%$

The answers are in the margin on page 383.

Changing Percents to Fractions

■ **Example 4**

Change 30% to a fraction in lowest terms.

Solution **Step 1** Change the percent number to a fraction.

$$30\% = \frac{30}{100}$$

Step 2 Reduce the fraction to lowest terms.

$$\frac{30}{100} = \frac{3}{10}$$

Practice check: Do exercises 25–28 in the margin.

Change the percent number to a fraction in lowest terms.

25. 15% **26.** 64% **27.** 22%

28. 85%

The answers are in the margin on page 383.

■ **Example 5**

Change 62.5% to a fraction in lowest terms.

Solution **Step 1** Change the percent number to a fraction.

$$62.5\% = \frac{62.5}{100}$$

It is easier to reduce fractions if both numerator and denominator of a fraction are whole numbers. Notice that the numerator has one decimal place.

Step 2 Since 62.5 has one decimal place, we multiply $\frac{62.5}{100}$ by $\frac{10}{10}$.

$$\frac{62.5}{100} \times \frac{10}{10} = \frac{62.5 \times 10}{100 \times 10} = \frac{625}{1000}$$

Not a whole number ↘ (62.5) Now a whole number ↙ (625)

Multiplying by 10 moves the decimal point 1 place to the right.

Step 3 Reduce.

$$\frac{625}{1000} = \frac{125}{200} = \frac{25}{40} = \frac{5}{8}$$

↑ Divide by 5. ↑ Divide by 5. ↑ Divide by 5.

Practice check: Do exercises 29–32 in the margin.

Answers to exercises 15–19

15. 0.79 **16.** 0.2 **17.** 0.05
18. 0.069 **19.** 0.1

Change the percent number to a fraction in lowest terms.

29. 58.6% **30.** 41.35%

31. 22.7% **32.** 6.484%

The answers are in the margin on page 383.

■ Example 6

Change $33\frac{1}{3}\%$ to a fraction in lowest terms.

Solution **Step 1** Change the percent number to a fraction.

$$33\frac{1}{3}\% = \frac{33\frac{1}{3}}{100}$$

Step 2 Change the numerator to an improper fraction.

$$\frac{33\frac{1}{3}}{100} = \frac{\frac{100}{3}}{100}$$

Step 3 Divide by 100.

$$\frac{\frac{100}{3}}{100} = \frac{100}{3} \div 100$$

$$= \frac{100}{3} \times \frac{1}{100}$$

$$= \frac{\overset{1}{\cancel{100}}}{3} \times \frac{1}{\underset{1}{\cancel{100}}} = \frac{1}{3}$$

$$33\frac{1}{3}\% = \frac{1}{3}$$

Practice check: Do exercises 33–37 in the margin.

Change the percent number to a fraction in lowest terms.

33. $16\frac{2}{3}\%$ **34.** $8\frac{1}{3}\%$

35. $91\frac{2}{3}\%$ **36.** $15\frac{1}{7}\%$

37. $34\frac{2}{3}\%$

The answers are in the margin on page 383.

■ Looking Back

You should now be able to change percents to decimals and a percent number to a fraction in lowest terms.

■ Problems 10–2

Change the percents to decimals.

1. 84% **2.** 98% **3.** 41% **4.** 79% **5.** 6%

6. 1% **7.** 500% **8.** 784% **9.** 39.6% **10.** 8.6%

11. 68.78% **12.** $77\frac{1}{3}\%$ **13.** $15\frac{2}{3}\%$ **14.** $5\frac{6}{7}\%$ **15.** $6.8\frac{2}{3}\%$

16. $10.2\frac{1}{2}\%$ **17.** $101\frac{2}{3}\%$ **18.** 46.8% **19.** 9% **20.** $17\frac{4}{5}\%$

21. $7.64\frac{2}{3}\%$ **22.** 339.4% **23.** 47% **24.** $69.84\frac{1}{3}\%$

Change the percent number to a fraction or a mixed number.

25. 16% **26.** 85% **27.** 71% **28.** 32% **29.** 4%

30. 21% **31.** 106% **32.** 236% **33.** 58% **34.** 99%

35. 62.5% **36.** 89.6% **37.** 7.7% **38.** 88.04% **39.** 36.57%

40. 10.5% **41.** 412.6% **42.** 362.95% **43.** 5.62% **44.** 13.302%

45. $33\frac{1}{3}\%$ **46.** $66\frac{2}{3}\%$ **47.** $21\frac{1}{5}\%$ **48.** $91\frac{2}{3}\%$ **49.** $56\frac{2}{5}\%$

50. $6\frac{1}{9}\%$ **51.** $83\frac{1}{3}\%$ **52.** $7\frac{3}{7}\%$ **53.** $41\frac{2}{3}\%$ **54.** 87%

55. $17\frac{1}{5}\%$ **56.** 66% **57.** 18.6% **58.** 7% **59.** 51.65%

60. $8\frac{1}{3}\%$ **61.** 5.108% **62.** 10%

Change the percents to decimals.

■ **Example 1**

Change $\frac{2}{3}\%$ to a decimal.

Solution **Step 1** Change the fraction to a decimal.

$$\frac{2}{3}\% = 0.66\frac{2}{3}\%$$

Step 2 Change $0.66\frac{2}{3}\%$ to a decimal.

$$0.66\frac{2}{3}\% = \frac{0.66\frac{2}{3}}{100} = 0.0066\frac{2}{3}$$

63. $\frac{1}{2}\%$ **64.** $\frac{1}{3}\%$ **65.** 0.2% **66.** 0.32% **67.** $\frac{7}{8}\%$ **68.** 0.104%

Change the percents to fractions.

■ **Example 2**

Change $\frac{3}{8}\%$ to a fraction.

Solution

$$\frac{3}{8}\% = \frac{\frac{3}{8}}{100} = \frac{3}{8} \times \frac{1}{100} = \frac{3}{800}$$

69. $\frac{1}{3}\%$ **70.** $\frac{5}{8}\%$ **71.** 0.62% **72.** 0.43%

The answers to odd-numbered exercises are in the back of the book.

Answers to exercises 20–24

20. $0.91\frac{2}{3}$ **21.** $0.08\frac{1}{3}$

22. $0.396\frac{1}{8}$ **23.** $0.41\frac{2}{3}$

24. $0.10\frac{1}{12}$

Answers to exercises 25–28

25. $\frac{15}{100} = \frac{3}{20}$ **26.** $\frac{64}{100} = \frac{16}{25}$

27. $\frac{22}{100} = \frac{11}{50}$ **28.** $\frac{85}{100} = \frac{17}{20}$

Answers to exercises 29–32

29. $\frac{293}{500}$ **30.** $\frac{827}{2000}$ **31.** $\frac{227}{1000}$

32. $\frac{1621}{25{,}000}$

Answers to exercises 33–37

33. $\frac{1}{6}$ **34.** $\frac{1}{12}$ **35.** $\frac{11}{12}$ **36.** $\frac{53}{350}$

37. $\frac{26}{75}$

10–3 CHANGING FRACTIONS AND DECIMALS TO PERCENT

Changing Decimals to Percent

■ **Example 1**

Change 0.621 to a percent.

Solution We need to change 0.621 to a fraction with 100 as a denominator.

$$0.621 = \frac{621}{1000} = \frac{621 \div 10}{1000 \div 10} = \frac{62.1}{100} = 62.1\%$$

Thus, 0.621 = 62.1%

Rule

To change a decimal to a percent, we move the decimal point two places to the right and attach a percent symbol.

Practice check: Do exercises 38–42 in the margin.

Change the decimal to a percent.

38. 0.79 **39.** 0.216

40. 0.034 **41.** 0.4681

42. 0.01698

The answers are in the margin on page 385.

Changing Fractions to Percent

To change a fraction to a percent, we will use the following steps:

Rule

To change a fraction to a percent:

Step 1 Change the fraction to at least a two-place decimal number.
Step 2 Change the decimal number to a percent.

■ **Example 2**

A class of 20 students has 13 women and 7 men. Thus $\frac{13}{20}$ of the class are women. What percent of the total class are women?

Solution **Step 1**

$$\frac{13}{20} = 0.65 \qquad \textit{Note: } \begin{array}{r} .65 \\ 20\overline{)13.00} \\ \underline{12\,0} \\ 1\,00 \\ \underline{1\,00} \end{array}$$

Step 2

$0.65 = 65\%$ Moving the decimal point two places to the right

65% of the class are women.

■ **Example 3**

Change $\frac{2}{5}$ to a percent.

Solution **Step 1**

$$\frac{2}{5} = 0.40 \qquad \textit{Note: } \begin{array}{r} .4 \\ 5\overline{)2.0} \\ \underline{2\,0} \end{array}$$

Step 2

$0.40 = 40\%$ Moving the decimal point two places to the right

Thus, $\frac{2}{5} = 40\%$

Practice check: Do exercises 43–46 in the margin.

Change the fractions to percents.

43. $\frac{1}{4}$ **44.** $\frac{7}{10}$

45. $\frac{23}{50}$ **46.** $\frac{1}{2}$

The answers are in the margin on page 387.

■ **Example 4**

Change $\frac{5}{8}$ to a percent.

Solution **Step 1**

$$\frac{5}{8} = 0.62\frac{1}{2}, \text{ or } 0.625 \qquad \textit{Note: } \begin{array}{r} .625 \\ 8\overline{)5.000} \\ \underline{4\,8} \\ 20 \\ \underline{16} \\ 40 \\ \underline{40} \end{array}$$

Step 2

$0.625 = 62.5\%$ Move the decimal point two places to the right.

Thus, $\frac{5}{8} = 62.5\%$, or $62\frac{1}{2}\%$.

Practice check: Do exercises 47–49 in the margin.

■ Example 5

Change $\frac{5}{7}$ to a percent.

Solution **Step 1** Change $\frac{5}{7}$ to at least a two-place decimal number.

$$\begin{array}{r} .71428 \\ 7\overline{)5.00000} \\ \underline{4\,9} \\ 10 \\ \underline{7} \\ 30 \\ \underline{28} \\ 20 \\ \underline{14} \\ 60 \\ \underline{56} \\ 4 \end{array}$$

The quotient is a nonterminating decimal.

When the decimal number is a nonterminating decimal, we can leave the answer in one of two forms:

1. As a number to the nearest hundredth, thousandth, etc., or
2. As a two-place decimal with a fraction.

Thus, $\frac{5}{7} \approx 0.714$, or $0.71\frac{3}{7}$.

Step 2 Change the decimal number to a percent.

$$\frac{5}{7} \approx 71.4\%, \quad \text{or} \quad 71\frac{3}{7}\%$$

■ Example 6

Change $\frac{1}{6}$ to a mixed number percent.

Solution **Step 1** Change $\frac{1}{6}$ to at least a two-place decimal number.

$$\begin{array}{r} .16\frac{4}{6}, \\ 6\overline{)1.00} \\ \underline{6} \\ 40 \\ \underline{36} \\ 4 \end{array} \quad \text{or} \quad 0.16\frac{2}{3}$$

Answers to exercises 38–42

38. 79% **39.** 21.6% **40.** 3.4% **41.** 46.81% **42.** 1.698%

Change the fractions to percents.

47. $\frac{3}{8}$ **48.** $\frac{5}{16}$

49. $\frac{43}{200}$

The answers are in the margin on page 387.

Change each fraction to a percent, rounded to the nearest hundredth.

50. $\frac{2}{3}$ **51.** $\frac{7}{12}$ **52.** $\frac{9}{12}$

53. $\frac{5}{6}$ **54.** $\frac{1}{12}$ **55.** $\frac{4}{9}$

The answers are in the margin on page 389.

Step 2 Change the decimal to a percent.

$$\frac{1}{6} \approx 0.16\frac{2}{3} = 16\frac{2}{3}\%$$

Practice check: Do exercises 50–55 in the margin.

■ Looking Back

You should now be able to change fractions and decimals to percent.

■ Problems 10–3

Review

Change the percents to decimals.

1. 67% **2.** 7% **3.** 8.9% **4.** 43.6% **5.** $14\frac{5}{8}\%$

Change each percent number to a fraction in lowest terms.

6. 91% **7.** $15\frac{1}{5}\%$ **8.** 72% **9.** 33.5% **10.** 8%

11. 32.76% **12.** $83\frac{1}{3}\%$ **13.** 10.104% **14.** 15% **15.** $22\frac{2}{7}\%$

16. $66\frac{2}{3}\%$

Change the fractions or mixed numbers to percents.

17. $\frac{3}{4}$ **18.** $\frac{17}{50}$ **19.** $\frac{9}{20}$ **20.** $\frac{6}{10}$ **21.** $\frac{1}{5}$ **22.** $\frac{2}{4}$

23. $\frac{15}{25}$ **24.** $\frac{3}{10}$ **25.** $\frac{3}{5}$ **26.** $\frac{1}{8}$ **27.** $\frac{3}{16}$ **28.** $\frac{5}{8}$

29. $\frac{27}{125}$ **30.** $\frac{29}{250}$ **31.** $4\frac{2}{5}$ **32.** $7\frac{5}{8}$ **33.** $5\frac{4}{5}$ **34.** $12\frac{1}{2}$

Change the fraction or mixed number to a percent, either (a) rounded to the nearest tenth, or (b) expressed as a mixed number.

35. $\frac{1}{6}$ **36.** $\frac{1}{3}$ **37.** $\frac{11}{12}$ **38.** $\frac{7}{9}$ **39.** $\frac{5}{6}$ **40.** $\frac{5}{13}$

41. $\frac{23}{64}$ **42.** $\frac{9}{17}$ **43.** $\frac{1}{8}$ **44.** $\frac{39}{46}$ **45.** $3\frac{1}{8}$ **46.** $18\frac{1}{3}$

Change the fractions to percents. If the quotient is nonterminating, write the percent as a mixed number.

47. $\frac{3}{5}$ **48.** $\frac{2}{3}$ **49.** $\frac{17}{20}$ **50.** $\frac{9}{16}$ **51.** $\frac{15}{32}$ **52.** $\frac{9}{25}$

53. $\frac{1}{12}$ **54.** $\frac{18}{125}$ **55.** $\frac{4}{5}$ **56.** $\frac{2}{3}$ **57.** $\frac{162}{200}$ **58.** $\frac{7}{16}$

59. $\frac{47}{50}$ **60.** $\frac{9}{10}$

Change the decimals to percents.

61. 0.68 **62.** 0.213 **63.** $0.68\frac{1}{2}$ **64.** 0.041

65. $0.13\frac{1}{3}$ **66.** $0.04\frac{1}{8}$ **67.** 0.7942 **68.** 0.0623

69. $0.347\frac{2}{3}$ **70.** $0.58\frac{1}{3}$ **71.** 0.83965 **72.** $0.08\frac{1}{3}$

73. 0.680 **74.** 0.204 **75.** 6 **76.** 73

77. 4.56 **78.** $1.08\frac{2}{3}$ **79.** 0.0026 **80.** $0.0031\frac{2}{3}$

Answers to exercises 43–46

43. 25% **44.** 70% **45.** 46% **46.** 50%

Answers to exercises 47–49

47. 37.5% **48.** 31.25% **49.** 21.5%

The answers to review and odd-numbered exercises are in the back of the book.

10–4 THE PERCENT QUANTITIES

Recognizing Percent Quantities

In order to solve percent problems, our attention is on three basic quantities:

1. The whole or *base, B*
2. The part compared to the whole, or *percentage, P*
3. The part compared to 100%, or *rate, R*

Knowing the value of any two of these quantities, we can solve for a third by using methods in section 10–5. In a percent problem, two of the quantities will always be known.

In the rest of this section we will learn to recognize these three basic quantities of a percent problem.

Rate, *R*

Of course, the rate in a percent problem is the easiest to recognize.

■ Examples

Which part of the percent problem means the same as *R*?

1. 66% of what number is 7?

   $R = 66$

2. What number is 15 percent of 8?

   $R = 15$

3. 35 is what percent of 70?

   $R =$ what? *R* is the unknown quantity in this case.

We can see that *R* is that number or word that is written with the word *percent* or the symbol %.

Practice check: Do exercises 56–59 in the margin.

Which part of the problem means the same as *R*?

56. 8 percent of 16 is what number?

57. 506 is what percent of 18?

58. 31 is 15% of what number?

59. What number is 305% of 55?

The answers are in the margin on page 391.

Base, *B*

The base *B* represents the whole. It usually comes right after the word *of.*

■ Examples

Which part of the problem means the same as *B*?

4. What number is 15 percent of 8?

   $B = 8$

5. 35 is what percent of 70?

   $B = 70$

6. 66% of what number is 7?

   $B =$ what? In this case, *B* is the unknown quantity.

Practice check: Do exercises 60–63 in the margin.

Which part of the problem means the same as *B*?

60. 8 percent of 16 is what number?

61. 506 is what percent of 18?

62. 31 is 15% of what number?

63. What number is 305% of 55?

The answers are in the margin on page 391.

Percentage, P

The part compared to the whole or percentage P is the quantity that is left after the rate R and base B have been identified.

■ Examples

Which part of the problem means the same as P?

7. 35 is what percent of 70?

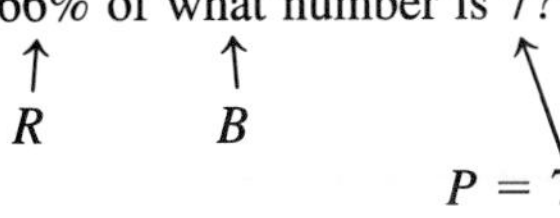

$P = 35$

8. 66% of what number is 7?

R B

$P = 7$

9. What number is 15 percent of 8?

R B

$P =$ what? In this case, the percentage is the unknown quantity.

Practice check: Do exercises 64–67 in the margin.

Rule

The three basic quantities of percent problems are

1. The part compared to 100% or rate (R) is the number that is written with the word *percent*, or symbol %.
2. The base number (B) comes right after the word *of*.
3. The part compared to the whole, or percentage (P) is the number that is left after R and B have been identified.

■ Example 10

Identify all three quantities in the percent problems.

a. 48% of 61 is what number?

b. What is 205% of 50?

c. What percent of 7 is 9?

d. 68 is 31% of what number?

Solution We first locate the percent (R) and the base (B). The remaining quantity is the portion (P).

a. $R = 48, B = 61, P =$ what?

b. $R = 205, B = 50, P =$ what?

c. $R =$ what? $B = 7, P = 9$

d. $R = 31, B =$ what? $P = 68$

Practice check: Do exercises 68–72 in the margin.

■ Looking Back

You should now be able to recognize the three basic quantities in a percent problem.

Answers to exercises 50–55

50. 66.67% **51.** 58.33%

52. 81.82% **53.** $83\frac{1}{3}\%$

54. $8\frac{1}{3}\%$ **55.** $44\frac{4}{9}\%$

Which part of the problem means the same as P?

64. 8 percent of 16 is what number?

65. 506 is what percent of 18?

66. 31 is 15% of what number?

67. What number is 305% of 55?

The answers are in the margin on page 391.

Identify all three quantities in the percent problems.

68. What is 6% of 35?

69. 26 is what percent of 106?

70. 8 is 39% of what number?

71. What percent of 34 is 17?

72. What is $66\frac{2}{3}\%$ of 27?

The answers are in the margin on page 391.

■ Problems 10–4

Review

Change the percents to decimals

1. $65\frac{1}{4}\%$

2. 18%

Change each percent to a fraction in lowest terms.

3. 225%

4. $83\frac{1}{3}\%$

Change the fractions or mixed numbers to percents rounded to the nearest tenth.

5. $\frac{2}{3}$

6. $\frac{5}{7}$

Identify all three basic quantities in the percent problems.

7. What is 39% of 67?

8. 15 is 6% of what number?

9. 106 is what percent of 58?

10. 60 is 30% of what number?

11. What percent of 72 is 206?

12. 215% of 66 is what?

13. What number is 80% of 32?

14. 617 is 31% of what number?

15. 28 is what percent of 84?

16. What is 150% of 9?

17. 26% of what number is 56?

18. What percent of 18 is 180?

19. 21 is 50% of what number?

20. What is 30% of 60?

21. 78% of 167 is what number?

22. 31 is 262% of what number?

23. 18 is 50% of what number?

24. What is 56% of 75?

25. What percent of 15 is 781?

26. 33 is what percent of 81?

The answers to review and odd-numbered exercises are in the back of the book.

Answers to exercises 56–59

56. $R = 8$ **57.** $R =$ what?
58. $R = 15$ **59.** $R = 305$

Answers to exercises 60–63

60. $B = 16$ **61.** $B = 18$
62. $B =$ what ? **63.** $B = 55$

Answers to exercises 64–67

64. $P =$ what? **65.** $P = 506$
66. $P = 31$ **67.** $P =$ what?

Answers to exercises 68–72

68. $R = 6, B = 35, P =$ what?
69. $R =$ what?, $B = 106, P = 26$
70. $R = 39, B =$ what?, $P = 8$
71. $R =$ what?, $B = 34, P = 17$
72. $R = 66\frac{2}{3}, B = 27, P =$ what?

10–5 THE THREE TYPES OF PERCENT PROBLEMS

In section 10–4, we learned to recognize the three basic quantities in a percent problem. In this section, we will learn to solve each of the three types of percent problems.

Rule

Indicating multiplication. Sometimes instead of using $\times$ for the multiplication sign, a raised dot or parentheses is used.

$$6 \times R = 6 \cdot R = 6(R)$$

Rule

The three types of percent problems:

1. Solving for P in $P = R\% \cdot B$ when B and R are given
2. Solving for $R\%$ in $R\% = \dfrac{P}{B}$ when P and B are given
3. Solving for B in $B = \dfrac{P}{R\%}$ when P and R are given

Type 1: Solving for P when B and R are given. Use $P = R\% \cdot B$

■ Example 1

What is 6% of 96?

Solution **Step 1** Identify the quantities.

$$R = 6, B = 96, P = \text{what?}$$

Step 2 Solve using $P = R\% \cdot B$.

$P = 6\% \cdot 96$ Replacing R by 6 and B by 96

$P = 0.06 \cdot 96$ Changing 6% to 0.06

$P = 5.76$ Multiplying 96 by 0.06

5.76 is 6% of 96.

■ Example 2

56% of 31 is what number?

Solution $R = 56, B = 31, P = $ what?

$$P = R\% \cdot B$$

$$P = 56\% \cdot 31 \quad \text{Replacing } R \text{ by 56 and } B \text{ by 31}$$

$$P = 0.56 \cdot 31 \quad \text{Changing 56\% to 0.56}$$

$$P = 17.36 \quad \text{Multiplying}$$

56% of 31 is 17.36.

Practice check: Do exercises 73–76 in the margin.

73. What is 25% of 16?

74. What is 13% of 35?

75. What is 120% of 44?

76. 0.8% of 66 is what number?

The answers are in the margin on page 395.

Type 2: Solving for R when P and B are given. Use $R\% = \frac{P}{B}$.

■ Example 3

7 is what percent of 28?

Solution $B = 28, P = 7, R = $ what?

$$R\% = \frac{P}{B}$$

$$R\% = \frac{7}{28} \quad \text{Replacing } P \text{ by 7 and } B \text{ by 28}$$

$$R\% = 0.25 \quad \text{Dividing 7 by 28}$$

$$R\% = 25\% \quad \text{Changing 0.25 to a percent}$$

7 is 25% of 28. *Note:*

```
       .25
  28)7.00
     5 6
     1 40
     1 40
```

Practice check: Do exercises 77–80 in the margin.

77. 9 is what percent of 54?

78. 29 is what percent of 15?

79. 3 is what percent of 270?

80. 4.3 is what percent of 16.8?

The answers are in the margin on page 395.

Type 3: Solving for B when P and R are given. Use $B = \frac{P}{R\%}$.

■ Example 4

0.5% of what number is 102?

Solution $R = 0.5, P = 102, B = $ what?

$$B = \frac{P}{R\%}$$

$$B = \frac{102}{0.5\%} \quad \text{Replacing } P \text{ by 102 and } R \text{ by 0.5}$$

$$B = \frac{102}{0.005} \quad \text{Changing 0.5\% to a decimal}$$

$$B = 20{,}400$$

0.5% of 20,400 is 102. *Note:*

$$\begin{array}{r} 20400 \\ 0.005\overline{)102000} \\ \underline{10} \\ 20 \\ \underline{20} \end{array}$$

Practice check: Do exercises 81–84 in the margin.

81. 32 is 8% of what number?

82. 250 is 4.5% of what number?

83. 150% of what number is 82?

84. 0.04% of what number is 2.5?

The answers are in the margin on page 395.

■ Looking Back

You should now be able to solve the three types of percent problems.

■ Problems 10–5

Review

Change the decimals to percents.

1. 0.375

2. 0.005

Identify all three basic quantities in the percent problems.

3. What is 63% of 37?

4. 40% of what number is 39?

5. 98 is what percent of 23?

6. What is 30% of 70?

7. What is 56% of 32?

8. What is 13% of 108?

9. 56 is 90% of what number?

10. 78 is 85% of what number?

11. 3 is what percent of 36?

12. 89 is what percent of 356?

13. 360 is what percent of 9?

14. 51 is what percent of 15?

15. What is 213% of 96?

16. What is $\frac{3}{8}$% of 416?

17. 180% of what number is 92?

18. 6% of what number is 29.22?

19. 366 is 55.2% of what number?

20. 88 is 0.8% of what number?

21. 0.7% of 18 is what number?

22. 0.18% of 35 is what number?

23. What percent of 48 is 18?

24. What percent of 45 is 120?

25. 78% of $46\frac{2}{5}$ is what number?

26. 135% of $24\frac{1}{3}$ is what number?

27. What percent of 39.5 is 5.25?

28. What percent of 47.6 is 21.42?

29. 0.6% of what number is 3.6?

30. $\frac{3}{5}$% of what number is 21?

31. $72\frac{1}{2}$ is what percent of $109\frac{1}{5}$?

32. $38\frac{1}{4}$ is what percent of $346\frac{1}{2}$?

33. What is $66\frac{2}{3}$% of $14\frac{1}{2}$?

34. What is $33\frac{1}{3}$% of $89\frac{1}{4}$?

35. $84\frac{1}{2}$ is 80% of what number?

36. 205 is $44\frac{3}{8}$% of what number?

The answers to review and odd-numbered exercises are in the back of the book.

10–6 APPLICATIONS OF TYPE 1 PROBLEMS

In this section, we will learn how to solve Type 1 application problems. These problems can be written in the form "What is R% of B?"

Rule

Type 1: Solve for P in $P = R\% \cdot B$ *when B and R are given.*

Example 1

The price of a pair of jeans is $17.00. If the state sales tax rate is 5% of the price, how much is the tax?

Solution Notice that the problem states "5% of the price." The price is $17.00. We can rewrite this problem as "What number is 5% of 17?"
Here $R = 5$, $B = 17$, and $P =$ what?

$$P = R\% \cdot B$$
$$P = 5\% \cdot 17$$
$$P = 0.05 \cdot 17$$
$$P = 0.85$$

The tax was $0.85 or 85¢.

Example 2

At Jason's, a coat was originally marked $110, but was discounted 15%. What was the amount of the discount?

Solution The discount is 15% of the original price, $110. We can rewrite the problem as "What number is 15% of 110?"
Here $R = 15$, $B = 110$, and $P =$ what?

$$P = R\% \cdot B$$
$$P = 15\% \cdot 110$$
$$P = 0.15 \cdot 110$$
$$P = 16.50$$

The amount of the discount is $16.50.

Practice check: Do exercises 85–87 in the margin.

Looking Back

You should now be able to solve Type 1 application problems.

Problems 10–6

Review

Change each percent to a fraction in lowest terms.

1. 18.25%

2. 145%

3. 54 is what percent of 80?

4. $\frac{1}{2}\%$ of 300 is what?

Answers to exercises 73–76

73. $P = 4$ **74.** $P = 4.55$
75. $P = 52.8$ **76.** $P = 0.528$

Answers to exercises 77–80

77. $16\frac{2}{3}\%$ **78.** $193\frac{1}{3}\%$

79. $1\frac{1}{9}\%$ **80.** 25.6%

Answers to exercises 81–84

81. $B = 400$ **82.** $B = 5555\frac{1}{2}$

83. $B = 54\frac{2}{3}$ **84.** $B = 6250$

85. A certain item marked $22.00 did not sell. The store owner decided to mark it down 25%. How much was the item reduced?

86. The Red Devils won 60% of the 46 games they played. How many games did they win?

87. In 1990, Jacksonville had a population of 562 people. The mayor estimated that the town would have 7% more people by 2000. How many more people would the town have? (Round to the nearest whole person.)

The answers are in the margin on page 397.

Solve. Write a statement for the answer.

5. **Welding.** The cost of aluminum for a given job is $2050. The supplier estimates an 8% rise in costs upon delivery. What is the rise in cost?

6. **Real Estate.** A real estate broker received a 7% commission on a $160,000 sale. What did he receive in commission?

7. **Clerical.** An 8% sales tax is added to the selling price of a power tool. If the selling price is $198.50, what is the sales tax? What is the final cost to the consumer?

8. **Wastewater Technology.** A 55-gallon drum contains water and 15% sodium hypochlorous acid. How many gallons of sodium hypochlorous acid are in the drum?

9. **Mason.** A mason had 35,000 bricks delivered on a building site. He found that 3% were broken. How many were broken?

10. **Printer.** Out of 10,500 printed brochures, 4% were rejected for various reasons. Find the total number of brochures that were rejected.

11. **Sales.** A salesperson receives 2.3% of her total sales as commission. During February, she sold $6530 worth of goods. How much money did she get for February?

12. **Plumber.** San-Tone Plumbing Company has 76.5% of its 230 employees as field people. How many employees are field people? Round to the nearest whole person.

13. **Accounting.** If 7.4% of your salary is withheld for Social Security, how much is withheld from your monthly salary of $1784.00? Round to the nearest cent.

14. **Electronics.** Radar-Electronics estimate that $3\frac{1}{2}$% of all their parts manufactured do not meet their quality control standards. If 22,000 parts will be manufactured in December, how many parts will be rejected?

15. **Police Science.** In 1998 a certain city had 125,400 major crimes committed. Of this amount 19% were auto thefts. How many automobiles were stolen during that year?

16. **Accounting.** After the first year, a certain new car depreciates $17\frac{1}{3}$%. If the new car cost $25,675, what was the first-year depreciation? What is the worth of the car after the first year? Round to the nearest dollar.

17. Water Management. Management decided to increase the flow of water by 75%. What is the increase in pipe area if the original pipe area is 32.5 square inches.

18. Statistician. This chart shows the personal income for various age groups in 1995. Determine the dollar amount for each age group if the total income for all age groups was \$1249.7 billion.

Answers to exercises 85–87

85. The item was reduced by \$5.50.
86. They won 27 games.
87. They would have 39 more people.

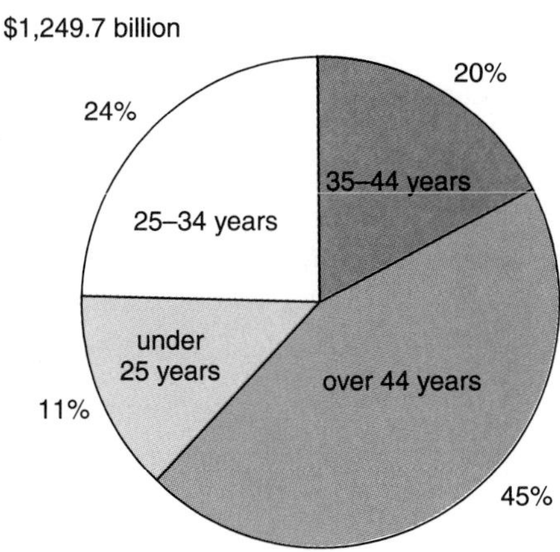

The answers to review and odd-numbered exercises are in the back of the book.

10–7 APPLICATIONS OF TYPE 2 PROBLEMS

In this section, we will learn how to solve Type 2 application problems. These problems can be written in the form "P is what % of B?"

Rule

Type 2: Solve for $R\%$ in $R\% = P/B$ when P and B are given.

■ Example 1

The Huskies played 18 games during the season and won 6 of them. What is the percent of the total games they won?

Solution We need to find R, the percent of the total games won. Notice that 6 is part of the 18, which is the whole total. We can rewrite this problem as "6 is what percent of 18?" So, $B = 18$ and $P = 6$.

We solve using $R\% = \frac{P}{B}$.

$$R\% = \frac{6}{18}$$

$$R\% = 0.33\frac{1}{3}$$

$$R\% = 33\frac{1}{3}\%$$

They won $33\frac{1}{3}\%$ of the games played. *Note:*

$$\begin{array}{r} .33\tfrac{6}{18} = .33\tfrac{1}{3} \\ 18\overline{)6.00} \\ \underline{5\,4} \\ 60 \\ \underline{54} \\ 6 \end{array}$$

■ Example 2

Bill got 17 questions correct and 5 incorrect. What percent of the questions did he get correct? Round to the nearest percent.

Solution We need to find R, the percent of the total questions that he got correct. In this case, the *whole* or *base* is the total number of questions on the test, or 22. Now we find that 17 is *part* of 22. We can rewrite the problem as "17 is what percent of 22?"
So, $B = 22$ and $P = 17$.

$$R\% = \frac{P}{B}$$

$$R\% = \frac{17}{22}$$

$$R\% = 0.77\frac{3}{11}$$

$$R\% = 77\frac{3}{11}\%$$

He got a score of 77% (rounded). *Note:* $\begin{array}{r} .77\tfrac{6}{22} = .77\tfrac{3}{11} \\ 22\overline{)17.00} \end{array}$

Practice check: Do exercises 88–90 in the margin.

88. Gloria Jackson saves $180 per month. If her monthly income is $1750, what percent of her monthly income does she save? Round to the nearest tenth of a percent.

89. 85 out of 325 employees received flu shots at Lincoln Industries. What percent of the employees received flu shots? Round to the nearest percent.

90. A college basketball team won 16 games and lost 27. What percent of the games did it lose? Round to the nearest tenth of a percent.

The answers are in the margin on page 401.

■ Looking Back

You should now be able to solve Type 2 application problems.

■ Problems 10–7

Review

Change the fraction to a percent.

1. $\frac{3}{8}$ **2.** $\frac{13}{16}$

Change the decimal to a percent.

3. 0.1875 **4.** 0.00625

Solve.

5. Ron earned $15.46 per hour as a checker in a grocery store. He received a 4.5% increase. How much was the increase to the nearest cent?

Solve. Write a statement for the answer.

6. **Generic.** Out of 2000 parts coming off the assembly line, 25 are defective. What percent of the parts are defective?

7. **Management.** A store manager paid $15.00 for an electric mixer. He added $7.50 for his profit. What percent of $15.00 was the increase?

8. **Wood Products.** A full flask of fluid weighs 8 grams. This includes 5 grams of formaldehyde. What percent of the weight of the formaldehyde is contained in the whole weight of the flask?

9. **Diesel Mechanics.** The mechanical efficiency of an engine is found by dividing the brake horsepower by the factory-specified indicated horsepower. What is the percent of mechanical efficiency if the braking horsepower of an engine is 152 and the indicated horsepower is 200?

10. **Automobile Mechanic.** A 5.25-gallon cooling system is filled with 2 gallons of antifreeze and the rest water. What percent of the solution is antifreeze? Round to the nearest tenth of a percent.

11. **Generic.** Betty earns $144.56 plus tips per week as a waitress. The tips during one week were $105.15. What percent of her total wages were tips? Round to the nearest whole percent.

12. **Sheet Metal Technicians.** A sheet metal company found that for every 120 sheets of metal, 6 sheets are wasted. Find the percent of sheets wasted.

13. **Forestry.** The conditions in a certain clear cut area require that 600 seedlings per acre survive. In order to maintain that requirement, the reseeding company needs to plant 16,000 seedlings per acre. What is the percent of seedlings that live through the first growing season in this area?

14. **Generic.** A certain astronaut weighed 176 pounds just before his flight to the space station. Upon his return to earth his weight showed that he had lost 11 pounds. What percent did he lose?

15. **Industrial Mechanics.** The indicated horsepower of a steam engine is 10.3 and the effective horsepower is 9.2 What percent of the indicated horsepower is the effective horsepower? Round to the nearest tenth of a percent.

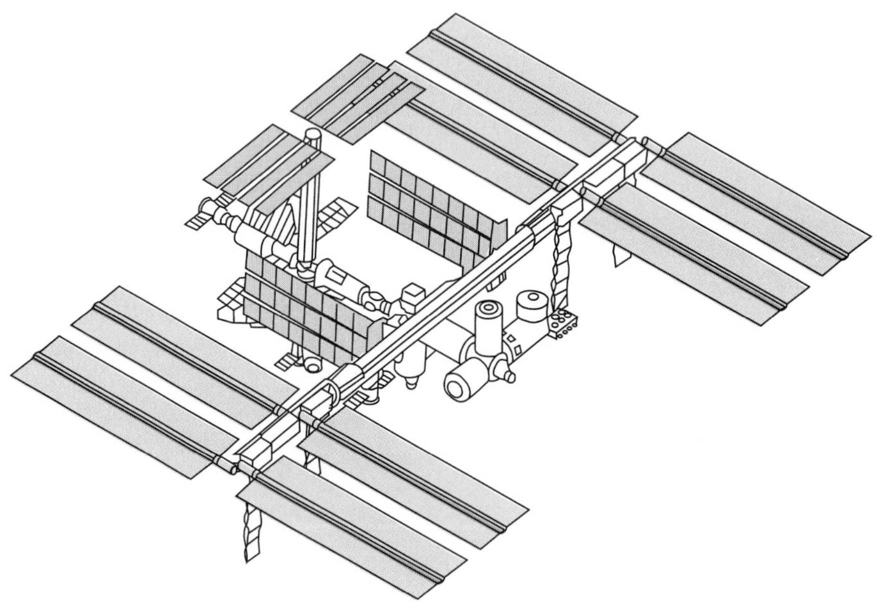

The answers to review and odd-numbered exercises are in the back of the book.

10–8 APPLICATIONS OF TYPE 3 PROBLEMS

In this section, we will learn how to solve Type 3 application problems. These problems can be written in the form "$R\%$ of what is B?

Rule

Type 3: Solve for B in $B = P/R\%$ *when P and R are given.*

■ Example 1

A certain car now costs $22,275. This is 60% of its original value. Find the original value, to the nearest dollar.

Solution We need to identify B, P, and R. It is easy to tell that $R = 60$. Now we must answer the question, Which is correct, $B = 22{,}275$, or $P = 22{,}275$? Read the problem again. We can rewrite the problem as "60% of the original price is $22,275."

So, $P = \$22{,}275$, and B = original price.

$$B = \frac{P}{R\%}$$

$$B = \frac{22{,}275}{60\%}$$

$$B = \frac{22{,}275}{0.60}$$

$$B = 37{,}125$$

The car's original price was \$37,125.

Answers to exercises 88–90

88. She saves 10.3% of her monthly income.
89. 26% of the employees received flu shots.
90. The team lost 62.8% of its games.

Example 2

Helen Santiani received 78% on a basic mathematics test. She got 21 questions correct. How many questions were on the test? Round to the nearest whole question.

Solution Here, 21 is part of the total number of questions on the test. 78% of what is 21?
So, $P = 21$, B = total number of questions, and $R = 78$.

$$B = \frac{P}{R\%}$$

$$B = \frac{21}{78\%}$$

$$B = \frac{21}{0.78}$$

$$B = 27 \quad \text{(rounded)}$$

There were 27 questions on the test.

Practice check: Do exercises 91and 92 in the margin.

91. In an election, 3469 people voted. This was 26.2% of all those registered to vote. How many people were eligible to vote? Round to the nearest whole voter.

92. In a period of 3 months, Bill Overwait lost 42.7 pounds. This was 22.6% of his original weight. What was his original weight to the nearest tenth of a pound?

The answers are in the margin on page 403.

Looking Back

You should now be able to solve Type 3 application problems.

Problems 10–8

Review

Solve. Write a statement for the answer.

1. **Generic.** In a body shop class, 13% of the 30 students received a grade of A. How many received an A? Round to the nearest whole student.

2. **Generic.** As an apprentice Sue receives \$450 per week. She banks \$25.00 of that amount. What percent of her paycheck does she save? Round to the nearest whole percent.

Solve. Write a statement for the answer.

3. **Generic.** John earned $85 per week. This was 45% of what he had in the bank. How much did he have in the bank? Round to the nearest cent.

4. **Marketing.** The mark-up on a sofa is $150. If this is 48% of the cost, how much did the sofa cost?

5. **Business.** A woman bought 15% of a business for $25,470. How much is the business worth?

6. **Sales.** A sales rep earned a 15% commission rate on total sales. If he earned $456 last week, what were his total sales?

7. **Nursing.** Of the total amount of internal secretions in a normal adult, about 8.6% or 450 $m\ell$ are bile secretions per day. What is the total amount of internal secretions?

8. **Real Estate.** A couple made a down payment of $15,000 on a home. This was $5\frac{2}{3}\%$ of the original price. What was the original price? Round to the nearest dollar.

9. **Accountant.** Betty received a raise of $0.25 per hour. This was 2% of her hourly wages. What were her hourly wages before the raise?

10. **Air Maintenance.** A tank is filled to 77% of its capacity with 235.7 gallons of high-octane fuel. What is the capacity of the tank to the nearest tenth of a gallon?

11. **Management.** Of all parts manufactured in a certain plant, $97\frac{1}{3}\%$ are not defective. How many parts would have to be made to fill an order of 370 parts? Round to the nearest whole part.

12. **Management.** A company estimates that 98.5% of its living grass plugs survive during shipping. If a customer orders 2200 plugs, how many should be shipped? Round to the nearest whole plug.

Mixed Practice

This set of problems contains a mixture of the three types of percent application problems. Solve. Write a statement for the answer.

13. **Landscaping.** The price of Lawn-Green fertilizer increased by 16%. If the old price was $19.95 for a 50-pound sack, what was the amount of the increase? Round to the nearest cent.

14. **Dairy.** An inspector found that a certain amount of milk contained 890 gallons of butterfat. If this was 3.2% of the total, how many gallons of milk were inspected? Round to the nearest gallon.

15. **Generic.** Juan earned $23,670 last year and saved $2100. What percent of his income did he save? Round to the nearest whole percent.

16. **Sales.** In one week Belle sold $2462 worth of Amboy Products. She gets to keep 15% of all sales. How much did she earn that week?

Answers to exercises 91 and 92

91. 13,240 people were eligible to vote.

92. Bill's original weight was 188.9 pounds.

17. **Construction.** A load of lumber weighed 9680 lb when dry. This figure represents 92.3% of the original weight of the lumber when it was moist. What did the load weigh before it was dried?

18. **Business.** Six months ago, lettuce sold for $0.98 per head. Within that time it was reduced in price by $0.55 per head. What percent of $0.98 was the decrease? Round to the nearest tenth of a percent.

19. **Agriculture.** A farmer found that out of 356 seed grains only 332 grains germinated. What percent germinated? Round to the nearest tenth of a percent.

20. **Food Processing.** A farmer's flat bed is loaded with 1120 lb of Royal Ann and Bing cherries. The load represents 30.1% of the weight of the truck. What is the weight of the empty truck?

21. **Management.** Flying Carpet Airlines earned $156,000 in 1999. They spent 5.6% of this on advertising. How much did they spend on advertising?

The answers to review and odd-numbered exercises are in the back of the book.

CHAPTER 10 OVERVIEW

Summary

10–1 You learned the meaning of *percent.*

10–2 You learned to change percents to decimals and fractions in lowest terms.

10–3 You learned to change fractions and decimals to percents.

10–4 You learned to recognize the three basic quantities in a percent problem.

10–5 You learned to solve the three types of percent problems using the percent proportion.

10–6 You learned to solve practical Type 1 problems.

10–7 You learned to solve practical Type 2 problems.

10–8 You learned to solve practical Type 3 problems.

Terms To Remember

	Page
Percent symbol	377
Base	388
Percentage	389
Rate	388

Rules

- When we divide by 100, we move the decimal point two places to the left.
- To change a percent to a decimal, we move the decimal point two places to the left and delete the percent symbol.
- To change a decimal to a percent, we move the decimal point two places to the right and attach a percent symbol.
- To change a fraction to a percent

  Step 1 Change the fraction to at least a two-place decimal number.

  Step 2 Change the decimal number to a percent.

- The three basic quantities of percent problems are

  1. The part compared to 100% or rate (R) is the number that is written with the word *percent,* or symbol %.

  2. The base number (B) comes right after the word *of.*

  3. The part compared to the whole, or percentage (P), is the number that is left after R and B have been identified.

- Indicating multiplication. Sometimes instead of using $\times$ for the multiplication sign, a raised dot or parentheses is used: $6 \times R = 6 \cdot R = 6(R)$
- The three types of percent problems

  1. Solving for P in $P = R\% \cdot B$ when B and R are given

  2. Solving for $R\%$ in $R\% = P/B$ when P and B are given

  3. Solving for B in $B = P/R\%$ when P and R are given

Type 1: Solve for P in $P = R\% \cdot B$ when B and R are given.

Type 2: Solve for $R\%$ in $R\% = P/B$ when P and B are given.

Type 3: Solve for B in $B = P/R\%$ when P and R are given.

Self-Test

10–1 Change the fraction to a number with a percent symbol.

1. $\frac{83}{100}$ **2.** $\frac{5.7}{100}$ **3.** $\frac{3\frac{1}{3}}{100}$

Change the percent numbers to reduced fractions or mixed numbers.

4. 56% **5.** 21.5% **6.** 0.09%

10–2 Change the percents to decimals.

7. 73% **8.** 8.3% **9.** $7\frac{5}{8}\%$

Change the percents to fractions in lowest terms.

10. 21% **11.** 68.4% **12.** $17\frac{1}{8}\%$

10–3 Change the fraction to a percents. If the quotient is nonterminating, write the percent as a mixed number.

13. $\frac{3}{8}$ **14.** $\frac{4}{9}$ **15.** $\frac{11}{16}$

Change the decimal to a percent.

16. 0.56 **17.** $0.98\frac{3}{4}$ **18.** $0.67\frac{3}{8}$

10–4 Using $P = R\% \cdot B$, $R\% = \frac{P}{B}$, or $B = \frac{P}{R\%}$, solve for the indicated letter.

19. $R = 10, B = 75, P = ?$ **20.** $R = 20, B = ?, P = 40$

21. $R = ?, B = 25, P = 5$

Identify all three basic quantities in the percent problems, R, B, and P.

22. $\frac{1}{2}$ = what percent of 6? **23.** 70 is 10% of what number?

10–5 Solve.

24. 36 is 20% of what number? **25.** What percent of 75 is 5?

10–6
10–7 **26.** A mechanic's yearly salary was $32,542.50. At the rate of 7.8% per year, how much does he pay in Social Security taxes?

10–8 **27.** Out of 1000 seat belts manufactured, a company found that 20 are defective. What percent of the output was found to be defective?

28. A person is liquidating his business. One article sold for \$385, thereby realizing a loss of 40%. What did the article cost originally?

Chapter Test

Directions: This test will aid in your preparation for a possible chapter test given by your instructor. The answers are in the back of the book. Go back and review in the appropriate section(s) if you missed any test items.

10–1 Change the fraction to a number with a percent symbol.

1. $\frac{71}{100}$ **2.** $\frac{8.6}{100}$ **3.** $\frac{7}{100}$ **4.** $\frac{\frac{2}{3}}{100}$

5. $\frac{91\frac{1}{3}}{100}$

Change the percent numbers to reduced fractions or mixed numbers.

6. 11% **7.** 8.1% **8.** 232% **9.** $66\frac{2}{3}\%$

10. 0.07%

10–2 Change the percents to decimals.

11. 66% **12.** 4% **13.** 3.8% **14.** 79.04%

15. $88\frac{1}{3}\%$ **16.** $6\frac{7}{9}\%$

Change the percent to a fraction in lowest terms.

17. 17% **18.** 5% **19.** 34.2% **20.** 6.8%

21. $66\frac{2}{3}\%$ **22.** $8\frac{1}{3}\%$

10–3 Change the fraction to a percent. If the quotient is nonterminating, write the percent as a mixed number.

23. $\frac{1}{4}$ **24.** $\frac{2}{5}$ **25.** $\frac{3}{8}$ **26.** $\frac{2}{3}$

27. $\frac{7}{13}$ **28.** $\frac{7}{12}$ **29.** $\frac{9}{16}$

Change the decimal to a percent.

30. 0.74 **31.** $0.22\frac{1}{2}$ **32.** $0.05\frac{1}{6}$ **33.** 0.309

34. $0.0506\frac{1}{3}$

10–4 Using $P = R\% \cdot B$, $R\% = \frac{P}{B}$, or $B = \frac{P}{R\%}$, solve for the indicated letter.

35. $P = 3, B = 15, R = ?$ **36.** $B = 8.5, R = 5, P = ?$

37. $P = 4.2, R = 7, B = ?$

Identify all three basic quantities in the percent problems, R, B, and P.

38. What is 4.6% of 38? **39.** 846 is 49% of what number?

40. What percent of 28 is 167? **41.** 150% of 38 is what number?

Solve

10–5 **42.** What is 63% of 39? **43.** 5 is what percent of 125?

44. 0.7% of what number is 4.9? **45.** What percent of 48 is 0.6?

10–6
10–7 **46.** Elsie Smart saved 6% of her paycheck. If her paycheck was $576, how much did she save?

10–8 **47.** Sami got 17 problems correct out of a possible 29. What was his score? Round to the nearest whole percent.

48. Jack Tolifson owned 51% of Multi-Corporation, which amounted to $105,000. What was the total worth of the corporation? Round to the nearest thousand dollars.

49. The Bullets won 89 games and lost 56. What percent of the total games played did the team lose? Round to the nearest tenth of a percent.

CHAPTER 11

Measurement

Choice: Either skip the Pretest and study all of Chapter 11 or take the Pretest and study only the section(s) you need to study.

Pretest

Directions: This Pretest will help you determine which sections to study in Chapter 11. If any of your answers are incorrect, or if you omitted any exercise, turn to the indicated section(s) and study the material; then take the Chapter Test.

11–2 Convert.

1. 4 ft to in. **2.** 21 in. to yd.

11–3 Convert.

3. 7 hm to m. **4.** 3.89 dkm to dm.

11–4 Perform the indicated operations.

5.
$$\begin{array}{r} 8\text{ ft }7\text{ in.} \\ +3\text{ ft }9\text{ in.} \\ \hline \end{array} = \underline{\ ?\ }\text{ ft}$$

6.
$$\begin{array}{r} 46\text{ cm }5\text{ mm} \\ -25\text{ cm }7\text{ mm} \\ \hline \end{array} = \underline{\ ?\ }\text{ cm}$$

7. $7 \times (5\text{ yd }2\text{ ft})$ **8.** $(46\text{ km }386\text{ mm }31\text{ cm}) \div 7$

11–5 **9.** Find the perimeter.

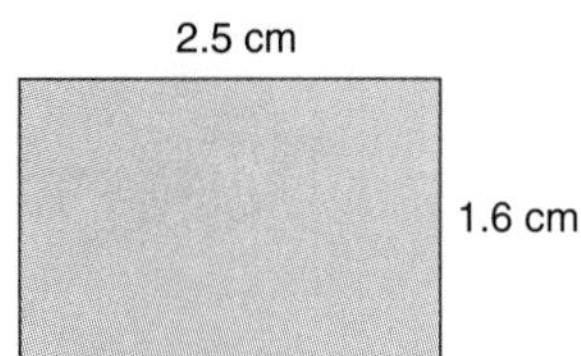

10. Find the circumference. (Use $\frac{22}{7}$ for π.)

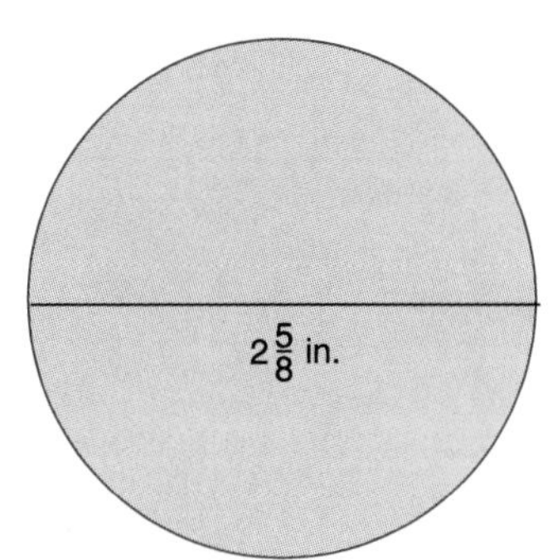

11–6 Find the area of each figure.

11. (Use 3.14 for π.)

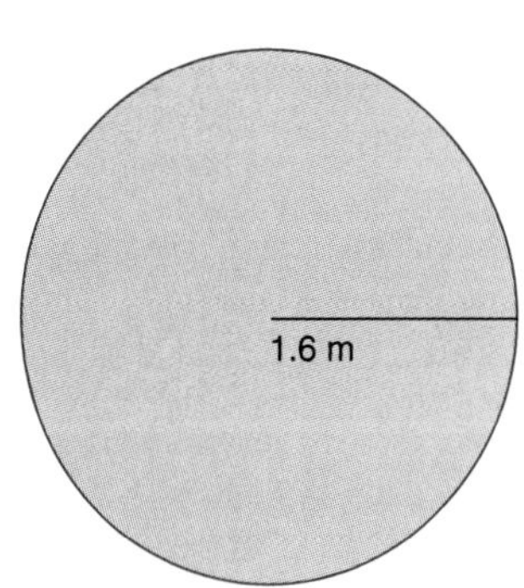

12.

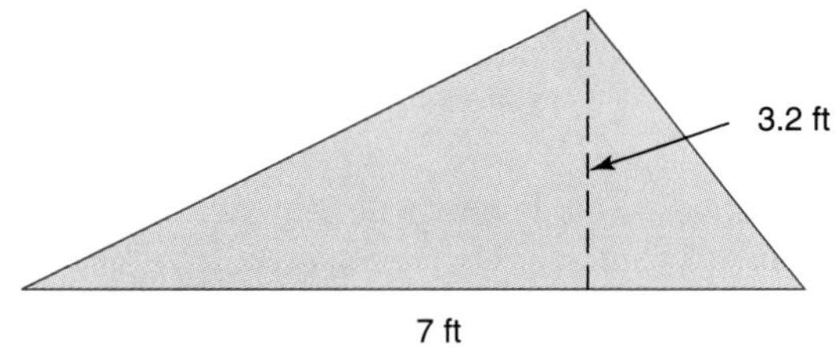

11–7 Convert.

13. 60 in.^{2} to ft^2

14. 8.9 ha to m^2

11–8 **15.** Find the volume.

16. Convert 3418 cℓ to ℓ.

11–9 **17.** Convert 580 dkg to g.

11–10 **18.** Convert 41°F to °C. **19.** Convert 85°C to °F.

11–11 Convert.

20. 32 in. to cm. **21.** 44 km to mi.

The answers are in the margin on page 412.

11–1 BODY MEASUREMENT—A LOOK INTO HISTORY

Caught without a measuring device like a ruler, people tend to use a part of their body or use pacing for a rough estimation of length. In early history measurement was first based on such measuring devices. For instance, Emperor Hadrian completed the Pantheon in Rome in 126 A.D. The measurement of the dome at the time was 66 cubits in diameter. A cubit was the length of one's forearm from the elbow to the tip of the middle finger.

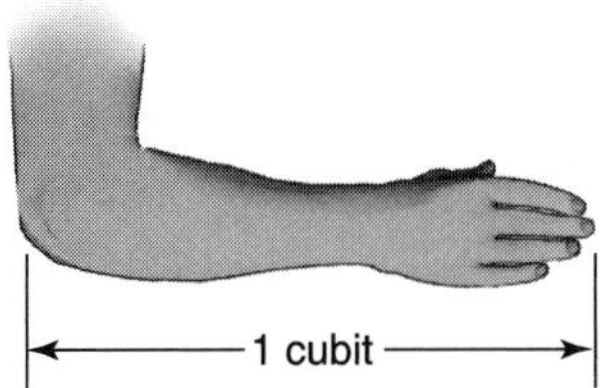

Figure 11–1

Since this measure varied from person to person, the Egyptians tried to standardize the cubit at approximately 21 inches in length. The Romans introduced their cubit of about 26 inches into Europe and England. The English replaced the cubit by the yard, which was standardized during the reign of Henry I. A yard was the distance between the tip of the king's nose and the end of his middle finger. King Edward I declared one inch equal to the distance of three barley corns laid end to end. Now, of course, all lengths of the units of measurements are standardized throughout the world.

Figure 11–2

In this chapter we shall work with two systems of measurement, American and metric.

Answers to Pretest

Circle the answers you missed or omitted. Cross them off after you've studied the corresponding section(s), then take the Chapter Test.

1. 48 in. **2.** $\frac{7}{12}$ yd **3.** 700 m

4. 389 dm **5.** $12\frac{1}{3}$ ft **6.** 20.8

7. 39 yd 2 ft

8. 6 km 626 m $61\frac{4}{7}$ cm

9. 8.2 cm **10.** $8\frac{1}{4}$ in.

11. 8.04 m^2 (rounded)

12. 11.2 ft^2 **13.** $\frac{5}{12}$ ft^2

14. 89,000 m^2 **15.** 0.54 cm^3

16. 34.18 ℓ **17.** 5800 g

18. 5°C **19.** 185°F

20. 81.28 cm **21.** 27.28 mi

11–2 LINEAR MEASUREMENT—THE AMERICAN SYSTEM

The major system of measurement in the world today is the metric system (see section 11–3). The U.S. Congress has decreed that the changover to the metric system will be voluntary in the United States, and so far the movement has been slow.

The metric system has 10 as a base, and changing from one unit to another is a matter of moving the decimal point. You are probably familiar with the following American units of linear measurement. If not, you should memorize them.

Rule

$$1 \text{ foot (ft)} = 12 \text{ inches (in.)}$$

$$1 \text{ yard (yd)} = 3 \text{ feet (ft)}$$

$$1 \text{ mile (mi)} = 5280 \text{ ft}$$

Conversion

To convert from one American unit to another, we use the following conversion ratios or factors:

$$\frac{1 \text{ foot}}{12 \text{ in.}} \quad \frac{12 \text{ in.}}{1 \text{ foot}} \quad \frac{1 \text{ yd}}{3 \text{ ft}} \quad \frac{3\text{ft}}{1 \text{ yd}} \quad \frac{1 \text{ mi}}{5280 \text{ ft}} \quad \frac{5280 \text{ ft}}{1 \text{ mi}}$$

Since the numerator is equal in distance to the denominator, each ratio is equal to 1.

■ **Example 1**

Convert 1 yard to inches.

Solution

$$\frac{1 \text{ yard}}{1} \times \frac{3 \text{ feet}}{1 \text{ yard}} \times \frac{12 \text{ inches}}{1 \text{ foot}}$$

We choose our conversion factors so that the following is true:

1. The numerators are in descending unit order (yards, feet, inches).
2. All unit labels will cancel out except inches.

$$\frac{1 \cancel{\text{yd}}}{1} \times \frac{3 \cancel{\text{ft}}}{1 \cancel{\text{yd}}} \times \frac{12 \text{ in.}}{1 \cancel{\text{ft}}} = 1 \times 3 \times 12 \text{ inches}$$

$$= 36 \text{ inches}$$

(*Note:* The value of the first factor is not changed, since the other factors are equal to 1.)

■ **Example 2**

Convert 1 mile to yards.

Solution

$$\frac{1 \text{ mile}}{1} \times \frac{1 \text{ yd}}{3 \text{ ft}} \times \frac{5280 \text{ ft}}{1 \text{ mi}}$$

We choose our conversion factors so that the following is true:

1. The numerators are in descending unit order (miles, yards, feet).
2. All unit labels will cancel out except yards.

$$\frac{1 \not{\text{mi}}}{1} \times \frac{1 \text{ yd}}{3 \not{\text{ft}}} \times \frac{5280 \not{\text{ft}}}{1 \not{\text{mi}}} = \frac{5280}{3} \text{ yd} = 1760 \text{ yd}$$

■ Example 3

Convert 6 yards to feet.

Solution

$$\frac{6 \not{\text{yd}}}{1} \times \frac{3 \text{ ft}}{1 \not{\text{yd}}} = 6 \times 3 \text{ ft} = 18 \text{ ft}$$

Practice check: Do exercises 1–4 in the margin.

Convert.

1. 4 yards to feet **2.** 6 ft to in.

3. 7 yd to in. **4.** 5 mi to ft

The answers are in the margin on page 415.

■ Example 4

Convert 3 inches to feet.

Solution $\frac{3 \text{ in.}}{1} \times \frac{1 \text{ ft}}{12 \text{ in.}}$

We choose our conversion factor so that the following is true:

1. The numerators are in ascending unit order (inches, feet).
2. All unit labels will cancel out except feet.

$$\frac{3 \not{\text{in.}}}{1} \times \frac{1 \text{ ft}}{12 \not{\text{in.}}} = \frac{3}{12} \text{ ft} = \frac{1}{4} \text{ ft}$$

■ Example 5

Convert 14 in. to yards.

Solution

$$\frac{14 \text{ in.}}{1} \times \frac{1 \text{ ft}}{12 \text{ in.}} \times \frac{1 \text{ yd}}{3 \text{ ft}} = \frac{14}{12 \times 3} \text{ yd} = \frac{14}{36} \text{ yd} = \frac{7}{18} \text{ yd}$$

Practice check: Do exercises 5–8 in the margin.

Convert.

5. 30 in. to ft **6.** 18 ft to yd

7. 24 in. to yd **8.** 4360 ft to mi

The answers are in the margin on page 415.

■ Looking Back

You should now be able to change one American unit of linear measurement to another.

■ Problems 11–2

Convert.

1. 3 ft to in.
2. 2 yd to in.
3. 4 mi to ft
4. 72 in. to ft
5. 3 mi to yd
6. 20 yd to ft
7. 33 ft to yd
8. 14 yd to in.
9. 18 in. to ft
10. 4 in. to ft
11. 57 ft to yd
12. 1.5 mi to yd
13. 24.1 yd to ft
14. 3 mi to ft
15. 13 ft to yd
16. $5\frac{1}{2}$ ft to yd
17. 24 ft to mi
18. 10,560 ft to mi
19. 18 yd to in.
20. 15 mi to ft
21. 18 in. to yd
22. 5280 yd to mi
23. 45 in. to ft
24. 561 ft to yd
25. 778 yd to in.
26. 32 mi to yd
27. 1201 in. to yd
28. 1860 ft to mi
29. 15.84 ft to mi
30. 12.8 ft to yd

Solve. Write a statement for the answer.

31. **Sheet Metal.** The heating duct space between the second and fourth floor of the newly constructed symphony hall is 57 feet. Find the number of yards.
32. **Welding.** An angle iron has length 24 ft. How long is this piece of metal in inches?

33. Generic. The Jackson family decided to walk across the Golden Gate Bridge, which spans 4200 feet. How many miles did they walk?

34. Electrician. Joe Jacobs won an electrical wiring contract that takes 75 ft of wiring. The inventory of his electrical wiring is only $8\frac{1}{4}$ yd. How many more yards does Joe need to complete the job?

35. Generic. Bill Jogger ran $7\frac{1}{4}$ miles.

a. Change this to feet.

b. If his step span averaged $2\frac{1}{2}$ ft, how many steps did Bill take in $7\frac{1}{4}$ miles?

The answers to odd-numbered exercises are in the back of the book.

Answers to exercises 1–4

1. 12 ft **2.** 72 in. **3.** 252 in. **4.** 26,400 ft

Answers to exercises 5–8

5. $\frac{5}{2}$ft, or $2\frac{1}{2}$ft **6.** 6 yd **7.** $\frac{2}{3}$ yd **8.** $\frac{109}{132}$ mi

11–3 LINEAR MEASUREMENT—THE METRIC SYSTEM

The basic metric linear measurement unit is the meter (m). The meter is about 39.37 inches in length, or a little longer than a yard. The meter has 100 centimeters (cm). Since a yard has 36 inches, we can see that a centimeter is smaller than an inch.

1 cm 1 in. (2.54 cm)

Practice check: Do exercises 9–11 in the margin. Use a metric ruler. Have someone help you.

9. Measure yourself.

a. Height = ______cm

b. Waist = ______cm

c. Head = ______m

d. Wrist = ______cm

10. Measure this book.

a. Width = ______cm

b. Length = ______cm

c. Thickness = ______cm

11. Measure your chair.

a. Width of seat = ______m

b. Length of seat = ______m

c. Height from floor to seat = ______m

The answers are in the margin on page 417.

The American system uses the mile to measure long distances. The metric system uses the kilometer (km). A kilometer is a little over $\frac{1}{2}$ mile in length—in other words, a mile contains not quite 2 kilometers. A kilometer is 1000 meters in length. All units of length are based on the meter and are either 10 times a meter, 100 times a meter, or $\frac{1}{10}$ of a meter, $\frac{1}{100}$ of a meter, $\frac{1}{1000}$ of a meter, etc.

Rule

Metric units of length:

$$1 \text{ kilometer } (km) = 1000 \text{ meters } (m)$$

$$1 \text{ hectometer } (hm) = 100 \text{ meters } (m)$$

$$1 \text{ dekameter } (dkm) = 10 \text{ meters } (m)$$

$$1 \text{ meter } (m) = 1 \text{ meter } (m)$$

$$1 \text{ decimeter } (dm) = \frac{1}{10} \text{ meter } (m)$$

$$1 \text{ centimeter } (cm) = \frac{1}{100} \text{ meter } (m)$$

$$1 \text{ millimeter } (mm) = \frac{1}{1000} \text{ meter } (m)$$

You should memorize the information in the chart on the previous page. It is helpful to relate the prefixes to the following words:

Milli relates to *millenium* or 1000 years. *Centi* relates to *century* or 100 years. *Deci* relates to *decimal* or tenths. *Deka* relates to *decade* or 10 years. *Hecto* relates to *hundred. Kilo* relates to *kilowatt* or 1000 watts.

Conversion

To convert from one metric unit of length to another we use the following conversion ratios or factors:

$$\frac{1\text{ km}}{1000\text{ m}} \quad \frac{1000\text{ m}}{1\text{ km}} \quad \frac{1\text{ hm}}{100\text{ m}} \quad \frac{100\text{ m}}{1\text{ hm}} \quad \frac{1\text{ dkm}}{10\text{ m}} \quad \frac{10\text{ m}}{1\text{ dkm}}$$

$$\frac{1\text{ m}}{10\text{ dm}} \quad \frac{10\text{ dm}}{1\text{ m}} \quad \frac{1\text{ m}}{100\text{ cm}} \quad \frac{100\text{ cm}}{1\text{ m}} \quad \frac{1\text{ m}}{1000\text{ mm}} \quad \frac{1000\text{ mm}}{1\text{ m}}$$

Since the numerator is equal in distance to the denominator, each ratio is equal to 1.

■ Example 1

Convert 2 km to m.

Solution

$$\frac{2\text{ km}}{1} \times \frac{1000\text{ m}}{1\text{ km}} \qquad \text{Multiplying by 1, or } \frac{1000\text{ m}}{1\text{ km}}$$

$$= \frac{2\cancel{\text{ km}}}{1} \times \frac{1000\text{ m}}{1\cancel{\text{ km}}} \qquad \text{Canceling}$$

$$= 2 \times 1000\text{ m} = 2000\text{ m}$$

Note: Canceling names for measurement is similar to canceling like factors in multiplication and division of fractions.

■ Example 2

Convert 4.2 dm to m.

Solution

$$\frac{4.2\text{ dm}}{1} \times \frac{1\text{ m}}{10\text{ dm}} \qquad \text{Multiplying by 1}$$

$$= \frac{4.2\cancel{\text{ dm}}}{1} \times \frac{1\text{ m}}{10\cancel{\text{ dm}}} \qquad \text{Canceling}$$

$$= \frac{4.2}{10}\text{m} = 0.42\text{ m}$$

Practice check: Do exercises 12–19 in the margin.

Convert.

12. 7.8 km to m

13. 3142 m to km

14. 18 cm to m

15. 529 m to cm

16. 0.6 hm to m

17. 13.608 m to mm

18. 157 m to dkm

19. 0.015 dm to m

The answers are in the margin on page 419.

■ Example 3

Convert 17 km to dkm.

Solution

$$\frac{17\text{ km}}{1} \times \frac{1000\text{ m}}{1\text{ km}} \times \frac{1\text{ dkm}}{10\text{ m}}$$

(*Note:* We choose the 1 factors so that all units will cancel out except dkm.)

$$= \frac{17\text{ }\cancel{\text{km}}}{1} \times \frac{\overset{100}{\cancel{1000}}\text{ }\cancel{\text{m}}}{1\text{ }\cancel{\text{km}}} \times \frac{1\text{ dkm}}{\cancel{10}\text{ }\cancel{\text{m}}} = 1700\text{ dkm}$$

■ Example 4

Convert 2.614 cm to hm.

Solution

$$\frac{2.614\text{ cm}}{1} \times \frac{1\text{ m}}{100\text{ cm}} \times \frac{1\text{ hm}}{100\text{ m}}$$

$$= \frac{2.614\text{ }\cancel{\text{cm}}}{1} \times \frac{1\text{ }\cancel{\text{m}}}{100\text{ }\cancel{\text{cm}}} \times \frac{1\text{ hm}}{100\text{ }\cancel{\text{m}}} = \frac{2.614}{10{,}000}\text{hm} = 0.0002614\text{ hm}$$

Practice check: Do exercises 20–25 in the margin.

Quick Conversions

We can make conversions quickly in the metric system simply by moving the decimal point. Notice below that each unit is ten times the unit to the right and $\frac{1}{10}$ the unit to the left.

Rule

km	hm	dkm	m	dm	cm	mm
1000	100	10	1	$\frac{1}{10}$	$\frac{1}{100}$	$\frac{1}{1000}$

One move to the right in the table is the same as moving the decimal point to the right 1 place. Three moves to the left in the table is the same as moving the decimal point 3 places to the left.

■ Example 5

Convert 21.72 cm to m.

Solution This is a movement of 2 places to the left in the table. We move the decimal 2 places to the left in 21.72. Thus, 21.72 cm = 0.2172 m.

■ Example 6

Convert 3.1007 hm to dm.

Solution This is a movement of 3 places to the right in the table. We move the decimal 3 places to the right in 3.1007. Thus, 3.1007 hm = 3100.7 dm

Practice check: Do exercises 26–33.

Answers to exercises 9–11

9. **a.** between 150 and 200 cm
   b. between 65 and 85 cm
   c. between 50 and 65 cm
   d. between 15 and 20 cm
10. **a.** 20.5 cm **b.** 28 cm
    c. 4 cm
11. **a.** between 30 and 35 cm
    b. between 30 and 35 cm
    c. between 40 and 45 cm

Convert.

20. 5.67 hm to dm

21. 42 km to cm

22. 6432 mm to dkm

23. 0.71 dkm to cm

24. 178.2 dm to km

25. 32.06 cm to hm

The answers are in the margin on page 419.

Make conversions quickly.

26. 1 km to m

27. 1 mm to cm

28. 1 hm to m

29. 856 m to km

30. 3.14 km to dkm

31. 4101.2 mm to dm

32. 15 cm to dkm

33. 0.07 dkm to mm

The answers are in the margin on page 419.

■ Looking Back

You should now be able to convert from one metric linear measurement unit to another.

■ Problems 11–3

Make conversions quickly, as often as possible.

1. 5 hm to m

2. 0.28 dkm to m

3. 7056 m to km

4. 559 hm to m

5. 325 dm to m

6. 1.07 km to m

7. 1 km to m

8. 1m to km

9. 5.8 km to m

10. 3581 m to dkm

11. 388 cm to m

12. 0.488 dkm to m

13. 4.67 mm to m

14. 33.61 dm to m

15. 1 dkm to m

16. 1 m to dkm

17. 7.8 cm to mm

18. 15km to hm

19. 1 hm to m

20. 1 m to hm

21. 23.9 km to m

22. 3.4 dkm to dm

23. 1 mm to cm

24. 1 hm to cm

25. 463 dkm to km

26. 5600 mm to dkm

27. 1 dm to mm

28. 1 km to dkm

29. 4.66 m to dkm

30. 78.4 m to dm

Solve. Use approximate values.

31. A car travels 85 km per hour. How many centimeters does it travel per hour?

32. The diameter (distance across) of a U.S. quarter is about

a. 2.1 mm **b.** 0.21 m **c.** 2.1 cm

33. The distance of a kilometer is about

a. a city block **b.** a half mile **c.** the distance from Boston to New York

d. the width of this book.

34. The width of a door is about

a. 1.5 m **b.** 150 dm **c.** 0.15 hm **d.** 0.015 km

35. The distance from Los Angeles to San Francisco is about 400 miles. This is about

a. 210 km **b.** 6400 m **c.** 640 km **d.** 2100 dkm

The answers to odd-numbered exercises are in the back of the book.

Answers to exercises 12–19

12. 7800 m **13.** 3.142 km
14. 0.18 m **15.** 52,900 cm
16. 60 m **17.** 13,608 mm
18. 15.7 dkm **19.** 0.0015 m

Answers to exercises 20–25

20. 5670 dm **21.** 4,200,000 cm
22. 0.6432 dkm **23.** 710 cm
24. 0.01782 km **25.** 0.003206 hm

Answers to exercises 26–33

26. 1000 m **27.** 0.1 cm
28. 100 m **29.** 0.856 km
30. 314 dkm **31.** 41.012 dm
32. 0.015 dkm **33.** 700 mm

11–4 DENOMINATE NUMBERS

Measurements such as those in sections 11–2 and 11–3 are called *denominate numbers.* A *denominate number* has two parts: a number and a unit of measurement.

Unit of measurement
↘
7 feet
↖
Number

Addition or subtraction of two or more denominate numbers must be done with like units of measurement.

Addition of Denominate Numbers

■ **Example 1**

Add 3 ft 4 in. to 7 ft 9 in.

Solution

$$\begin{array}{r r} 3\text{ ft} & 4\text{ in.} \\ +7\text{ ft} & 9\text{ in.} \\ \hline 10\text{ ft} & 13\text{ in.} \end{array}$$

This must be less than a foot, so we change 13 in. to 12 in. + 1 in., or 1 ft + 1 in.

$$10\text{ ft } 13\text{ in.} = 10\text{ ft} + \underbrace{1\text{ ft} + 1\text{ in.}}_{13\text{ in.}}$$

$$= 11\text{ ft } 1\text{ in.} \quad \text{or} \quad 11\frac{1}{12}\text{ ft}$$

■ **Example 2**

Add 5 km 360 m to 7 km 895 m.

Solution

$$\begin{array}{r r} 5\text{ km} & 360\text{ m} \\ +7\text{ km} & 895\text{ m} \\ \hline 12\text{ km} & 1255\text{ m} \end{array}$$

This must be less than a kilometer, so we change 1255 m to 1000 m + 255 m, or 1 km + 255 m.

$$12\text{ km } 1255\text{ m} = 12\text{ km} + \underbrace{1\text{ km} + 255\text{ m}}_{1255\text{ m}}$$

$$= 13\text{ km } 255\text{ m}, \quad \text{or} \quad 13.255\text{ km}$$

Practice check: Do exercises 34–37 in the margin.

Add.

34. 7 ft 9 in.
16 ft 5 in.
= ______ft

35. 13 yd 2 ft 8 in.
22 yd 1 ft 8 in.
= ______yd

36. 7 km 476 m
18 km 743 m
= ______km

37. 43 m 76 cm
108 m 53 cm
= ______m

The answers are in the margin on page 423.

Subtraction of Denominate Numbers

■ **Example 3**

Subtract 3 ft 9 in. from 8 ft 2 in.

Solution

$$\begin{array}{r r} 8\text{ ft} & 2\text{ in.} \\ -3\text{ ft} & 9\text{ in.} \\ \hline \end{array}$$

We can't subtract 9 in. from 2 in., so

1. "Borrow" 1 ft from the 8 ft.
2. Change the 1 ft to 12 in.
3. Add the 12 in. to the 2 in.

$$\begin{array}{r} 7\ \ 14 \\ \not{8}\text{ ft } \not{2}\text{ in.} \\ -3\text{ ft } 9\text{ in.} \\ \hline 4\text{ ft } 5\text{ in.,} \end{array} \quad \text{or} \quad 4\frac{5}{12}\text{ ft}$$

■ **Example 4**

Subtract 9 cm 9 mm from 15 cm 7 mm.

Solution

$$\begin{array}{r} 15 \text{ cm } \ 7 \text{ mm} \\ - \ 9 \text{ cm } \ 9 \text{ mm} \\ \hline \end{array}$$

We can't subtract 9 mm from 7 mm, so
1. "Borrow" 1 cm from the 15 cm.
2. Change the 1 cm to 10 mm.
3. Add the 10 mm to the 7 mm.

$$\begin{array}{rr} 14 & 17 \\ \not{15} \text{ cm} & \not{7} \text{ mm} \\ -9 \text{ cm} & 9 \text{ mm} \\ \hline 5 \text{ cm} & 8 \text{ mm}, \end{array} \quad \text{or} \quad 5.8 \text{ cm}$$

Practice check: Do exercises 38–41 in the margin.

Multiplication of Denominate Numbers

■ Example 5

Multiply $5 \times (6 \text{ ft } 4 \text{ in.})$.

Solution

$$\begin{array}{rr} 6 \text{ ft} & 4 \text{ in.} \\ & \times 5 \\ \hline 30 \text{ ft} & 20 \text{ in.} \end{array} \quad \text{Multiplying each denominate number by 5}$$

$$= 31 \text{ ft } 8 \text{ in.}, \quad \text{or} \quad 31\frac{2}{3} \text{ ft}$$

■ Example 6

Multiply $7 \times (5 \text{ km } 360 \text{ m})$.

Solution

$$\begin{array}{rr} 5 \text{ km} & 360 \text{ m} \\ & \times 7 \\ \hline 35 \text{ km} & 2520 \text{ m} \end{array}$$

$$= 37 \text{ km } 520 \text{ m}, \text{ or } 37.52 \text{ km}$$

You can also multiply a denominate number by another denominate number, but more on this later.

Practice check: Do exercises 42–45 in the margin.

Division of Denominate Numbers

The process of dividing a denominate number by a whole number is similar to that of dividing whole numbers.

■ Example 7

Divide (9 yd 2 ft 7 in.) by 4.

Subtract.

38. $\begin{array}{r} 7 \text{ ft } 9 \text{ in.} \\ -3 \text{ ft } 5 \text{ in.} \\ \hline \end{array}$ = ______ft

39. $\begin{array}{r} 5 \text{ yd } 1 \text{ ft} \\ -2 \text{ yd } 2 \text{ ft} \\ \hline \end{array}$ = ______yd

40. $\begin{array}{r} 8 \text{ km } 382 \text{ m} \\ -5 \text{ km } 147 \text{ m} \\ \hline \end{array}$ = ______km

41. $\begin{array}{r} 81 \text{ cm } 4 \text{ mm} \\ -32 \text{ cm } 8 \text{ mm} \\ \hline \end{array}$ = ______cm

The answers are in the margin on page 423.

Multiply.

42. $6 \times (3 \text{ ft } 7 \text{ in.})$

43. $15 \times (6 \text{ yd } 2 \text{ ft})$

44. $8 \times (15 \text{ km } 55 \text{ m})$

45. $22 \times (13 \text{ m } 8 \text{ mm})$

The answers are in the margin on page 423.

Solution **Step 1** Divide yards.

$$\begin{array}{rlll}
 & \underline{\text{2 yd}\qquad\qquad\qquad\qquad} & & \\
4) & \text{9 yd} & \text{2 ft} & \text{7 in.} \\
 & \underline{-\text{8 yd}} & & \\
 & \text{1 yd} = & \underline{\text{3 ft}} & \\
 & & \text{5 ft} & \text{Adding 2 ft and 3 ft}
\end{array}$$

Step 2 Divide feet.

$$\begin{array}{rllll}
 & \text{2 yd} & \text{1 ft} & & \\
4) & \text{9 yd} & \text{2 ft} & \text{7 in.} & \\
 & \underline{\text{8 yd}} & & & \\
 & \text{1 yd} = & \underline{\text{3 ft}} & & \\
 & \hookrightarrow & \text{5 ft} & & \\
 & & \underline{-\text{4 ft}} & & \\
 & & \text{1 ft} = & \underline{\text{12 in.}} & \\
 & & & \text{19 in.} & \text{Adding 7 in. and 12 in.}
\end{array}$$

Step 3 Divide inches.

$$\begin{array}{rllll}
 & \text{2 yd} & \text{1 ft} & \text{4 in.} & \\
4) & \text{9 yd} & \text{2 ft} & \text{7 in.} & \\
 & \underline{\text{8 yd}} & & & \\
 & \text{1 yd} = & \underline{\text{3 ft}} & & \\
 & & \text{5 ft} & & \\
 & & \underline{-\text{4 ft}} & & \\
 & & \text{1 ft} = & \underline{\text{12 in.}} & \\
 & & \hookrightarrow & \text{19 in.} & \\
 & & & \underline{-\text{16 in.}} & \\
 & & & \text{3 in.} & \\
 & & & \uparrow & \text{When divided by 4 we get } \tfrac{3}{4} \text{ in.}
\end{array}$$

$$(\text{9 yd 2 ft 7 in.}) \div 4 = \text{2 yd 1 ft } 4\frac{3}{4} \text{ in.}$$

■ Example 8

Divide (15 km 330 m 29 cm) by 2.

Solution **Step 1** Divide kilometers.

$$\begin{array}{rlll}
 & \text{7 km} & & \\
2) & \text{15 km} & \text{330 m} & \text{29 cm} \\
 & \underline{-\text{14 km}} & & \\
 & \text{1 km} = & \underline{\text{1000 m}} & \\
 & & \text{1330 m} &
\end{array}$$

Step 2 Divide meters.

```
         7 km     665 m
      2)15 km     330 m      29 cm
       | 14 km
       |  1 km = 1000 m
       └──────→ 1330 m
               −1330 m
                   0 m  =    0 cm
                            29 cm
```

Step 3 Divide the centimeters.

```
   7 km     665 m     14 cm
2)15 km     330 m     29 cm
 | 14 km
 |  1 km = 1000 m
 |         1330 m
 |         1330 m
 |            0 m  =   0 cm
 └─────────────────→  29 cm
                      28 cm
                       1 cm  ←── When divided by 2 we get ½ cm.
```

$$(15 \text{ km } 330 \text{ m } 29 \text{ cm}) \div 2 = 7 \text{ km } 665 \text{ m } 14\frac{1}{2} \text{ cm}$$

Practice check: Do exercises 46–49 in the margin.

Answers to exercises 34–37

34. 24 ft 2 in., or $24\frac{1}{6}$ ft
35. 36 yd 1 ft 4 in.
36. 26 km 219 m, or 26.219 km
37. 152 m 29 cm, or 152.29 m

Answers to exercises 38–41

38. 4 ft 4 in., or $4\frac{1}{3}$ ft
39. 2 yd 2 ft, or $2\frac{2}{3}$ yd
40. 3 km 235 m, or 3.235 km
41. 48 cm 6 mm, or 48.6 cm

Answers to exercises 42–45

42. 21 ft 6 in., or $21\frac{1}{2}$ ft
43. 90 yd 30 ft = 100 yd
44. 120 km 440 m, or 120.44 km
45. 286 m 176 mm, or 286.176 m

Divide.

46. (6 yd 1 ft 7 in.) ÷ 3
47. (17 yd 2 ft 9 in.) ÷ 5
48. (31 km 898 m 72 cm) ÷ 6
49. (32 m 53 cm 8 mm) ÷ 7

The answers are in the margin on page 425.

■ Looking Back

You should now be able to add, subtract, multiply, and divide with denominate numbers.

■ Problems 11–4

Review

Convert.

1. 62 ft to yd
2. 39 yd to in.
3. 3 km to m
4. 8.24 km to dkm
5. 5 in. to ft
6. 789 ft to yd
7. 5204.3 mm to dm
8. 18 cm to dkm

Add.

9. 3 ft 7 in.
5 ft 6 in.

10. 8 ft 9 in.
10 ft 5 in.

11. 7 km 356 m
15 km 805 m
= ______km

12. 28 km 760 m
108 km 455 m
= ______km

13. 6 yd 2 ft
8 yd 1 ft
15 yd 2 ft
= ______yd

14. 22 ft 9 in.
6 ft 5 in.
14 ft 7 in.
= ______ft

15. 17 cm 5 mm
82 cm 9 mm
15 cm 1 mm
= ______cm

16. 31 m 52 cm
86 m 31 cm
5m 76 cm
= ______m

Subtract.

17. 17 ft 9 in.
9 ft 4 in.
= ______ft

18. 29 ft 11 in.
18 ft 5 in.
= ______ft

19. 15 km 132 m
7 km 58 m
= ______km

20. 208 m 31 cm
184 m 29 cm
= ______m

21. 24 yd 1 ft
13 yd 2 ft
= ______yd

22. 31 ft 3 in.
17 ft 7 in.
= ______ft

23. 17 mi 308 ft
12 mi 3640 ft
= ______mi

24. 8 mi 2480 ft
2 mi 4575 ft
= ______mi

Answers to exercises 46–49

46. 2 yd $6\frac{1}{3}$ in.
47. 3 yd 1 ft 9 in.
48. 5 km 316 m $45\frac{1}{3}$ cm
49. 4 m 64 cm $8\frac{2}{7}$ mm

25. 69 cm 6 mm
31 cm 7 mm
= ______cm

26. 66 m 14 cm
18 m 56 cm
= ______m

Multiply.

27. 5 × (5 yd 2 ft)

28. 3 × (22 ft 6 in.)

29. 2 × (9 m 58 cm)

30. 7 × (15 km 256 m)

31. 8 × (6 yd 2 ft 5 in.)

32. 6 × (15 yd 1 ft 8 in.)

33. 9 × (22 km 171 m 31 cm)

34. 5 × (33 km 506 m 79 cm)

Divide.

35. (4 ft 10 in.) ÷ 2

36. (10 yd 2 ft) ÷ 2

37. (33 km 342 m) ÷ 3

38. (56 m 49 cm) ÷ 7

39. (32 ft 8 in.) ÷ 6

40. (15 yd 1 ft) ÷ 9

41. (58 cm 7 mm) ÷ 5

42. (17 m 79 cm) ÷ 11

43. (6 yd 2 ft 4 in.) ÷ 3

44. (9 yd 1 ft 8 in.) ÷ 4

45. (48 km 761 m 32 cm) ÷ 7

46. (81 m 98 cm 9 mm) ÷ 5

The answers to review and odd-numbered exercises are in the back of the book.

11–5 PERIMETER AND CIRCUMFERENCE

Perimeter

Bill McDonald's garden was shaped as shown. To keep out dogs and cats, he decided to put a fence around it. To figure how much fencing to buy, Bill found the distance around the garden to be 67 ft.

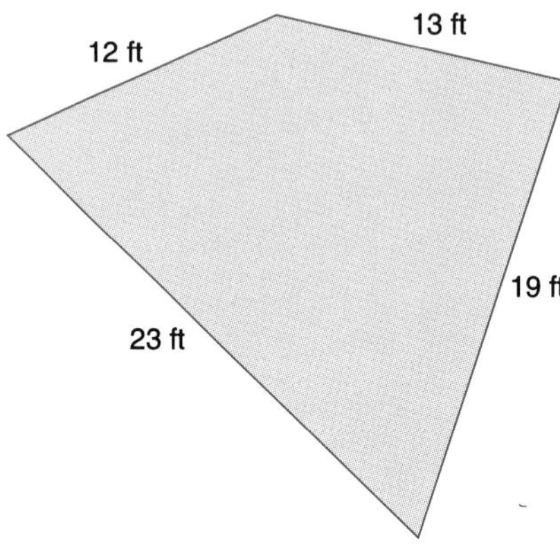

Figure 11–3

The distance around a closed geometric figure such as the one here is called the *perimeter*.

Rectangles and squares are two examples of closed geometric figures. A square, however, is a special type of rectangle. The different parts of each figure are named below. (*Note:* A square is a rectangle with the length of all sides equal.)

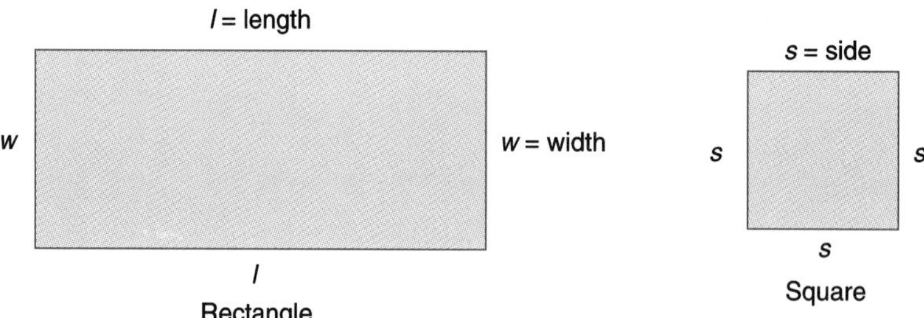

Figure 11–4

The formulas for the perimeters of the rectangles above are as follows:

Rule

Perimeter Formulas

Let P *be the perimeter.*

Square: *If* s *is the length of one side, then* $P = 4 \times s$, or $P = 4s$.

Rectangle: *If* l *is the length and* w *is the width, then* $P = 2 \times (l + w)$, *or* $P = 2 \cdot (l + w)$.

■ **Example 1**

Find the perimeter of each of the following:

a.

15 in.

9 in.

Rectangle

Figure 11–5

b.

7 ft

Square

Figure 11–6

Solution **a.** Here, $l = 15$ in and $w = 9$ in.

$$P = 2(l + w)$$

$= 2(15 \text{ in.} + 9 \text{ in.})$ Substituting for l and w

$= 2\,(24 \text{ in.})$ Adding

$= 48$ in. Multiplying

$$P = 4 \text{ ft}$$

b. Here, $s = 7$ ft.

$$P = 4 \times s$$

$= 4 \times 7$ ft Substituting for s

$= 28$ ft Multiplying

$$P = 9 \text{ yd } 1 \text{ ft, or } 9\frac{1}{3} \text{ yd}$$

Practice check: Do exercise 50 in the margin.

Circumference

The *circumference* is the measured distance around a circle. The different parts of a circle are named below.

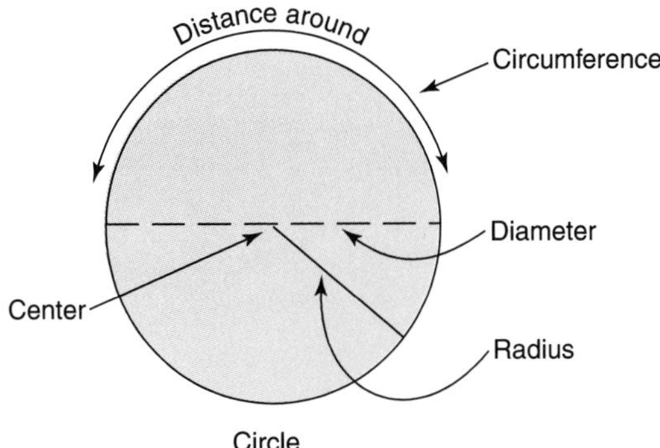

Figure 11–7

50. Find the perimeter of each of the following:

a.

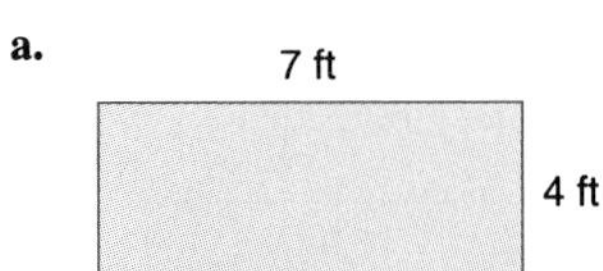

b.

c.

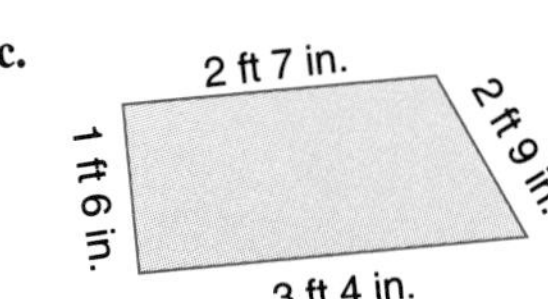

d.

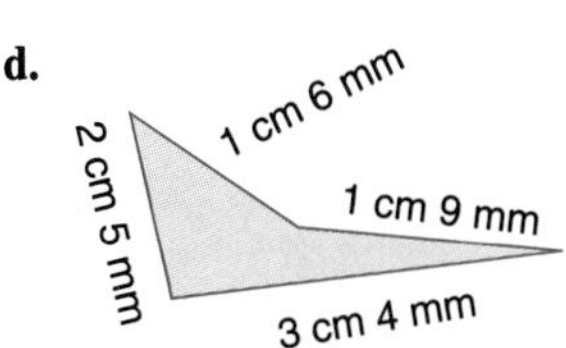

The answers are in the margin on page 429.

(*Note:* The radius is half the diameter.)

Rule

Circumference Formula

Let C be the circumference.
Circle: If r is the radius and d is the diameter, then $C = \pi \cdot d$, or $C = 2 \cdot \pi \cdot r$.

There is no exact decimal or fraction for π (pronounced pie), so we use approximate values of 3.14 or $\frac{22}{7}$.

■ **Example 2**
Find the circumference.

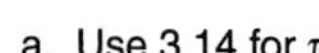

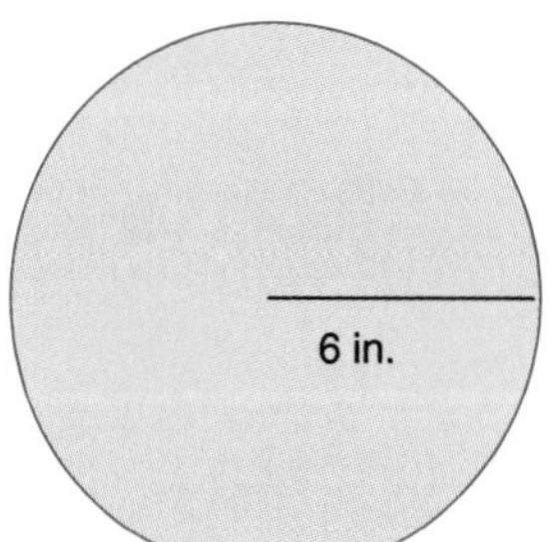

Figure 11–8

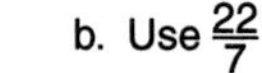

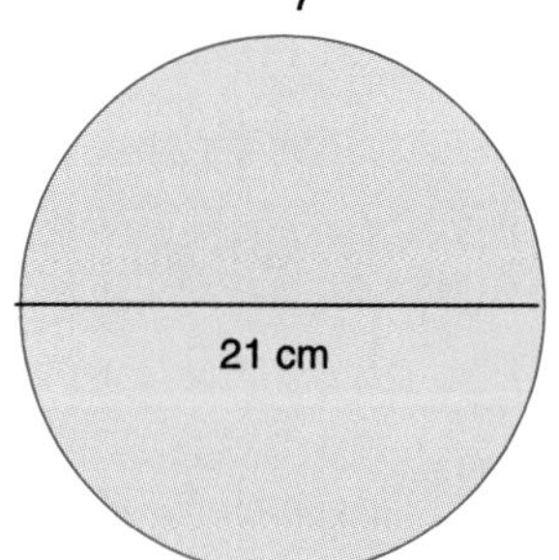

Figure 11–9

Solution **a.** Here, $r = 6$ in.

$$C = 2 \cdot \pi \cdot r$$
$$\approx 2 \times 3.14 \times 6 \text{ in.}$$
$$C \approx 37.68 \text{ in.}$$

b. Here, $d = 21$ cm.

$$C = \pi \cdot d$$
$$\approx \frac{22}{7} \times 21 \text{ cm}$$
$$\approx 22 \times 3 \text{ cm}$$
$$C = 66 \text{ cm}$$

(*Note:* $\approx$ means *approximately equal to.*)

Practice check: Do exercises 51–53 in the margin.

Find the circumference.

51.

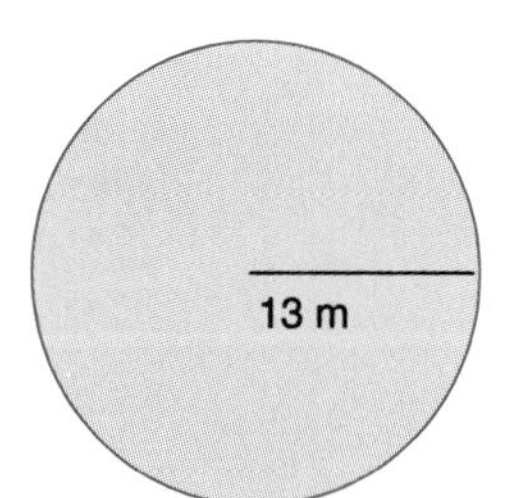

Use 3.14 for π.

52.

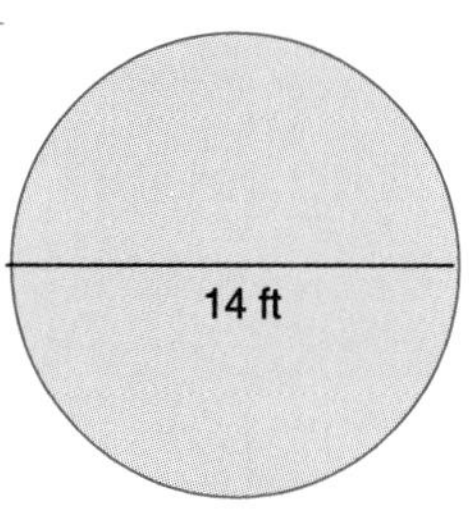

Use $\frac{22}{7}$ for π.

53.

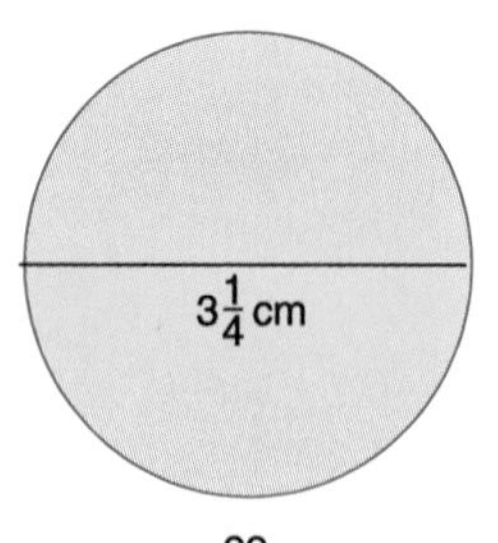

Use $\frac{22}{7}$ for π.

The answers are in the margin on page 431.

■ Looking Back

You should now be able to find the perimeter of a closed geometric figure and the circumference of a circle.

■ Problems 11–5

Answer to exercise 50

a. 22 ft, or $7\frac{1}{3}$ yd
b. 96 cm
c. 10 ft 2 in.
d. 9 cm 4 mm, or 9.4 cm

Review

Divide.

1. (40 km 655 m) ÷ 5

2. (15 ft 9 in.) ÷ 3

3. (56 yd 2 ft 3 in.) ÷ 4

4. (70 km 532 m 46 cm) ÷ 7

Find the perimeter of each figure.

5.

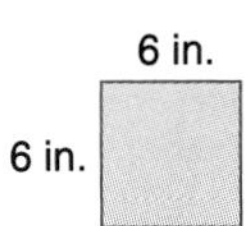

6.

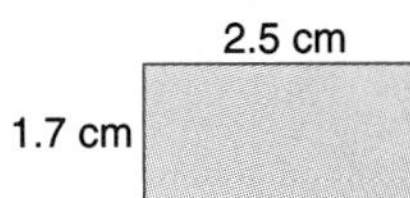

7.

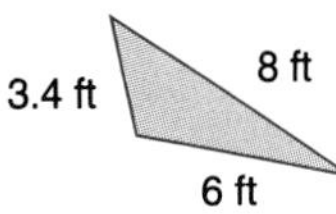

8.

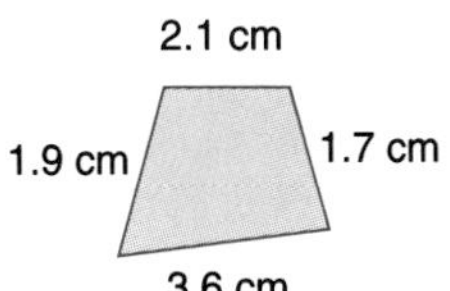

9.

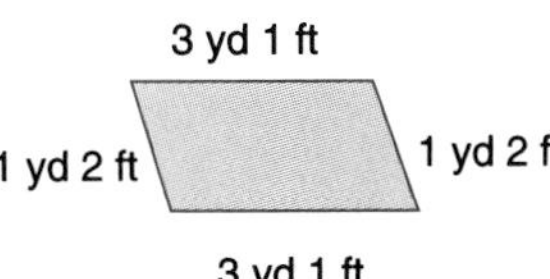

10.

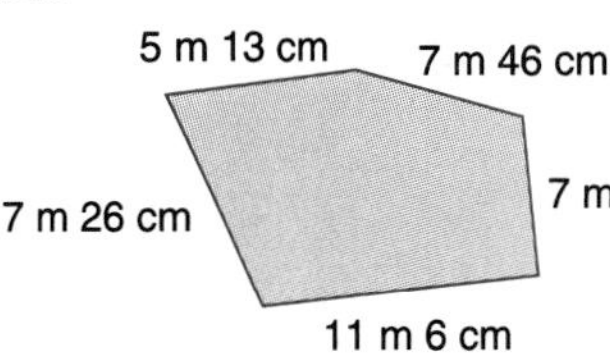

Find the perimeter of each square.

11. 6.7 dkm on each side

12. 33.4 ft on each side

Find the perimeter of each rectangle.

13. 10.5 in. by 8.4 in.

14. 51.2 cm by 4.2 m *Hint:* Change 51.2 cm to m.

Find the circumference of each circle.

15.

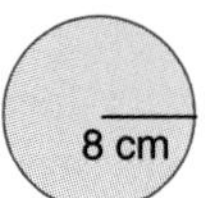

16.

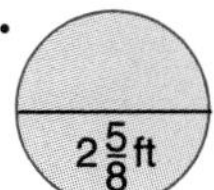

17.

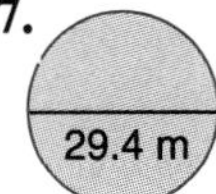

18.

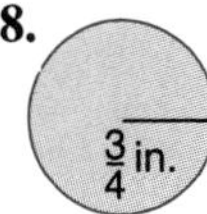

Find the perimeter of the following figures.

19.

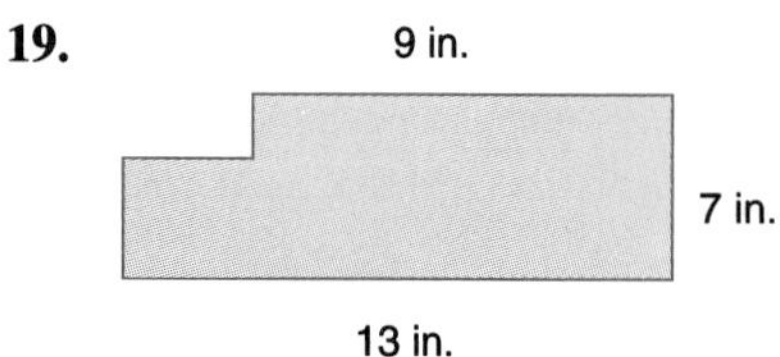

20.

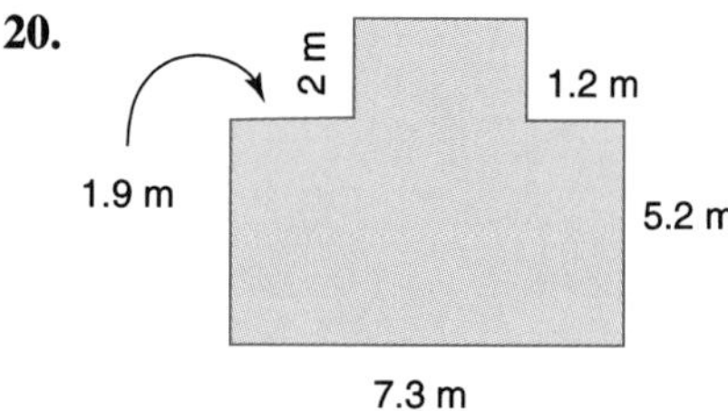

21.

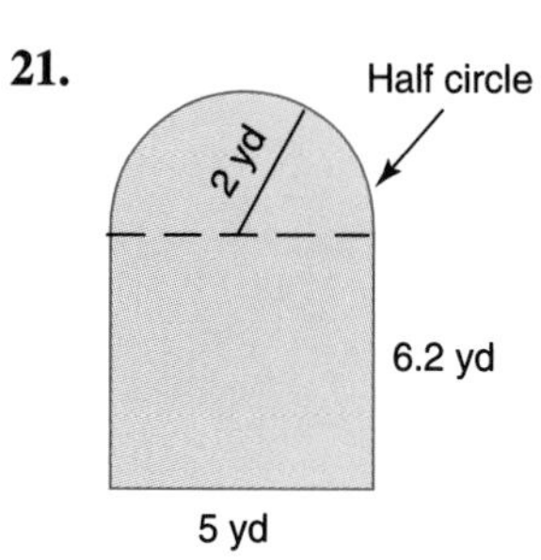

22.

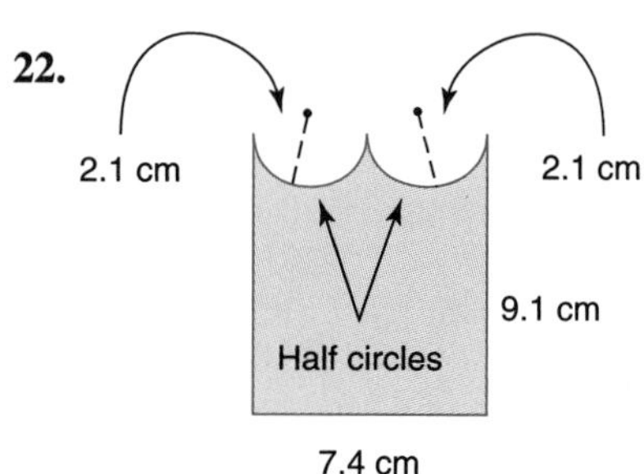

Solve. Write a statement for the answer.

23. Fire Fighter. The men in Engine 3 received permission to expand their equipment storage to an area of 4.5 meters long and 3.7 meters wide. Find the perimeter.

24. Carpentry. Jim Strom needs to replace the baseboard in his bathroom. If the bathroom measures 10 ft 4 in. by 5 ft 3 in., how many feet of baseboard should Jim buy now?

25. Carpentry. How many feet of molding would be needed to edge a circular table with a diameter of $4\frac{1}{2}$ ft?

26. Landscaping. G. Thumb needs to enclose a triangular garden with a fence whose measurements are pictured below.

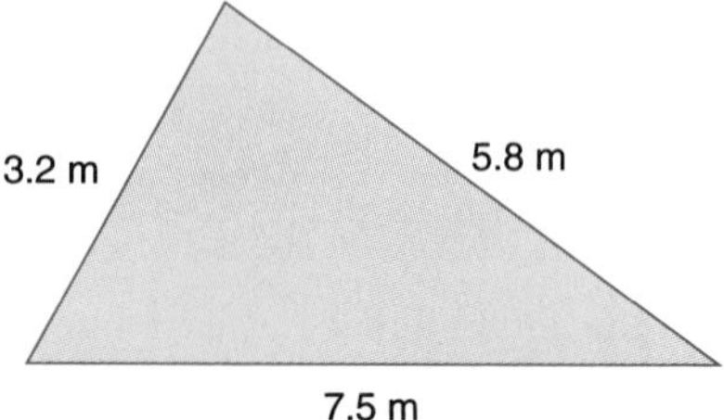

How many meters of fencing does he need? If fencing cost $1.50 per meter, how much would the fencing cost?

27. Generic. Judy's project is to frame a picture for her home. How many meters of framing will she need to buy if the picture is to measure 45 cm by 72 cm and 4 cm are allowed for each comer?

28. Plumbing. McNair Plumbing has to install PCV drainage pipe completely around a building 74 ft long and 26 ft wide.

a. How many feet of pipe are needed?

b. If this grade of plastic pipe sells for $2.24 per linear foot, how much did this pipe cost?

Answers to exercises 51–53

51. $C \approx 81.64$ m
52. $C \approx 44$ ft
53. $C \approx \frac{143}{14}$ cm, or $10\frac{3}{14}$ cm

29. Landscaping. H. P. Gardener can edge a lawn in 25 minutes. If this rectangular-shaped lawn measures 22 m by 18 m, how many centimeters per minute is this?

30. Generic. In a baseball game, Sam hit a home run with the bases loaded. If there are 90 ft between bases, how many total feet did the baserunners travel?

31. Machinist. A pulley with 5 spokes is 26 in. in diameter. Using the rim end point of one of the spokes as a constant, how many revolutions did the pulley turn if the total traveled by the end point of the spoke was one mile?

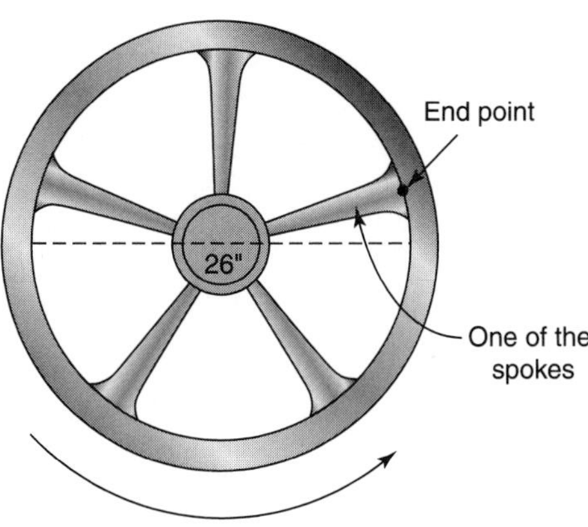

32. Machinist. A certain machine cuts at the rate of 14 inches per minute. Jack's job is to cut out circular disks that have a radius of 5.6 inches. How many pieces would he cut in a 7-hour work shift?

The answers to review and odd-numbered exercises are in the back of the book.

11–6 AREAS OF CLOSED FIGURES

Dana Simmon's Saturday project was to tile a closet floor. The tiles were 1 ft square, and the dimensions of the closet were 5 ft by 3 ft. How many square tiles were needed to cover all of the floor?

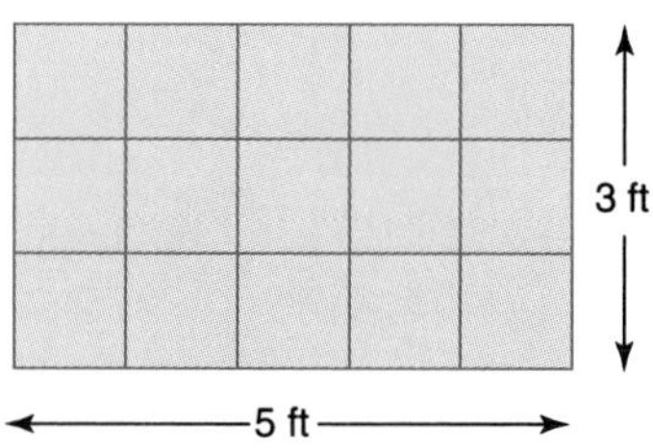

Figure 11–10

It took 15 tiles to cover the floor.

Rule

The measurement of the space covered in closed geometric figure is called area.

Two units of area measurement are the *square inch* and the *square centimeter*, shown below.

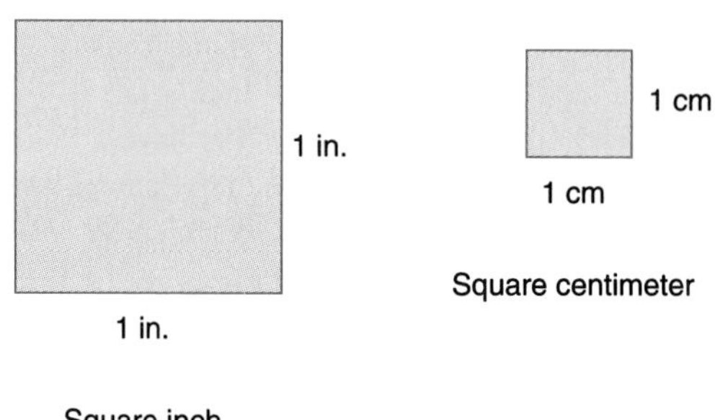

Figure 11–11

Some other units are *square feet, square miles, square meters.*

Rectangles

■ **Example 1**

Find the area of the rectangle.

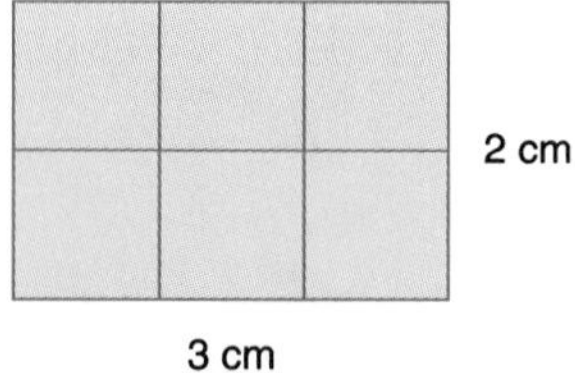

Figure 11–12

Solution The rectangular region is filled with 6 square centimeters, so the area is 6 square centimeters (sq cm).

Practice check: Do exercises 54 and 55 in the margin.

We can find the area of any rectangular region by multiplying the length by the width.

Rule

Rectangular Area Formula

Let A *be the area. If* l *is the length and* w *is the width, then* A = l × w, *or* A = l · w.

■ Example 2

Find the area of a rectangle 7 yd by 6 yd.

Solution

$$A = l \times w$$

$$A = 7 \text{ yd} \times 6 \text{ yd}$$

$$A = 42 \text{ yd}^2$$

(*Note:* We read 42 yd^2 as 42 square yards.)
The area of a square can be found by multiplying the length of the side by itself.

Practice check: Do exercises 56–60 in the margin.

Triangles

A triangle is a three-sided geometric figure with parts as shown below.

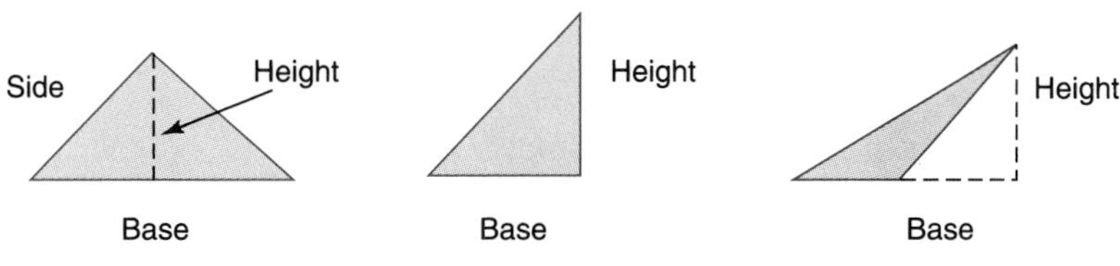

Figure 11–13

Consider half the area of the rectangle below. The area of the triangle formed is $\frac{1}{2} \times l \times w$ —the area of the triangle is half the area of the rectangle.

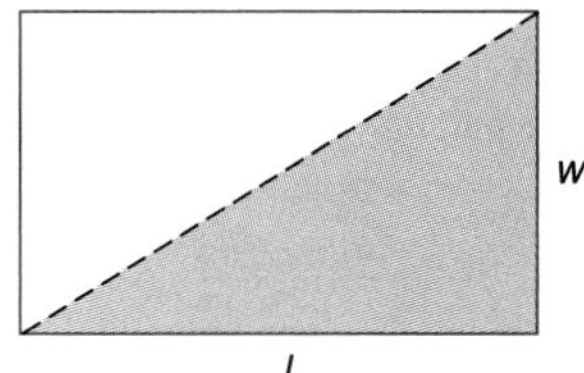

Figure 11–14

By changing length to base (b) and width to height (h), the formula for the area of this triangle is $A = \frac{1}{2} \times b \times h$. This is also the area formula for all types of triangles.

Find the areas.

54.

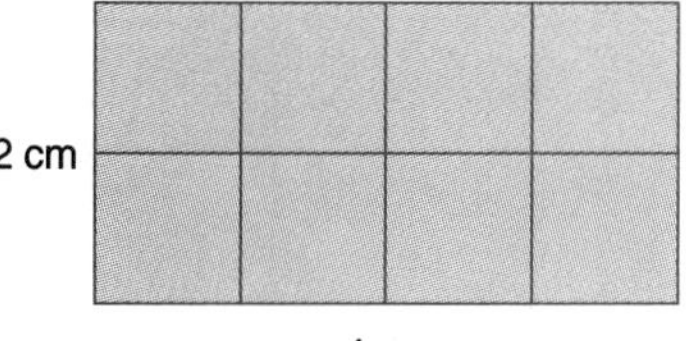

55.

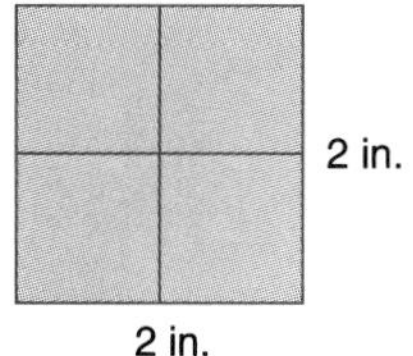

The answers are in the margin on page 435.

Find the area of the following rectangular regions.

56. 4.5 m by 3.2 m

57. $3\frac{1}{2}$ in. by $\frac{3}{4}$ in.

Find the area of the square whose side length is

58. 10 cm **59.** $2\frac{1}{2}$ ft **60.** 11.6 yd

The answers are in the margin on page 435.

Rule

Triangle Area Formula

Let A *be the area. If* b *is the base, and* h *is the height, then* $A = \frac{1}{2} \times b \times h$, *or* $A = \frac{1}{2} \cdot b \cdot h$.

Example 3

Find the area of the triangles.

a.

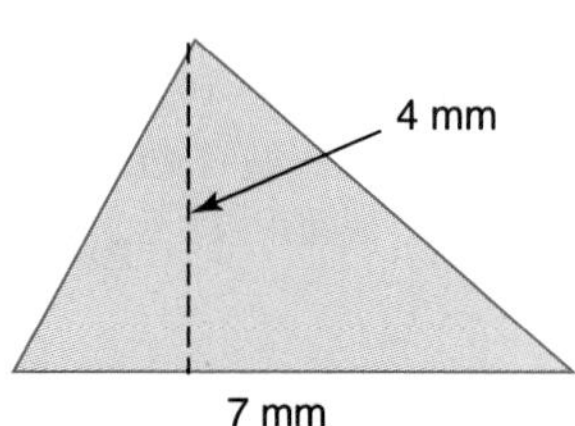

b.

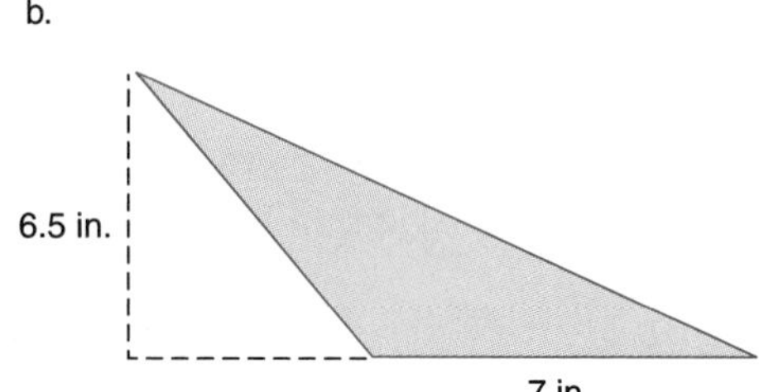

Figure 11–15

Solution

a. $A = \frac{1}{2} \times 7 \text{ mm} \times 4 \text{ mm}$

$A = \frac{1}{2} \times 28 \text{ mm}^2$

$A = 14 \text{ mm}^2$

b. $A = \frac{1}{2} \times 7 \text{ in.} \times 6.5 \text{ in.}$

$A = 0.5 \times 45.5 \text{ in}^2.$

$A = 22.75 \text{ in}^2.$

Practice check: Do exercises 61–63 in the margin.

Find the areas.

61.

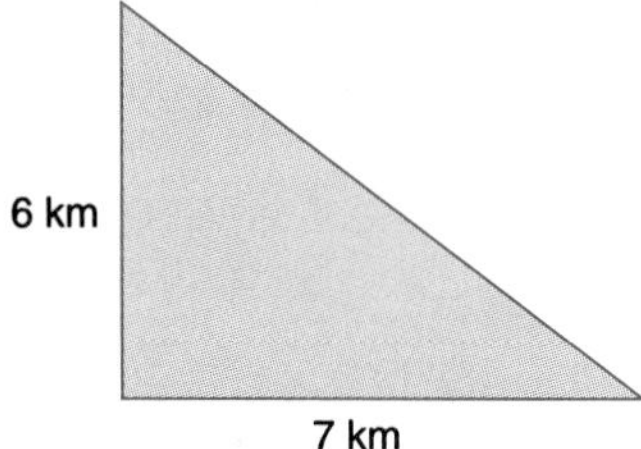

62.

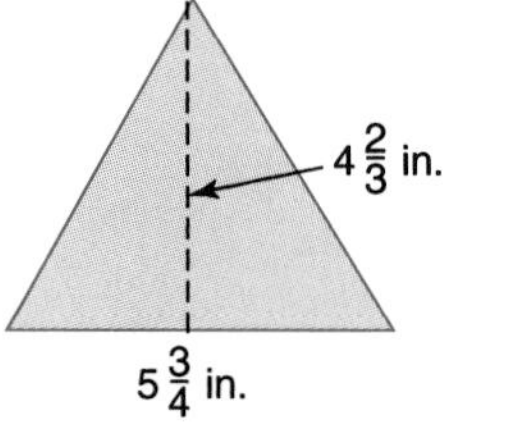

63. $b = 3.02$ cm, $h = 0.6$ cm

The answers are in the margin on page 437.

Circles

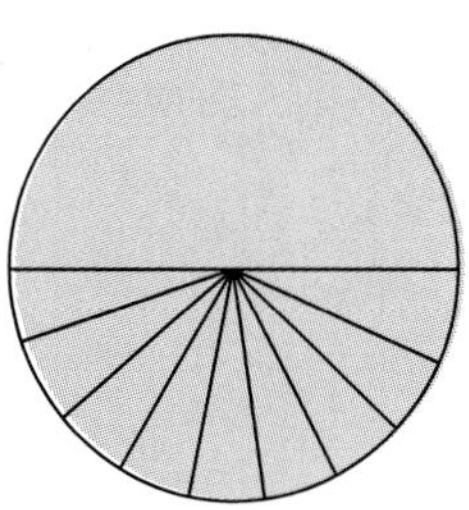

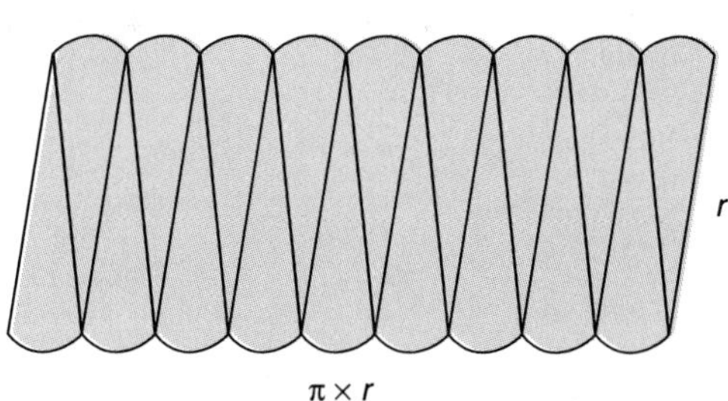

Figure 11–16

Suppose we cut one-half the circular region shown into smaller triangular parts and arrange them as shown. Then do the same for the other half and fit them into the spaces that are left. We now have a rectangular region whose length is about $\pi \times r$ (half the circumference) and whose width is r. The area is approximately $(\pi \cdot r) \cdot r$, which also is the area of the circle.

Rule

Circle Area Formula

Let A *be the area. If* r *is the radius, then* $A = \pi \cdot r \cdot r$, *or* $A = \pi r^2$

■ Example 4

Find the area of the circles.

a.

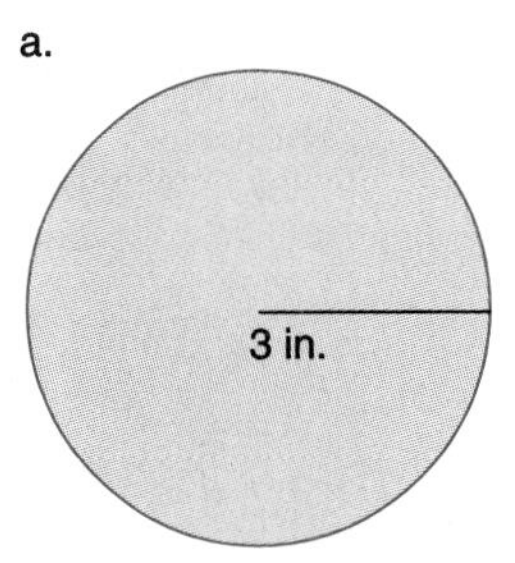

b.

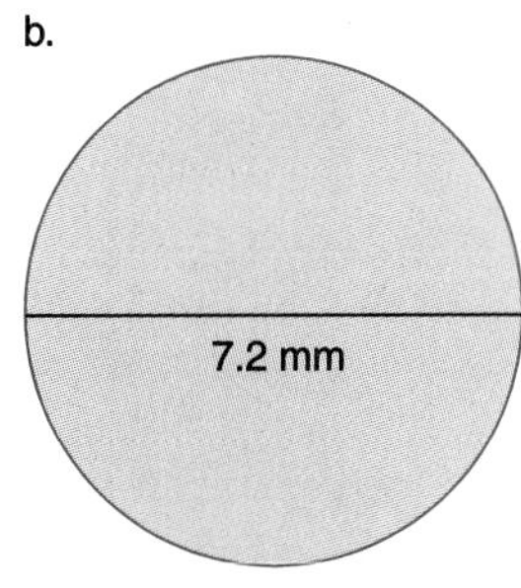

Figure 11–17

Solution **a.** $A \approx 3.14 \cdot 3 \text{ in.} \cdot 3 \text{ in.}$

$A \approx 3.14 \cdot 9 \text{ in}^2$

$A \approx 28.26 \text{ in}^2$

b. *Note:* $r = \frac{1}{2} \cdot d$

$r = \frac{1}{2} \cdot 7.2 \text{ mm} = 3.6 \text{ mm}$

$A \approx 3.14 \cdot 3.6 \text{ mm} \cdot 3.6 \text{ mm}$

$A \approx 3.14 \cdot 12.96 \text{ mm}^2$

$A \approx 40.6944 \text{ mm}^2$

$A \approx 40.7 \text{ mm}^2$

Practice check: Do exercises 64–66 in the margin.

■ Looking Back

You should now be able to find the areas of rectangles, triangles, and circles.

■ Problems 11–6

Find the area of each figure.

1.

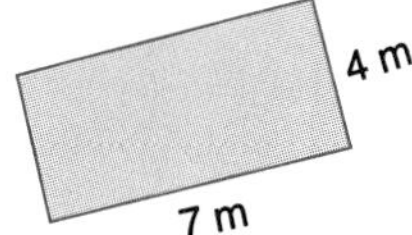

2.

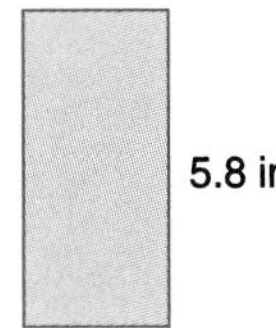

3.

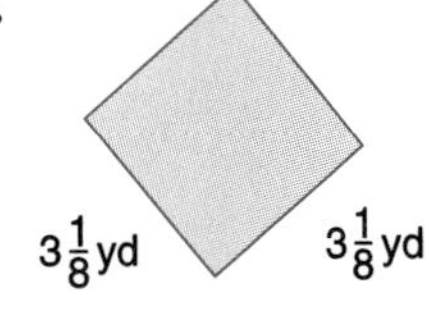

Answers to exercises 54 and 55

54. 8 square centimeters
55. 4 square inches

Answers to exercises 56–60

56. 14.4 m^2 **57.** $2\frac{5}{8} \text{ in}^2$

58. 100 cm^2 **59.** $6\frac{1}{4} \text{ ft}^2$

60. 134.56 yd^2

Find the areas.

64.

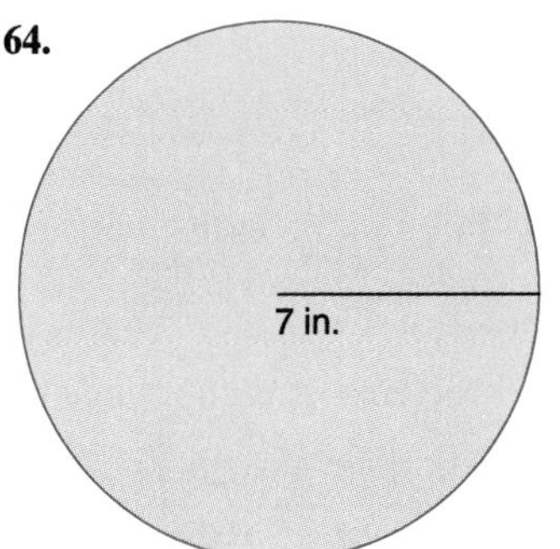

65.

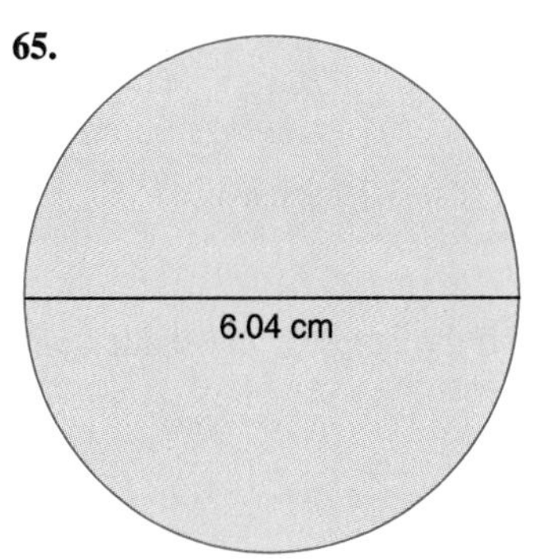

66. $d = 3\frac{1}{2}$ ft Use $\frac{22}{7}$ for π.

The answers are in the margin on page 437.

4.
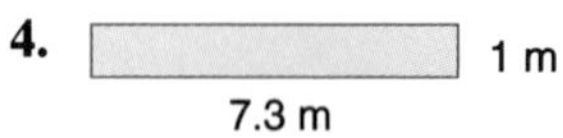

5.
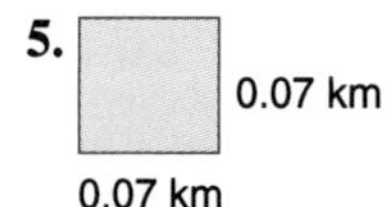

6.
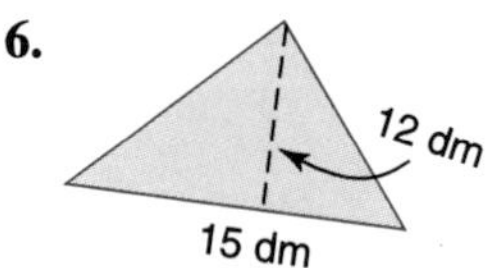

7. $1\frac{1}{4}$ ft, $3\frac{1}{2}$ ft

8. 7.8 in., 8 in.

9.
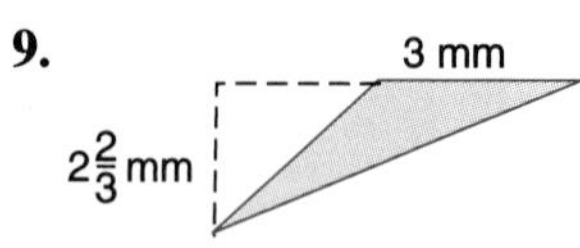

10. 0.6 cm, 0.2 cm

11. 5 yd

12.
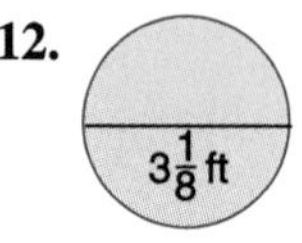

13. 14 km

14.
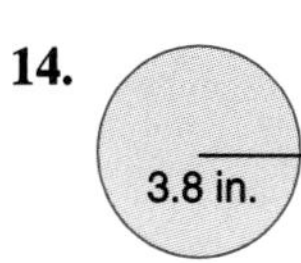

Find the area of each rectangle.

15. 5 ft by 90 ft

16. 18.6 m by 10 m

17. $3\frac{1}{5}$ dkm on each side

18. 14.04 in. on each side

Find the area of each triangle.

19. $b = 5$ yd, $h = 20$ yd

20. $b = 6\frac{3}{4}$ cm, $h = 4$ cm

Find the area of each circle.

21. $r = 3$ ft
Use 3.14 for π.

22. $d = 2\frac{1}{5}$ yd
Use $\frac{22}{7}$ for π.

23. $d = 5$ hm
Use 3.14 for π.

24. $r = 2.8$ mm
Use $\frac{22}{7}$ for π.

Find the areas of the shaded parts.

25. 2 ft, 5 ft, 5 ft, 6 ft, 6 ft, 10 ft

26. 8.2 m, 8.5 m, 4 m, 2 m, 2 m

27.
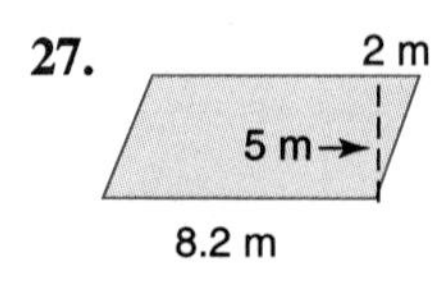

28.

25 yd

110 yd

29.

4.5 cm

2 cm

4.5 cm

30.

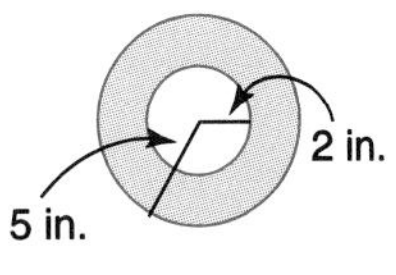

Answers to exercises 61–63

61. 21 km^2 **62.** $13\frac{5}{12}$ in^2

63. 0.906 cm^2

Answers to exercises 64–66

64. $A \approx 154$ m^2

65. $A \approx 28.64$ cm^2 (rounded)

66. $A \approx 9\frac{5}{8}$ ft^2

Solve. Write a statement for the answer.

31. Construction. How many square tiles, each 12 in. on a side, will cover a floor 15 ft by 10 ft?

32. Construction. A living room rectangular floor measures 21 ft by 15 ft. How many square yards of carpeting are needed to cover the floor? (*Hint*: First change feet to yards.)

33. Construction. The city code requires plastic under each new house construction. One new house measures 15 m by 9 m.

a. How many square meters of plastic are needed?

b. If each roll contains 50 m^2, how many rolls are needed?

c. How many square meters will be left over?

34. Interior Decorator. Dan Wright contracted a decorating service to wallpaper one wall of a bathroom. It measures 3.12 m by 4.34 m. A window in the wall measures 0.9 m by 1.2 m.

a. How much area of the wall needs to be papered?

b. If a roll will cover $6\frac{2}{3}$ sq m, how many rolls are needed? Round to the next greatest roll.

35. Landscaping. A plot of land measures 70 ft by 150 ft. The house on the lot measures 50 ft by 25 ft, and the driveway measures 15 ft by 40 ft. All the rest of the lot is planted in lawn. If 50 lb of fertilizer covers 2500 square feet, how much fertilizer is needed for the lawn?

36. Forestry. A forest ranger can see from her tower for a distance of 32 miles in all directions. How many square miles can she see in all directions?

The answers to odd-numbered exercises are in the back of the book.

11–7 AREA UNIT CONVERSION

American Units

Sometimes it is necessary to convert from one American unit of area measurement to another. We use the measurements in the following table to form the factors of one.

Rule

1 square foot (ft^2) = 144 square inches (in^2)

1 square yard (yd^2) = 9 square feet (ft^2)

1 square mile (mi^2) = 640 acres

To convert quickly, you should memorize this table.

■ Example 1

Change 22 ft^2 to in^2.

Solution

$$\frac{22\ ft^2}{1} \times \frac{144\ in^2}{1\ ft^2} \quad \text{Multiplying by 1}$$

$$= \frac{22\ \cancel{ft^2}}{1} \times \frac{144\ in^2}{1\ \cancel{ft^2}} \quad \text{Canceling}$$

$$= 22 \times 144\ in^2 = 3168\ in^2$$

■ Example 2

Change 27 ft^2 to yd^2.

Solution

$$\frac{27\ \cancel{ft^2}}{1} \times \frac{1\ yd^2}{9\ \cancel{ft^2}} = \frac{27}{9}\ yd^2 = 3\ yd^2$$

Practice check: Do exercises 67–71 in the margin.

Convert.

67. 30 ft^2 to in^2

68. 3 ft^2 to yd^2

69. 15 yd^2 to ft^2

70. 36 in^2 to ft^2

71. 15 mi^2 to acres

The answers are in the margin on page 441.

Metric Units

To convert metric area units quickly, study and memorize the following table:

Rule

1 square meter (m^2) = 10,000 square centimeters (cm^2)
(100 cm × 100 cm)

1 square kilometer (km^2) = 1,000,000 square meters (m^2)
(1000 m × 1000 m)

1 hectare (ha) = 10,000 m^2 (approximately 2.47 acres)
(100 m × 100 m)

$$1 \text{ are (a)} = 100 \text{ m}^2 \quad \text{(pronounced "air")}$$
$$(10 \text{ m} \times 10 \text{ m})$$
$$1 \text{ centare (ca)} = 1 \text{ m}^2 \quad \text{(approximately 1 sq. yd)}$$
$$(1 \text{ m} \times 1 \text{ m})$$

■ Example 3

Change 2.5 a to m^2.

Solution

$$\frac{2.5 \text{ a}}{1} \times \frac{100 \text{ m}^2}{1 \text{ a}}$$
$$= \frac{2.5 \not{\text{a}}}{1} \times \frac{100 \text{ m}^2}{1 \not{\text{a}}} = 2.5 \times 100 \text{ m}^2 = 250 \text{ m}^2$$

■ Example 4

Change 4000 m^2 to ha.

Solution

$$\frac{4000 \text{ m}^2}{1} \times \frac{1 \text{ ha}}{10{,}000 \text{ m}^2} = \frac{4000}{10{,}000} \text{ha} = \frac{2}{5} \text{ha}$$

Practice check: Do exercises 72–75 in the margin.

Convert.

72. 10.8 ha to m^2

73. 0.5 km^2 to m^2

74. 5800 m^2 to a

75. 5.6 ca to m^2

The answers are in the margin on page 441.

■ Looking back

You should now be able to convert American and metric area units.

■ Problems 11–7

Review

1. 15 yd to in.

2. 36 ft to yd

3. 4 mi to yd

4. 8 in. to ft

5. 3640 yd to mi

6. 18 in. to yd

7. 8.9 km to m

8. 535 dm to m

9. 4.66 m to dkm

10. 496 dkm to km

11. 0.015 mm to dm

12. 5.3 cm to mm

Convert.

13. 32 yd^2 to ft^2

14. 46.1 in^2 to ft^2

15. 20 ft^2 to in^2

16. 40 mi^2 to acres

17. 432 in^2 to ft^2

18. 48 in^2 to ft^2

19. 88 yd^2 to ft^2

20. 720 in^2 to ft^2

21. 84 ft^2 to yd^2

22. 1470 acres to mi^2

23. 5.6 ha to m^2

24. 92.6 a to m^2

25. 20.06 km^2 to m^2

26. 0.042 m^2 to cm^2

27. 41.4 ha to m^2

28. 0.7 km^2 to m^2

29. 7400 m^2 to ca

30. 20,000 m^2 to ha

31. 314.7 m^2 to km^2

32. 1700 m^2 to a

33. 18.43 cm^2 to m^2

34. 6 m^2 to ha

35. 3 mi^2 to ft^2

36. 5 mi^2 to yd^2

37. 3.2 yd^2 to in^2

38. 2592 in^2 to yd^2

39. 6 ha to a

40. 7.2 ca to ha

41. 1 km^2 to ha

42. 1500 a to ha

Solve.

43. Surveyor. After surveying a plot of land the surveyor finds the measurement to be 215 ft by 102 ft. If the plot is in the form of a rectangle, find the area of the plot in square yards.

44. Landscaping. A homeowner contracted Bill's Landscaping Service to put in a tile patio that is 150 m^2. If the tiles measure 20 cm by 20 cm, how many tiles will it take to complete the patio?

45. Catering. A standard round banquet table has diameter 72 in.

a. Find the area of the top. (Use 3.14 for π.)

b. How many persons can sit around the table at once if 2 ft for each person is allowed?

46. Agriculture. Ray Cook planted 10 ha in winter wheat.

a. State the number of m^2.

b. State the number of acres.

Answers to exercises 67–71

67. 4320 in^2 **68.** $\frac{1}{3}$ yd^2

69. 135 ft^2 **70.** $\frac{1}{4}$ ft^2

71. 9600 acres

Answers to exercises 72–75

72. 108,000 m^2 **73.** 500,000 m^2
74. 58 a **75.** 5.6 m^2

The answers to review and odd numbered exercises are in the back of the book.

11–8 VOLUME AND CAPACITY

Volume

Volume is the measurement of space inside of a solid object. To measure area we used unit squares; to measure volume we use unit cubes. A cube is like a child's block or a die (the singular of *dice*). Two units of volume measurement are the cubic inch and the cubic centimeter, shown below.

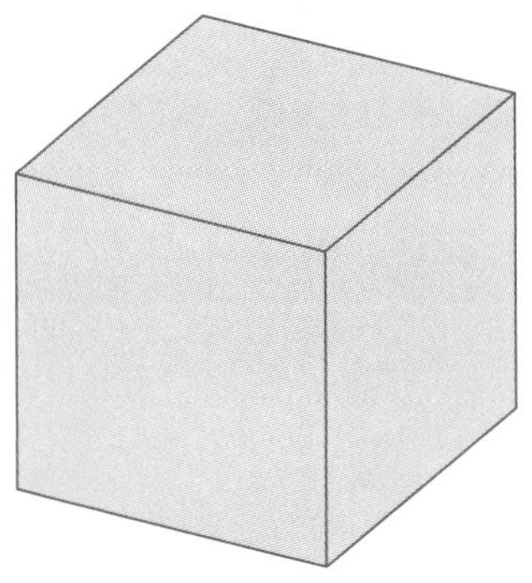

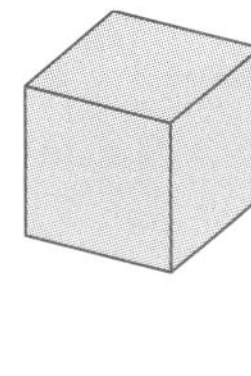

Figure 11–18

■ Example 1

Count the cubes that fill this rectangular solid.

Figure 11–19

Solution 1 There are 3 horizontal layers: top, middle, and bottom. Each layer has 15 cubes. Thus, there are 3 × 15, or 45 cubes.

Solution 2 The rectangular solid is 5 cubes long, 3 cubes wide, and 3 cubes high. We find the product of these numbers is 5 · 3 · 3, or 45 cubes.

Practice check: Do exercises 76 and 77 in the margin.

By now you have discovered that the volume of a rectangular solid is the product of its length, width, and height.

Rule

Rectangular Solid Volume Formula

Let V *be the volume. If* l *is the length,* w *is the width, and* h *is the height, then* V = l · w · h.

■ Example 2

Find the volume of this solid.

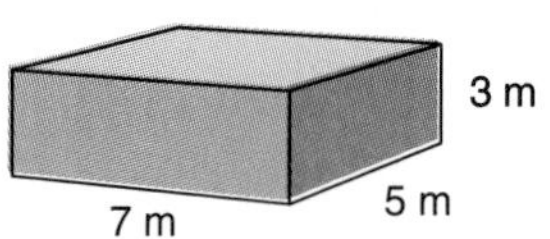

Figure 11–20

Solution

$$V = l \cdot w \cdot h$$

$$V = 7\text{ m} \cdot 5\text{ m} \cdot 3\text{ m}$$

$$V = 105\text{ m}^3$$

(*Note:* We read 105 m^3 as 105 cubic meters.)

Practice check: Do exercises 78–80 in the margin.

Count the cubes that fill the rectangular solid.

76.

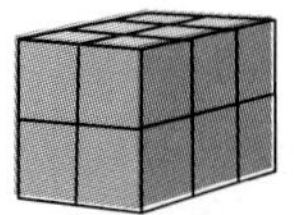

77.

The answers are in the margin on page 444.

Find the volume of this solid.

78.

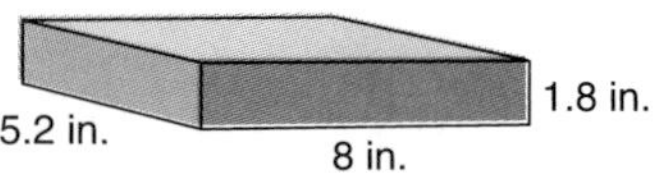

Find the volume of the rectangular solids with the given dimensions.

79. $l = 8$ cm, $w = 0.5$ cm, $h = 4.6$ cm

80. $l = 3\frac{1}{2}$ ft, $w = 6$ ft, $h = 5\frac{3}{8}$ ft

The answers are in the margin on page 444.

Capacity

Words like *gallons, quarts, liters, teaspoons,* and *milliliters* are measures of capacity. The American system has two separate units for measuring capacity: dry and liquid measure. A person can become confused between ounces that measure fluid and ounces that measure weight.

The metric system is easier to understand. All forms of capacity measurement are based on the liter, which is a little larger than a quart.

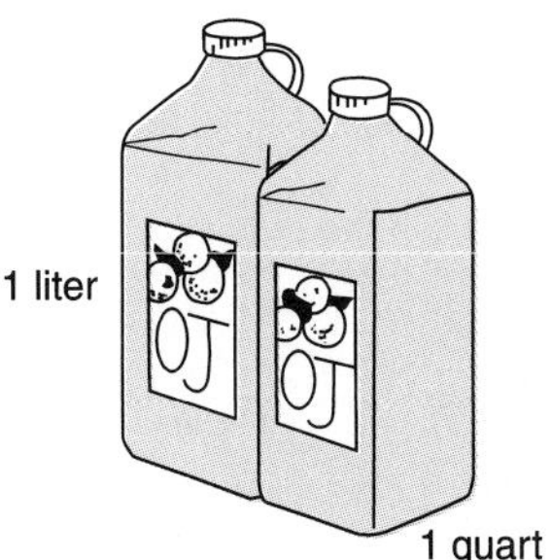

Figure 11–21

$$1 \text{ liter} = 1.06 \text{ quarts}$$

$$1 \text{ quart} = 0.946 \text{ liter}$$

The tip of your little finger including all of your fingernail is about 1 cubic centimeter (cc). (Check this out by measuring.) There are 1000 cc, or 1000 milliliters (mℓ), in 1 liter. The metric prefixes are also used with the liter as with the meter. The table below helps when converting.

Rule

Metric Units of Capacity

$$1 \text{ kiloliter (k}\ell) = 1000 \text{ liters}$$

$$1 \text{ hectoliter (h}\ell) = 100 \text{ liters}$$

$$1 \text{ dekaliter (dk}\ell) = 10 \text{ liters}$$

$$1 \text{ liter } (\ell) = 1 \text{ liter}$$

$$1 \text{ deciliter (d}\ell) = \frac{1}{10} \text{ liter}$$

$$1 \text{ centiliter (c}\ell) = \frac{1}{100} \text{ liter}$$

$$1 \text{ milliliter (m}\ell) = \frac{1}{1000} \text{ liter} = (1 \text{ cc})$$

■ **Example 3**

Convert 7 ℓ to mℓ.

Solution $\frac{7\,\ell}{1} \times \frac{1000 \text{ m}\ell}{1\,\ell} = 7000 \text{ m}\ell$

Answers to exercises 76 and 77

76. 12 cubes **77.** 72 cubes

Answers to exercises 78–80

78. 74.88 in^3 **79.** 18.4 cm^3

80. $112\frac{7}{8}$ ft^3

Example 4

Convert 346 $c\ell$ to dkℓ.

Solution $$\frac{346\ c\ell}{1} \times \frac{1}{10\ c\ell} \times \frac{1\ \text{dk}\ell}{10} = \frac{346}{1000}\ \text{dk}\ell$$

Note: To convert quickly, we move the decimal point 3 places to the left in Example 4, since there is movement of 3 places to the left in the table below.

Rule

kℓ	hℓ	dkℓ	ℓ	dℓ	cℓ	mℓ
1000	100	10	1	$\frac{1}{10}$	$\frac{1}{100}$	$\frac{1}{1000}$

Practice check: Do exercises 81–86 in the margin.

Convert.

81. 186 m ℓ to ℓ

82. 78 dℓ to ℓ

83. 15 dkℓ to ℓ

84. 9 kℓ to dkℓ

85. 7 dkℓ to cℓ

86. 1.07 dk ℓ to dℓ

The answers are in the margin on page 446.

Looking Back

You should now be able to find the volume of a rectangular box and to convert from one metric capacity unit to another.

Problems 11–8

Review

Find the perimeter of each figure.

1.

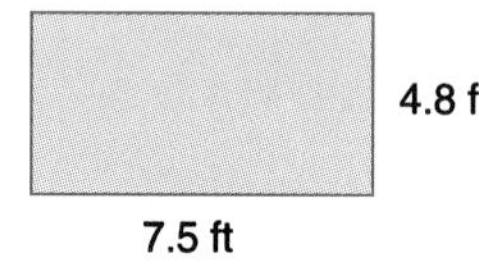

2.

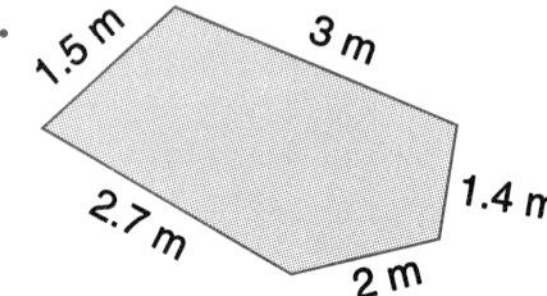

3.

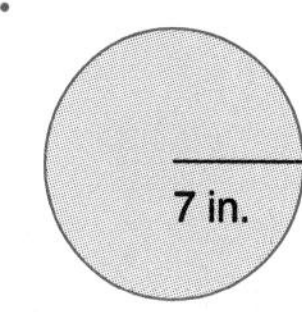

4.

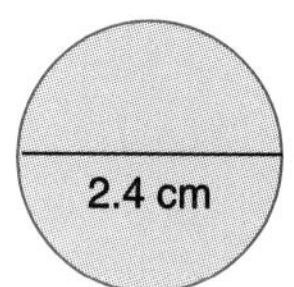

5. Find the area of the figure in exercise 1.

6. Find the area of the circle in exercise 4.

Find the volumes.

7.
4 m
6 m 5 m

8.
2.2 in.
1.5 in. 3 in.

9.

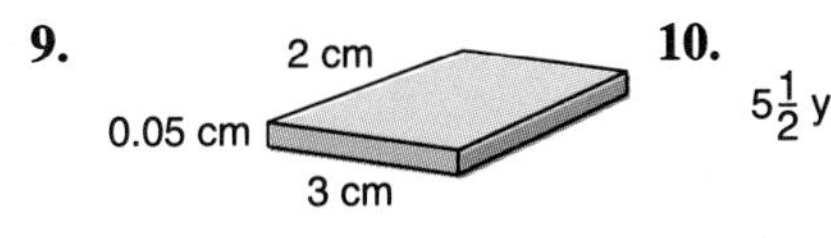

10.
$5\frac{1}{2}$ yd
6 yd $1\frac{3}{4}$ yd

11.
13 cm
9 cm 85 mm

12.

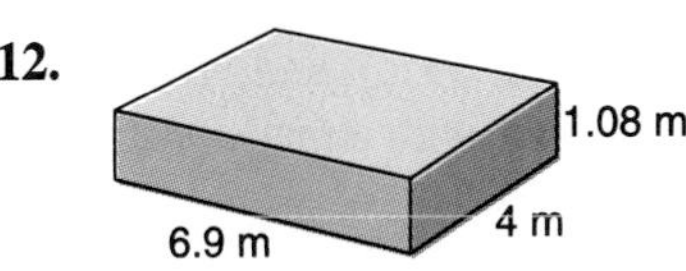

Convert.

13. 3.2 hℓ to ℓ

14. 438 ℓ to dkℓ

15. 3042 ℓ to kℓ

16. 0.082 dkℓ to ℓ

17. 536 cℓ to ℓ

18. 5.07 ℓ to mℓ

19. 4 mℓ to cc

20. 1.7 cc to mℓ

21. 132 ℓ to dℓ

22. 6 dℓ to ℓ

23. 5.62 hℓ to dℓ

24. 18 kℓ to dkℓ

25. 139.2 mℓ to dkℓ

26. 29.8 dℓ to kℓ

27. 269 dkℓ to dℓ

28. 32 dℓ to cℓ

Solve. Write a statement for the answer.

29. Construction. A footing for a building is 2 yd by 3 yd by 5 yd. How many cubic yards of concrete are needed for the footing?

30. Generic. Find the volume of an aquarium that is 45 cm long, 20 cm wide, and 35 cm high. How many liters does the aquarium hold?

31. Design. Find the cubic yards of a tank that measures 8 ft by 6 in. by 2 ft. (*Hint*: First change the measurements to yd.)

32. Nursing. How many flu shots of 4 mℓ each can be obtained from one liter of vacine?

Answers to exercises 81–86

81. 0.186 ℓ **82.** 7.8 ℓ **83.** 150 ℓ
84. 900 dkℓ **85.** 7000 cℓ
86. 107 dℓ

33. Landscape Management. A truck bed measuring 5 m by 2.5 m by 3 m filled with beauty bark is dumped on a driveway. How many trips will it take one worker to spread the bark if his wheelbarrow holds only 0.3 cubic meters?

The answers to review and odd-numbered exercises are in the back of the book.

11–9 WEIGHT

Tomatoes, potatoes, nails, meat, honey, and coal are some items that are priced according to their weight. We are familiar with American units of weights such as pounds, ounces, and tons.

In the metric system of weights the basic unit is the gram. The gram is the weight of 1 cc of water, which is about the weight of a paper clip. A kilogram (kg) contains 10 grams (g) and weighs approximately 2.2 pounds.

Rule

Metric Units of Weight:

1 kilogram (kg) = 1000 grams (g)
1 hectogram (hg) = 100 grams
1 dekagram (dkg) = 10 grams
1 gram = 1 gram
1 decigram (dg) = 0.1 gram
1 centigram (cg) = 0.01 gram
1 milligram (mg) = 0.001 gram

■ **Example 1**

Convert 15 g to mg.

Solution $$\frac{15\cancel{g}}{1} \times \frac{1000 \text{ mg}}{1\cancel{g}} = 15{,}000 \text{ mg}$$

■ **Example 2**

Convert 4061 dg to hg.

Solution $$\frac{4061 \cancel{dg}}{1} \times \frac{1 \cancel{g}}{10 \cancel{dg}} \times \frac{1 \text{ hg}}{100 \cancel{g}} = \frac{4061}{1000} \text{hg} = 4.061 \text{ hg}$$

Convert.

87. 7 dkg to g **88.** 5.6 mg to g

89. 608 dg to g **90.** 4 kg to cg

91. 9 dkg to dg

92. 0.045 dkg to mg

The answers are in the margin on page 449.

Note: To convert quickly, we move the decimal point 3 places to the left in Example 2, since there is a movement of 3 places to the left in table below.

Rule

kg	hg	dkg	g	dg	cg	mg
1000	100	10	1	0.1	0.01	0.001

Practice check: Do exercises 87–92 in the margin.

■ Looking Back

You should now be able to convert from one metric weight to another.

■ Problems 11–9

Review

Convert.

1. 43 yd^2 to ft^2
2. 32 ft^2 to in^2
3. 348 in^2 to ft^2
4. 98 ft^2 to yd^2

5. 6 in^2 to yd^2
6. 4.8 yd^2 to in^2
7. 7.8 ha to m^2
8. 31.6 km^2 to m^2

9. 38.6 ha to m^2
10. 5420 m^2 to ca
11. 531.4 m^2 to km^2
12. 14.09 cm^2 to m^2

Convert.

13. 69 kg to g
14. 32 hg to kg
15. 1 mg to g
16. 1 g to mg
17. 456 dkg to g

18. 89 cg to g
19. 1 g to cg
20. 1 cg to g
21. 0.464 kg to hg
22. 0.094 dg to g

23. 1 kg to g
24. 1 g to kg
25. 0.0047 cg to hg
26. 15 mg to cg
27. 346 hg to g

28. 64 cg to g
29. 1 dkg to g
30. 1 cg to g
31. 9 kg to cg
32. 46 dg to g

33. 5 g to hg
34. 999 dg to g
35. 1 cg to hg
36. 1 kg to dg
37. 3.2 g to kg
38. 3474 mg to kg

Solve. Write a statement for the answer.

39. Butcher. A package of ground round is marked 500 g. How many kilograms is this?

40. Retail Management. A certain hand-held calculator weighs 350 g. How many kg do 5 calculators weigh?

41. Generic. How much does 5.68 kg cost at 41¢ per kilogram?

42. Grocery Management. A sack of tomatoes weighs 7 kg 332 g. How much does half a sack of tomatoes weigh?

The answers to review and odd-numbered exercises are in the back of the book.

11–10 TEMPERATURE

The two temperature scales commonly used in the United States are Fahrenheit, for American measurement, and Celsius (or Centigrade), for metric measurement. The figure below shows how the two scales compare.

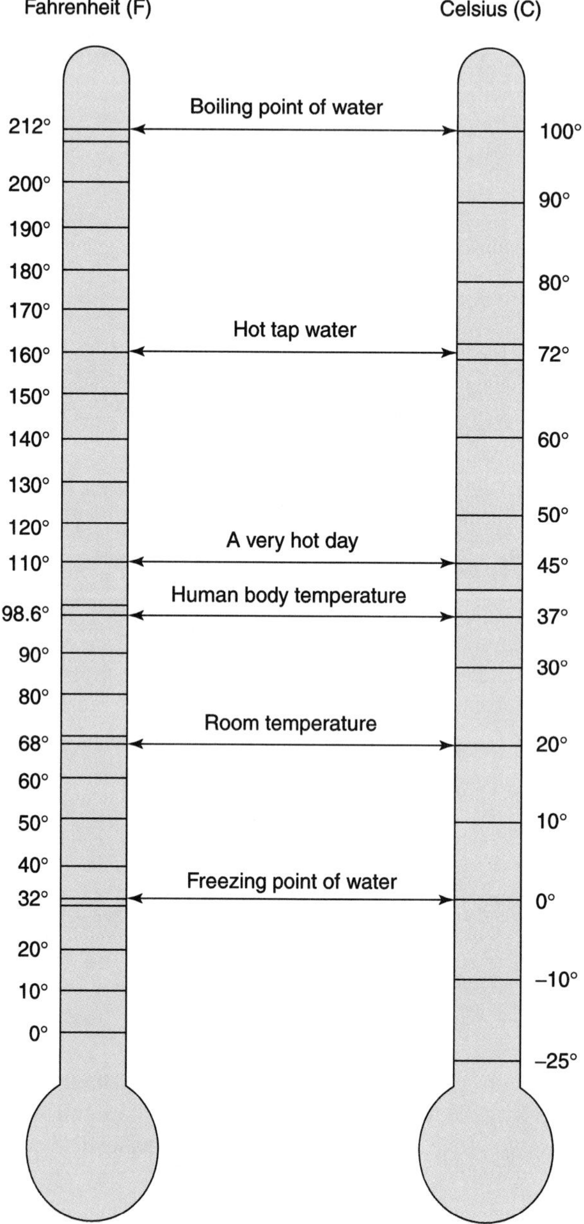

Figure 11–22

In 1714, Gabriel Fahrenheit, a German scientist, developed a temperature scale with the freezing point of water at 32° and the boiling point at 212°. This is the scale with which we are most familiar.

The metric system has adopted the Celsius (pronounced sell-see-us) scale, developed in 1742 by Aders Celsius, a Swedish astronomer. He conveniently divided his scale into 100 degrees between the freezing and boiling points of water.

By using a straight edge or ruler on the scales, we can easily estimate when converting from one measure of temperature to another.

Answers to exercises 87–92

87. 70 g **88.** 0.0056 g
89. 60.8 g **90.** 400,000 cg
91. 900 dg **92.** 450 mg

■ Example 1

Estimate the equivalent of 50°F in Celsius.

Solution 50°F is approximately 10°C.

■ Example 2

Estimate the equivalent of 60°C in Fahrenheit.

Solution 60° is approximately 139°F.

Practice check: Do exercises 93–98 in the margin.

Estimate the equivalent in Celsius.

93. 85°F **94.** 10°F

95. 145°F

Estimate the equivalent in Fahrenheit.

96. 15°C **97.** −5°C

98. 95°C

The answers are in the margin on page 451.

To get an exact conversion we use these formulas:

Rule

Temperature Conversion Formula

From Fahrenheit to Celsius, use $C = \frac{5}{9} \cdot (F - 32)$.
From Celsius to Fahrenheit use $F = \frac{9}{5} \cdot C + 32$.

■ Example 3

Convert 85°F to Celsius.

Solution

$$C = \frac{5}{9} \cdot (F - 32)$$

$$C = \frac{5}{9} \cdot (85 - 32) \quad \text{Substituting 85 for } F$$

$$C = \frac{5}{9} \cdot (53) \quad \text{Subtracting first}$$

$$C \approx 29.4 \quad \text{Rounding to the nearest tenth}$$

$$85°F \approx 29.4°C$$

■ Example 4

Convert 85°C to Fahrenheit.

Solution

$$F = \frac{9}{5} \cdot C + 32$$

$$F = \frac{9}{5} \cdot 85 + 32 \quad \text{Substituting 85 for } C$$

$$F = 153 + 32 \quad \text{Multiplying first}$$

$$F = 185$$

$$85°C = 185°F$$

Make the following conversions. Round to the nearest tenth of a degree when necessary.

99. 76°F to Celsius

100. 58°C to Fahrenheit

101. 132°F to Celsius

102. 24°C to Fahrenheit

The answers are in the margin on page 452.

Practice check: Do exercises 99–102 in the margin.

■ Looking Back

You should now be able to convert from one measure of temperature to another.

■ Problems 11–10

Review

Find the volume.

1.

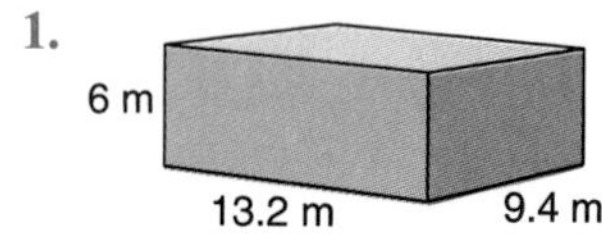

2.

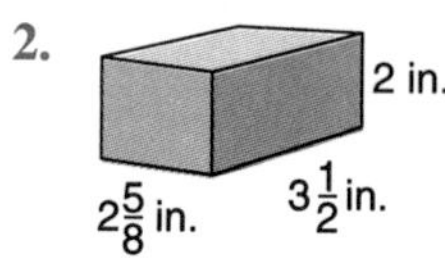

Convert.

3. 4.8 hℓ to ℓ

4. 342 cℓ to ℓ

5. 6.01 hℓ to dℓ

6. 83 dℓ to cℓ

7. 1.4 mℓ to cc

8. 24 kℓ to dkℓ

Estimate the equivalent in Celsius or Fahrenheit by using the scales on page 449.

9. 167°F

10. 55°F

11. 203°F

12. 5°F

13. 88°F

14. 170°F

15. 99°C

16. 35°C

17. 65°C

18. −7°C

19. 52°C

20. 17°C

Convert to Celsius or Fahrenheit. Use the formulas on page 449.

21. 129°F

22. 171°F

23. 43°F

24. 87°F

25. 70°C

26. 105°C

27. 50°C

28. 35°C

Solve.

29. A dangerously high fever is 105°F. How many degrees Celsius is this?

30. The temperature in Death Valley on July 10, 1913, was 134°F. How many degrees Celsius is this?

Answers to exercises 93–98

93. 30°C **94.** −12°C **95.** 62°C **96.** 59°F **97.** 22°F **98.** 207°F

31. The temperature is 3°C and clear. What would you wear to go for a walk?

32. Would you go ice skating or swimming if the outside temperature were 29°C?

The answers to review and odd-numbered exercises are in the back of the book.

11–11 CONVERSIONS BETWEEN THE AMERICAN AND METRIC SYSTEMS

In certain jobs it is often necessary to convert between the American and metric systems. The following tables will help in making those conversions.

Rule

Linear Measurement

1 in. = 2.54 cm	1 cm = 0.39 in.
1 ft = 30.5 cm	1 m = 3.28 ft
1 yd = 0.91 m	1 m = 1.09 yd
1 mi = 1.61 km	1 km = 0.62 mi

Weight Measurement

1 oz = 28.4 g	1 g = 0.035 oz
1 lb = 0.45 kg	1 kg = 2.2 lb

Volume Measurement

1 pint (pt) = 0.47 ℓ	1 ℓ = 2.1 pt
1 quart (qt) = 0.95 ℓ	1 ℓ = 1.06 qt
1 gallon (gal) = 3.8 ℓ	1 ℓ = 0.26 gal

■ **Example 1**

Convert 56 in. to centimeters.

Answers to exercises 99–102

99. $C \approx 24.4°$ **100.** $F = 136.4°$
101. $C \approx 55.6°$ **102.** $F = 75.2°$

Solution Notice that 1 in. = 2.54 cm. Since we want inches to cancel out, we choose $\frac{2.54 \text{ cm}}{1 \text{ in.}}$ as the conversion factor.

$$\frac{56 \cancel{\text{in.}}}{1} \times \frac{2.54 \text{ cm}}{1 \cancel{\text{in.}}} = 14.24 \text{ cm}$$

■ Example 2

Convert 5 g to oz.

Solution 1 g = 0.035 oz

$$\frac{5\cancel{g}}{1} \times \frac{0.035 \text{ oz}}{1\cancel{g}} = 0.175 \text{ oz}$$

■ Example 3

Convert 26 gal to ℓ.

Solution 1 gal = 3.8 ℓ

$$\frac{26 \cancel{\text{gal}}}{1} \times \frac{3.8\ \ell}{1\cancel{\text{gal}}} = 98.8\ \ell$$

Practice check: Do exercises 103–106 in the margin.

Convert.

103. 24 yd to m **104.** 2.3 km to mi

105. 8.6 lb to kg **106.** 32 ℓ to gal

The answers are in the margin on page 455.

■ Looking Back

You should now be able to convert between American and metric measurements.

■ Problems 11–11

Review

Convert.

1. 56 kg to g **2.** 0.032 dg to g **3.** 18 kg to cg **4.** 1 cg to g

Convert to Celsius or Fahrenheit.

5. 154°F **6.** 77°C

Convert.

7. 29 in. to cm **8.** 39 m to yd **9.** 380 km to mi

10. 152 yd to m

11. 3 ft to cm

12. 5 cm to in.

13. 7 oz to g

14. 11 kg to lb

15. 531 lb to kg

16. 152 g to oz

17. 22 qt to ℓ

18. 9 ℓ to gal

19. 180 ℓ to gal

20. 246 pt to ℓ

21. 3.8 mi to km

22. 46.4 m to yd

23. 88.9 m to ft

24. 106.04 in. to cm

25. 5.5 g to oz

26. 7.84 lb to kg

27. 346.5 pt to ℓ

28. 1.08 ℓ to gal

29. $2\frac{1}{2}$ in. to cm

30. $3\frac{2}{3}$ m to ft

31. $6\frac{3}{5}$ mi to km

32. $10\frac{1}{4}$ m to yd

33. $\frac{3}{8}$ g to oz

34. $\frac{3}{4}$ lb to kg

35. $5\frac{1}{8}$ ℓ to pt

36. $4\frac{5}{7}$ qt to ℓ

Solve. Write a statement for the answer.

37. Retail Grocery Management. The manager at Sunrise Grocery ordered 20 sacks of Idaho potatoes each 50 lb in weight. How many kilograms did the complete order weigh?

38. Trucking. The distance from Portland to Seattle is 150 miles. How many kilometers is this?

39. Trucking. Jack Filler used 197 liters of diesel in his rig before filling his tank. How many gallons did he use?

40. Interior Decorating. An interior decorator estimates that it would take 6 gallons of paint to finish the interior rooms of a home. The total cost would be $93.00.

a. What is the cost per gallon?

b. What is the cost per liter?

41. Carpentry. A chair measures 41 cm from the floor to the seat. How many inches does this measure?

The answers to review and odd-numbered exercises are in the back of the book.

CHAPTER 11 OVERVIEW

Summary

11–1 You learned about the history of measurement.

11–2 You learned to change one American unit of linear measurement to another.

11–3 You learned to change one metric unit of linear measurement to another.

11–4 You learned to add, subtract, multiply, and divide denominate numbers.

11–5 You learned to find the perimeter of a closed geometric figure and the circumference of a circle.

11–6 You learned to find the areas of rectangles, triangles, and circles.

11–7 You learned to convert between American and metric units of area.

11–8 You learned to find the volume of a rectangular box and to convert from one metric capacity unit to another.

11–9 You learned to convert from one metric weight unit to another.

11–10 You learned to convert from one measurement of temperature to another.

11–11 You learned to make conversions between the American and metric systems.

Terms To Remember

	Page		***Page***
Are	439	Deciliter	443
Area	432	Decimeter	415
Celsius	448	Dekagram	446
Centare	439	Dekaliter	443
Centigram	446	Dekameter	415
Centiliter	443	Denominate nos.	419
Centimeter	415	Fahrenheit	448
Decigram	446	Gram	446

Answers to exercises 103–106

103. 21.84 m **104.** 1.43 mi
105. 3.87 kg **106.** 8.32 gal

Rules

- Metric Units of Length

$$\begin{aligned}
1 \text{ kilometer (km)} &= 1000 \text{ meters (m)} \\
1 \text{ hectometer (hm)} &= 100 \text{ meters (m)} \\
1 \text{ dekameter (dkm)} &= 10 \text{ meters (m)} \\
1 \text{ meter (m)} &= 1 \text{ meter (m)} \\
1 \text{ decimeter (dm)} &= \frac{1}{10} \text{ meter (m)} \\
1 \text{ centimeter (cm)} &= \frac{1}{100} \text{ meter (m)} \\
1 \text{ millimeter (mm)} &= \frac{1}{1000} \text{ meter (m)}
\end{aligned}$$

- American Units of Length

$$\begin{aligned}
1 \text{ foot (ft)} &= 12 \text{ inches (in.)} \\
1 \text{ yard (yd)} &= 3 \text{ feet (ft)} \\
1 \text{ mile (mi)} &= 5280 \text{ ft}
\end{aligned}$$

- Quick conversions

km	hm	dkm	m	dm	cm	mm
1000	100	10	1	$\frac{1}{10}$	$\frac{1}{100}$	$\frac{1}{1000}$

- Perimeter Formulas

  Let P be the perimeter.

  Square: If s is the length of one side, then $P = 4 \times s$, or $P = 4s$.

  Rectangle: If l is the length and w is the width, then $P = 2 \times (l + w)$, or $P = 2 \cdot (l + w)$.

- Circumference Formula

  Let C be the circumference.

  Circle: If r is the radius and d is the diameter, then $C = \pi \cdot d$, or $C = 2 \cdot \pi \cdot r$.

- The measurement of the space covered in a closed geometric figure is called *area*.
- Rectangular Area Formula

  Let A be the area. If l is the length and w is the width, then $A = l \times w$, or $A = l \cdot w$.

- Triangle Area Formula

  Let A be the area. If b is the base and h is the height, then $A = \frac{1}{2} \times b \times h$, or $A = \frac{1}{2} \cdot b \cdot h$.

- Circle Area Formula

  Let A be the area. If r is the radius, then $A = \pi \cdot r \cdot r$, or $A = \pi r^2$.

- American Area Units

$$1 \text{ square foot (ft)} = 144 \text{ square inches (in}^2)$$
$$1 \text{ square yard (yd}^2) = 9 \text{ square feet (ft}^2)$$
$$1 \text{ square mile (mi}^2) = 640 \text{ acres}$$

- Metric Area Units

1 square meter (m^2) =	10,000 square centimeters (cm^2) (100 cm × 100 cm)	
1 square kilometer (km^2) =	1,000,000 square meters (m^2) (1000 m × 1000 m)	
1 hectare (ha) =	10,000 m^2 (100 m × 100 m)	(approximately 2.47 acres)
1 are (a) =	100 m^2 (10 m × 10 m)	(pronounced "air")
1 centare (ca) =	1 m^2 (1 m × 1 m)	(approximately 1 sq. yd)

- Rectangular Solid Volume Formula

  Let V be the volume. If l is the length, w is the width, and h is the height, then $V = l \cdot w \cdot h$.

- Metric Units of Capacity

1 kiloliter (kℓ) = 1000 liters	1 deciliter (dℓ) = $\frac{1}{10}$ liter
1 hectoliter (hℓ) = 100 liters	1 centiliter (cℓ) = $\frac{1}{100}$ liter
1 dekaliter (dkℓ) = 10 liters	1 milliliter (mℓ) = $\frac{1}{1000}$ liter
1 liter (ℓ) = 1 liter	

- Quick Conversions

kℓ	hℓ	dkℓ	ℓ	dℓ	cℓ	mℓ
1000	100	10	1	$\frac{1}{10}$	$\frac{1}{100}$	$\frac{1}{1000}$

- Metric Units of Weight

1 kilogram (kg) = 1000 grams (g)	1 decigram (dg) = 0.1 gram
1 hectogram (hg) = 100 grams	1 centigram (cg) = 0.01 gram
1 dekagram (dkg) = 10 grams	1 milligram (mg) = 0.001 gram
1 gram = 1 gram	

- Quick Conversion

kg	hg	dkg	g	dg	cg	mg
1000	100	10	1	0.1	0.01	0.001

- Temperature Conversion Formula
  From Fahrenheit to Celsius, use $C = \frac{5}{9} \cdot (F - 32)$.
  From Celsius to Fahrenheit, use $F = \frac{9}{5} \cdot C + 32$.
- American and Metric Conversion

Linear Measurement

1 in. = 2.54 cm 1 cm = 0.39 in. 1 yd = 0.91 m 1 m = 1.09 yd

1 ft = 30.5 cm 1 m = 3.28 ft 1 mi = 1.61 km 1 km = 0.62 mi

Weight Measurement

1 oz = 28.4 g 1 g = 0.035 oz 1 lb = 0.45 kg 1 kg = 2.2 lb

Volume Measurement

1 pint (pt) = 0.47 ℓ 1 ℓ = 2.1 pt 1 gallon (gal) = 3.8 ℓ 1 ℓ = 0.26 gal

1 quart (qt) = 0.95 ℓ 1 ℓ = 1.06 qt

Self Test

11–2 Convert.

1. 7 ft to in. **2.** 38 ft to yd **3.** 24 in. to yd

11–3 Convert.

4. 8 hm to m **5.** 6.59 mm to m **6.** 5.9 dkm to dm

11–4 Perform the indicated operations.

7.
```
  9 ft  5 in.
+2 ft  7 in.
```

8.
```
  32 dm   8 cm
−17 dm   9 cm
```

9. 5 × (6 yd 1 ft) **10.** (13 m 9 dm 8 cm) ÷ 4

11–5 Find the perimeter or circumference.

11. Use 3.14 for π.

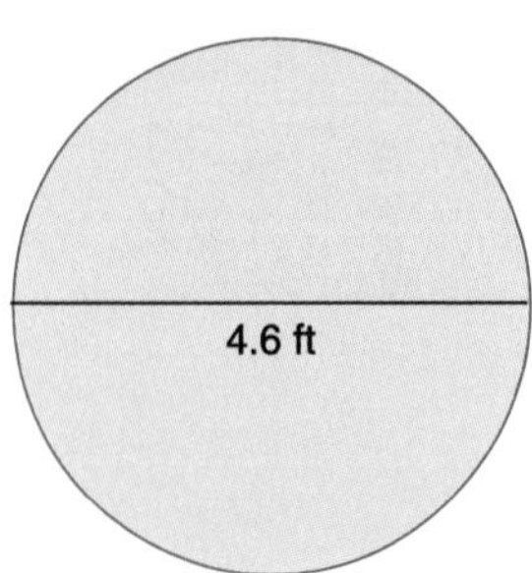

12.

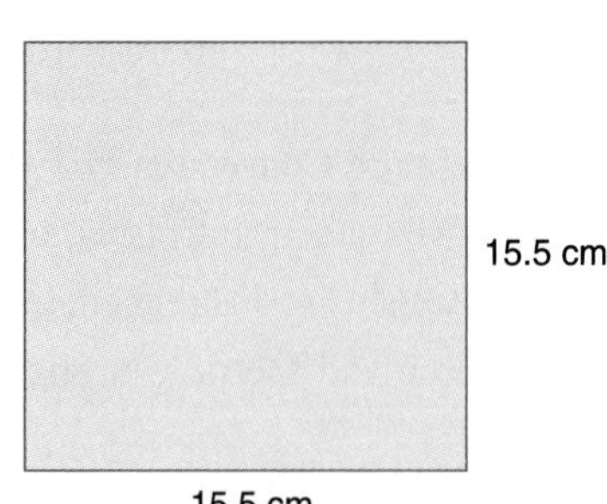

11–6 Find the area of each figure.

13.

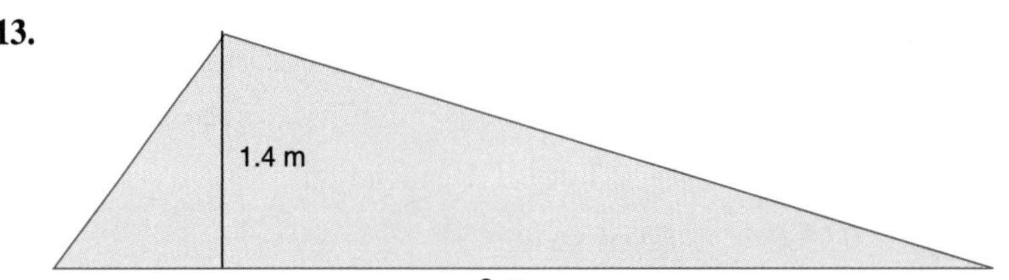

14.

0.7 in.

6.8 in.

11–7 Convert.

15. 50 in^2 to ft^2 **16.** 0.6 km^2 to m^2 **17.** 78 ft^2 to yd^2

11–8 Find the volume.

18.

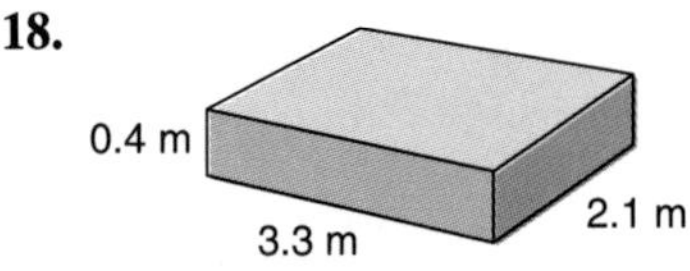

Convert.

19. 546 ℓ to dk ℓ

Solve.

20. Twenty oil drums contain a total of 78.2 hectoliters. How many liters does each drum hold?

11–9 Convert.

21. 2.8 kg to g

22. 3904 mg to kg

11–10 23. Convert 56°F to °C

24. Convert 105°C to °F

11–11 Convert.

25. 5.6 mi to km

26. 2.08 ℓ to gal

Chapter Test

Directions: This test will aid in your preparation for a possible chapter test given by your instructor. The answers are in the back of the book. Go back and review in the appropriate section(s) if you missed any test items.

11–2 Convert.

1. 5 ft to in.

2. 26 ft to yd

3. 14 yd to in.

4. 10,560 ft to mi

5. 24 in. to yd

6. 30 yd to ft

7. Death Valley is 282 ft below sea level. How many yards is this?

11–3 8. 6 hm to m

9. 3479 m to km

10. 5.79 mm to m

11. 0.477 dkm to m

12. 4.8 dkm to dm

13. 7800 mm to dkm

11–4 Perform the indicated operations.

14.
$$\begin{array}{r} 7 \text{ ft } 6 \text{ in} \\ +5 \text{ ft } 9 \text{ in} \\ \hline \end{array}$$
= ______ ft

15.
$$\begin{array}{r} 8 \text{ km } 489 \text{ m} \\ +13 \text{ km } 634 \text{ m} \\ \hline \end{array}$$
= ______ km

16.
$$\begin{array}{r} 39 \text{ cm } 4 \text{ mm} \\ -13 \text{ cm } 8 \text{ mm} \\ \hline \end{array}$$
= ______ cm

17.
$$\begin{array}{r} 13 \text{ mi } 3460 \text{ ft} \\ -9 \text{ mi } 4742 \text{ ft} \\ \hline \end{array}$$
= ______ mi

18. $6 \times (7 \text{ yd } 1 \text{ ft})$

19. $8 \times (13 \text{ km } 360 \text{ m})$

20. $(39 \text{ km } 538 \text{ m } 48 \text{ cm}) \div 8$

21. $(12 \text{ yd } 1 \text{ ft } 5 \text{ in.}) \div 5$

11–5 Find the perimeter.

22.

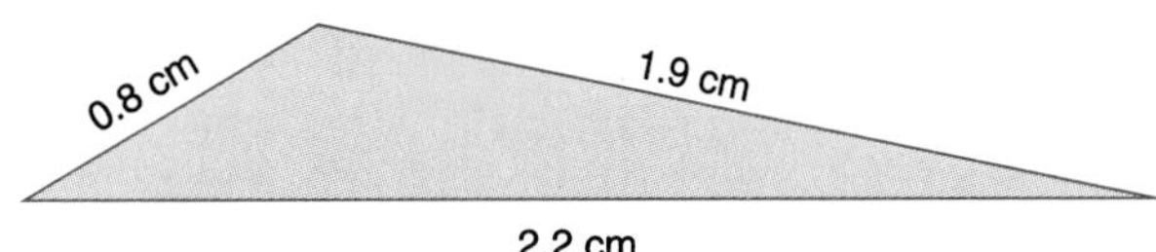

23. Find the circumference. Use 3.14 for π.

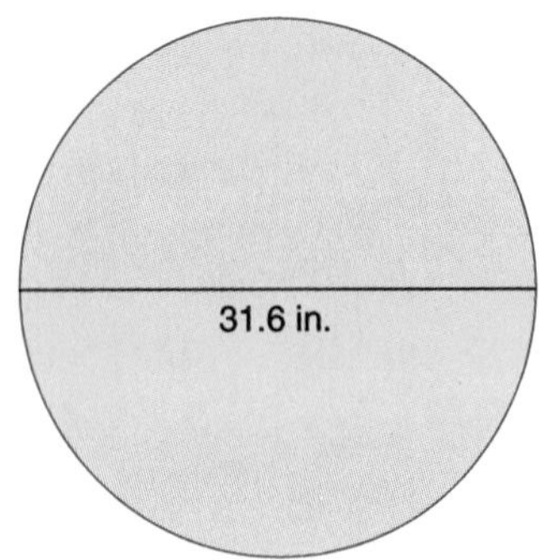

11–6 Find the area of each figure.

25.

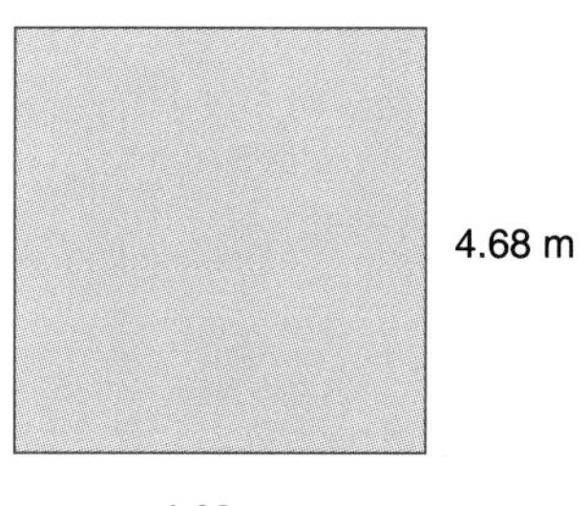

26. Use 3.14 for π.

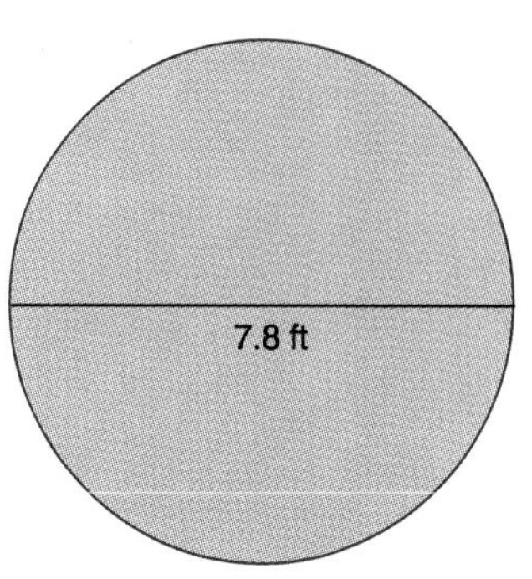

27.

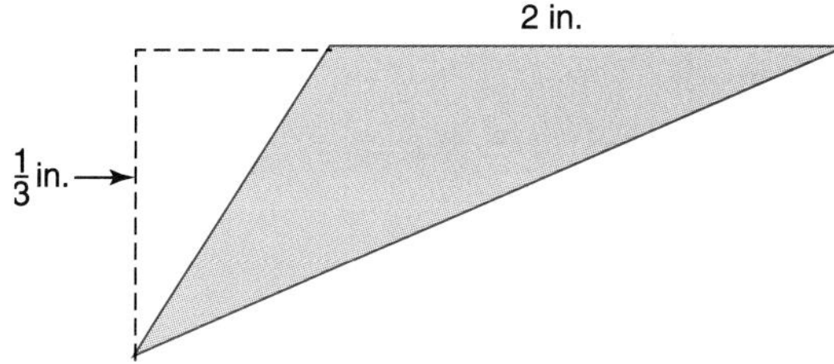

28. How many square tiles, each 12 in. on a side, will cover a floor 21 ft by 15 ft?

11–7 Convert.

29. 42 yd^2 to ft^2

30. 120 in^2 to ft^2

31. 50 mi^2 to acres

32. 58 ft^2 to yd^2

33. 6.7 ha to m^2

34. 0.9 km^2 to m^2

35. 23.9 cm^2 to m^2

36. 13 m^2 to ha

11–8 Find the volume

37.

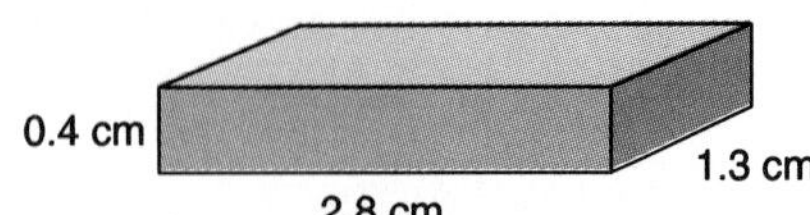

Convert.

38. 4.6 hℓ to ℓ

39. 4076 cℓ to ℓ

40. 9.84 hℓ to dℓ

41. 25 kℓ to dk ℓ

42. How many cubic yards of dirt must be removed to make a hole $3\frac{1}{2}$ yd by 5 yd by $2\frac{1}{8}$ yd?

11–9 Convert.

43. 84 kg to g

44. 306 dkg to g

45. 11 kg to cg

46. 4898 mg to kg

11–10 Convert by estimation. Use the scales on page 448.

47. 147°F to °C

48. 84°C to °F

Convert. Use the formulas on page 449.

49. 98°F to °C

50. 21°C to °F

51. Normal body temperature is 98.6°F. How many degrees Celsius is this?

11–11 Convert.

52. 78 ft to cm

53. 8 kg to lb

54. 109 qt to ℓ

55. 5.67 m to in.

56. The marked speed limit on the freeway is 60 mph. What is the speed limit in kilometers per hour?

CHAPTER 12

Algebra Preview

Choice: Either skip the Pretest and study all of Chapter 12 or take the Pretest and study only the section(s) you need to study.

Pretest

Directions: This Pretest will help you determine which sections to study in Chapter 12. If any of your answers are incorrect, or if you omitted any exercise, turn to the indicated sections and study the material; then take the Chapter Test.

12–1 Write the following numbers in words.

1. -56.4

2. $+5\frac{1}{8}$

3. Name point A on the number line.

Place $<$ or $>$ symbols between the following numbers to make a true statement.

4. $-29 \quad 0$

5. $-\frac{1}{8} \quad -\frac{7}{8}$

6. $-2.05 \quad -3.05$

7. Find the indicated absolute value. $\left|-\frac{2}{3}\right|$

12–2 Add.

8. $5 + (-2)$

9. $-\frac{5}{9} + 16$

10. $-246 + (-56.07)$

12–3 11. What is the opposite of -5.2?

Subtract.

12. $15 - 39$

13. $-2\frac{7}{15} - \left(-13\frac{2}{5}\right)$

14. $-9.314 - (18.6)$

12–4 Multiply.

15. -6×5

16. $3\left(-2\frac{1}{3}\right)$

17. $-6.04(-0.47)$

Divide.

18. $-18 \div 9$

19. $-\frac{7}{10} \div \frac{14}{15}$

20. $-9.19 \div (-0.21)$

12–5 Solve.

21. $x - 8 = 20$

22. $-4y - 7 = 1$

Solve. Write a statement for your answer.

23. If 13 is subtracted from 3 times a number, the result is 20. What is the number?

24. If 30% of the yearly wage of a worker is $5100, find his total annual income.

25. The area of an automobile showroom is 6750 ft. If the length is 90 ft, what is the width?

26. Two spherical vats weigh a total of 205 lb. The first vat weighs 95 lb more than the second. Find the weight of each.

The answers are in the margin on page 469.

12–1 INTRODUCTION TO SIGNED NUMBERS

Up to this point in the text we have learned to use the arithmetic of whole numbers, fractions, and decimals. These numbers, however useful, will not give answers to all practical problems that you may face. For instance, suppose the temperature is 15°C. If the temperature drops by 20°C, you can see by the figure that the new temperature is 5°C below zero. (Beginning at 15° count 20° downward.)

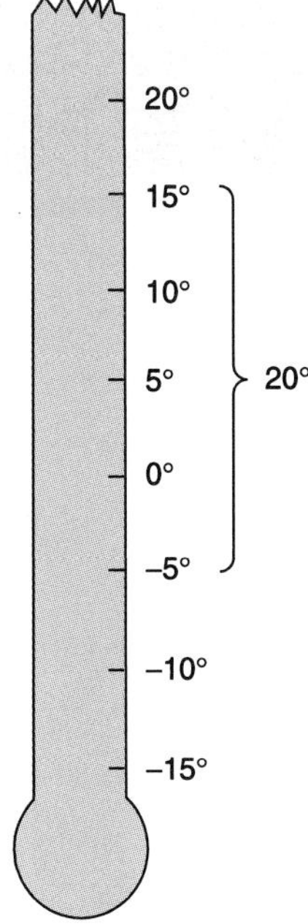

Figure 12–1

All numbers less than zero are called negative numbers and are expressed with a negative sign ($-$) preceding the number. In the preceding example, 5°C below zero is written as -5°C and is read, "negative five degrees Celsius."

All numbers greater than zero are called positive numbers and may be written with or without a positive sign ($+$). Thus 15°C above zero may be written as $+15$°C, or 15°C.

■ Example 1

Write the following numbers in words:

a. -39 **b.** $+56$ **c.** -18.2 **d.** $\frac{3}{8}$

Solution **a.** negative thirty-nine

b. positive fifty-six, or fifty-six

c. negative eighteen and two-tenths

d. positive three-eighths, or three-eighths

Practice check: Do exercises 1–4 in the margin.

Write the following numbers in words.

1. $+152$ **2.** -46

3. $-5\frac{1}{4}$ **4.** 29.04

The answers are in the margin on page 471.

The Number Line

Negative and positive numbers can be pictured on a number line as shown in the figure.

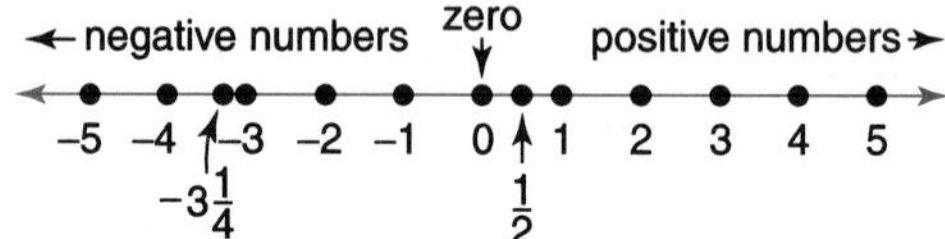

Figure 12–2

Negative numbers are to the left of zero, and positive numbers are to the right of zero.

■ Example 2

Name point *A* on the number line.

a.

A 0 1

Figure 12–3

b.

-24 -23 A

Figure 12–4

Solution **a.** -4 **b.** -18

■ **Example 3**

Locate $-\frac{1}{3}$ on a number line.

Solution Divide the space between 0 and -1 into 3 equal parts.

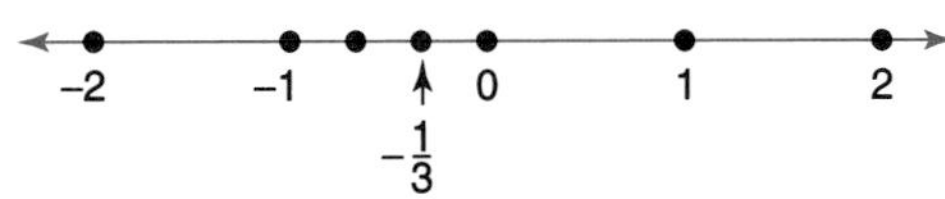

Figure 12–5

The mark closest to 0 is $-\frac{1}{3}$.

■ **Example 4**

Locate $-3\frac{3}{5}$ on a number line.

Solution Divide the space between -3 and -4 into 5 equal parts.

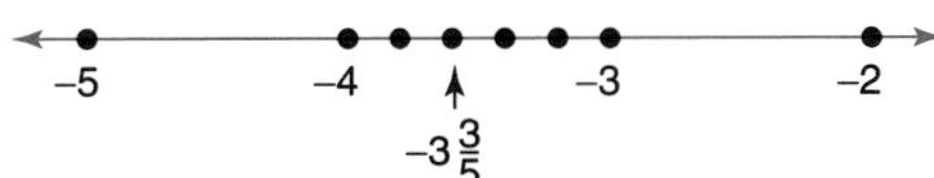

Figure 12–6

The third mark to the left of -3 is $-3\frac{3}{5}$.

Practice check: Do exercises 5–8 in the margin.

Order and Signed Numbers

On a number line a number to the left is less than a number to its right.

■ **Examples**

5. $-3 < 2$ **6.** $-4 < -2$

7. $2\frac{1}{2} > -2\frac{1}{2}$ **8.** $-\frac{1}{3} > -\frac{2}{3}$

Practice check: Do exercises 9–14 in the margin.

Absolute Value

Consider the following number line.

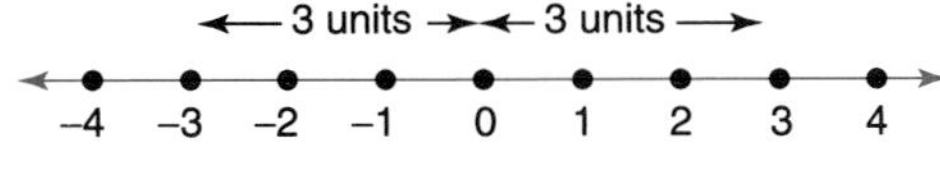

Figure 12–7

Answers to Pretest

1. negative fifty-six and four tenths
2. plus five and one eighth
3. -20 **4.** $<$ **5.** $>$
6. $>$ **7.** $\frac{2}{3}$ **8.** 3 **9.** $15\frac{4}{9}$
10. -302.07 **11.** 5.2 **12.** -24
13. $10\frac{14}{15}$ **14.** -27.914
15. -30 **16.** -7 **17.** 2.8388
18. -2 **19.** $-\frac{3}{4}$ **20.** 43.76
21. 28 **22.** -2 **23.** 11
24. $17,000 **25.** 75 ft
26. 55 lb and 150 lb

Name point A on the number line.

5.

A 0 1

6.

A −16 −15

7. Locate $-1\frac{2}{3}$ on a number line.

8. Locate $-6\frac{3}{4}$ on a number line.

The answers are in the margin on page 471.

Place $<$ and $>$ symbols between the following numbers to make true statements.

9. 5 -6 **10.** 0 13

11. -4 -1 **12.** -8 -12

13. $-3\frac{1}{3}$ 0 **14.** $-\frac{3}{4}$ $-\frac{1}{2}$

The answers are in the margin on page 471.

The distance from 0 to +3 is 3 units.

The distance from −3 to 0 is 3 units.

Since distance is always expressed as a nonnegative number (0, or a positive number), you can see from the number line example that the following statements are true:

The distance from 0 to +23 is 23 units.

The distance between −13 and 0 is 13 units.

The distance from $-3\frac{1}{3}$ to 0 is $3\frac{1}{3}$ units.

Rule

Definition of Absolute Value

The absolute value of a number is the distance that number is from zero and this value is never negative.

Absolute value is indicated by vertical bars on each side of the number. So, $|-6|$ means "the absolute value of −6."

Examples

9. $|-15| = 15$ **10.** $\left|2\frac{2}{3}\right| = 2\frac{2}{3}$ **11.** $|0| = 0$

Practice check: Do exercises 15–24 in the margin.

Find.

15. $|-19|$ **16.** $|11|$

17. $|0|$ **18.** $\left|-66\frac{2}{3}\right|$

19. $|32.4|$ **20.** $|-532|$

21. $\left|\frac{3}{8}\right|$ **22.** $|-0.008|$

23. $|3470|$ **24.** $\left|-7\frac{5}{8}\right|$

The answers are in the margin on page 473.

To help you understand addition and subtraction of signed numbers, we will make use of the number line and absolute value in sections 12–2 and 12–3.

Looking Back

You should now be able to

1. Write positive and negative numbers in words.
2. Name and locate positive and negative numbers on a number line.
3. Decide which signed number is larger or smaller than another.
4. Find the absolute value of numbers.

Problems 12–1

Write the following numbers in words.

1. +7 **2.** −18 **3.** 47 **4.** +15

5. −77 **6.** −43.8 **7.** $5\frac{3}{8}$ **8.** −563.03

9. $-29\frac{4}{7}$ **10.** $+\frac{2}{19}$

Name point A on the following number lines:

11.

12.

13.

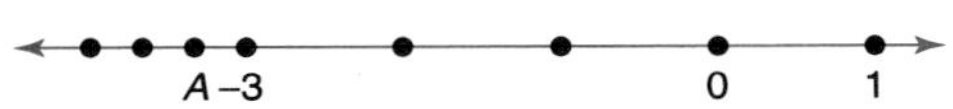

14.

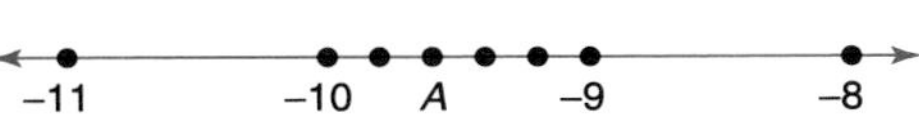

Answers to exercises 1–4

1. positive one hundred fifty-two, or one hundred fifty-two
2. negative forty-six
3. negative five and one-fourth
4. positive twenty-nine and four hundredths, or twenty-nine and four hundredths

Answers to exercises 5–8

5. -3 **6.** -19

7.

−3 −2 $-1\frac{2}{3}$ −1 0

8.

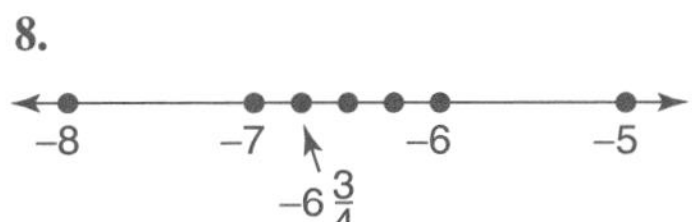

Answers to exercises 9–14

9. $5 > -6$ **10.** $0 < 13$
11. $-4 < -1$ **12.** $-8 > -12$

13. $-3\frac{1}{3} < 0$ **14.** $-\frac{3}{4} < -\frac{1}{2}$

Locate the following numbers on a number line.

15. $2\frac{2}{3}$ **16.** $4\frac{3}{5}$ **17.** $-1\frac{1}{2}$ **18.** $-3\frac{3}{4}$

Place < or > symbols between the following numbers to make a true statement.

19. $-6 \quad 10$ **20.** $-7 \quad 13$ **21.** $39 \quad 0$ **22.** $-39 \quad 0$

23. $-4 \quad -8$ **24.** $-13 \quad -21$ **25.** $3 \quad -5$ **26.** $7 \quad -13$

27. $-5 \quad -1$ **28.** $-10 \quad -6$ **29.** $0 \quad 5$ **30.** $0 \quad -8$

31. $-13 \quad -12$ **32.** $-35 \quad -25$ **33.** $-2\frac{1}{8} \quad -3\frac{2}{3}$ **34.** $-1\frac{1}{4} \quad -\frac{3}{4}$

35. $-\frac{1}{5} \quad -\frac{3}{5}$ **36.** $-\frac{3}{4} \quad -\frac{5}{8}$ **37.** $-0.53 \quad -0.49$ **38.** $-1.04 \quad -2.04$

Find. Think of a number line and the distance from zero.

39. $|-16|$ **40.** $|-5|$ **41.** $|23|$ **42.** $|90|$

43. $|108|$ **44.** $|379|$ **45.** $|-5060|$ **46.** $|-8467|$

Answers to exercises 15–24

15. 19 **16.** 11 **17.** 0 **18.** $66\frac{2}{3}$

19. 32.4 **20.** 532 **21.** $\frac{3}{8}$

22. 0.008 **23.** 3470 **24.** $7\frac{5}{8}$

47. $\left|32\frac{1}{2}\right|$ **48.** $\left|\frac{7}{8}\right|$ **49.** $\left|-18\frac{7}{8}\right|$ **50.** $\left|-\frac{4}{19}\right|$

51. $|-0.04|$ **52.** $|-13.641|$ **53.** $|0|$ **54.** $|6.4|$

Find two numbers whose absolute value is

55. 3 **56.** 8 **57.** $56\frac{5}{8}$ **58.** $\frac{9}{16}$

59. 109.7 **60.** 0.643

12–2 ADDITION OF SIGNED NUMBERS

We can use the number line to explain addition of signed numbers by using the following steps.

Rule

Addition of Signed Numbers Using the Number Line
Draw and direct an arrow from the point of the first number as follows:

1. *Direct the arrow to the right if the second number is positive.*
2. *Direct the arrow to the left if the second number is negative.*
3. *Stay at the first number if the second number is 0.*

Draw the arrow the unit length of the second number.
The sum is at the tip of the arrow.

■ **Example 1**

Add $2 + (-3)$.

Solution

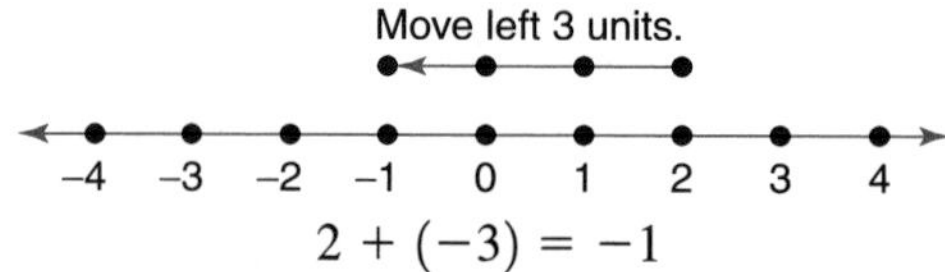

$$2 + (-3) = -1$$

Figure 12–8

■ **Example 2**

Add $-3 + (-2)$.

Solution

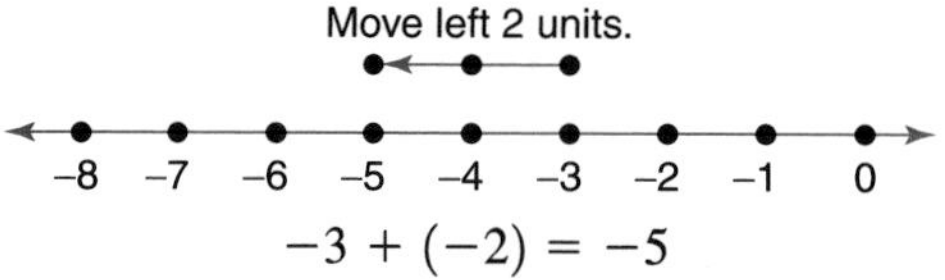

$$-3 + (-2) = -5$$

Figure 12–9

■ **Example 3**

Add $-5 + 7$.

Solution

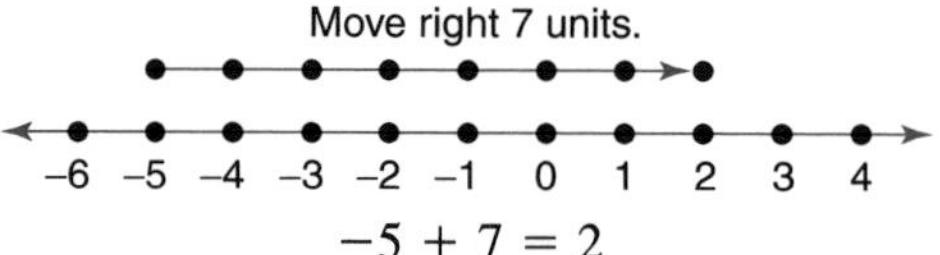

$$-5 + 7 = 2$$

Figure 12–10

■ **Example 4**

Add $-4 + 4$.

Solution

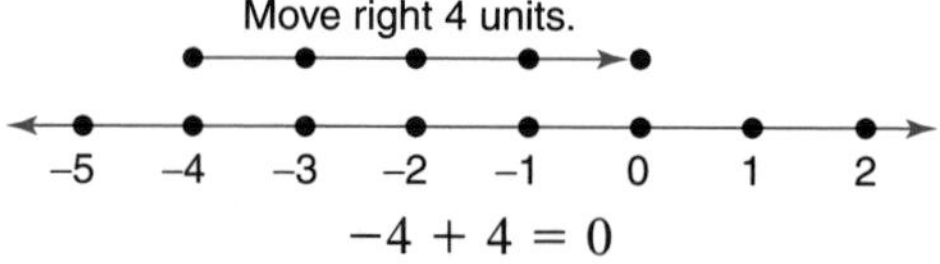

$$-4 + 4 = 0$$

Figure 12–11

■ **Example 5**

Add $-3 + 0$.

Solution

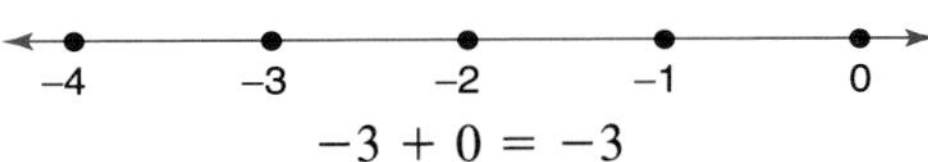

$-3 + 0 = -3$

Figure 12–12

Practice check: Do exercises 25–30 in the margin.

Add. Use a number line.

25. $5 + 2$ **26.** $3 + (-1)$

27. $-2 + (-5)$ **28.** $-5 + 3$

29. $-2 + 0$ **30.** $3 + (-3)$

The answers are in the margin on page 477.

Since it is impractical to find sums for examples such as $-1689 + 389$ on a number line, the short method as described below is more useful for quick calculations.

Addition with the Same Sign

If the signs of the two numbers to be added are both positive or both negative use the following steps.

Rule

A Short Method for Adding Two Numbers with the Same Sign

Step 1 *Find the sum of the absolute values of the numbers.*

Step 2 *a. Make the sum positive if the numbers to be added are positive.*
b. Make the sum negative if the numbers to be added are negative.

■ **Examples**

Add. **6.** $5 + 8 = 13$ **7.** $-5 + (-8) = -13$

8. $-13\frac{1}{6} + \left(-2\frac{2}{3}\right) = -15\frac{5}{6}$

Note:

$$\begin{array}{rr} 13\frac{1}{6} = & 13\frac{1}{6} \\ +2\frac{2}{6} = & 2\frac{4}{6} \\ \hline & 15\frac{5}{6} \end{array}$$

Practice check: Do exercises 31–36 in the margin.

Add using the short method.

31. $15 + 43$ **32.** $24\frac{3}{4} + 9\frac{7}{18}$

33. $-8 + (-7)$ **34.** $-26 + (-17)$

35. $-43\frac{3}{8} + \left(-\frac{9}{16}\right)$

36. $-4.05 + (-32.532)$

The answers are in the margin on page 477.

Addition with Different Signs

If the signs of the two numbers to be added are different, use the following steps.

Rule

A Short Method for Adding Two Numbers with Different Signs

Step 1 *Subtract the smaller absolute value from the larger absolute value.*

Step 2 *a. Make the difference positive if the larger absolute value is a positive number.*

b. Make the difference negative if the larger absolute value is a negative number.

■ Example 9

Add $-9 + 4$

Solution *Think:* $|-9| = 9$ and $|4| = 4$. From this we see that $|-9| > |4|$.

Step 1 Subtract. $9 - 4 = 5$

Step 2b Make the answer negative. $-9 + 4 = -5$

Note: We make the answer negative, since -9 is negative and $|-9| > |4|$.

■ Examples

Add. **10.** $7 + (-5) = 2$ **11.** $\frac{1}{2} + \left(-\frac{3}{4}\right) = -\frac{1}{4}$

Explanation The answer in Example 10 is positive, since 7 is positive and $|7| > |-5|$. The answer in Example 11 is negative, since $-\frac{3}{4}$ is negative and $\left|-\frac{3}{4}\right| > \left|\frac{1}{4}\right|$.

Practice check: Do exercises 37–44 in the margin.

Add using the short method.

37. $6 + (-9)$ **38.** $-8 + 5$

39. $-15 + 23$ **40.** $89 + (-104)$

41. $\frac{4}{9} + \left(-\frac{1}{6}\right)$ **42.** $-8\frac{7}{8} + 3\frac{5}{24}$

43. $-2.5 + 0.15$

44. $10.26 + (-4.319)$

The answers are in the margin on page 479.

■ Looking back

You should now be able to add signed numbers using the number line and using the short method.

■ Problems 12–2

Review

Write the following numbers in words.

1. $+9$ **2.** -53 **3.** -35.18 **4.** $6\frac{9}{17}$

5. Name point *A* on the number line.

Find the absolute value of the following numbers.

6. $+82$ **7.** -13 **8.** 0 **9.** $-16\frac{1}{2}$

10. 34.04

Answers to exercises 25–30

25. 7 **26.** 2 **27.** −7 **28.** −2 **29.** −2 **30.** 0

Answers to exercises 31–36

31. 58 **32.** $34\frac{5}{36}$ **33.** −15 **34.** −43 **35.** $-43\frac{15}{16}$ **36.** −36.582

Find.

11. $|-8|$ **12.** $|13|$ **13.** $|-309|$ **14.** $\left|\frac{6}{7}\right|$

15. $|0|$ **16.** $|-3.43|$ **17.** $\left|-52\frac{1}{8}\right|$ **18.** $|0.058|$

Add. Use a number line.

19. 3 + 3 **20.** 5 + 2 **21.** 7 + (−2) **22.** 3 + (−1)

23. −6 + 4 **24.** 3 + (−7) **25.** −4 + (−3) **26.** −6 + (−2)

27. −3 + 0 **28.** 5 + (−5)

Add. Use the short method.

29. 18 + 39 **30.** 32 + 98 **31.** −9 + (−6) **32.** −5 + (−7)

33. 7 + (−4) **34.** −6 + 8 **35.** 3 + (−7) **36.** −9 + 5

37. $15\frac{1}{4} + \frac{5}{8}$

38. $3\frac{7}{10} + 8\frac{7}{8}$

39. $32 + (-18)$

40. $-74 + 104$

41. $-37 + (-29)$

42. $-146 + (-39)$

43. $-67 + 43$

44. $346 + (-589)$

45. $\frac{3}{5} + \left(-\frac{1}{4}\right)$

46. $\left(-\frac{3}{7}\right) + \frac{9}{14}$

47. $-\frac{7}{18} + 19$

48. $11\frac{3}{8} + \left(-5\frac{9}{16}\right)$

49. $-4\frac{3}{10} + \left(-8\frac{2}{5}\right)$

50. $-22\frac{3}{4} + \left(-13\frac{5}{18}\right)$

51. $-22\frac{3}{5} + 14\frac{7}{15}$

52. $6\frac{3}{5} + \left(-10\frac{1}{2}\right)$

53. $-\frac{7}{12} + \frac{1}{6}$

54. $\frac{5}{18} + \left(-\frac{3}{4}\right)$

55. $-481.6 + (-32.08)$

56. $-72.41 + (-3.682)$

57. $3.0305 + (-2.6732)$

58. $-1.9565 + 4.9$

59. $-5.05 + 3.9253$

60. $10.43 + (-17)$

12–3 SUBTRACTION OF SIGNED NUMBERS

Opposites of Signed Numbers

Subtraction is the opposite of addition. To subtract signed numbers you will need to know the meaning of opposite numbers. Study the number line below.

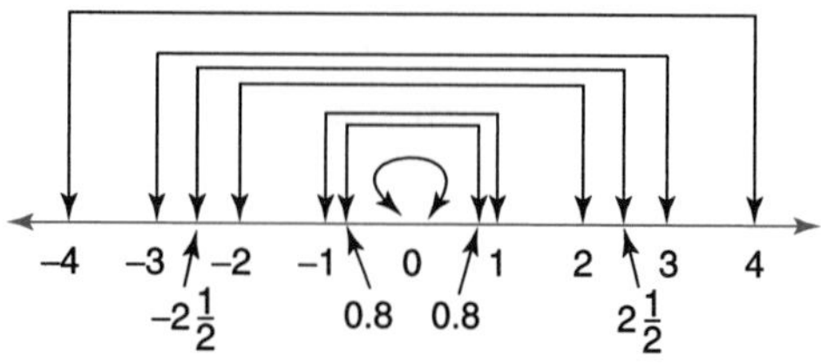

Figure 12–13

Every number has an opposite. Each of the following pairs of numbers are opposites to each other.

1 and −1, −3 and 3, 0.8 and −0.8, −22 and 22, −64.4 and 64.4, $8\frac{7}{8}$ and $-8\frac{7}{8}$, 0 and 0

Notice that 0 is the opposite of itself.

Practice check: Do exercises 45–54 in the margin.

Answers to exercises 37–44

37. −3 **38.** −3 **39.** 8

40. −15 **41.** $\frac{5}{18}$ **42.** $-5\frac{2}{3}$

43. −2.35 **44.** 5.941

Find the opposite of the following numbers.

45. 5 **46.** −5

47. −18 **48.** $23\frac{1}{7}$

49. 0 **50.** −432.08

51. $-\frac{3}{4}$ **52.** 2034

53. −0.009 **54.** 32

The answers are in the margin on page 481.

Number Line Subtraction

We can subtract signed numbers on a number line by using the following steps.

Rule

Subtraction of Signed Numbers Using the Number Line
Draw and direct an arrow from the point of the first number as follows:

1. *Direct the arrow to the left if the second number is positive.*
2. *Direct the arrow to the right if the second number is negative.*
3. *Stay at the first number if the second number is 0.*

Draw the arrow the unit length of the second number.
The difference is at the tip of the arrow.

Note: In subtraction the arrow is in opposite direction from the arrow in addition. Subtraction is the opposite of addition.

■ **Example 1**

Subtract 5 − 2.

Solution

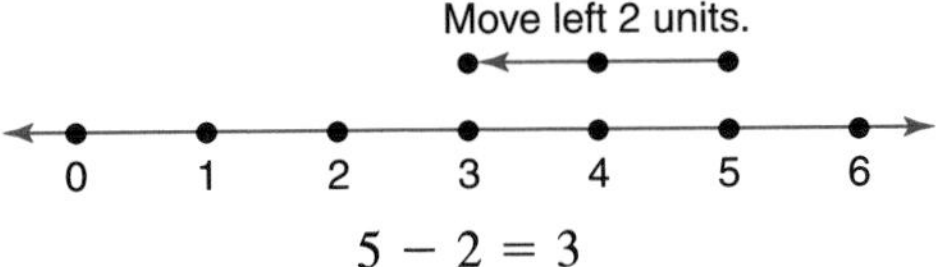

Figure 12–14

Remember to draw the arrow in the opposite direction because subtraction is the opposite of addition.

Example 2

Subtract $2 - 5$.

Solution

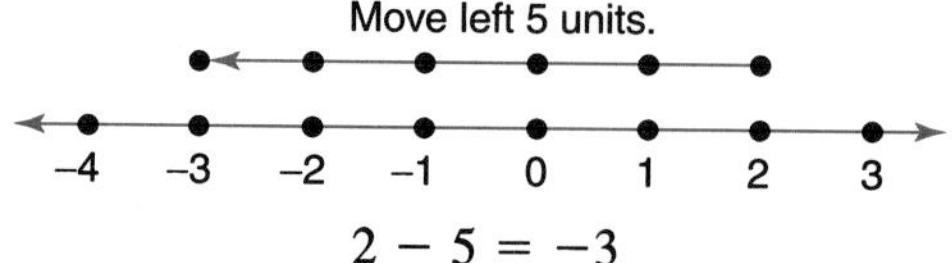

$$2 - 5 = -3$$

Figure 12–15

Example 3

Subtract $-2 - (-3)$.

Solution

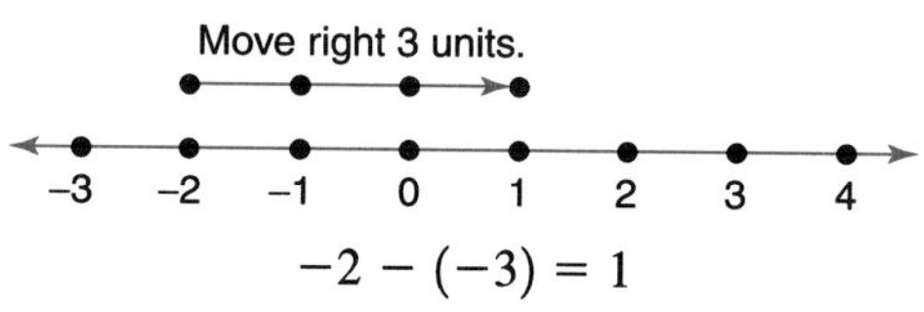

$$-2 - (-3) = 1$$

Figure 12–16

Explanation Since -3 is negative, the arrow moves in a positive direction.

Example 4

Subtract $-4 - 6$.

Solution

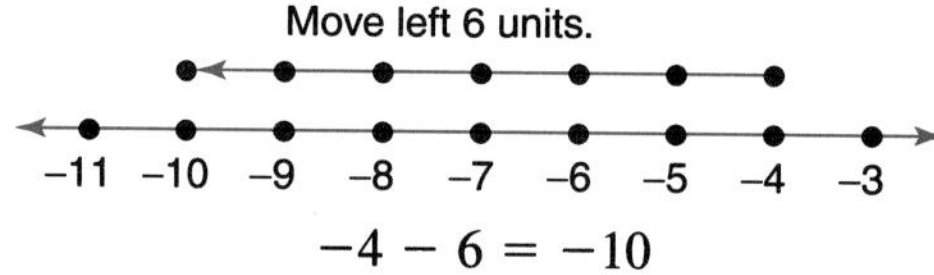

$$-4 - 6 = -10$$

Figure 12–17

Practice check: Do exercises 55–61 in the margin.

Subtract using a number line.

55. $5 - 3$ **56.** $3 - 7$

57. $5 - 6$ **58.** $1 - (-3)$

59. $-3 - (5)$ **60.** $5 - (-3)$

61. $-4 - (-4)$

The answers are in the margin on page 482.

Again, as with addition of signed numbers, the short method as described below should be used for quick calculations.

Short Method for Subtraction

To subtract signed numbers you simply add the opposite of the number to be subtracted. To find the difference follow these steps.

Rule

A Short Method for Subtracting Signed Numbers

Step 1 *Change the subtraction sign to an addition sign.*

Step 2 *Replace the number to be subtracted by its opposite.*
Step 3 *Find the difference by adding.*

Answers to exercises 45–54

45. -5 **46.** 5 **47.** 18

48. $-23\frac{1}{7}$ **49.** 0 **50.** 432.08

51. $\frac{3}{4}$ **52.** -2034 **53.** 0.009

54. -32

■ Example 5

Subtract $3 - 9$.

Solution $3 - 9 = ?$

Step 1	$3 +$	Step 1 above
Step 2	$3 + (-9)$	The opposite of 9 is -9. See Step 2 above.
Step 3	$3 + (-9) = -6$	Step 3 above

■ Examples

Subtract.

6. $3 - (-4) = 3 + (4) = 7$ The opposite of -4 is 4.
7. $-3 - (-3) = -3 + (3) = 0$
8. $13 - (-5) = 13 + (5) = 18$
9. $-29 - 32 = -29 + (-32) = -61$ The opposite of 32 is -32
10. $-\frac{3}{4} - \left(-\frac{5}{16}\right) = -\frac{3}{4} + \left(+\frac{15}{16}\right) = \frac{3}{16}$
11. $53.8 - 68.2 = 53.8 + (-68.2) = -14.4$

Practice check: Do exercises 62–68 in the margin.

Subtract using the short method.

62. $15 - 8$ **63.** $46 - 64$

64. $26 - (-19)$

65. $-39 - (-56)$

66. $-13 - (54)$ **67.** $\frac{3}{12} - \left(\frac{5}{8}\right)$

68. $-301.6 - (-47.5)$

The answers are in the margin on page 482.

■ Looking Back

You should now be able to subtract signed numbers using the number line and the short method.

■ Problems 12–3

Review

1. Write $-13\frac{5}{8}$ in words.

2. Name point A on the number line.

Answers to exercises 55–61

55. 2 **56.** -4 **57.** -1 **58.** 4
59. -8 **60.** 8 **61.** 0

Answers to exercises 62–68

62. 7 **63.** -18 **64.** 45 **65.** 17
66. -67 **67.** $-\frac{3}{8}$ **68.** -254.1

3. Find the absolute value of 568.9.

Add. Use the number line.

4. $5 + (-4)$ **5.** $-3 + (-4)$

Add using the short method.

6. $8 + (-9)$ **7.** $-73 + (-104)$ **8.** $-\frac{5}{8} + \frac{7}{12}$

Find the opposite of the following numbers.

9. 6 **10.** 15 **11.** -6 **12.** -16

13. $26\frac{3}{8}$ **14.** $3\frac{7}{9}$ **15.** -0.041 **16.** -3.610

Subtract using a number line.

17. $7 - 4$ **18.** $6 - 2$ **19.** $4 - 6$ **20.** $2 - 7$

21. $1 - 4$ **22.** $3 - 9$ **23.** $2 - (-4)$ **24.** $6 - (-8)$

25. $-5 - (-4)$ **26.** $-1 - (-3)$ **27.** $3 - (-4)$ **28.** $8 - (-2)$

The factors are both negative.

$-1 \times -2 = 2$
$-2 \times -2 = 4$
$-3 \times -2 = 6$
etc.

The product is positive.

This pattern shows that the product of two negative numbers is positive. This is true for all numbers.

Rule

To find the product of two numbers having the same sign, we multiply their absolute values and make the product positive.

Examples

Multiply.

5. $-9(-8) = 72$

6. $6 \times 7 = 42$

7. $-\frac{2}{7}\left(-\frac{1}{2}\right) = \frac{1}{7}$

8. $-31(-6) = 186$

Practice check: Do exercises 75–80 in the margin.

Find the product.

75. $5 \cdot 9$

76. $-6(-8)$

77. $-21(-251)$

78. $108 \cdot 32$

79. $-\frac{2}{5}\left(-\frac{5}{7}\right)$

80. $-3.61(-8.04)$

The answers are in the margin on page 487.

Dividing with Signed Numbers

We can use the rules for multiplication of signed numbers to discover the rules for dividing signed numbers.
Consider the following examples: We know that since

$$-2 \times -5 = 10, \quad \text{then} \quad 10 \div -2 = -5$$

and since

$$3 \times -2 = -6, \quad \text{then} \quad -6 \div 3 = -2$$

From these examples we can see that the quotient of two numbers with different signs is negative.

On the other hand, suppose the numbers to be divided are both negative as in $-12 \div -6$?
Since

$$-6 \times 2 = -12, \quad \text{then} \quad -12 \div -6 = 2$$

This example illustrates that the quotient of two numbers with the same sign is positive. Notice that the rules for multiplying and dividing with signed numbers are the same.

Rule

When multiplying or dividing with two signed numbers the answer is

a. Positive if the signs are the same.
b. Negative if the signs are different.

■ Examples

Divide.

9. $-8 \div 4 = -2$

10. $36 \div -9 = -4$

11. $-42 \div -2 = 21$

12. $\frac{7}{8} \div -\frac{5}{12} = -2\frac{1}{20}$

13. $-5.8 \div -0.02 = 290$

Practice check: Do exercises 81–87 in the margin.

Divide.

81. $18 \div 9$

82. $-81 \div 3$

83. $-2432 \div 4$

84. $-64 \div (-16)$

85. $-\frac{8}{9} \div \frac{4}{3}$

86. $-1\frac{3}{5} \div \left(-8\frac{2}{3}\right)$

87. $-76.38 \div (-19)$

The answers are in the margin on page 489.

■ Looking Back

You should now be able to multiply and divide signed numbers.

Exercise 12–4

Review

1. Name point *A* on the number line.

2. Evaluate $|-16.9|$.

3. Add $-82 + 56$.

4. Add $-\frac{7}{6} + \left(-\frac{5}{24}\right)$.

5. What is the opposite of -0.015?

6. Subtract $-348 - (-976)$.

7. Subtract $-6.43 - (27.6)$.

8. Subtract $3\frac{3}{4} - 7\frac{1}{8}$.

Answers to exercises 69–74

69. -36 **70.** -377 **71.** -2176

72. $-\frac{1}{9}$ **73.** $-3\frac{3}{5}$ **74.** -4.2658

Answers to exercises 75–80

75. 45 **76.** 48 **77.** 5271

78. 3456 **79.** $\frac{2}{7}$ **80.** 29.0244

Multipy.

9. 3×4 **10.** 8×7 **11.** -6×8 **12.** -5×4

13. $-7(-9)$ **14.** $-4(-6)$ **15.** $9(-8)$ **16.** $3(-5)$

17. 16×32 **18.** 66×90 **19.** $-43(-21)$ **20.** $-19(-84)$

21. $114(-39)$ **22.** $346(-104)$ **23.** $-200(-32)$ **24.** $-400(-85)$

25. $-76(5)$ **26.** $-124(9)$ **27.** $-\frac{7}{8} \times \frac{12}{21}$ **28.** $\frac{2}{3}\left(-\frac{9}{10}\right)$

29. $9.8(-0.5)$ **30.** -13.04×2.6 **31.** $-\frac{7}{12}\left(-\frac{4}{21}\right)$ **32.** $-\frac{9}{10}\left(-\frac{35}{18}\right)$

33. $-5.12(-8.4)$ **34.** $-0.007(-1.72)$

Divide.

35. $15 \div 3$ **36.** $6 \div 2$ **37.** $-16 \div 8$ **38.** $-35 \div 7$

39. $-42 \div (-3)$ **40.** $-75 \div (-5)$ **41.** $234 \div 26$ **42.** $5447 \div 13$

43. $-1617 \div (-7)$ **44.** $-344 \div (-4)$ **45.** $1736 \div (-31)$

46. $1770 \div (-59)$ **47.** $-2060 \div (-20)$ **48.** $-1976 \div (-19)$

49. $-1888 \div 8$ **50.** $-1836 \div 6$ **51.** $-\frac{7}{8} \div \frac{5}{12}$ **52.** $\frac{6}{7} \div \left(-\frac{9}{10}\right)$

53. $2.1 \div (-3.5)$ **54.** $0.0072 \div (-0.9)$ **55.** $-\frac{7}{8} \div \left(-\frac{7}{6}\right)$ **56.** $-\frac{15}{16} \div \frac{5}{14}$

57. $0.6552 \div (-0.021)$ **58.** $46.5 \div (-2.5)$ **59.** $4\frac{3}{4} \div \left(-\frac{7}{16}\right)$ **60.** $-2\frac{7}{8} \div 5\frac{1}{2}$

12–5 BASIC EQUATIONS

In this chapter you will learn the fundamentals of solving equations. Equations are used to solve problems in higher mathematics and everyday mathematics including technical, occupational, and vocational problems involving mathematics.

Members of an Equation

Every equation has a left member and a right member separated by an equal sign.

■ **Example 1**

$$\underbrace{4 + x}_{\text{Left member}} \overset{\text{Equal sign}}{=} \underbrace{6}_{\text{Right member}}$$

Answers to exercises 81–87

81. 2 **82.** −27 **83.** −608

84. 4 **85.** $-\frac{2}{3}$ **86.** $\frac{12}{65}$ **87.** 4.02

Variables in Equations

Equations contain one or more letters called variables.

■ Example 2

$$3x = 6$$

↑ Variable (points to x)

Solutions to Equations

To find a solution for an equation means to find a number or numbers to replace the variable or variables in order to make the left member equal to the right member. When this is done, the replaced number or numbers are the solutions.

■ Example 3

Find the solution for $x + 2 = 3$.

Solution If x is replaced by 1, the equation is true.

Proof

$$1 + 2 = 3$$
$$3 = 3$$

The solution is 1.

■ Example 4

Find the solution for $5 - x = 3$.

Solution If x is replaced by 2, the equation is true.

Proof

$$5 - 2 = 3$$
$$3 = 3$$

The solution is 2.

■ Example 5

Find the solution for $6x = 18$.

Solution If x is replaced by 3, the equation is true.

Proof

$$6(3) = 18$$
$$18 = 18$$

The solution is 3.

■ Example 6

Find the solution for $8x = 2$.

Solution If x is replaced by $\frac{1}{4}$, the equation is true.

Proof

$$8\left(\frac{1}{4}\right) = 2$$

$$2 = 2$$

The solution is $\frac{1}{4}$.

■ Example 7

Find the solution for $5x - 2 = -17$.

Solution If x is replaced by -3, the equation is true.

Proof

$$5(-3) - 2 = -17$$

$$-15 - 2 = -17$$

$$-17 = -17$$

The solution is -3.

Practice check: Do exercises 88 and 89 in the margin.

Find the solutions.

88. $7 = x + 10$

89. $16 = -8x$

The answers are in the margin on page 493.

Solving Basic Equations

To solve equations, the object is to isolate the variable with its coefficient equal to 1 on either the left member or the right member of the equation.

Rule

To simplify a basic equation

1. *Multiply or divide both sides of the equation by the same nonzero number.*
2. *Add or subtract the same number to both sides of the equation.*

■ Example 8

Solve $a + 9 = 12$

$a + 9 - 9 = 12 - 9$ Subtract 9 from both sides.

$a + 0 = 3$ Simplifying

$a = 3$

Proof

$$3 + 9 = 12$$

$$12 = 12$$

The solution is 3.

■ Example 9

Solve $11 + y = 20$

$11 - 11 + y = 20 - 11$ Subtract 11 from both sides.

$0 + y = 9$ Simplifying

$y = 9$

Proof

$$11 + 9 = 20$$
$$20 = 20$$

The solution is 9.

■ Example 10

Solve

$$x - 7 = 20$$
$$x - 7 + 7 = 20 + 7 \quad \text{Add 7 to both sides.}$$
$$x + 0 = 27 \quad \text{Simplifying}$$
$$x = 27$$

Proof

$$27 - 7 = 20$$
$$20 = 20$$

The solution is 27.

Note: Since the variable x is similar to the multiplication sign (×), the raised · is used in algebra to signify the operation of multiplication.

■ Example 11

Solve

$$9 \cdot b = 27$$
$$\frac{9b}{9} = \frac{27}{9} \quad \text{Divide by 9.}$$
$$b = 3 \quad \text{Simplify.}$$

Proof

$$9 \cdot 3 = 27$$
$$27 = 27$$

The solution is 3.

Practice check: Do exercises 90–92 in the margin.

Solve.

90. $14 = 5 + x$

91. $x - 15 = 30$

92. $35 = 7a$

The answers are in the margin on page 493.

■ Example 12

Solve

$$3a + 8 = 11$$
$$3a + 8 - 8 = 11 - 8 \quad \text{Subtract 8 from both sides.}$$
$$3a + 0 = 3 \quad \text{Simplify.}$$
$$3a = 3$$
$$\frac{3a}{3} = \frac{3}{3} \quad \text{Divide both sides by 3.}$$
$$a = 1 \quad \text{Simplify.}$$

Proof

$$3 \cdot 1 + 8 = 11$$
$$3 + 8 = 11$$
$$11 = 11$$

The solution is 1.

■ **Example 13**

Solve $5x - 2 = 2x + 10$

$5x - 2 + 2 = 2x + 10 + 2$ Add 2 to both sides.

$5x = 2x + 12$ Simplify.

$5x - 2x = 2x - 2x + 12$ Subtract $2x$ from both sides.

$3x = 12$ Simplify.

$\frac{3x}{3} = \frac{12}{3}$ Dividing both sides by 3

$x = 4$ Simplify.

Proof $5 \cdot 4 - 2 = 2 \cdot 4 + 10$

$20 - 2 = 8 + 10$

$18 = 18$

The solution is 4.

■ **Example 14**

Solve $-16x + 6 = 150$

$-16x + 6 - 6 = 150 - 6$ Subtract 6 from both sides.

$-16x = 144$ Simplify.

$\frac{-16x}{-16} = \frac{144}{-16}$ Divide both sides by -16.

$x = -9$ Simplify.

Proof $-16 \cdot (-9) + 6 = 150$

$144 + 6 = 150$

$150 = 150$

The solution is -9.

Practice check: Do exercises 93–95 in the margin.

Solve.

93. $7b + 6 = 20$

94. $3x + 4 = x - 8$

95. $-34 = -9q - 7$

The answers are in the margin on page 495.

Applied Problems

■ **Example 15**

Solve

After a steel rod is milled by 2 inches, the length is 16 inches. What is the length before the milling?

Solution Let x be the length before the milling.

Let $x - 2$ be the final length after the milling.

Since 16 inches is the final length after the milling, the equation is

$$x - 2 = 16$$

$$x - 2 + 2 = 16 + 2 \qquad \text{Add 2 to both sides.}$$

$$x = 18$$

The length before the milling is 18 in.

Answers to exercises 88 and 89

88. -3 **89.** -2

Answers to exercises 90–92

90. $9 = x$ or $x = 9$
The solution is 9.

91. $x - 15 + 15 = 30 + 15$
$x + 0 = 45$
$x = 45$
The solution is 45.

92. $35 = 7a$

$$\frac{35}{7} = \frac{7a}{7}$$

$5 = a$ or $a = 5$
The solution is 5.

■ Example 16

Solve
Four washers used as a spacer on a bolt measure 1.2 cm in length. Find the thickness of each washer.

Solution Let n be the thickness of one washer.
Let $4n$ be the length of 4 washers.
Since 1.2 cm is the length of 4 washers, the equation is

$$4n = 1.2$$

$$\frac{4n}{4} = \frac{1.2}{4} \qquad \text{Divide by 4.}$$

$$n = 0.3$$

Proof

$$4 \cdot 0.3 = 1.2$$

$$1.2 = 1.2$$

Each washer thickness is 0.3 cm.

Practice check: Do exercises 96 and 97 in the margin.

Solve.

96. The sum of a number and 4 is 6. Find the number.

97. Alvistax Rental charges \$21.00 per day and \$0.10 per mile for compacts. If James rented a car for 1 day for a total charge of \$28.00, how many miles did he drive the compact?

The answers are in the margin on page 495.

■ Looking Back

You should now be able to solve simple equations and applied problems using equations.

■ Problems 12–5

Review

1. Evaluate $\left|-\frac{2}{3}\right|$.

2. Place $<$ or $>$ between the numbers to make a true statement.

$-18 \qquad -21$

3. Add $9 + (-8)$.

4. Add $-\frac{2}{3} + \left(-\frac{5}{6}\right)$.

5. Find the opposite of -0.35.

6. Subtract $-48 - (-80)$.

7. Subtract $-4\frac{1}{8} - 16\frac{3}{4}$.

8. Multiply $5(-3)$.

9. Divide $-8\frac{8}{9} \div 6\frac{2}{3}$

10. Divide $39.3 \div (-3)$

Solve.

11. $x + 15 = 22$

12. $3 + x = 3$

13. $21 + y = -4$

14. $a - 4 = 8$

15. $10x = 100$

16. $7y = -56$

17. $4b - 6 = 6$

18. $-12x - 3 = 21$

Solve. Write a statement for your answer.

19. Generic. If 4 is subtracted from a number, the result is 9. What is the number?

20. Generic. The sum of three times a number and 4 is 28. Find the number.

21. Generic. If 5 is increased by 4 times a number, the result is -7. Find the number.

22. Generic. When 8 is subtracted from two times a number, the result is -16. What is the number?

23. Generic. When five times a number is subtracted from -4, the result is 26. What is the number?

Answers to exercises 93–95

93. The solution is 2.
94. The solution is -3..
95. The solution is 3.

Answers to exercises 96 and 97

96. The number is 2.
97. James drove 70 mi.

24. Generic. Dick is 5 years older than his sister Carol. The sum of their ages six years ago was 15. How old are they now?

25. Plumbing. Find the wall thickness of the tubing in the figure.

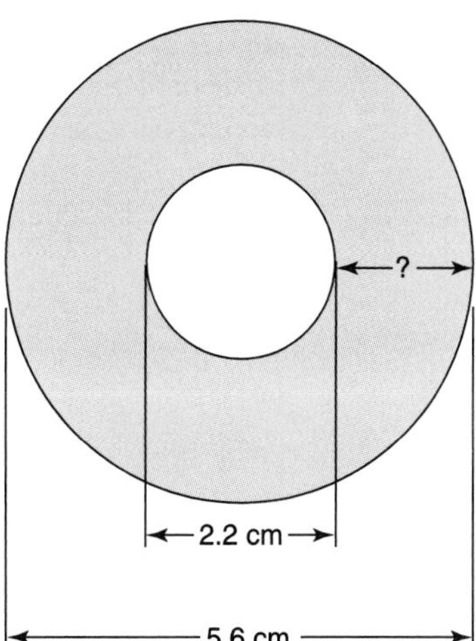

26. Waste Water. One tank holds three times the capacity of another. If both hold a total of 600 gallons, how many gallons do each of the two hold?

27. Accounting. If 70% of the total income of a commercial airplane construction firm is $250,045, what is the total income of the firm?

28. Construction. Juan Fencing was contracted to construct a rectangular wood fence around a yard. The yard is 3 times as long as it is wide and has a perimeter of 352 feet. Find the dimensions of the yard. (Perimeter means distance around a rectangle.)

29. Machine Shop. A shop floor has area 600 sq ft. If the width is 24 ft, how long is the shop?

30. Plumbing. A circular pipe has a circumference 17.27 cm. What is the diameter? Use $C = \pi d$.

31. Accounting. A steel worker keeps 80% of his annual pay due to deductions. If his income is $15,000, what is his total yearly wage?

32. Sheet Metal. Find the missing dimension in the figure.

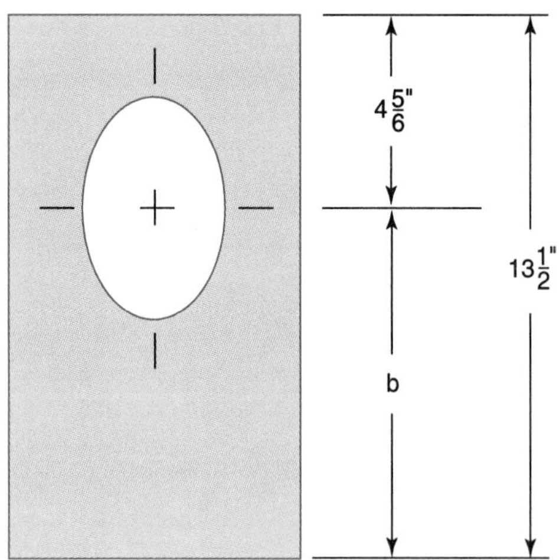

33. Carpentry. A carpenter needs to cut a 9-ft board into 3 pieces. The first board is 2 ft longer than the second, and the third is 1 ft longer than the second. What are the lengths of the three pieces?

CHAPTER 12 OVERVIEW

Summary

12–1 You learned to write positive and negative numbers in words, to locate positive and negative numbers on a number line, to decide which signed number is larger or smaller than another, and to find the indicated absolute value of a number.

12–2 You learned to add signed numbers using the number line and the short method.

12–3 You learned to subtract signed numbers using the number line and the short method.

12–4 You learned to multiply and divide signed numbers.

12–5 You learned to solve simple equations and applied problems using equations.

Terms To Remember

	Page
Negative numbers	467
Positive numbers	467
Absolute value	475
Opposites	478

Rules

- Definition of Absolute Value

  The absolute value of a number is the distance that number is from zero, and this value is never negative.

- Addition of Signed Numbers Using the Number Line

  Draw and direct an arrow from the point of the first number as follows:

  1. Direct the arrow to the right if the second number is positive.

  2. Direct the arrow to the left if the second number is negative.

  3. Stay at the first number if the second number is 0.

  Draw the arrow the unit length of the second number.

  The sum is at the tip of the arrow.

- A Short Method for Adding Two Numbers with the Same Sign

  Step 1 Find the sum of the absolute values of the numbers.

  Step 2 **a.** Make the sum positive if the numbers to be added are positive.

  b. Make the sum negative if the numbers to be added are negative.

- A Short Method for Adding Two Numbers with Different Signs

  Step 1 Subtract the smaller absolute value from the larger absolute value.

  Step 2 **a.** Make the difference positive if the larger absolute value is a positive number.

  b. Make the difference negative if the larger absolute value is a negative number.

- Subtraction of Signed Numbers Using the Number Line

  Draw and direct an arrow from the point of the first number as follows:

  1. Direct the arrow to the left if the second number is positive.

  2. Direct the arrow to the right if the second number is negative.

  3. Stay at the first number if the second number is 0.

  Draw the arrow the unit length of the second number.

  The difference is at the tip of the arrow.

- A Short Method for Subtracting Signed Numbers

  Step 1 Change the subtraction sign to an addition sign.

  Step 2 Replace the number to be subtracted by its opposite.

  Step 3 Find the difference by adding.

- To find the product of two numbers with different signs, we multiply their absolute values and make the product negative.
- To find the product of two numbers having the same sign, we multiply their absolute values and make the product positive.
- When multiplying or dividing with two signed numbers the answer is
  a. Positive if the signs are the same.
  b. Negative if the signs are different.

Self-Test

12–1 Write the number in words.

1. -201.17

2. Name point A on the number line.

3. Place < or > between the numbers to make a true statement.

$$-16 \quad -18$$

4. Find the indicated absolute value.

$$|5.1|$$

12–2 **5.** Add.

$$1\frac{5}{6} + \left(-17\frac{5}{12}\right)$$

12–3 **6.** Subtract.

$$-2.165 - (-0.876)$$

12–4 **7.** Multiply.

$$-3\left(-2\frac{1}{3}\right)$$

8. Divide.

$$-84 \div 0.15$$

12–5 **9.** Solve.

$$11x - 2 = 31$$

Solve. Write a statement for your answer.

10. When 1 is added to twice a number, the result is 5.

11. Two shims have a total combined thickness of 0.672 cm. One shim is 1.4 times as thick as the other. What is the thickness of each?

Chapter Test

12–1 Write the number in words.

1. -18.1

2. $+31\frac{5}{7}$

3. Name point *A* on the number line.

Place < or > between the numbers to make a true statement.

4. $-8 \quad 15$

5. $-3\frac{1}{6} \quad -5\frac{5}{6}$

6. $-0.43 \quad -0.39$

7. Find the indicated absolute value.

$$|-0.64|$$

12–2 Add.

8. $7 + (-3)$ **9.** $5\frac{5}{6} + \left(-11\frac{1}{6}\right)$ **10.** $-16.5 + (-9.71)$

12–3 **11.** What is the opposite of 562?

Subtract.

12. $98 - 462$ **13.** $-14\frac{5}{12} - \left(-6\frac{3}{8}\right)$ **14.** $-0.005 - 3.6$

12–4 Multiply.

15. $-10 \cdot 6$ **16.** $-\frac{7}{8}\left(\frac{8}{21}\right)$ **17.** $-0.05(-1.5)$

Divide.

18. $-21 \div 7$ **19.** $\frac{6}{35} \div \left(-\frac{3}{7}\right)$ **20.** $-0.039 \div (-7.5)$

12–5 Solve.

21. $a + 16 = 9$ **22.** $6x + 9 = -9$

Solve. Write a statement for your answer.

23. If 5 is added to a number the result is 18. What is the number?

24. If 2% of a company's profit is $18,500, find the total profit.

25. The area of a plumbing establishment is 5260 sq ft. If the width is 50 ft, what is the length?

26. Two construction workers' combined monthly take home pay is $3025. If one earns $175 more than the other, find the monthly take home pay of each.

Test Your Memory

Chapter 1

1–1 **1.** Name the smallest counting number.

2. Name the second digit in 467,105.

1–2 **3.** Write an expanded number for 5069.

1–3 **4.** Write a word name for 306,400,986.

5. What is the meaning of 8 in the standard number 785,404,259?

1–4 **6.** Place the $<$ or $>$ symbol between 706 and 760 to make a true statement.

1–5 **7.** Round 6254 to the nearest hundred.

Chapter 2

2–2 **8.** Add.

$$\begin{array}{r} 3760 \\ \underline{2239} \end{array}$$

2–3 **9.** Add.

$$\begin{array}{r} 8439 \\ \underline{6498} \end{array}$$

2–4 **10.** Add and check: 56 + 4802 + 98 + 349.

2–5 **11.** Estimate the sum to the nearest thousand, then find the exact sum and compare.

$$\begin{array}{r} 6043 \\ 3518 \\ 9384 \\ \underline{1940} \end{array}$$

2–6 **12.** The Baker family used the following amounts of heating fuel oil during 1999: 175, 208, 229, 155, 167 and 118 gallons. How many gallons were used in 1999?

2–8 **13.** Subtract.

$$\begin{array}{r} 3982 \\ \underline{-1601} \end{array}$$

2–9 **14.** Subtract.

$$\begin{array}{r} 4236 \\ \underline{-2871} \end{array}$$

2–10 **15.** Subtract.

$$\begin{array}{r} 7001 \\ \underline{-5219} \end{array}$$

2–11 **16.** Cid Knapp found that his family used 7516 kilowatt hours of electricity during January and 7608 kilowatt hours during February. How many more kilowatt hours were used during February?

Chapter 3

Multiply

3–2 **17.** $\begin{array}{r} 914 \\ \times 1000 \\ \hline \end{array}$ **18.** 7000×5

3–4 **19.** $\begin{array}{r} 5342 \\ \times 6 \\ \hline \end{array}$ **20.** $\begin{array}{r} 708 \\ \times 80 \\ \hline \end{array}$

3–5 **21.** $\begin{array}{r} 2106 \\ \times 43 \\ \hline \end{array}$

3–6 **22.** How far can a car travel on one tankful if the tank holds 11 gallons of gasoline and the car averages 27 miles per gallon?

Divide.

3–8 **23.** $4\overline{)8408}$

3–9 **24.** $4\overline{)8144}$

3–10 **25.** $278\overline{)167{,}716}$

3–11 **26.** Sara Lee bought a microwave oven for $270. If she decides to pay for it in 9 equal installments, what is the amount of each payment?

Chapter 4

4–1 **27.** Write a fractional number for the shaded part.

28. Find $40\frac{2}{3}$ on a number line.

4–2 **29.** Find the factors of 24.

4–3 **30.** Find the prime factorization of 24.

4–4 **31.** Reduce $\frac{36}{96}$ to lowest terms.

4–5 **32.** Multiply: $\dfrac{5}{6} \times \dfrac{1}{2} \times \dfrac{8}{15}$.

4–6 **33.** Jack Painter used $\frac{4}{7}$ of a gallon to paint a room. His helper did another room using $\frac{1}{2}$ as much paint. What part of a gallon did Jack's helper use?

4–7 **34.** State whether $\frac{16}{16}$ is a proper or improper fraction.

4–8 **35.** Divide and simplify: $\frac{\frac{3}{8}}{\frac{9}{16}}$.

Chapter 5

5–3 **36.** Add: $\frac{7}{15} + \frac{9}{10}$.

5–4 **37.** Sue Williams bought $\frac{7}{8}$ yd of print material and $\frac{1}{2}$ yd of plain material. How much material did Sue buy?

5–5 **38.** Subtract: $\frac{11}{12} - \frac{5}{8}$.

5–6 **39.** Decide which fraction is larger, $\frac{3}{8}$ or $\frac{15}{31}$.

5–7 **40.** A piece $\frac{5}{12}$ yd was cut from a board $\frac{3}{4}$ yd long. How long was the piece that was left?

Chapter 6

6–1 **41.** Change $\frac{19}{4}$ to a mixed number.

6–2 **42.** Multiply: $4\frac{2}{3} \times 2\frac{4}{7}$.

6–3 **43.** Divide: $1\frac{1}{9} \div 1\frac{2}{3}$.

6–4 **44.** Add: $8\frac{3}{8} + 5\frac{1}{6}$.

6–5 **45.** Subtract.

$$\begin{array}{r} 11\frac{3}{8} \\ -5\frac{7}{16} \\ \hline \end{array}$$

6–6 **46.** A jet traveled from Boston to Seattle in $4\frac{3}{4}$ hours. It took $5\frac{1}{3}$ hours on the return flight. How much faster was the first flight?

Chapter 7

7–1 **47.** Write a mixed number for 23.054.

48. Write an expanded number for 5.406.

7–2 **49.** Write a word name for 3.502.

7–3 **50.** Which number is larger, 0.502 or 0.5010?

7–4 **51.** Add: 4.68 + 0.004 + 32 + 8.46.

7–5 **52.** Estimate the sum to the nearest tenth: 56.413 + 9.064.

7–6 **53.** The weather bureau reported the following rainfall measurement for the first 5 days in February: 0.06, 1.32, 2.1, 0.52, and 0.007 inches. How many inches fell during those days?

7–7 **54.** Subtract: 51.6 − 25.98.

7–8 **55.** Subtract: 18 − 7.605.

7–9 **56.** John Swift ran the 100-yd dash in 11.6 seconds on Tuesday. On Wednesday, it took him 10.9 seconds. How much faster was he on Wednesday?

Chapter 8

8–2 **57.** Multiply: 3.69×0.05.

8–3 **58.** Write the product directly: $5.07 \times 10{,}000$.

8–5 **59.** A certain engine consumes 6.81 gallons of fuel per hour. How many gallons would be used if the engine ran for 24 hours?

8–6 **60.** Write the quotient directly: $23.5 \div 10{,}000$.

8–7 **61.** Divide: $5.19\overline{)0.07266}$.

8–8 **62.** Round to the nearest tenth: $22.5\overline{)78.64}$.

8–9 **63.** Express $0.66\frac{2}{3}$ as a common fraction.

8–10 **64.** A car travels 237.1 miles in 5.2 hours. What is its average speed per hour? Round to the nearest tenth of a mile.

Chapter 9

9–1 **65.** Express as a ratio in reduced form: 5 dimes to 18 cents, in cents.

9–2 **66.** Express as a ratio in whole number terms: $3\frac{1}{5}$ quarts to $3\frac{3}{4}$ quarts.

9–3 **67.** Find the unit cost or rate.

Cost	Quantity
$14.06	7.6 pounds

9–5 **68.** Show by the cross product method that the following statement is true:

$$\frac{\frac{3}{2}}{6} = \frac{2}{8}.$$

9–6 **69.** A map has a scale of 10 miles for every $\frac{1}{4}$ inch. How many inches represent 624 miles?

Chapter 10

10–2 **70.** Change 5.9% to a decimal.

71. Change $33\frac{1}{3}\%$ to a fraction in lowest terms.

10–3 **72.** Change $\frac{5}{8}$ to a percent.

73. Change 0.0641 to a percent.

10–4 **74.** 0.8% of what number is 6.4?

10–6 **75.** A certain item marked \$35.00 was reduced by 15%. How much was the item reduced?

10–7 **76.** 68 out of 432 employees are women. What percent of the employees are men? Round to the nearest whole percent.

10–8 **77.** The mark-up of a motorcycle is \$432. If this is 30% of the cost, how much did the motorcycle cost?

Chapter 11

11–2 **78.** Convert 38.6 yd to ft.

11–3 **79.** Convert 680.5 mm to dkm.

11–4 **80.** (53 m 84 cm 7 mm) ÷ 6.

11–5 **81.** Find the perimeter of a square that is 0.78 ft on each side.

82. Find the circumference of a circle with a radius of 7.5 cm. Use 3.14 for π.

11–6 **83.** Find the area of a rectangle 14.7 m by 10 m.

84. Find the area of a circle with diameter $3\frac{1}{3}$ in. Use $\frac{22}{7}$ for π.

11–7 **85.** Convert 66 yd^2 to ft^2.

86. Convert 389.4 m^2 to km^2.

11–8 **87.** Find the volume.

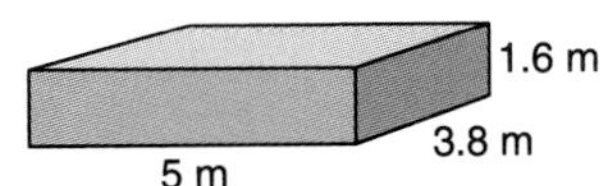

88. Convert 347 cℓ to ℓ.

11–9 **89.** Convert 56 cg to g.

11–10 **90.** Convert 116°C to °F.

11–11 **91.** Convert 104 mi to km.

Chapter 12

12–1 **92.** Locate $-5\frac{5}{8}$ on a number line.

12–3 **93.** Subtract: $-8 - (-2.3)$.

12–4 **94.** Divide: $-15 \div (-3)$.

12–5 **95.** Solve: $7 = x + 15$.

Write an equation and solve.

96. A number increased by 2.5 results in 5.5. Find the number.

97. A salesman's weekly earnings exceeded her sales commission by $320. What were her sales commissions if her earnings were $1010?

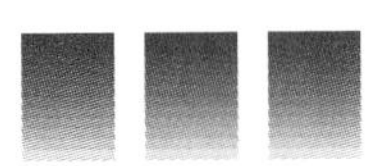

ANSWERS TO ODD-NUMBERED AND REVIEW PROBLEMS

Chapter 1

Problems 1–1

1. 52 **3.** 0 is included in the set of whole numbers.
5. 50, 51, 52, 53, 54, 55, 56, 57, 58, 59, 60, 61, 62, 63 **7.** 100
9. 7 **11.** 9, 8 **13.** 4 **15.** 5

Problems 1–2

1. 15 **2.** 6 **3.** 4000 + 500 + 20 + 6
5. 6000 + 800 + 20 + 1 **7.** 70 + 8 **9.** 200 + 10 + 1
11. 9000 + 90 **13.** 6000 + 600 + 1 **15.** 200 + 6 **17.** 1000 + 20 + 2
19. 3564 **21.** 9613 **23.** 43 **25.** 358
27. 6073 **29.** 2008 **31.** 2050 **33.** 6700
35. 20,000 + 6000 + 500 + 70 + 7 **37.** 20,000 + 3000 + 400 + 6
39. 30,000 + 9 **41.** 80,000 + 9000 **43.** 79,726 **45.** 50,609
47. 17,005 **49.** 20,006

Problems 1–3

1. 105,690, 105,691, 105,692 **2.** 2, 0, 9, 5, 4, 1, 3 **3.** 5000 + 300 + 4
4. 7405 **5.** millions **7.** forty-three
9. nine hundred sixty-eight **11.** two thousand, three hundred fifty-eight
13. one hundred thirty-one thousand, fifteen **15.** 352
17. 7,000,000 **19.** 1 **21.** 6 **23.** 4 hundred million
25. four trillion, six hundred ninety-two billion, eight hundred million dollars
27. two hundred ninety-nine trillion, six hundred eighty billion, four hundred eleven million, five hundred seventy-one thousand, four hundred ten
29. 28,000,005,117,000

Problems 1–4

1. 6 **2.** 222 **3.** 7000 + 500 + 2 **4.** 5069
5. five hundred nineteen million, six hundred four thousand, five hundred thirty-three
6. 926,580 **7.** $<$ **9.** $<$ **11.** $>$
13. $>$ **15.** $>$ **17.** $>$ **19.** $>$
21. $>$

Problems 1–5

1. 263 **2.** 1 **3.** 5000 + 800 + 60
4. 95,078 **5.** three hundred six thousand, twenty-two.

6. 7 ten billion **7.** $<$ **8.** $<$ **9.** 20
11. 840 **13.** 230 **15.** 300 **17.** 4900
19. 700 **21.** 8000 **23.** 1000 **25.** 78,000
27. 418,000 **29.** 846,502,000 **31.** 12,938,000

Chapter 2
Problems 2–1

1. 5 **2.** 4 **3.** 9 **4.** 7
5. 10 **6.** 2 **7.** 13 **8.** 7
9. 8 **10.** 11 **11.** 14 **12.** 3
13. 6 **14.** 9 **15.** 7 **16.** 12
17. 15 **18.** 4 **19.** 6 **20.** 9
21. 6 **22.** 12 **23.** 8 **24.** 5
25. 11 **26.** 14 **27.** 5 **28.** 7
29. 10 **30.** 6 **31.** 13 **32.** 10
33. 1 **34.** 13 **35.** 16 **36.** 7
37. 8 **38.** 11 **39.** 3 **40.** 12
41. 15 **42.** 8 **43.** 9 **44.** 12
45. 2 **46.** 14 **47.** 17 **48.** 9
49. 10 **50.** 16 **51.** 1 **52.** 11
53. 18 **54.** 10 **55.** 5 **56.** 10
57. 8 **58.** 12 **59.** 6 **60.** 5
61. 13 **62.** 16 **63.** 3 **64.** 9
65. 17 **66.** 1 **67.** 11 **68.** 2
69. 9 **70.** 15 **71.** 9 **72.** 5
73. 7 **74.** 7 **75.** 12 **76.** 12
77. 9 **78.** 11 **79.** 13 **80.** 14
81. 4 **82.** 11 **83.** 10 **84.** 8
85. 15 **86.** 10 **87.** 18 **88.** 7
89. 8 **90.** 3 **91.** 12 **92.** 14
93. 6 **94.** 13 **95.** 11 **96.** 8
97. 14 **98.** 16 **99.** 10 **100.** 0

Problems 2–2

1. 7 **2.** 13 **3.** 8 **4.** 11
5. 14 **6.** 12 **7.** 15 **8.** 12
9. 11 **10.** 14 **11.** 13 **12.** 13
13. 16 **14.** 11 **15.** 15 **16.** 12
17. 17 **18.** 11 **19.** 15 **20.** 13
21. 29 **23.** 87 **25.** 96 **27.** 79
29. 693 **31.** 779 **33.** 789 **35.** 499
37. 5671 **39.** 6497 **41.** 1899 **43.** 12,968
45. 66,389 **47.** 62,991

Problems 2–3

1. 53 **3.** 84 **5.** 100 **7.** 859
9. 1303 **11.** 1010 **13.** 910 **15.** 9820
17. 8541 **19.** 18,020 **21.** 10,000 **23.** 91,855
25. 494,014 **27.** 1,000,000

Problems 2–4

1. 27 **3.** 23 **5.** 177 **7.** 202
9. 1152 **11.** 1531 **13.** 9223 **15.** 10,152
17. 1671 **19.** 7858 **21.** 3560 **23.** 8361
25. 15,757 **27.** 6,653,464

Problems 2–5

1. 12,589 **2.** 9480 **3.** 5711 **4.** 2700
5. 6580, 6600, 7000 **6.** 3820, 3800, 4000 **7.** 220, 216 **9.** 220, 219
11. 1700, 1696 **13.** 1700, 1725 **15.** 17,000, 17,486 **17.** 16,000, 16,654
19. 1730, 1700, correct **21.** 11,680, 11,700, correct
23. 101,920, 101,900, 102,000, 101,912

Problems 2–6

1. $246 **3.** 3953 gal **5.** $5302 **7.** 1102 tickets
9. 49,013 persons

Problems 2–7

1. 9 **2.** 9 **3.** 9 **4.** 8
5. 1 **6.** 8 **7.** 7 **8.** 0
9. 9 **10.** 7 **11.** 7 **12.** 7
13. 6 **14.** 3 **15.** 8 **16.** 2
17. 6 **18.** 7 **19.** 6 **20.** 0
21. 5 **22.** 4 **23.** 1 **24.** 3
25. 2 **26.** 5 **27.** 6 **28.** 5
29. 1 **30.** 4 **31.** 4 **32.** 3
33. 2 **34.** 1 **35.** 6 **36.** 5
37. 0 **38.** 4 **39.** 3 **40.** 1
41. 5 **42.** 4 **43.** 8 **44.** 3
45. 5 **46.** 1 **47.** 4 **48.** 3
49. 9 **50.** 2 **51.** 2 **52.** 0
53. 1 **54.** 2 **55.** 8 **56.** 9
57. 3 **58.** 2 **59.** 1 **60.** 8
61. 3 **62.** 6 **63.** 1 **64.** 9
65. 7 **66.** 9 **67.** 5 **68.** 7
69. 5 **70.** 8 **71.** 6 **72.** 4
73. 8 **74.** 6 **75.** 4 **76.** 3

77. 9	**78.** 7	**79.** 5	**80.** 2
81. 9	**82.** 7	**83.** 5	**84.** 8
85. 6	**86.** 4	**87.** 3	**88.** 0

Problems 2–8

1. 7	**2.** 2	**3.** 2	**4.** 4
5. 3	**6.** 2	**7.** 4	**8.** 3
9. 3	**10.** 2	**11.** 2	**12.** 1
13. 6	**14.** 5	**15.** 6	**16.** 4
17. 5	**18.** 5	**19.** 4	**20.** 3
21. 42	**23.** 31	**25.** 28	**27.** 211
29. 322	**31.** 338	**33.** 3211	**35.** 4126
37. 3701	**39.** 16,112	**41.** 97,989	**43.** 43,221
45. 34,103			

Problems 2–9

1. 9	**2.** 9	**3.** 8	**4.** 9
5. 7	**6.** 8	**7.** 7	**8.** 6
9. 4	**10.** 6	**11.** 5	**12.** 4
13. 6	**14.** 5	**15.** 5	**16.** 4
17. 9	**18.** 5	**19.** 14	**21.** 28
23. 14	**25.** 292	**27.** 78	**29.** 165
31. 1185	**33.** 869	**35.** 5873	**37.** 8985
39. 682,629			

Problems 2–10

1. 3213	**2.** 3322	**3.** 38	**4.** 16
5. 354	**6.** 159	**7.** 1979	**8.** 739
9. 12	**11.** 35	**13.** 282	**15.** 46
17. 8884	**19.** 2167	**21.** 1518	**23.** 2569
25. 2588	**27.** 4677	**29.** 2359	**31.** 1539
33. 32,756	**35.** 438,530	**37.** 284,372	

Problems 2–11

1. 3585	**2.** 1638	**3.** 2887	**4.** 1289
5. 4878	**6.** 3394	**7.** 1679	**8.** 1092
9. $631	**11.** 32 in.	**13.** 9 amps	**15.** $685
17. 160 screws			

Chapter 3
Problems 3–1

1. 3	**2.** 40	**3.** 0	**4.** 12
5. 18	**6.** 27	**7.** 35	**8.** 2

9. 6 **10.** 12 **11.** 7 **12.** 36
13. 24 **14.** 0 **15.** 16 **16.** 0
17. 56 **18.** 63 **19.** 5 **20.** 32
21. 30 **22.** 0 **23.** 24 **24.** 0
25. 28 **26.** 10 **27.** 36 **28.** 64
29. 72 **30.** 0 **31.** 45 **32.** 24
33. 9 **34.** 30 **35.** 4 **36.** 18
37. 25 **38.** 0 **39.** 0 **40.** 0
41. 5 **42.** 45 **43.** 8 **44.** 21
45. 32 **46.** 16 **47.** 16 **48.** 14
49. 0 **50.** 0 **51.** 9 **52.** 6
53. 40 **54.** 3 **55.** 42 **56.** 4
57. 63 **58.** 0 **59.** 4 **60.** 0
61. 56 **62.** 35 **63.** 48 **64.** 28
65. 0 **66.** 8 **67.** 12 **68.** 10
69. 0 **70.** 18 **71.** 1 **72.** 54
73. 24 **74.** 21 **75.** 9 **76.** 20
77. 15 **78.** 7 **79.** 49 **80.** 42
81. 81 **82.** 20 **83.** 6 **84.** 0
85. 0 **86.** 6 **87.** 72 **88.** 18
89. 2 **90.** 27 **91.** 15 **92.** 12
93. 36 **94.** 8 **95.** 54 **96.** 14
97. 8 **98.** 0 **99.** 48 **100.** 0

Problems 3–2

1. 70 **3.** 60 **5.** 705,000 **7.** 10,500
9. 8910 **11.** 98,000 **13.** 4800 **15.** 3500
17. 300 **19.** 60 **21.** 140 **23.** 46,100
25. 1200 **27.** 8000 **29.** 120
31. 15,000 impressions **33.** 20,000 bricks **35.** 900 lb

Problems 3–3

1. 3000 **2.** 23,000 **3.** 720 **4.** 36,800
5. 42,000 **6.** 150 **7.** 504 **9.** 236
11. 387 **13.** 2420 **15.** 4794 **17.** 4510
19. 2262 **21.** 2454 **23.** 1536 **25.** 16,080
27. 30,648 **29.** 19,628 **31.** 42,161 **33.** 17,776
35. 72,016 **37.** 69,916 **39.** 174,492 **41.** 390,215
43. 86,432 **45.** 2,809,230 **47.** $2286 **49.** $177.10

Problems 3–4

1. 534 **2.** 2805 **3.** 5472 **4.** 16,648
5. 52,024 **6.** 56,035 **7.** 500 **9.** 6000

11. 3200 **13.** 18,000 **15.** 480,000 **17.** 1,500,000
19. 2240 **21.** 14,450 **23.** 58,000 **25.** 272,400
27. 2,284,000 **29.** 15,189,000 **31.** 24,040,000 **33.** 5,720,240,000
35. 392,173,600,000 **37.** 24,000 lb **39.** $1024

Problems 3–5

1. 2100 **2.** 12,000 **3.** 2118 **4.** 54,288
5. 5040 **6.** 3,256,000 **7.** 2108 **9.** 2769
11. 16,327 **13.** 13,950 **15.** 39,424 **17.** 92,768
19. 132,734 **21.** 147,712 **23.** 190,242 **25.** 156,180
27. 237,930 **29.** 2,304,469 **31.** 817,563 **33.** 125,048
35. 3,015,514 **37.** 11,641,088 **39.** 12,378,540 **41.** 403,449,624
43. 506,334,936 **45.** 2,053,749,248 **47.** 54,886,320,412
49. 35,443,547,008 **51.** 4,300,588,884,935

Problems 3–6

1. 61,224 **2.** 45,063 **3.** 1462 **4.** 14,898
5. 14,428 **6.** 3,173,346 **7.** 17,506,412 **8.** 36,719
9. 563,066 **10.** 300,780 **11.** 245,412 **12.** 4,024,824
13. 56,296,080 **14.** 242,552,518 **15.** 4875 lb **17.** 10,660 pages
19. $7416 **21.** 7500 sq ft **23.** 121,500 gal **25.** 15,145 gal

Problems 3–7

1. 3 **2.** 8 **3.** not possible **4.** 3
5. 3 **6.** 3 **7.** 5 **8.** 2
9. 6 **10.** 6 **11.** 7 **12.** 6
13. 3 **14.** 0 **15.** 4 **16.** not possible
17. 8 **18.** 7 **19.** 5 **20.** 8
21. 6 **22.** 0 **23.** 6 **24.** 0
25. 4 **26.** 2 **27.** 4 **28.** 8
29. 9 **30.** 0 **31.** 5 **32.** 4
33. 9 **34.** 5 **35.** 4 **36.** 2
37. 5 **38.** not possible **39.** not possible **40.** 0
41. 5 **42.** 5 **43.** 2 **44.** 7
45. 4 **46.** 8 **47.** 2 **48.** 0
49. 0 **50.** 9 **51.** 3 **52.** 5
53. 3 **54.** 6 **55.** 1 **56.** 9
57. not possible **58.** 4 **59.** 0 **60.** 7
61. 1 **62.** 7 **63.** 6 **64.** 7
65. 1 **66.** 4 **67.** 4 **68.** 5
69. not possible **70.** 6 **71.** 1 **72.** 6
73. 8 **74.** 3 **75.** 1 **76.** 5
77. 3 **78.** 7 **79.** 7 **80.** 7

81. 4	**82.** 2	**83.** 0	**84.** 6
85. 8	**86.** 9	**87.** 2	**88.** 9
89. 5	**90.** 2	**91.** 9	**92.** 8
93. 9	**94.** 2	**95.** 8	**96.** 0
97. 8	**98.** not possible	**99.** 1	

Problems 3–8

1. 31	**3.** 14	**5.** 12	**7.** 31
9. 12	**11.** 11	**13.** 213	**15.** 232
17. 221	**19.** 3421	**21.** 1231	**23.** 1122
25. 21,324	**27.** 18 amps	**29.** 62 mi	

Problems 3–9

1. 93	**3.** 93	**5.** 411 R 2	**7.** 359
9. 4865	**11.** 7680 R 3	**13.** 4697 R 4	**15.** 80 R 4
17. 3369	**19.** 15,400 R 5	**21.** 78,341 R 1	**23.** 834,926 R 4
25. $180	**27.** 1035 bricks	**29.** 44 acres remainder 4	

Problems 3–10

1. 115 R 1	**2.** 400 R 4	**3.** 500 R 4	**4.** 2144 R 8
5. 3834	**6.** 16,035 R 5	**7.** 2 R 8	**9.** 1 R 47
11. 8 R 4	**13.** 7	**15.** 17 R 7	**17.** 34 R 9
19. 82	**21.** 87 R 1	**23.** 144 R 14	**25.** 274
27. 405 R 66	**29.** 1726	**31.** 426 R 14	**33.** 402 R 31
35. 3 R 10	**37.** 70	**39.** 680 R 250	**41.** 721 R 406
43. 521 R 200	**45.** 637	**47.** 29 R 2010	**49.** 7845 R 1867

Problems 3–11

1. 4351 R 4	**2.** 14600 R 4	**3.** 60 R 14	**4.** 1369 R 16
5. 1360 R 123	**6.** 6711 R 697	**7.** $15	**9.** $89

11. 14 mi with 200 ft left over **13.** 62 fires

15. 6 weeks with 44 pounds left over

Chapter 4
Problems 4–1

1. $\frac{2}{5}$ **3.** $\frac{3}{8}$ **5.**

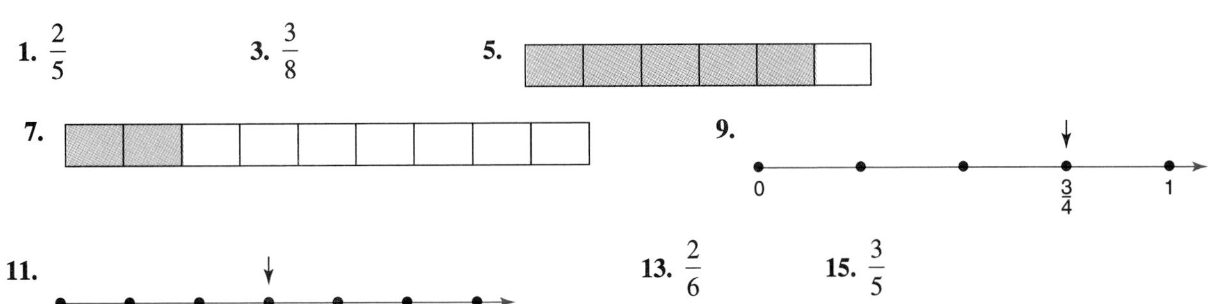

13. $\frac{2}{6}$ **15.** $\frac{3}{5}$

17. $\frac{4}{9}$ **19.** numerator (7), denominator (9)

21. numerator (172), denominator (468) **23.** 0 **25.** 1

27. 1 **29.** not possible **31.** 1 **33.** 1

35. $\frac{6}{6} = 1$ **37.** not possible **39.** 1 **41.** $\frac{8}{3}$, or $2\frac{2}{3}$

43.

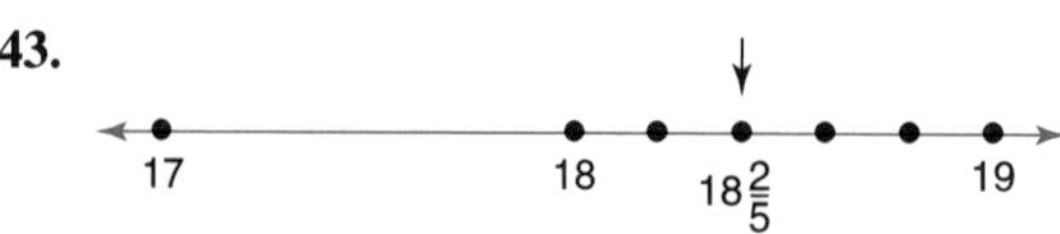

45. $21\frac{5}{7}$

Problems 4–2

1. $\frac{4}{12}$

2.

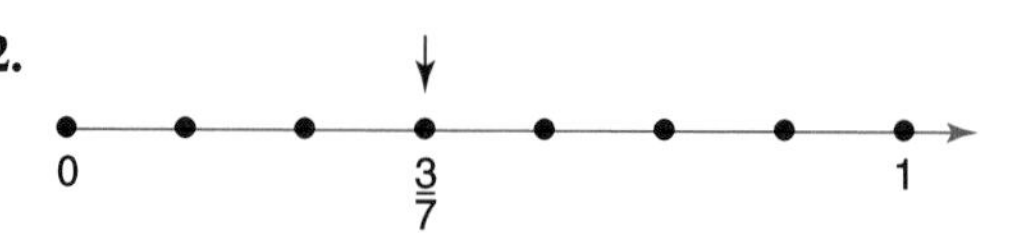

3. 29

4. 1 **5.** 56 **6.** 0 **7.** 1

8. not a number **9.** $1\frac{3}{6}$, or $\frac{9}{6}$ **10.** $1\frac{4}{6}$ **11.** by 3

13. by 2, 3 **15.** by 2, 5 **17.** by 3 **19.** by 3

21. by 2 **23.** none **25.** by 2 **27.** by 3

29. none **31.** by 2, 3 **33.** by 3, 5 **35.** none

37. by 2, 5 **39.** by 2, 3 **41.** 1, 3, 5, 15 **43.** 1, 19

45. 1, 2, 11 **47.** 1, 5, 25 **49.** 1 **51.** 1, 19

53. 1, 2 **55.** 1, 2, 4, 11, 22, 44 **57.** 1, 43

59. 1, 2, 3, 5, 6, 10, 15, 30 **61.** 1, 2, 3, 6, 9, 18, 27, 54

63. 1, 67 **65.** 1, 2, 19, 38 **67.** 1, 2, 4, 5, 10, 20, 25, 50, 100

69. 1, 2, 4, 8, 11, 22, 44, 88

Problems 4–3

1.

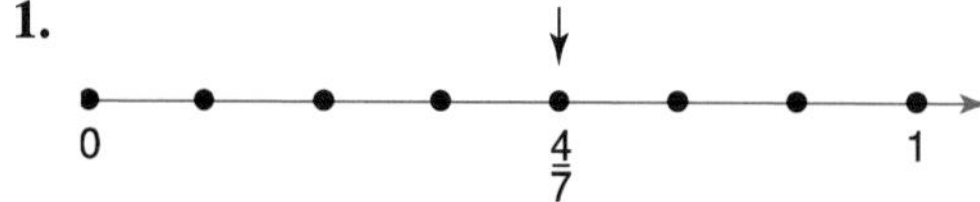

2. 1 **3.** 183

4. $\frac{8}{5}$, or $1\frac{3}{5}$ **5.** The sum of the digits must be divisible by 3.

6. 1, 2, 4, 11, 22, 44 **7.** composite **9.** neither

11. prime **13.** composite **15.** prime **17.** composite

19. prime **21.** prime **23.** composite **25.** composite

27. composite **29.** composite **31.** 2×2 **33.** 2×5

35. 13 **37.** $2 \times 2 \times 5$ **39.** $2 \times 2 \times 2 \times 3$ **41.** 5×7

43. 47 **45.** $2 \times 3 \times 3 \times 3$ **47.** 2×29 **49.** 2×37

51. $3 \times 3 \times 3 \times 3$ **53.** 3×31 **55.** $3 \times 3 \times 11$ **57.** 5×23

59. $3 \times 3 \times 3 \times 5$ **61.** 2×143 **63.** $2 \times 3 \times 3 \times 7$ **65.** 11×11

67. $5 \times 5 \times 19$ **69.** $2 \times 2 \times 3 \times 3 \times 29$ **71.** 5389

Problems 4–4

1. 31

2.

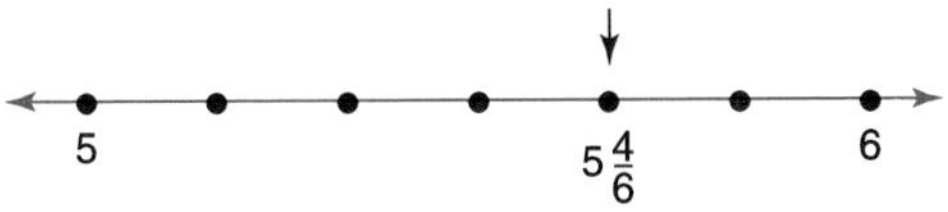

3. Yes. If the last digit is an even number, the number is divisible by 2.

4. 1, 5, 7, 35 **5.** 2, 3, 5, 7, 11 **6.** $2 \times 2 \times 11$ **7.** 6

9. 28 **11.** 24 **13.** 30 **15.** 51

17. 48 **19.** 56 **21.** 36 **23.** 40

25. 9 **27.** 20 **29.** 76 **31.** $\frac{1}{2}$

33. $\frac{1}{6}$ **35.** $\frac{3}{10}$ **37.** $\frac{3}{4}$ **39.** $\frac{3}{4}$

41. $\frac{1}{4}$ **43.** $\frac{2}{7}$ **45.** $\frac{29}{40}$ **47.** $\frac{4}{9}$

49. $\frac{3}{4}$ **51.** $\frac{1}{3}$ **53.** $\frac{2}{3}$ **55.** $\frac{1}{2}$

57. $\frac{2}{5}$ **59.** $\frac{32}{45}$ **61.** 120 **63.** 26

65. 84 **67.** 105 **69.** $\frac{2}{5}$ **71.** $\frac{10}{49}$

73. $\frac{7}{8}$ **75.** $\frac{1}{5}$

Problems 4–5

1. not possible **2.** $2\frac{1}{3}$

3. No. If the sum of the digits is divisible by 3, the number is divisible by 3.

4. $3 \times 5 \times 7$ **5.** 28 **6.** $\frac{3}{4}$ **7.** $\frac{8}{15}$

9. $\frac{15}{56}$ **11.** $\frac{25}{48}$ **13.** $\frac{3}{35}$ **15.** $\frac{7}{18}$

17. $\frac{1}{7}$ **19.** $\frac{7}{18}$ **21.** $\frac{7}{8}$ **23.** $\frac{2}{3}$

25. $\frac{3}{5}$ **27.** $\frac{1}{2}$ **29.** $\frac{81}{175}$ **31.** $\frac{4}{15}$

33. $\frac{6}{35}$ **35.** $\frac{9}{80}$ **37.** $\frac{81}{320}$ **39.** $\frac{1}{12}$

41. $\frac{1}{48}$ **43.** $\frac{2}{33}$ **45.** $\frac{1}{8}$ **47.** $\frac{4}{15}$

49. $\frac{3}{98}$ **51.** $\frac{1}{6}$ **53.** $\frac{4}{27}$

Problems 4–6

1. $\frac{3}{9}$ **2.** $33\frac{2}{7}$ **3.** 1, 2, 4, 8, 16 **4.** 18

5. proper **7.** improper **9.** proper **11.** proper

13. improper **15.** improper **17.** improper **19.** $\frac{9}{7}$

21. none **23.** $\frac{1}{56}$ **25.** $\frac{1}{117}$ **27.** 67

29. $\frac{1}{2106}$ **31.** $\frac{3}{20}$ **33.** $\frac{656}{109}$ **35.** $\frac{35}{32}$

37. $\frac{32}{35}$ **39.** $\frac{5}{4}$ **41.** $\frac{2}{3}$ **43.** $\frac{20}{21}$

45. $\frac{1}{5}$ **47.** $\frac{32}{27}$ **49.** $\frac{6}{5}$ **51.** $\frac{21}{8}$

53. 9 **55.** 9

Problems 4–7

1. prime **2.** 21 **3.** $\frac{2}{3}$ **4.** improper

5. $\frac{14}{14}$ **6.** $\frac{7}{8}$ **7.** $\frac{32}{35}$ **8.** 3

9. $\frac{9}{20}$ **10.** $\frac{12}{25}$ **11.** $\frac{21}{10}$ **12.** $\frac{1}{6}$

13. $\frac{4}{7}$ **14.** $\frac{3}{7}$ **15.** 5 **16.** $\frac{5}{18}$

17. 1 **19.** $\frac{6}{13}$ **21.** 1 **23.** $\frac{7}{15}$

25. $\frac{5}{3}$ **27.** $\frac{100}{21}$

Problems 4–8

1. $\frac{3}{40}$ **3.** 6 cakaws **5.** $\frac{5}{6}$ km **7.** $\frac{4}{15}$ of the calls

9. 12 rows **11.** $\frac{2}{15}$ sq yd **13.** 6 cups **15.** 6 dishes

17. $\frac{3}{2}$ times more **19.** $\frac{2}{5}$ of the cordless telephones used nickel cadmium

Chapter 5
Problems 5–1

1. $\frac{7}{12}$ **2.** $\frac{11}{10}$ **3.** $\frac{2}{3}$ **4.** $\frac{2}{45}$

5. $\frac{5}{6}$ **7.** 1 **9.** $\frac{7}{8}$ **11.** $\frac{6}{5}$

13. $\frac{1}{2}$ **15.** $\frac{7}{5}$ **17.** 1 **19.** $\frac{13}{2}$

21. $\frac{7}{8}$ **23.** $\frac{79}{91}$ **25.** $\frac{39}{40}$

Problems 5–2

1. $\frac{5}{6}$ **2.** $\frac{1}{4}$ **3.** $\frac{27}{10}$ **4.** $\frac{7}{8}$

5. $\frac{1}{2}$ **6.** 1 **7.** $\frac{2}{5}$ **8.** $\frac{12}{13}$

9. 18 **11.** 20 **13.** 24 **15.** 30

17. 36 **19.** 140 **21.** 120 **23.** 48

25. 96 **27.** 3150 **29.** $\frac{8}{18}, \frac{3}{18}$ **31.** $\frac{8}{12}, \frac{3}{12}$

33. $\frac{16}{60}, \frac{25}{60}$ **35.** $\frac{8}{20}, \frac{15}{20}$ **37.** $\frac{14}{42}, \frac{24}{42}$ **39.** $\frac{117}{144}, \frac{40}{144}$

41. $\frac{15}{24}, \frac{12}{24}, \frac{42}{24}$ **43.** $\frac{60}{240}, \frac{75}{240}, \frac{108}{240}$ **45.** $\frac{232}{432}, \frac{117}{432}$ **47.** $\frac{150}{960}, \frac{195}{960}, \frac{468}{960}$

Problems 5–3

1. $\frac{2}{3}$ **2.** $\frac{3}{14}$ **3.** $\frac{12}{25}$ **4.** 1

5. $\frac{1}{11}$ **6.** $\frac{12}{25}$ **7.** $\frac{11}{12}$ **9.** $\frac{11}{10}$

11. $\frac{37}{30}$ **13.** $\frac{23}{24}$ **15.** $\frac{3}{4}$ **17.** $\frac{31}{40}$

19. $\frac{153}{80}$ **21.** $\frac{23}{32}$ **23.** $\frac{17}{12}$ **25.** $\frac{15}{16}$

27. $\frac{51}{48}$ **29.** $\frac{73}{48}$

Problems 5–4

1. $\frac{2}{3}$ **2.** $\frac{19}{16}$ **3.** $\frac{5}{9}$ **4.** $\frac{9}{20}$

5. $\frac{37}{33}$ **6.** 1 **7.** $\frac{5}{8}$ **8.** $\frac{18}{7}$

9. $\frac{5}{12}$ **11.** $\frac{9}{20}$ **13.** $\frac{11}{60}$ **15.** $\frac{11}{48}$

17. $\frac{3}{32}$ **19.** $\frac{4}{15}$ **21.** $\frac{31}{66}$ **23.** $\frac{17}{36}$

25. $\frac{3}{16}$ **27.** $\frac{7}{20}$ **29.** $\frac{1}{6}$ **31.** 0

Problems 5–5

1. $\frac{5}{18}$ **2.** $\frac{19}{56}$ **3.** $\frac{21}{32}$ **4.** $\frac{3}{2}$

5. $\frac{5}{9}$ **6.** $\frac{41}{36}$ **7.** $\frac{2}{3}$ **8.** $\frac{1}{40}$

9. $\frac{10}{7}$ 10. $\frac{43}{24}$ 11. $\frac{7}{8}$ 13. $\frac{1005}{1000}$

15. $\frac{7}{9}$ 17. $\frac{9}{7}$ 19. $\frac{7}{2}$ 21. $\frac{7}{10}$

23. $\frac{9}{10}$

Problems 5–6

1. $\frac{3}{4}$ cup 3. $\frac{5}{6}$ lb 5. 2 times more 7. $\frac{21}{2}$ more

9. $\frac{25}{36}$ lb 11. $\frac{5}{16}$ ft 13. $\frac{3}{8}$ in. 15. $\frac{8}{27}$ of votes cast

17. A gain by $\frac{3}{16}$ of a point 19. $\frac{5}{16}$ in.

Chapter 6
Problems 6–1

1. $\frac{7}{5}$ 2. $\frac{5}{2}$ 3. $\frac{9}{5}$ 4. $\frac{3}{8}$

5. $\frac{251}{180}$ 6. $\frac{35}{36}$ 7. $2\frac{3}{5}$ 9. $6\frac{5}{9}$

11. $17\frac{1}{2}$ 13. 11 15. $42\frac{3}{5}$ 17. $26\frac{2}{5}$

19. $6\frac{3}{10}$ 21. $104\frac{16}{51}$ 23. $\frac{8}{3}$ 25. $\frac{107}{7}$

27. $\frac{139}{9}$ 29. $\frac{57}{16}$ 31. $\frac{72}{13}$ 33. $\frac{25}{16}$

35. $\frac{405}{8}$ 37. $\frac{60{,}013}{1000}$ 39. $\frac{2024}{133}$

Problems 6–2

1. $\frac{1}{2}$ 2. $\frac{11}{12}$ 3. $\frac{2}{3}$ 4. $\frac{31}{15}$

5. $\frac{5}{48}$ 6. $\frac{7}{8}$ 7. $\frac{2}{3}$ 9. 5

11. 0 13. 14 15. 24 17. $\frac{238}{9}$, or $26\frac{4}{9}$

19. 2 21. $6\frac{7}{8}$ 23. $8\frac{2}{5}$ 25. 17

27. $\frac{35}{3}$, or $11\frac{2}{3}$ 29. 0 31. $\frac{49}{48}$, or $1\frac{1}{48}$ 33. 390

Problems 6–3

1. $\frac{21}{20}$, or $1\frac{1}{20}$ **2.** $\frac{11}{15}$ **3.** $\frac{9}{8}$, or $1\frac{1}{8}$ **4.** $\frac{1}{3}$

5. $\frac{13}{10}$, or $1\frac{3}{10}$ **6.** 5 **7.** $\frac{3}{2}$ **9.** $\frac{4}{25}$

11. $\frac{2}{9}$ **13.** 0 **15.** $\frac{23}{44}$ **17.** 32

19. $\frac{2}{9}$ **21.** not possible **23.** $\frac{45}{176}$ **25.** 2

27. $\frac{11}{16}$ **29.** $\frac{288}{85}$, or $3\frac{33}{85}$ **31.** $6\frac{5}{8}$ **33.** $\frac{9}{31}$

35. $\frac{47}{160}$ **37.** $\frac{65}{4}$, or $16\frac{1}{4}$

Problems 6–4

1. $\frac{9}{16}$ **2.** $\frac{59}{18}$, or $3\frac{5}{18}$ **3.** $3\frac{3}{8}$ **4.** 12

5. $\frac{3}{7}$ **6.** 0 **7.** $\frac{4}{45}$ **8.** $\frac{13}{16}$

9. $15\frac{1}{2}$ **11.** $20\frac{5}{7}$ **13.** $32\frac{5}{6}$ **15.** $70\frac{5}{36}$

17. $32\frac{15}{16}$ **19.** $104\frac{4}{7}$ **21.** $127\frac{9}{10}$ **23.** $186\frac{23}{30}$

25. $373\frac{3}{10}$ **27.** $18\frac{13}{40}$ **29.** $22\frac{11}{12}$ **31.** $206\frac{1}{24}$

33. $1906\frac{37}{48}$

Problems 6–5

1. $\frac{47}{14}$, or $3\frac{5}{14}$ **2.** $\frac{25}{34}$ **3.** $\frac{14}{27}$ **4.** $\frac{546}{25}$, or $21\frac{21}{25}$

5. $25\frac{5}{7}$ **6.** $24\frac{2}{3}$ **7.** $22\frac{7}{10}$ **8.** $16\frac{29}{72}$

9. $3\frac{7}{9}$ **11.** 3 **13.** $16\frac{4}{13}$ **15.** $4\frac{11}{48}$

17. $3\frac{6}{7}$ **19.** $33\frac{4}{7}$ **21.** $17\frac{11}{18}$ **23.** $16\frac{39}{40}$

25. $2\frac{28}{31}$ **27.** 5 **29.** $1\frac{1}{12}$ **31.** $24\frac{11}{27}$

33. $42\frac{12}{17}$ **35.** $33\frac{5}{18}$ **37.** $30\frac{1}{35}$ **39.** $7\frac{13}{36}$

41. $\frac{125}{192}$ **43.** $9\frac{173}{216}$

Problems 6–6

1. $11\frac{1}{4}$ pounds **3.** 20 mph **5.** $15\frac{3}{8}$ gal **7.** $25\frac{25}{32}$ sq yd

9. $34\frac{11}{36}$ **11.** $2\frac{58}{243}$, or approximately 2 hours **13.** 24 post holes

15. $1\frac{3}{8}$ hours **17.** $1\frac{7}{24}$ in. **19.** $11\frac{5}{12}$ in. **21.** $3\frac{3}{16}$

Chapter 7
Problems 7–1

1. hundredths **3.** ten millions **5.** hundred thousands

7. $20 + 4 + \frac{4}{10} + \frac{6}{100}$ **9.** $9 + \frac{6}{10}$

11. $\frac{3}{10} + \frac{0}{100} + \frac{0}{1000} + \frac{4}{10{,}000}$ **13.** $3000 + 90 + \frac{5}{10}$

15. $5000 + 2 + \frac{7}{1000}$ **17.** $10{,}000 + 500 + 1 + \frac{9}{10{,}000}$

19. 0.35 **21.** 735.03 **23.** 9.802 **25.** 807.609

27. 5001.1007 **29.** 0.5 **31.** 5.07 **33.** 31.0000001

35. 356.00004 **37.** 19.009 **39.** $\frac{7}{1000}$ **41.** $\frac{2}{100{,}000}$

43. $\frac{1}{10{,}000}$ **45.** $68\frac{9}{1{,}000{,}000}$ **47.** $579\frac{4}{10{,}000}$

Problems 7–2

1. ten thousandths **2.** thousands **3.** 0.0009 **4.** $\frac{7}{1000}$

5. $\frac{5}{10} + \frac{3}{100}$ **6.** $500 + 4 + \frac{6}{10} + \frac{9}{100}$ **7.** 0.7009

8. 607.0501 **9.** eight tenths **11.** ninety-one hundredths

13. two hundred forty-eight and nineteen hundredths

15. sixty-three thousandths

17. one hundred ninety-seven and six hundred forty-nine thousandths

19. three and four thousandths **21.** twenty two and seventeen hundredths

23. sixty-eight and two hundred four ten thousandths **25.** 0.8

27. 0.06 **29.** 0.307 **31.** 0.835 **33.** 86.00673

Problems 7–3

1. thousandths **2.** trillionths **3.** $\frac{7}{100} + \frac{4}{1000}$

4. $7000 + 50 + \frac{6}{10} + \frac{9}{1000} + \frac{7}{10{,}000}$ **5.** 0.0508 **6.** 55.608

7. nine-hundred forty-two thousandths **8.** three hundred and fifty-six thousandths

9. 800.029 **10.** 350.0406 **11.** 0.04 **13.** 5.6

15. 0.8 **17.** 0.005 **19.** 1.749 **21.** 0.31

Problems 7–4

1. ten millionths **2.** $20 + 6 + \frac{5}{10} + \frac{7}{100}$ **3.** 31.85
4. 15.016 **5.** $\frac{31}{10,000}$ **6.** thirty-two and forty-two thousandths
7. 0.9 **8.** 9.704 **9.** 1.53 **11.** 69.3
13. 3.77 **15.** 38.541 **17.** 36.879 **19.** 8.507
21. 10.023 **23.** 15.677 **25.** 19.897 **27.** 51.559
29. 13.1349 **31.** 101.4551 **33.** 39.494 **35.** 62.324
37. 76.926

Problems 7–5

1. 806.04 **2.** $17\frac{5}{100}$ **3.** 24.69 **4.** 0.81
5. 27.324 **6.** 25.652 **7.** 0.1, 0.09, 0.089
9. 6.4, 6.41, 6.409 **11.** 53.0, 53.03, 53.031 **13.** 14.389, 14.3
15. 86.3922, 86.4 **17.** 18.0593, 18.05 **19.** 60.2993, 60.3

Problems 7–6

1. 53.00003 **2.** twenty-three and four hundred sixteen hundred thousandths
3. 5.62 **4.** 23.4656 **5.** 38.164; 38.2 **6.** 76.06
7. 0.31 **9.** 3.203 **11.** 322.4031 **13.** 22.017
15. 402.3639 **17.** 1.635 **19.** 0.7 **21.** 0.6254
23. 19.077 **25.** 76.867

Problems 7–7

1. $400 + 70 + \frac{6}{100} + \frac{1}{1000}$ **2.** 21.006
3. fifty eight and six hundred nine thousandths **4.** 0.003
5. 0.4361 **7.** 2.2349 **9.** 2.2704 **11.** 1.2437
13. 4.9724 **15.** 10.255 **17.** 1.452 **19.** 6.231
21. 83.265 **23.** 19.148

Problems 7–8

1. \$82.79 **3.** 3.35 ft **5.** 1346.4 miles **7.** 36.5141
9. 13.9 yards **11.** 4.075 in. **13.** 25.5 hours **15.** 0.013 in.
17. 1876 ft **19.** 32.1 kw

Chapter 8
Problems 8–1

1. 3.0 **3.** 14.28 **5.** 133.128 **7.** 21.7548
9. 21 **11.** 0.16 **13.** 0.008 **15.** 6403.45

17. 2606.65 **19.** 0.0017 **21.** 81 in. **23.** 79.625 lb
25. $331.88 **27.** 20.5 hr **29.** 470.4 mℓ

Problems 8–2

1. 0.63 **3.** 365 **5.** 0.588 **7.** 38.944
9. 22.1543 **11.** 0.0032 **13.** 1850.4996 **15.** 361.0516
17. 1.14 **19.** 4228.2 **21.** $154.15 **23.** 2.744 ft
25. 725.86 sq ft **27.** 60,196.5 cu in. **29.** $400.11

Problems 8–3

1. 48.6 **3.** 3.18 **5.** 170 **7.** 7187
9. 0.207 **11.** 47.58 **13.** 93.07616 **15.** 17.892
17. 31.8 **19.** 9.146952 **21.** 9 in. **23.** 5250 lb
25. **a.** 750 ma **b.** 7500 ma **c.** 75 ma

Problems 8–4

1. 2.88 **3.** 7.5 **5.** 0.5005 **7.** 11.86636
9. 0.7958 **11.** $3056.25 **13.** 318.09 lb **15.** $1116.30

Problems 8–5

1. 12.5 **3.** 20.46 **5.** 0.0541 **7.** 9.146
9. 0.0376 **11.** 0.2168 **13.** 0.009013 **15.** 3
17. 0.7 **19.** 0.69 **21.** 0.0987 **23.** 3.98676
25. 39 **27.** 60 **29.** 0.938 in. **31.** 0.0655 ft
33. 0.083 grains **35.** 12.26 hr

Problems 8–6

1. 862 **3.** 20 **5.** 320 **7.** 2.59
9. 8.301 **11.** 6.4 **13.** 766 **15.** 130.8
17. 11,784 **19.** 31.2 **21.** 1845 **23.** 4060 rev
25. $4700 **27.** 7 hr **29.** 56 dinners

Problems 8–7

1. 14.5 **3.** $28.1\frac{2}{3}$ **5.** $1934.78\frac{6}{23}$ **7.** 22.75
9. $0.44\frac{3}{8}$ **11.** 1579.4 **13.** 6.0 **15.** 1521.74
17. 4.46 **19.** 5485.71 **21.** 8 rev **23.** 20 pieces
25. 3.18 in. per hr

Problems 8–8

1. 141.954 **2.** 20.93301 **3.** 341 **4.** 6.2
5. 43.75 **6.** 0.07 **7.** 108.52 **8.** 1.41
9. 0.06 **11.** $0.33\frac{1}{3}$ **13.** 17.5 **15.** 0.85
17. 0.84 **19.** 37.69 **21.** 3.125 **23.** 132.92
25. 2.5 **27.** 0.875 **29.** $\frac{1}{5}$ **31.** $\frac{7}{20}$
33. $6\frac{1}{3}$ **35.** $\frac{1}{4}$ **37.** $\frac{2}{3}$ **39.** $16\frac{7}{8}$
41. $3\frac{5}{6}$ **43.** $\frac{5}{12}$ **45.** $9\frac{3}{8}$ **47.** $46\frac{11}{22}$

Problems 8–9

1. $297.16 **3.** 3.58 hr **5.** $50.24 **7.** $15.84
9. 12,416.9 lb **11.** 3,134.28 mi **13.** 32.6 gal; $46 **15.** 7.3 hr
17. 1.583 in.

Chapter 9
Problems 9–1

1. $\frac{50}{45}$ **3.** $\frac{2 \text{ dimes}}{16 \text{ cents}} = \frac{20 \text{ cents}}{16 \text{ cents}} = \frac{20}{16}$ **5.** $\frac{5}{7}$
7. $\frac{3}{15}$ **9.** $\frac{2 \text{ feet}}{2 \text{ yards}} = \frac{2 \text{ feet}}{2 \times 3 \text{ feet}} = \frac{2}{6}$ **11.** $\frac{2}{5}$
13. $\frac{9}{5}, \frac{9}{14}, \frac{14}{5}$ **15.** $\frac{28}{43}$ **17.** $\frac{16}{52} = \frac{4}{13}$ **19.** $\frac{36}{81} = \frac{4}{9}$
21. $\frac{39}{75} = \frac{13}{25}$ **23.** $\frac{18}{12} = \frac{3}{2}$
25. $\frac{6}{10} = \frac{18}{?}; \frac{6}{10} \times \frac{3}{3} = \frac{18}{30};$ It served 30 persons.
27. $\frac{5}{13} = \frac{15}{?}; \frac{5}{13} \times \frac{3}{3} = \frac{15}{39};$ 39 feet **29.** **a.** yes **b.** yes **c.** no

Problems 9–2

1. $\frac{10}{3}$ **3.** $\frac{300}{11}$ **5.** $\frac{5}{12}$ **7.** 4 to 1
9. 1 to 500

Problems 9–3

1. $0.54 per pound **3.** 21.4 cents per cup **5.** $1.25 per pint
7. 44.7 km per hr **9.** 42.7 mi per hr
11. 1.5 males to 1 female; 2.5 total to 1 female
13. $3\frac{1}{16}$ lb of steel to 1 lb aluminum **15.** $3.27 per pound

Problems 9–4

1. \$0.89 **2.** \$0.639 **3.** 25 miles per hour
4. 47 feet per second **5.** yes **7.** yes
9. no **11.** no **13.** no

Problems 9–5

1. $6 = 6$; true **3.** $192 = 192$; true **5.** $351 \neq 360$; false
7. $0.462 = 0.462$; true **9.** $20 = 20$; true **11.** $n = 6$
13. 12.5, or $12\frac{1}{2}$ **15.** $r = 54$ **17.** $n = 108$ **19.** $r = 70$
21. $n = \frac{8}{15}$ **23.** $n = \frac{7}{96}$ **25.** $\frac{9}{17}$
27. $n = 9.1$ **29.** $r = 76$

Problems 9–6

1. \$72.86 **3.** 12 lb **5.** $187\frac{1}{2}$ ft **7.** 5.2 gallons
9. 0.57 lb per ft **11.** 36 servings **13.** \$23.15 **15.** 130.4 lb

Chapter 10
Problems 10–1

1. 41% **3.**

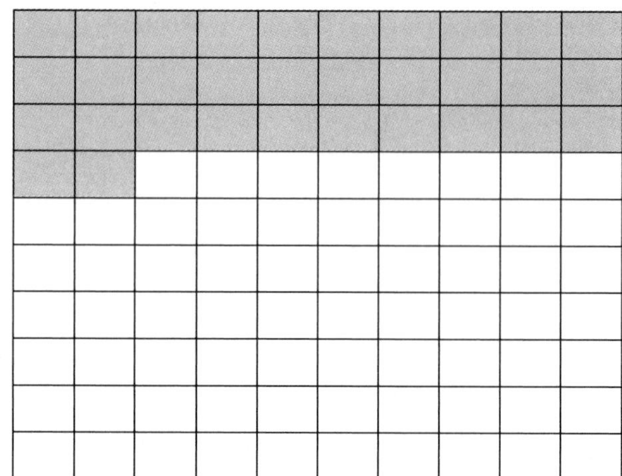

5. 62% **7.** 56% **9.** 156% **11.** 15.67%
13. $\frac{1}{2}\%$ **15.** 29% **17.** 360% **19.** 2681%
21. $\frac{10}{100}$ **23.** $\frac{6.1}{100}$ **25.** $\frac{33\frac{1}{3}}{100}$ **27.** $\frac{57.84}{100}$
29. $\frac{62\frac{1}{2}}{100}$ **31.** $\frac{315}{100}$ **33.** $\frac{83\frac{1}{3}}{100}$ **35.** $\frac{0.004}{100}$

Problems 10–2

1. 0.84 **3.** 0.41 **5.** 0.06 **7.** 5.0
9. 0.396 **11.** 0.6878 **13.** $0.15\frac{2}{3}$ **15.** $0.068\frac{2}{3}$

17. $1.01\frac{2}{3}$ **19.** 0.09 **21.** $0.0764\frac{2}{3}$ **23.** 0.47

25. $\frac{16}{100} = \frac{4}{25}$ **27.** $\frac{71}{100}$ **29.** $\frac{4}{100} = \frac{1}{25}$ **31.** $\frac{106}{100} = \frac{53}{50}$

33. $\frac{58}{100} = \frac{29}{50}$ **35.** $\frac{62.5}{100} = \frac{25}{40} = \frac{5}{8}$ **37.** $\frac{7.7}{100} = \frac{77}{1000}$ **39.** $\frac{3657}{10{,}000}$

41. $\frac{4126}{1000} = \frac{2063}{500} = 4\frac{63}{500}$ **43.** $\frac{5.62}{100} = \frac{281}{5000}$ **45.** $\frac{1}{3}$

47. $\frac{53}{250}$ **49.** $\frac{141}{250}$ **51.** $\frac{5}{6}$ **53.** $\frac{5}{12}$

55. $\frac{43}{250}$ **57.** $\frac{93}{500}$ **59.** $\frac{1033}{2000}$ **61.** $\frac{1277}{25{,}000}$

63. 0.005 **65.** 0.002 **67.** 0.00875 **69.** $\frac{1}{300}$

71. $\frac{31}{5000}$

Problems 10–3

1. 0.67 **3.** 0.089 **5.** $0.14\frac{5}{8}$ **7.** $\frac{19}{125}$

9. $\frac{67}{200}$ **11.** $\frac{819}{2500}$ **13.** $\frac{1263}{12{,}500}$ **15.** $\frac{39}{175}$

17. 75% **19.** 45% **21.** 20% **23.** 60%

25. 60% **27.** $18\frac{3}{4}$%, or 18.75% **29.** 21.6%

31. 440% **33.** 580% **35.** 16.7%, $16\frac{2}{3}$% **37.** 91.7%, $91\frac{2}{3}$%

39. 83.3%, $83\frac{1}{3}$% **41.** 35.9%, $35\frac{15}{16}$% **43.** 12.5%, $12\frac{1}{2}$% **45.** 312.5%, $312\frac{1}{2}$%

47. 60% **49.** 85% **51.** $46\frac{7}{8}$% **53.** $8\frac{1}{3}$%

55. 80% **57.** 81% **59.** 94% **61.** 68%

63. $68\frac{1}{2}$% **65.** $13\frac{1}{3}$% **67.** 79.42% **69.** $34.7\frac{2}{3}$%

71. 83.965% **73.** 68% **75.** 600% **77.** 456%

79. 0.26%

Problems 10–4

1. $0.65\frac{1}{4}$ **3.** $\frac{9}{4}$ **5.** 66.7%

7. P = what?, $R = 39$, $B = 67$ **9.** $P = 106$, R = what?, $B = 58$

11. R = what?, $B = 72$, $P = 206$ **13.** P = what?, $R = 80$, $B = 32$

15. $P = 28$, R = what?, $B = 84$ **17.** $R = 26$, B = what?, $P = 56$

19. $P = 21$, $R = 50$, B = what? **21.** $R = 78$, $B = 167$, P = what?

23. $P = 18$, $R = 50$, B = what? **25.** R = what?, $B = 15$, $P = 781$

Problems 10–5

1. $37\frac{1}{2}\%$ **2.** $\frac{1}{2}\%$ **3.** $R = 63\%, B = 37, P =$ what?

4. $R = 40, B =$ what?, $P = 39$ **5.** $R =$ what?, $P = 98, B = 23$

7. 17.92 **9.** $62\frac{2}{9}$ **11.** $8\frac{1}{3}\%$ **13.** 4000%

15. 204.48 **17.** 51.1 **19.** 663 **21.** 0.126

23. 37.5% **25.** 36.192 **27.** 13.29% **29.** 600

31. 66.4% **33.** 9.67 **35.** 105.6

Problems 10–6

1. $\frac{73}{400}$ **2.** $\frac{29}{20}$ **3.** 67.5% **4.** 1.5

5. $164 **7.** $214.38 **9.** 1050 bricks **11.** $150.19

13. $132.02 **15.** 23,826 cars **17.** 24.4 sq in.

Problems 10–7

1. 37.5% **2.** $81\frac{1}{4}\%$ **3.** $18\frac{3}{4}\%$ **4.** 0.625%

5. $0.70 per hr **7.** 50% **9.** 76% **11.** 73%

13. 3.75%, or $3\frac{3}{4}\%$ **15.** 89.3%

Problems 10–8

1. 4 students **2.** 6% **3.** $188.89 **5.** $169,800

7. 5,232.6 mℓ **9.** $12.50 **11.** 380 parts **13.** $3.19

15. 9% **17.** 10,487.54 lb **19.** 93.3% **21.** $8736

Chapter 11

Problems 11–2

1. 36 in. **3.** 21,120 ft **5.** 5280 yd **7.** 11 yd

9. $1\frac{1}{2}$ **11.** 19 yd **13.** 72.3 ft **15.** $\frac{13}{3}$ yd, or $4\frac{1}{3}$ yd

17. $\frac{1}{220}$ mi **19.** 648 in. **21.** $\frac{1}{2}$ yd **23.** $3\frac{3}{4}$ ft

25. 28,008 in. **27.** $33\frac{13}{36}$ yd **29.** $\frac{3}{1000}$ mi **31.** 19 yd

33. 0.8 mi **35.** **a.** 38,280 ft **b.** 15,312 steps

Problems 11–3

1. 500 m **3.** 7.056 km **5.** 32.5 m **7.** 1000 m

9. 5800 m **11.** 3.88 m **13.** 0.00467 m **15.** 10 m

17. 78 mm **19.** 100 m **21.** 23,900 m **23.** 0.1 cm

25. 4.63 km **27.** 100 mm **29.** 0.466 dkm **31.** 8,500,000 cm
33. b **35.** c

Problems 11–4

1. $20\frac{2}{3}$ yd **2.** 1404 in. **3.** 3000 m **4.** 824 dkm
5. $\frac{5}{12}$ ft **6.** 263 yd **7.** 52.043 dm **8.** 0.018 dkm
9. 9 ft 1 in. **11.** 23.161 km **13.** $30\frac{2}{3}$ yd **15.** 115.5 cm
17. $8\frac{5}{12}$ ft **19.** 8.074 km **21.** $10\frac{2}{3}$ yd **23.** $4\frac{487}{1320}$ mi
25. 37.9 cm **27.** 28 yd 1 ft **29.** 19 m 16 cm **31.** 54 yd 1 ft 4 in.
33. 199 km 541 m 79 cm **35.** 2 ft 5 in., or $2\frac{5}{12}$ ft
37. 11 km 114 m, or 11.114 km **39.** 5 ft $5\frac{1}{3}$ in.
41. 11 cm $7\frac{2}{5}$ mm **43.** 2 yd $9\frac{1}{3}$ in. **45.** 6 km 965 m $90\frac{2}{7}$ cm

Problems 11–5

1. 8 km 131 m, or 8.131 m **2.** 5 ft 3 in., or $5\frac{1}{4}$ ft **3.** 14 yd $6\frac{3}{4}$ in.
4. 10 km 76 m $4\frac{4}{7}$ cm **5.** 24 in. **7.** 17.4 ft
9. 10 yd **11.** 2.8 dkm **13.** 37.8 in. **15.** 50.24 cm
17. 92.4 m **19.** 40 in. **21.** 23.68 yd **23.** 8.4 m
25. $14\frac{1}{7}$ ft **27.** 2.18 m **29.** 32 cm per min **31.** 64.6 rev.

Problems 11–6

1. 28 m^2 **3.** $\frac{625}{64}$ yd^2, or $9\frac{49}{64}$ yd^2 **5.** 0.0049 km^2
7. $\frac{35}{8}$ ft^2 **9.** 4 mm^2 **11.** 78.5 yd^2 **13.** 154 km^2
15. 450 ft^2 **17.** $10\frac{6}{25}$ dkm^2 **19.** 50 yd^2 **21.** 28.26 ft^2
23. 19.625 hm^2 **25.** 70 ft^2 **27.** 41 m^2 **29.** 16.25 cm^2
31. 150 squares **33.** **a.** 135 m^2 **b.** 3 rolls **c.** 15 m^2 **35.** 173 lb

Problems 11–7

1. 540 in. **2.** 12 yd **3.** 7040 yd **4.** $\frac{2}{3}$ ft
5. $2\frac{3}{44}$ mi **6.** $\frac{1}{2}$ yd **7.** 8900 m **8.** 53.5 m

9. 0.466 dkm **10.** 4.96 km **11.** 0.00015 dm **12.** 53 mm
13. 288 ft^2 **15.** 2880 in^2 **17.** 3 ft^2 **19.** 792 ft^2
21. $9\frac{1}{3}$ yd^2 **23.** 56,000 m^2 **25.** 20,060,000 m^2 **27.** 414,000 m^2
29. 7400 ca **31.** 0.0003147 km^2 **33.** 0.001843 m^2 **35.** 83,635,200 ft^2
37. 4147.2 in^2 **39.** 600 a **41.** 1000 ha **43.** $2436\frac{2}{3}$ yd^2
45. **a.** 4069.4 in^2 **b.** 9 persons

Problems 11–8

1. 24.6 ft **2.** 10.6 m **3.** 43.9 in. **4.** 7.536 cm
5. 36 ft^2 **6.** 4.5 cm^2 (rounded) **7.** 120 m^3
9. 0.3 cm^3 **11.** 994.5 cm^3 **13.** 320 ℓ **15.** 3.042 kℓ
17. 5.36 ℓ **19.** 4 cc **21.** 1320 dℓ **23.** 5620 dℓ
25. 0.01392 dkℓ **27.** 26,900 dℓ **29.** 30 yd^3 **31.** $\frac{8}{27}$ yd^3
33. 125 trips

Problems 11–9

1. 387 ft^2 **2.** 4608 ft^2 **3.** $2\frac{5}{12}$ ft^2 **4.** $10\frac{8}{9}$ yd^2
5. $\frac{1}{216}$ yd^2 **6.** 6220.8 in^2 **7.** 78,000 m^2 **8.** 31,600,000 m^2
9. 386,000 m^2 **10.** 5420 ca **11.** 0.0005314 km^2 **12.** 0.001409 m^2
13. 69,000 g **15.** 0.001 g **17.** 4560 g **19.** 100 cg
21. 4.64 hg **23.** 1000 g **25.** 0.00000047 hg **27.** 34,600 g
29. 10 g **31.** 900,000 cg **33.** 0.05 hg **35.** 0.0001 hg
37. 0.0032 kg **39.** 0.5 kg **41.** \$2.33

Problems 11–10

1. 744.48 m^3 **2.** $18\frac{3}{18}$ in^3 **3.** 480 ℓ **4.** 3.42 ℓ
5. 6010 dℓ **6.** 830 cℓ **7.** 1.4 cc **8.** 2400 dkℓ
9. 75°C **11.** 95°C **13.** 30°C **15.** 210°F
17. 150°F **19.** 125°F **21.** 54°C (rounded) **23.** 6.1°C (rounded)
25. 158°F **27.** 122°F **29.** 40.6°C (rounded)
31. probably a light jacket

Problems 11–11

1. 56,000 g **2.** 0.0032 g **3.** 1,800,000 cg **4.** 0.01 g
5. 68°C **6.** 171°F **7.** 73.66 cm **9.** 235.6 mi
11. 91.5 cm **13.** 198.8 g **15.** 238.95 kg **17.** 20.9 ℓ
19. 46.8 gal **21.** 6.118 km **23.** 291.592 ft **25.** 0.19 oz

27. 162.855 ℓ **29.** 6.35 cm **31.** 10.304 km **33.** 0.013 oz
35. 10.76 pt **37.** 450 kg **39.** 51.2 gal **41.** 16 in.

Chapter 12
Problems 12–1

1. positive seven **3.** forty-seven **5.** negative seventy-seven
7. five and three eighths **9.** negative twenty-nine and four sevenths

11. -2 **13.** $-3\frac{1}{3}$  **15.**

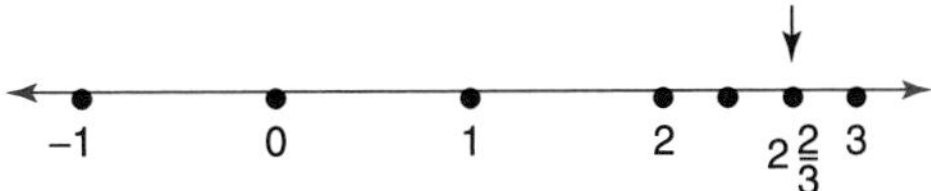

 **17.**

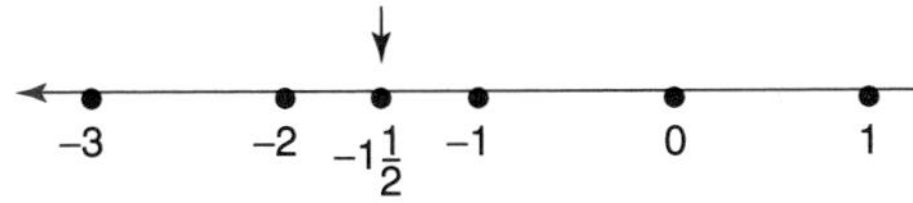

19. $<$ **21.** $>$

23. $>$ **25.** $>$ **27.** $<$ **29.** $<$
31. $<$ **33.** $>$ **35.** $>$ **37.** $<$
39. 16 **41.** 23 **43.** 108 **45.** 5060
47. $32\frac{1}{2}$ **49.** $18\frac{7}{8}$ **51.** 0.04 **53.** 0
55. $-3, 3$ **57.** $-56\frac{5}{8}, 56\frac{5}{8}$ **59.** $-109.7, 109.7$

Problems 12–2

1. positive nine **2.** negative fifty-three
3. negative thirty-five and eighteen hundredths
4. six and nine seventeenths **5.** $-6\frac{2}{5}$ **6.** 82
7. 13 **8.** 0 **9.** $16\frac{1}{2}$ **10.** 34.04
11. 8 **12.** 13 **13.** 309 **14.** $\frac{6}{7}$
15. 0 **16.** 3.43 **17.** $52\frac{1}{8}$ **18.** 0.058
19. 6 **21.** 5 **23.** -2 **25.** -7
27. -3 **29.** 57 **31.** -15 **33.** 3
35. -4 **37.** $15\frac{7}{8}$ **39.** 14 **41.** -66
43. -24 **45.** $\frac{7}{20}$ **47.** $18\frac{11}{18}$ **49.** $-12\frac{7}{10}$
51. $-8\frac{2}{15}$ **53.** $-\frac{5}{12}$ **55.** -513.68 **57.** 0.3573
59. -1.1247

Problems 12–3

1. negative thirteen and five eighths **2.** $-3\frac{1}{2}$ **3.** 568.9

4. 1 **5.** -7 **6.** -1 **7.** -177

8. $-\frac{1}{24}$ **9.** -6 **11.** 6 **13.** $-26\frac{3}{8}$

15. 0.041 **17.** 3 **19.** -2 **21.** -3

23. 6 **25.** -1 **27.** 7 **29.** 11

31. -16 **33.** -159 **35.** 55 **37.** 39

39. -167 **41.** -65 **43.** $-\frac{5}{12}$ **45.** $11\frac{2}{15}$

47. 64.45 **49.** 2.396

Problems 12–4

1. -27 **2.** 16.9 **3.** -26 **4.** $-1\frac{3}{8}$

5. 0.015 **6.** 628 **7.** -34.03 **8.** $-3\frac{3}{8}$

9. 12 **11.** -48 **13.** 63 **15.** -72

17. 512 **19.** 903 **21.** -4446 **23.** 64,000

25. -380 **27.** $-\frac{1}{2}$ **29.** -4.9 **31.** $\frac{1}{9}$

33. 43.008 **35.** 5 **37.** -2 **39.** 14

41. 9 **43.** 231 **45.** -56 **47.** 103

49. -236 **51.** $-2\frac{1}{10}$ **53.** 0.6 **55.** $\frac{3}{4}$

57. -31.2 **59.** $-10\frac{6}{7}$

Problems 12–5

1. $\frac{2}{3}$ **2.** $>$ **3.** 1 **4.** $-\frac{3}{2}$

5. 0.35 **6.** 32 **7.** $-20\frac{7}{8}$ **8.** -15

9. $-\frac{4}{3}$ or $-1\frac{1}{3}$ **10.** -13.1 **11.** 7 **13.** -25

15. 10 **17.** 3 **19.** 13 **21.** -3

23. -6 **25.** The thickness is 1.7 cm. **27.** $357,207

29. The length is 25 ft.

31. The worker's yearly pay was $18,750.

33. The first board is 4 ft; the second is 2 ft; the third is 3 ft.

ANSWERS TO SELF-TESTS

Chapter 1

1. 52	**2.** 0	**3.** 88	
4. 41, 42, 43, 44, 45, 46, 47, 48		**5.** 0	**6.** 7000 + 40 + 8
7. 6048	**8.** eighteen thousand, four hundred six		**9.** 4 ten thousands
10. <	**11.** 3500		

Chapter 2

1. 14	**2.** 14	**3.** 16	**4.** 16
5. 13	**6.** 11	**7.** 11	**8.** 12
9. 99	**10.** 9488	**11.** 91	**12.** 14,195
13. 180	**14.** 8701	**15.** 190, 185	**16.** 94 hr
17. 9	**18.** 5	**19.** 4	**20.** 3
21. 9	**22.** 8	**23.** 6	**24.** 9
25. 4422	**26.** 1659	**27.** 1477	**28.** 2639
29. 1385 ft or 1 roll + 385 ft			

Chapter 3

1. 80	**2.** 806,000	**3.** 8900	**4.** 24,000
5. 20,000	**6.** 350	**7.** 17,084	**8.** 21,063
9. 24,280	**10.** 3,542,000	**11.** 49,896	**12.** 5,949,086
13. 175,197	**14.** 42,520,350	**15.** $764,400	**16.** 25,228 sq meters
17. 2131	**18.** 228 R 1	**19.** 138 R 2	**20.** 71 R 5
21. 7822 R 2	**22.** 36 R 54	**23.** 910 R 49	**24.** 33 R 204
25. 75	**26.** 409 R 180	**27.** 741 R 231	**28.** 423 miles
29. 15 hours			

Chapter 4

1. $\frac{3}{8}$

2.

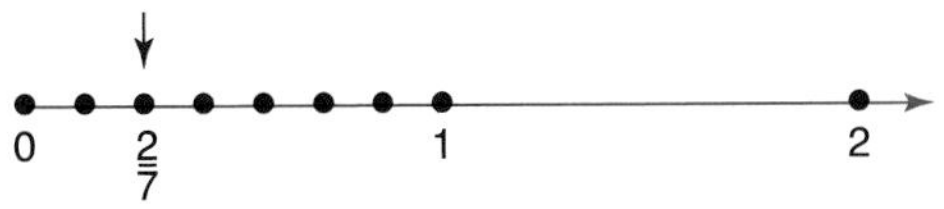

3. 16

4. 562	**5.** $21\frac{4}{5}$	**6.** 3	**7.** 1, 3, 11, 33
8. prime	**9.** $36 = 2 \times 2 \times 3 \times 3$		**10.** $\frac{3}{4}$

11. $\frac{1}{9}$ **12.** $\frac{10}{9}$, or $1\frac{1}{9}$ **13.** 1 **14.** $\frac{3}{14}$

15. $\frac{3}{2}$ of a turn

Chapter 5

1. $\frac{8}{9}$ **2.** 1 **3.** $\frac{3}{4}$ **4.** 40

5. 6 **6.** 36 **7.** $\frac{29}{18}$ **8.** $\frac{19}{16}$

9. $\frac{41}{30}$ **10.** $\frac{193}{21}$ **11.** $\frac{2}{5}$ **12.** $\frac{2}{9}$

13. $\frac{1}{3}$ **14.** $\frac{13}{28}$ **15.** $\frac{5}{9}$ **16.** $\frac{1}{24}$

17. $\frac{17}{48}$ **18.** $\frac{5}{9}$ **19.** $\frac{3}{8}$ **20.** $\frac{15}{27}$

21. $\frac{730}{1000}$ in. or $\frac{73}{100}$ in. **22.** $\frac{15}{16}$ in. **23.** yes, $\frac{1}{16}$ in.

24. $\frac{17}{32}$ in.

Chapter 6

1. $2\frac{2}{3}$ **2.** $2\frac{3}{19}$ **3.** $\frac{45}{7}$ **4.** $\frac{283}{9}$

5. $\frac{5}{8}$ **6.** 0 **7.** 11 **8.** $22\frac{2}{3}$

9. $6\frac{7}{8}$ **10.** $1\frac{5}{11}$ **11.** 0 **12.** $\frac{2}{3}$

13. 35 **14.** $16\frac{3}{8}$ **15.** $79\frac{13}{15}$ **16.** $71\frac{31}{36}$

17. $5\frac{1}{3}$ **18.** $6\frac{1}{3}$ **19.** $27\frac{1}{3}$ **20.** $16\frac{31}{36}$

21. $11\frac{2}{7}$ **22.** 8 pieces, 4 left over **23.** $4\frac{3}{4}$ lb

24. $1\frac{13}{16}$ ft **25.** $2\frac{5}{32}$ in.

Chapter 7

1. ten-millionths **2.** 23.017 **3.** $9\frac{5}{10{,}000}$ **4.** 906.0057

5. fifty-six and seven hundred nine thousandths **6.** 0.7199

7. 103.8 **8.** 53.889 **9.** 3.56 **10.** 28.497

11. 83.310 **12.** 22.318 **13.** 8.29 **14.** 11.21

15. 28.399 **16.** 1.821 in. **17.** 0.00325 in. **18.** 5.9375 cm
19. $9629.14

Chapter 8

1. 1.50101 **2.** 36.81 **3.** 18.4059 **4.** 1.8
5. 1.96 **6.** 0.3234 **7.** 24.5616 **8.** 89,600
9. 6 **10.** 36.8 **11.** 5.696 **12.** 0.4131
13. 7.2 **14.** 0.0055 **15.** 2.46 **16.** 0.006
17. 6.98 **18.** 0.015 **19.** 1815 **20.** $15.92\frac{42}{49}$
21. 797.2 **22.** 0.0428 **23.** 0.25 **24.** 3.625
25. $\frac{1}{125}$ **26.** $5\frac{251}{300}$ **27.** $39\frac{3}{8}$ **28.** $66.58
29. 1.092 in. **30.** 7.6 lb **31.** 0.337

Chapter 9

1. $\frac{7}{12}$ **2.** $\frac{3}{1}$ **3.** $\frac{29}{3}$ **4.** $\frac{1690}{7}$
5. 17 mpg **6.** 252 gal per min **7.** no **8.** $0.02 = 0.02$
9. $n = 7$ **10.** $36\frac{2}{3}$ lb

Chapter 10

1. 83% **2.** 5.7% **3.** $3\frac{1}{3}\%$ **4.** $\frac{14}{25}$
5. $\frac{43}{200}$ **6.** $\frac{9}{10{,}000}$ **7.** 0.73 **8.** 0.083
9. 0.07625 **10.** $\frac{21}{100}$ **11.** $\frac{171}{250}$ **12.** $\frac{137}{800}$
13. 37.5% **14.** $44\frac{4}{9}\%$ **15.** 68.75% **16.** 56%
17. $98\frac{3}{4}\%$ **18.** $67\frac{3}{8}\%$ **19.** 7.5 **20.** 200
21. 20% **22.** $R = ?, P = \frac{1}{2}, B = 6$
23. $R = 10, B = ?, P = 70$ **24.** 180 **25.** $6\frac{2}{3}\%$
26. $2538.32 **27.** 2% **28.** $962.50

Chapter 11

1. 84 in. **2.** $12\frac{2}{3}$ yd **3.** $\frac{2}{3}$ yd **4.** 800 m
5. 0.00659 m **6.** 590 dm **7.** 12 ft **8.** 14.9 dm

9. 31 yd 2 ft **10.** 3 m 4 km $9\frac{1}{2}$ cm **11.** 14.4 ft **12.** 62 cm

13. 5.6 m^2 **14.** 4.76 in^2 **15.** 0.347 ft^2 **16.** 600 m^2

17. $8\frac{2}{3}$ yd^2 **18.** 2.8 m^3 **19.** 54.6 dkℓ **20.** 391 ℓ

21. 2800 g **22.** 0.003904 kg **23.** $13\frac{1}{3}$ °C **24.** 221°F

25. 9.02 km **26.** 0.5408 gal

Chapter 12

1. negative two hundred one and seventeen hundredths **2.** −56

3. > **4.** 5.1 **5.** $-15\frac{7}{12}$ **6.** −1.289

7. 7 **8.** −560 **9.** 3 **10.** The number is 2.

11. 0.28 cm; 0.392 cm

ANSWERS TO CHAPTER TESTS

Chapter 1 Test

1. Whole numbers begin with 0. Natural numbers begin with 1. **2.** 60
3. 1 **4.** 1052
5. 372, 373, 374, 375, 376, 377, 378, 379, 380, 381, 382 **6.** 8
7. 700 + 80 + 4 **8.** 2000 + 40 **9.** 5671 **10.** 8060
11. two hundred fifteen million, four hundred eight thousand, three
12. 28,000,400,682 **13.** 5 hundred million
14. > **15.** < **16.** 20 **17.** 650
18. 400 **19.** 300 **20.** 1000 **21.** 2300
22. 6000 **23.** 1000 **24.** 10,000

Chapter 2 Test

1. 13 **2.** 14 **3.** 7 **4.** 15
5. 12 **6.** 11 **7.** 14 **8.** 13
9. 13 **10.** 16 **11.** 11 **12.** 15
13. 17 **14.** 18 **15.** 9 **16.** 12
17. 78 **18.** 513 **19.** 7946 **20.** 111
21. 1133 **22.** 11,565 **23.** 183 **24.** 1227
25. 11,791 **26.** 6190 **27.** 180,178 **28.** 2200, 2214
29. 26,000, 25,672 **30.** 5060 shingles **31.** 12,258 passengers
32. 9 **33.** 9 **34.** 8 **35.** 9
36. 9 **37.** 8 **38.** 8 **39.** 9
40. 5 **41.** 7 **42.** 8 **43.** 8
44. 3223 **45.** 1111 **46.** 3576 **47.** 2658
48. 34 **49.** 3849 **50.** 2377 **51.** 2536 wall boards
52. 587 fires

Chapter 3 Test

1. 70 **2.** 717,000 **3.** 6700 **4.** 40,000
5. 720 **6.** 20,000 **7.** 16,143 **8.** 32,064
9. 49,210 **10.** 2,454,000 **11.** 53,958 **12.** 8,313,678
13. 185,152 **14.** 56,647,080 **15.** 43,000 sheets **16.** 3243
17. 264 R 2 **18.** 58 R 2 **19.** 5763 R 1 **20.** 45
21. 1377 R 21 **22.** 84 **23.** 263 R 782 **24.** 808 R 116
25. 31 mpg

Chapter 4 Test

1. $\frac{10}{20}$

2. 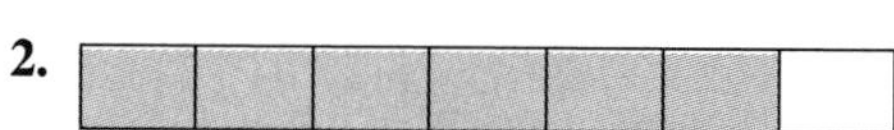

3. 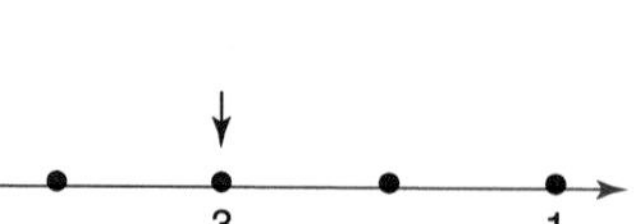

4. $\frac{4}{9}$

5. 37

6. 1

7. 0

8. not possible

9. 1

10. 682

11. $\frac{11}{5}$ or $2\frac{1}{2}$

12. 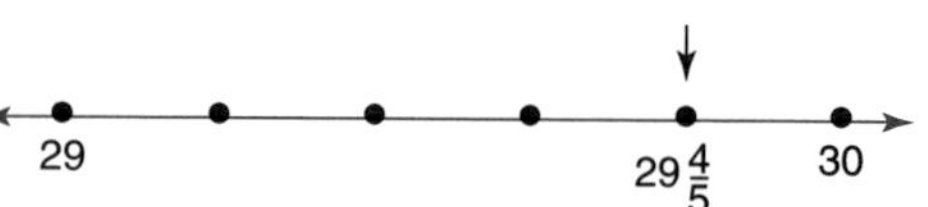

13. $56\frac{2}{8}$

14. by 2, 5

15. by 3

16. 1, 2, 4, 13, 26, 52

17. neither

18. prime

19. composite

20. $2 \times 2 \times 2 \times 7$

21. 24

22. $\frac{2}{3}$

23. $\frac{8}{15}$

24. $\frac{4}{15}$

25. $\frac{1}{21}$

26. proper

27. proper

28. improper

29. improper

30. $\frac{8}{6}$

31. None

32. $\frac{1}{206}$

33. $\frac{22}{39}$

34. $\frac{2}{9}$

35. $\frac{15}{32}$ in.

Chapter 5 Test

1. $\frac{3}{4}$

2. $\frac{1}{6}$

3. $\frac{2}{3}$

4. $\frac{7}{8}$

5. $\frac{8}{9}$

6. $\frac{2}{9}$

7. $\frac{3}{5}$

8. 1

9. $\frac{6}{7}$

10. 15

11. 36

12. 42

13. 60

14. 18

15. 30

16. $\frac{4}{35}$

17. $\frac{1}{2}$

18. $\frac{53}{60}$

19. $\frac{1}{36}$

20. $\frac{47}{48}$

21. $\frac{5}{42}$

22. $\frac{4}{5}$

23. $\frac{5}{6}$

24. $\frac{1}{12}$

25. $\frac{13}{16}$ lb

26. $\frac{17}{24}$ cups

27. $\frac{19}{15}$ miles

28. $\frac{3}{16}$ in.

29. James, $\frac{1}{30}$ of an hour

Chapter 6 Test

1. $4\frac{1}{2}$ **2.** $3\frac{5}{17}$ **3.** $\frac{61}{8}$ **4.** $\frac{206}{7}$
5. $\frac{5}{7}$ **6.** $\frac{80}{3}$, or $26\frac{2}{3}$ **7.** 0 **8.** $\frac{684}{5}$, or $136\frac{4}{5}$
9. $\frac{10}{7}$, or $1\frac{3}{7}$ **10.** $\frac{55}{2}$, or $27\frac{1}{2}$ **11.** $5\frac{3}{8}$ **12.** $\frac{21}{25}$
13. 0 **14.** $\frac{7}{6}$, or $1\frac{1}{6}$ **15.** $\frac{3}{10}$ **16.** not possible
17. $38\frac{2}{3}$ **18.** $13\frac{7}{16}$ **19.** $66\frac{1}{8}$ **20.** $43\frac{29}{48}$
21. $16\frac{1}{42}$ **22.** 12 **23.** $13\frac{1}{3}$ **24.** $36\frac{4}{9}$
25. $7\frac{13}{36}$ **26.** $4\frac{4}{33}$ **27.** $7\frac{4}{7}$ **28.** $18\frac{37}{41}$
29. $4\frac{6}{13}$ **30.** 46 cents **31.** $147\frac{1}{9}$ lb **32.** $300
33. $20\frac{3}{16}$ in.

Chapter 7 Test

1. millionths **2.** hundred thousands **3.** 15.008
4. $13\frac{15}{10{,}000}$ **5.** $6 + \frac{8}{10} + \frac{4}{1000}$ **6.** 4090.8005
7. forty-one and six hundred five thousandths **8.** 8.021
9. 0.680 **10.** 29.6 **11.** 19.0537 **12.** 82.019
13. 9.37 **14.** 24.326 **15.** 746.06 **16.** 94.3
17. 30.228 **18.** 0.0059 **19.** 8.64 **20.** 58.466
21. 9.43 **22.** 0.067 **23.** 6.498 **24.** 0.5625 in.
25. $30.13 **26.** 1.365 in. **27.** yes **28.** 30.39 in.
29. $1.61

Chapter 8 Test

1. 0.019 **2.** 38.88 **3.** 2.34 **4.** 58.4835
5. 0.072 **6.** 34.8 **7.** 0.389 **8.** $9.13
9. $12.27 **10.** 0.0028 **11.** 0.1045 **12.** 7.01
13. 1724 **14.** 276.0 **15.** 0.75 **16.** 23.875
17. $17\frac{2}{3}$ **18.** $\frac{9}{40}$ **19.** $0.38
20. 28 days, or 27 days with 1.4 gallons left
21. 51 mph **22.** 21 lb

Chapter 9 Test

1. $\frac{8}{7}$ **2.** $\frac{15}{17}$ **3.** $\frac{4}{13}$ **4.** $\frac{6}{5}$

5. $\frac{3}{1}$ **6.** no **7.** $\frac{4628}{113}$ **8.** $\frac{3}{4}$

9. 76.8 cents per lb **10.** 36.8 mph **11.** $26.39 per week

12. $\frac{2}{3} = \frac{2}{3}$ **13.** $\frac{35}{1} = \frac{35}{1}$ **14.** $112 = 112$; true

15. $4.32 = 4.32$; true **16.** $4 = 4$; true **17.** $n = 8$

18. $r = 50$ **19.** $y = 16.8$ **20.** $1416\frac{2}{3}$ gal **21.** 15.12 in.

22. 12 cups

Chapter 10 Test

1. 71% **2.** 8.6% **3.** 7% **4.** $\frac{2}{3}\%$

5. $91\frac{1}{3}\%$ **6.** $\frac{11}{100}$ **7.** $\frac{81}{1000}$ **8.** $\frac{58}{25}$

9. $\frac{2}{3}$ **10.** $\frac{7}{10{,}000}$ **11.** 0.66 **12.** 0.04

13. 0.038 **14.** 0.7904 **15.** $0.88\frac{1}{3}$ **16.** $0.06\frac{7}{9}$

17. $\frac{17}{100}$ **18.** $\frac{1}{20}$ **19.** $\frac{171}{500}$ **20.** $\frac{17}{250}$

21. $\frac{2}{3}$ **22.** $\frac{1}{12}$ **23.** 25% **24.** 40%

25. $37\frac{1}{2}\%$ **26.** $66\frac{2}{3}\%$ **27.** $53\frac{11}{13}\%$ **28.** $58\frac{1}{3}\%$

29. $56\frac{1}{4}\%$ **30.** 74% **31.** $22\frac{1}{2}\%$ **32.** $5\frac{1}{6}\%$

33. 30.9% **34.** $5.06\frac{1}{3}\%$ **35.** $R = 20$ **36.** $P = 0.425$

37. $B = 60$ **38.** $P =$ what?, $B = 38$, $R = 4.6$

39. $P = 846$, $R = 49$, $B = ?$ **40.** $R = ?$, $B = 28$, $P = 167$

41. $R = 150$, $B = 38$, $P = ?$ **42.** 24.57 **43.** 4%

44. 700 **45.** 1.25% **46.** $34.56 **47.** 59%

48. $206,000 **49.** 38.6%

Chapter 11 Test

1. 60 in. **2.** $8\frac{2}{3}$ yd **3.** 504 in. **4.** 2 mi

5. $\frac{2}{3}$ yd **6.** 90 ft **7.** 94 yd **8.** 600 m

9. 3.479 **10.** 0.00579 m **11.** 4.77 m **12.** 480 dm

13. 0.78 dkm **14.** $13\frac{1}{4}$ ft **15.** 22.123 km **16.** 25.6 cm

17. $3\frac{1999}{2640}$ **18.** 44 yd **19.** 106.880 km **20.** 4 km 942 m 31 cm

21. 2 yd 1 ft $5\frac{4}{5}$ in. **22.** 4.9 cm **23.** 99.224 in. **24.** 164 ft

25. 21.9 m^2 (rounded) **26.** 47.76 ft^2 **27.** $1\frac{1}{3}$ in^2 **28.** 315 tiles

29. 378 ft^2 **30.** $\frac{5}{6}$ ft^2 **31.** 32,000 acres **32.** $6\frac{4}{9}$ yd^2

33. 67,000 m^2 **34.** 900,000 m^2 **35.** 0.00239 m^2 **36.** 0.0013 ha

37. 1.456 cm^3 **38.** 460 ℓ **39.** 40.76 ℓ **40.** 9840 dℓ

41. 2500 dkℓ **42.** $37\frac{3}{16}$ yd^3 **43.** 84,000 g **44.** 3060 g

45. 1,100,000 cg **46.** 0.004898 kg **47.** 63°C **48.** 188°F

49. 36.7°C (rounded) **50.** 69.8°F **51.** 37°C **52.** 2379 cm

53. 17.6 lb **54.** 103.55 ℓ **55.** 223.2 in. (rounded) **56.** 100 km/hr

Chapter 12 Test

1. negative eighteen and one tenth **2.** positive thirty-one and five sevenths

3. −34 **4.** $<$ **5.** $>$ **6.** $<$ **7.** 0.64

8. 4 **9.** $-5\frac{1}{3}$ **10.** −26.21 **11.** −562

12. −364 **13.** $-8\frac{1}{24}$ **14.** −3.605 **15.** −60

16. $-\frac{1}{3}$ **17.** 0.075 **18.** −3 **19.** $-\frac{2}{5}$

20. 0.0052 **21.** −7 **22.** −3 **23.** 13

24. $925,000 **25.** 105 ft **26.** $1425 and $1600

TEST YOUR MEMORY ANSWERS

Test Your Memory—Chapters 1–3

1. 6 **2.** 80,506 **3.** 500,000 + 7000 + 300 + 20

4. twenty-three million, five hundred seven thousand, seven **5.** 645,772

6. 75,360, 75,400, 75,000 **7.** 857 **8.** 15,340

9. 373 **10.** 5512 **11.** 24,000 **12.** 49,458

13. 321 **14.** 51 **15.** 215 **16.** 9200

17. 64 R 2 **18.** 10,127 **19.** 62 **20.** 1269

21. 25,880 **22.** 1,510,626 **23.** 1223 R 2 **24.** 1818

25. 3,545,000 **26.** 268 **27.** 6608 **28.** 12,188

29. 22,766 **30.** 1617 **31.** 13,886,688 **32.** 519 R 4

33. 660 R 209 **34.** 624,869 **35.** 2,417,812 **36.** 1314 R 316

37. 49,294,080 **38.** 592 R 268 **39.** 309 drill bits **40.** 784 ft

41. 32 therms **42.** 16,427 lb **43.** 59 in. **44.** $147

45. 20 gal **46.** 24 mph **47.** $10,640

Test Your Memory—Chapters 1–6

1. 3, 1, 0, and 4 **2.** 20,000 + 500 + 3

3. Three hundred fifty-one million, sixty-eight thousand, nine **4.** 7 ten thousands

5. $<$ **6.** $>$ **7.** $<$ **8.** 30

9. 560 **10.** 700 **11.** 2900 **12.** 8000

13. 57,000 **14.** 65 **15.** 9595 **16.** 7065

17. 240 **18.** 91,692 **19.** 7234 **20.** 264

21. 2685 **22.** 2586 **23.** 40,864 **24.** 1360

25. 1,824,000 **26.** 187,859 **27.** 4,020,621 **28.** 7812

29. 74 **30.** 2169 **31.** 508 R 30 **32.** 69

33. 58 **34.**

0 $\frac{3}{8}$ 1

35. $\frac{5}{3}$, or $1\frac{2}{3}$ **36.** $2 \times 2 \times 7$ **37.** $\frac{7}{12}$ **38.** $\frac{4}{45}$

39. $\frac{1}{2}$ mile **40.** $\frac{14}{9}$ **41.** $\frac{2}{3}$ **42.** $\frac{5}{4}$

43. $\frac{7}{8}$ **44.** 1 **45.** $\frac{46}{35}$ **46.** $\frac{89}{60}$

47. $\frac{19}{28}$ **48.** $\frac{1}{3}$ **49.** $\frac{18}{35}$ **50.** $\frac{1}{24}$

51. $\frac{5}{8}$ **52.** $\frac{1}{10}$ of an hour longer **53.** $1\frac{2}{5}$

54. $7\frac{1}{2}$ **55.** $\frac{25}{8}$ **56.** $\frac{93}{7}$ **57.** $\frac{21}{4}$, or $5\frac{1}{4}$

58. $\frac{35}{2}$, or $17\frac{1}{2}$ **59.** $\frac{4}{5}$ **60.** $\frac{1}{8}$ **61.** $3\frac{7}{8}$

62. $27\frac{5}{12}$ **63.** $3\frac{19}{24}$ **64.** $6\frac{7}{18}$ **65.** $4\frac{3}{4}$

66. $17\frac{11}{16}$ **67.** $36\frac{1}{10}$ **68.** $5\frac{9}{16}$ **69.** 198 miles **70.** $3\frac{7}{8}$ ft

Test Your Memory—Chapters 1–9

1. 0 **2.** 1 **3.** 7000 + 100 + 9

4. eighteen million, six hundred one thousand, four hundred ninety-two

5. six million **6.** 840 > 804 **7.** 8700 **8.** 9219

9. 12,937 **10.** 4515 **11.** 22,000; 22,002 **12.** 1086 gal

13. 2192 **14.** 1385 **15.** 1674 **16.** 1778 dresses

17. 821,000 **18.** 24,000 **19.** 29,827 **20.** 36,300

21. 173,016 **22.** 338 mi **23.** 2204 **24.** 3047

25. 435 R 267 **26.** $778 **27.** $\frac{5}{8}$

28.

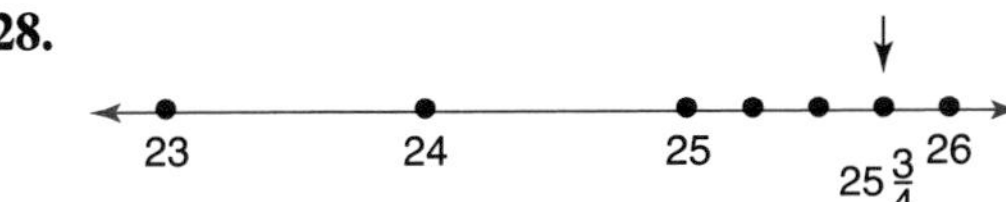

29. 1, 2, 3, 6, 7, 14, 21, 42

30. $2 \times 3 \times 7$ **31.** $\frac{2}{3}$ **32.** $\frac{1}{25}$ **33.** $\frac{5}{16}$ cu in.

34. improper **35.** $\frac{5}{6}$ **36.** $\frac{19}{15}$ **37.** $1\frac{3}{8}$

38. $\frac{5}{36}$ **39.** $\frac{5}{8}$ **40.** $\frac{15}{32}$ in. **41.** $8\frac{1}{3}$

42. $3\frac{1}{8}$ **43.** $\frac{3}{2}$ **44.** $7\frac{33}{50}$ **45.** $11\frac{15}{32}$

46. $\frac{23}{24}$ of an hour **47.** $32\frac{23}{500}$ **48.** $8 + \frac{6}{10} + \frac{1}{100} + \frac{8}{1000}$

49. seven and fifty-seven thousandths **50.** 0.703 **51.** 65.755

52. 75.6 **53.** 151.1 cu yd **54.** 30.91 **55.** 7.291

56. 0.7 sec **57.** 0.1904 **58.** 61,800 **59.** 548.64 gal

60. 0.00486 **61.** 0.031 **62.** 4.4 **63.** $\frac{1}{3}$

64. 26.1 mph **65.** $\frac{1}{6}$ **66.** yes **67.** $\frac{15}{8}$

68. $14.40 per yd **69.** $n = 64$ **70.** $\frac{232}{580} = \frac{500}{?}$; $1250

Test Your Memory—Chapters 1–12

1. 1 **2.** 6 **3.** 5000 + 60 + 9

4. three hundred six million, four hundred thousand, nine hundred eighty-six

5. 8 ten million **6.** $706 < 760$ **7.** 6300 **8.** 5999

9. 14,937 **10.** 5305 **11.** 21,000; 20,885 **12.** 1052 gal

13. 2381 **14.** 1365 **15.** 1782

16. 92 more kilowatt hours **17.** 914,000 **18.** 35,000

19. 32,052 **20.** 56,640 **21.** 90,558 **22.** 297 mi

23. 2102 **24.** 2036 **25.** 603 R 82

26. \$30 is the amount for each payment. **27.** $\frac{9}{9}$

28.

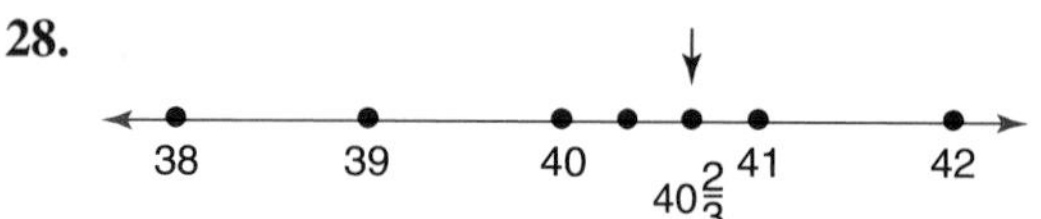

29. 1, 2, 3, 4, 6, 8, 12, 24

30. $2 \cdot 2 \cdot 2 \cdot 3$ **31.** $\frac{3}{8}$ **32.** $\frac{2}{9}$ **33.** $\frac{2}{7}$ of a gallon

34. improper fraction **35.** $\frac{2}{3}$ **36.** $1\frac{11}{30}$ **37.** $1\frac{3}{8}$ yd

38. $\frac{7}{24}$ **39.** $\frac{15}{31}$ **40.** $\frac{1}{3}$ yd was left. **41.** $4\frac{3}{4}$

42. 12 **43.** $\frac{2}{3}$ **44.** $13\frac{13}{24}$ **45.** $5\frac{15}{16}$

46. $\frac{7}{12}$ hour faster **47.** $23\frac{27}{500}$

48. $5 + \frac{4}{10} + \frac{6}{1000}$ **49.** three and five hundred two thousandths

50. 0.502 **51.** 45.144 **52.** 65.5 **53.** 4.007 in.

54. 25.62 **55.** 10.395 **56.** 0.7 seconds **57.** 0.1845

58. 50,700 **59.** 163.44 gal **60.** 0.00235 **61.** 0.014

62. 3.5 **63.** $\frac{2}{3}$ **64.** 45.6 mph **65.** $\frac{25}{9}$

66. $\frac{64}{75}$ **67.** \$1.85 per pound **68.** $12 = 12$ **69.** 15.6 in.

70. 0.059 **71.** $\frac{1}{3}$ **72.** 62.5% **73.** 6.41%

74. 800 **75.** \$5.25 **76.** 84% **77.** \$1440

78. 115.8 ft **79.** 0.06805 dkm **80.** 8 m 97 cm $4\frac{1}{2}$ mm

81. 3.12 ft **82.** 47.1 cm **83.** 147 m^2 **84.** $8\frac{46}{63}$ in^2, or 8.7 in^2

85. 594 ft^2 **86.** 0.0003894 km^2 **87.** 30.4 m^3 **88.** 3.47 ℓ

89. 0.56 g **90.** 240.8°F **91.** 167.4 km

92. 

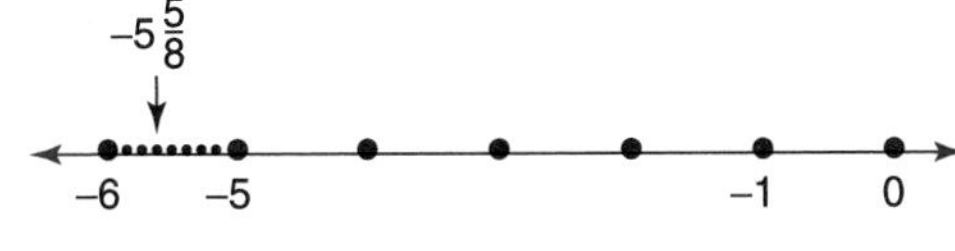

93. -5.7

94. 5 **95.** -8 **96.** 3

97. Her sales commission was $690.

INDEX

VOLUME, CAPACITY, AND WEIGHT EQUIVALENTS

1 gallon = 231 cubic inches
1 cubic foot = $7\frac{1}{2}$ gallons (approximately)
1 bushel = $1\frac{1}{4}$ cubic feet
= 2,150.42 cubic inches
1 cubic foot of water weighs $62\frac{1}{2}$ pounds (approximately)

DRY MEASURE

1 quarter = 2 pints = 67.2 cu in. = 1.101 liters
1 peck = 8 quarts = 537.6 cu in. = 8.81 liters
1 bushel = 4 pecks = 1.244 cu ft = 35.24 liters
1 cord of wood = 128 cu ft = 4 ft × 4 ft × 8 ft = 3.625 m^3

FORMULAS

w
l

Rectangle
Perimeter Area
$P = 2l + 2w$ $A = lw$

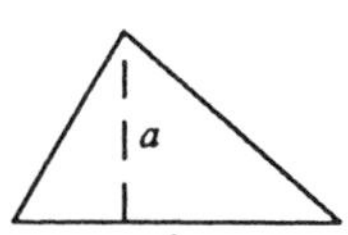

Triangle
Area
$A = \frac{1}{2}ab$

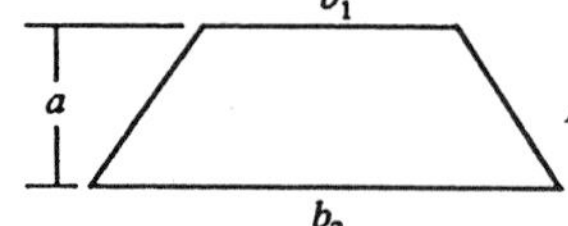

Trapezoid
Area
$A = \frac{1}{2}a(b_1 + b_2)$

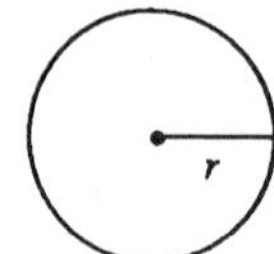

Circle
Circumference Area
$C = 2\pi r$ $A = \pi r^2$
$\pi \approx 3.1416$

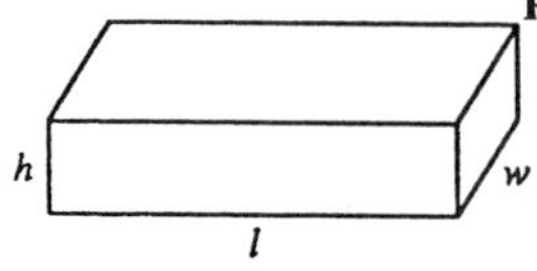

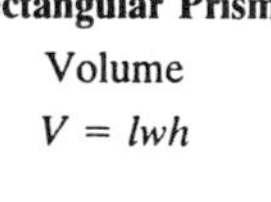

Rectangular Prism
Volume
$V = lwh$

Sphere
Surface Area Volume
$S = 4\pi r^2$ $V = \frac{4}{3}\pi r^3$

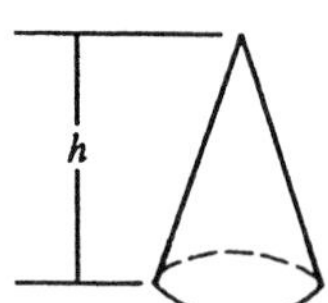

Cone
Volume
$V = \frac{1}{3}\pi r^2 h$

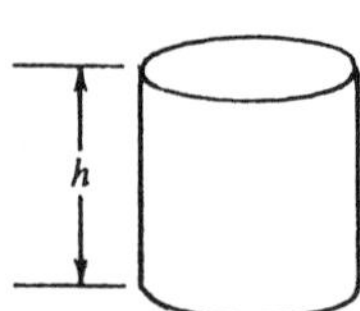

Cylinder
Volume
$V = \pi r^2 h$